THE CYANIDE HANDBOOK

McGraw-Hill Book Company

Publishers of Books for

Electrical World	The Engineering and Mining Journal
Engineering Record	Engineering News
Railway Age Gazette	American Machinist
Signal Engineer	American Engineer
Electric Railway Journal	Coal Age
Metallurgical and Chemical Engineering	Power

THE
Cyanide Handbook

BY

J. E. CLENNELL, B.Sc. (LONDON)

*Associate of the Institution of Mining and Metallurgy; Associate
of the Chemical, Metallurgical and Mining Society of South
Africa; Author of "Chemistry of Cyanide Solutions,"
etc.; Cyanide Chemist to the Creston-Colorada
Company*

SECOND EDITION
REVISED AND ENLARGED

McGRAW–HILL BOOK COMPANY, Inc.

239 WEST 39TH STREET, NEW YORK

6 BOUVERIE STREET, LONDON, E. C.

1915

ISBN 978-1-4276-1941-9

INTRODUCTION

NOTWITHSTANDING the fact that so much has been written about the cyanide process since its successful establishment in South Africa in 1890, there still appear to be some aspects of the subject which are only slightly touched upon, or quite inadequately treated, by the many able exponents of the process, and many facts of great value to the practical worker can only be gathered by long and painful search through technical literature. The various mechanical developments of the process have been dealt with by different writers too numerous to mention, and books have been published in which this side of the question is handled in a fairly adequate manner. The chemical aspect of the process has, however, been treated only in a fragmentary and imperfect manner, although valuable contributions to our knowledge of this subject have appeared from time to time, chiefly in the form of papers read before such bodies as the Chemical and Metallurgical Society of South Africa, the Institution of Mining and Metallurgy, and the Society of Chemical Industry.

The aim of the present work is to bring together such scattered information in a handy and accessible form, which will appeal to all workers interested in the process, whether as managers of plants, foremen, chemists, or assayers. Depending as it does on the application of chemical principles, a very important section of the work connected with a cyanide plant is carried on in the laboratory; this department will therefore receive a larger share of attention than is usually given to it in books of this description.

The discussion of the chemical properties, reactions, methods of manufacture, and analysis of cyanide itself, which have been scarcely touched upon in other similar books, will here be dealt with in sufficient detail. An outline will be given of the latest and most approved practice in modern cyanide plants, but without going into minute details of working, which are liable to constant change.

PREFACE TO THE SECOND EDITION

In the present edition every effort has been made to bring the information on all important sections of the subject up to date, so far as is possible in such a rapidly developing branch of metallurgy as the Cyanide Process.

It has not been thought desirable to omit any considerable part of the descriptions of older methods, as such processes are generally still in use, and new developments frequently involve the adoption of systems previously discarded, sometimes without change or with only slight modifications. New sections and illustrations will be found dealing with such important matters as improved systems of crushing and grinding, hydraulic classification, filter-pressing, vacuum filtration, agitating and aërating devices, zinc and aluminum dust precipitation, etc., and the latest information on the application of cyanide to the treatment of refractory ores has also been incorporated.

The writer desires to express his thanks for valuable aid and encouragement from well-known workers in the field, notably Messrs. Chas. Butters, G. H. Clevenger, J. J. Denny, E. M. Hamilton, Edwin Burt, J. V. N. Dorr, C. W. Merrill and E. L. Oliver, and the description of their inventions and discoveries adds greatly to the value of this new edition.

J. E. C.

September, 1914

CONTENTS

PART VI. SPECIAL MODIFICATIONS OF THE CYANIDE PROCESS

PART VII. ASSAYING

PART IX. METALLURGICAL TESTS

PART I

GENERAL

SECTION I

EARLY HISTORY OF THE CYANIDE PROCESS

(A) Discovery of Cyanogen and its Chief Compounds

Early Use of Cyanogen Compounds. — The cyanide process is a technical application of certain chemical reactions displayed by a group of substances known as the "cyanogen compounds." The beginnings of its history must therefore be traced in the beginnings of our knowledge of those compounds. Mixtures containing one or other of these substances, such as could easily be obtained from natural sources, have probably been known and used from very ancient times without any accurate knowledge of their composition. Thus it is stated by Hoefer[1] that prussic acid was known to the ancient Egyptians, who used it for poisoning those who had been guilty of divulging the sacred mysteries of the priests. An extract of laurel leaves seems to have been used for similar purposes in the Middle Ages.

Discovery of Prussian Blue. — In 1704 the substance now known as Prussian blue was accidentally discovered by Diesbach[2] whilst endeavoring to prepare a red coloring matter or " lake " by mixing a decoction of cochineal with alum, green vitriol, and fixed alkali: Using alkali, which had previously been used for the distillation of animal matter, instead of a red precipitate, he obtained a fine blue color. The discovery was communicated to Dippel, who improved and simplified the process; but it was kept secret until 1724, when Woodward[3] published a method of making Prussian blue by fusing dried blood and alkali, and treating the extract with alum, green vitriol, and spirit of salt. It was soon found that other animal matters might be used instead of blood, and many chemists

[1] "Histoire de Chimie," I, 226.
[2] Macquer, "Dictionnaire de Chimie," I, 265. See also "Miscellanea Berolinensis ad Incrementum Scientiarum," I, 377 (1710).
[3] "Philosophical Transactions," XXXIII, 15–24 (1724).

occupied themselves with the preparation and examination of this remarkable coloring matter.

Macquer's Researches on Prussian Blue. — The first attempt at a scientific examination of it appears to be that of Macquer[1] (1752). He came to the conclusion that it consists of iron charged with an excess of inflammable matter furnished by the "phlogisticated alkali" used for precipitating it. He also showed that the peculiar principle which imparted the blue color might be removed from Prussian blue and transferred to an alkali, which latter thus lost its alkaline properties and yielded a solution from which pure Prussian blue could be obtained again by addition of an iron salt and an acid. He had thus obtained a ferrocyanide.

Prussic Acid. — Other investigators followed, notably Bergman and Guyton, the latter of whom gave the name "acide prussique" to the substance to which the color of Prussian blue was supposed to be due. In 1782 important researches on this substance were made by Scheele,[2] who succeeded in isolating prussic acid by distilling the extract (obtained by fusing animal matter and alkali) with excess of sulphuric acid; he showed that pure Prussian blue could be obtained from the distillate by adding to it an iron salt and a few drops of sulphuric acid. He concluded that the "coloring material" was composed of aerial acid (*i.e.*, carbonic acid) and an inflammable substance. He also showed that the neutral salt obtained by the action of alkali on Prussian blue (*i.e.*, ferrocyanide of potassium or sodium) is a triple salt consisting of alkali, a little iron, and the coloring material.

Scheele's Observations on the Cyanides of Gold and Silver. — But the most interesting observations from our present point of view are those which he made on the cyanides and double cyanides of gold and silver, and attention may be drawn to the following remarkable passages:

"The calx of iron is not the only one which has the property of fixing the co   material, but this also happens with gold, silver, copper, and   s with several metallic calces, for if the solutions of the same are precipitated with the 'liquor præcipitans' and again dissolved by pouring in more liquor, the solution remains clear in

---

[1] "Memoire de l'Academie des Sciences": see Macquer's "Dictionnaire de Chimie," I, 265–274 (1778).

[2] "Sammtliche Physische und Chemische Werke," II, 321–348 ("Versuche über die Färbende Materie im Berlinerblau").

the open air, and the calx is not again precipitated by the aerial acid." [1]

Translated into modern language this passage means that when solutions of gold, silver, and copper salts are precipitated by a cyanide solution and the precipitate redissolved by excess of cyanide, a solution is formed which remains clear on exposure to air, and from which the metallic oxide is not thrown down by carbonic acid.

"If some of this liquor" (*i.e.*, the alkaline extract of animal matter) "be dropped into a completely saturated gold solution, the gold is thrown down as a white precipitate, but if much of the liquor be added, it is again dissolved. This solution is colorless like water; the precipitate is, however, insoluble in acids. Silver is thrown down white, like cheese; if more liquor be added the precipitate is again dissolved." [2]

He had thus prepared the double cyanides of gold and silver, on the formation of which the extraction of the precious metals by cyanide entirely depends.

Berthollet on Composition of Prussic Acid. — Berthollet[3] (1787) repeated and extended the observations of Scheele, showed that the views previously held, that ammonia and carbonic acid are contained in prussic acid, were erroneous, and gave the correct composition of the acid as follows: "Hence I conclude that hydrogen and nitrogen exist in prussic acid, that they are combined with carbon, and that when oxygen is added the necessary conditions for forming carbonate of ammonia are present." He was not able to determine the proportions of the constituents, however.

Berthollet also seems to have obtained chloride of cyanogen, but made no exact investigation of it. He says (*loc. cit.*, p. 35): "If oxygenated muriatic acid" (*i.e.*, chlorine) "be mixed with prussic acid prepared by Scheele's method, the former is converted into muriatic acid and the latter acquires a much sharper odor, and appears to have become more volatile: its affinity for alkalis is still further diminished. It forms no Prussian blue with iron solutions, but a green precipitate which becomes blue on exposure to light or mixture with sulphurous acid."

---

[1] *Ibid.*, p. 347.

[2] *Ibid.*, p. 339.

[3] "Annales de Chimie," I, 30 (1790): "Extrait d'un Mémoire sur l'Acide Prussique."

Artificial Formation of Cyanides. — Clouet,[1] in 1791, published a method of obtaining cyanogen compounds which is the basis of many methods that have been used or proposed for the manufacture of cyanide. He passed ammonia gas through a porcelain tube containing charcoal heated to redness, and received the issuing gases in vessels containing a solution of ferrous sulphate. He found that the product exhibited the properties recognized by chemists as belonging to the coloring matter of Prussian blue.

Cyanides in Nature. — Vauquelin[2] (1803) showed that prussic acid is present ready formed in certain vegetable substances occurring in nature.

Solubility of Gold in Cyanides. — Hagen, in 1805 ("Untersuchungen," I, 665), is reported to have made the statement that gold is dissolved, not only by free chlorine and aqua regia, but by a solution of prussiate of potash. I have not been able to confirm this reference, but, if correct, it would seem to be the earliest statement of the solubility of metallic gold in cyanide solutions. The existence of a soluble "prussiate" of gold was of course well known, having been discovered, as already mentioned, by Scheele; but in all references to this substance which I have been able to find it is formed by acting on salts of gold and not on the metal itself.

Proust on Cyanides and Ferrocyanides. — In the following year (1806) Proust published an extensive investigation of the properties of the prussiates,[3] showed that iron in the "triple prussiates" (*i.e.*, ferrocyanides) exists in the condition of the "black oxide" (*i.e.*, in the ferrous condition), and requires the addition of a solution of the red oxide (*i.e.*, a ferric salt) in order to give Prussian blue. With regard to the reactions of cyanides and ferrocyanides with gold, he remarks:

" Gold $\begin{cases} \text{Triple prussiate,} & \text{Nothing;} \\ \text{Simple } \text{ "} & \text{White precipitate;} \end{cases}$

which becomes a fine yellow. On heating the mixture, the precipitate does not explode; it is a true prussiate of gold. Heated in a retort, it gives water, empyreumatic oil pretty abundantly,

---

[1] "Annales de Chimie" [1] XI, 30–35: "Memoire sur la Composition de la Matière Colorante du Bleu de Prusse."

[2] "Annales de Chimie et de Physique," XLV, 206.

[3] "Annales de Chimie," LX, 185–224, 225–252: "Faits pour Servir à l'Histoire des Prussiates."

a carbonaceous gas which burns with a blue flame, and leaves a residue of gold mixed with charcoal powder."

Cyanogen Obtained. — Proust also tried the effect of heating cyanide of mercury, and found that it gave off inflammable matter which burnt with a red and blue flame with a yellowish border. He formed the erroneous conclusion, however, that the gas consisted of a mixture of prussic acid and "gaseous oxide."

Researches on Prussic Acid. — Further researches were made by Ittner[1] (1809), who succeeded in obtaining prussic acid by the action of hydrochloric acid on the cyanide of mercury. By carrying out this process and condensing the distillate in a mixture of ice and salt, Gay-Lussac,[2] in 1811, succeeded in obtaining anhydrous prussic acid, both in the solid and liquid form, and determined its boiling-point, 26.5° C, melting-point, − 15° C, and specific gravity, .70583.

Electrolysis of Ferrocyanides. — R. Porrett,[3] in 1814, published a paper "on the nature of the salts termed triple prussiates." He electrolyzed the "triple prussiate of soda" (*i.e.*, sodium ferrocyanide), and observed that ferrous oxide and prussic acid are obtained at the positive pole and only soda at the negative pole. Hence he concluded that the iron and the elements of prussic acid must together constitute a peculiar acid, which he succeeded in isolating by treating a solution of barium ferrocyanide with the exactly equivalent quantity of sulphuric acid. He also discovered (in 1808) the acid of the class of substances now known as "thiocyanates" or "sulphocyanides"; and suggested the use of these compounds for the quantitative estimation of copper, when added together with a reducing agent.

Gay-Lussac: Analysis of Prussic Acid and Cyanogen. — In 1815 Gay-Lussac[4] published his researches on prussic acid, in which he describes the analysis of prussic acid by exploding the vapor with oxygen in a vessel standing over mercury, and also by passing the vapor over heated iron. He determined the quantitative composition of mercuric cyanide, and showed that the gas produced by heating this compound consists only of carbon and

---

[1] Ittner, "Beiträge sur Geschichte der Blausäure."

[2] Gay-Lussac, "Note sur l'Acide Prussique": "Annales de Chimie," LXXVII, 128.

[3] Porrett, "Philosophical Transactions," XXVI, 527.

[4] Gay-Lussac, "Recherches sur l'Acide Prussique": "Annales de Chimie," XCV, 156.

nitrogen, in the proportion (approximately) of 6 parts of the former to 7 of the latter. He proposed the name "Cyanogène" for the new gas, and having shown that prussic acid is a compound of equal volumes of cyanogen and hydrogen, he suggested for it the name "acide hydrocyanique."

The discovery of cyanogen is of great importance in the history of chemistry, being the first example of the isolation of a "compound radical," that is to say, of a group of elements associated together and acting in a number of chemical changes as though they were a single element.

Davy's Researches on Cyanogen. — The observations of Gay-Lussac were confirmed and extended by Sir H. Davy,[1] in 1816, who electrolyzed anhydrous hydrocyanic acid, showed that no water is formed by the combustion of cyanogen in oxygen, and also decomposed cyanogen by the electric spark, and found that carbon was separated and a volume of nitrogen liberated equal to that of the cyanogen taken. He also prepared iodide of cyanogen by heating iodine with mercuric cyanide.

From this time onward many chemists occupied themselves with the study of cyanogen and its compounds, and we can here only allude to a few of the more important discoveries.

Cyanic Acid. — In 1818, Vauquelin found that one of the products obtained when cyanogen gas is dissolved in water is an acid, which is decomposed by stronger acids into carbonic acid and an ammonia salt. The name cyanic acid was given to this substance; it was found to contain cyanogen and oxygen, and gave salts, known as cyanates, containing different metals in combination with these elements.

Synthesis of Urea. — F. Wöhler[2] noted that when cyanogen gas is passed into solution of ammonia various bodies are formed, including "a characteristic crystalline substance, which, however, does not appear to be ammonium cyanate." In 1828[3] he identified this substance as urea, and showed that it can be formed by direct transformation of ammonium cyanate, which has the same percentage composition. This observation is very interesting, as being the first example of the formation of an organic body from

[1] Davy, "Gilb. Ann.," LIV, 383. "Journ. of Science and the Arts," I, 238: "On the Prussic Basis and Acid."
[2] Wöhler, "Pogg. Ann.," III, 177 (1825).
[3] "Pogg. Ann.," XII, 253.

purely inorganic materials, for it had already been shown by Clouet (1791) that cyanides could be formed from ammonia passed over red-hot carbon. It is also one of the earliest observed examples of isomerism, or of two substances having identical chemical composition but different properties, the reaction being

$$NH_4 \cdot CNO = CO(NH_2)_2.$$

In 1831, J. Pelouze[1] showed that potassium formate is one of the products obtained by evaporating an aqueous solution of potassium cyanide, this being another instance of the artificial production of organic compounds.

Manufacture of Cyanides. — In 1834, F. and E. Rodgers[2] described various methods of manufacturing cyanides, including that universally practised until a few years ago, which is given by them in the following terms: "Cyanuret of potassium may be prepared by exposing a mixture of anhydrous carbonate of potash and anhydrous ferrocyanuret of potassium to a moderate red heat in a covered crucible for about 20 minutes."

Action of Acids on Ferrocyanides. — In 1835, Everitt[3] gave the correct explanation of the reaction taking place when potassium ferrocyanide is distilled with dilute sulphuric acid, the equation (expressed in modern symbols) being:

$$3H_2SO_4 + K_4FeCy_6 = 3KHSO_4 + KFeCy_3 + 3HCy.$$

obtaining the white or yellowish insoluble salt ($KFeCy_3$) now known as Everitt's salt.

Haloid Cyanogen Compounds. — Bromide of cyanogen was discovered and investigated by Serullas[4] (1827), who also studied and determined the composition of the analogous chloride of cyanogen (first observed by Berthollet: see above), and rediscovered and investigated the iodide of cyanogen, first obtained by Davy (1816).

Other Investigators. — Further important researches on cyanogen and its compounds were made by Liebig, Bunsen, Berzelius, Kuhlmann, Berthelot, Joannis, Playfair, and many others, some of which will be referred to later in dealing with the chemistry

[1] Pelouze, "Annales de Chimie et de Physique," XLVIII, 395.
[2] Rodgers, "Philosophical Magazine," (3) IV, 91: "On Certain Metallic Cyanurets."
[3] Everitt, "Philosophical Magazine," (3) VI, 97.
[4] Serullas, "Annales de Chimie et de Physique," (2) XXVII, 184; XXXV, 291, 337; XXXVIII, 370.

of cyanide. We shall now proceed to describe those observations which more immediately led up to the discovery of the cyanide process.

(B) Early Observations on the Solubility of the Precious Metals in Cyanide

Elkington's Patent. — The earliest definite statement of the solubility of metallic gold and silver in cyanide solutions appears to be that contained in the patent taken out Sept. 25, 1840, by J. R. and H. Elkington (Brit. patent, No. 8447, of 1840), which was based on information supplied to them by Dr. Alexander Wright, of Birmingham, who in his turn appears to have been led to the discovery by the observations of Scheele, already quoted. The following are extracts from Elkington's specification:

" We take oxide of gold prepared by any of the known methods, or metallic gold in fine division, and dissolve the same in a solution of prussiate of potash or soda; to about two ounces of gold converted into oxide we add two pounds of prussiate of potash dissolved in one gallon of water, and boil the same for half an hour; the solution is then ready for use." . . ." We claim, therefore, in respect of this branch of our invention, the method of coating, covering, or plating the metals above mentioned with gold by the use, first, of an oxide of gold or metallic gold in fine division dissolved in prussiate of potash or any soluble prussiate, and also by the use of the oxide of gold dissolved in any analogous salt; and secondly, by combining the action of a galvanic current, with the use of a solution of gold as above described, giving the preference to a solution of gold in the prussiate of potash as above particularly described."

It will be obvious, from the quotations already given from early writers (Proust, Porrett, and others), that the expression " prussiate of potash " signifies the cyanide and not the ferrocyanide of potassium; and in fact it appears to be used in this sense as late as 1880 (see Clark's specification, U. S. Patent, No. 229,586, quoted below).

Bagration: Solubility of Gold in Cyanide. — In 1843, Prince Pierre Bagration[1] published his paper " On the Property which Potassium Cyanide and Ferrocyanide Possess of Dissolving

[1] Bagration, "Bull. de l'Acad. Imple de St. Pet." (1843), II, 136. See abstract in "Journ. für. präkt. Chem.," XXXI, 367 (1844).

Metals." In using gold plates as anodes in electro-gilding operations, he had observed that the gold continued to dissolve in the cyanide solution, even when the electric current was interrupted. He therefore tried the experiment of suspending a plate of gold so as to be partially immersed in a solution of potassium cyanide, and found that " after the lapse of about 3 days the part which dipped into the liquid was almost completely dissolved. The strongest action had taken place in the upper part where the solution and the plate were in contact with the atmospheric air." He found that the action was promoted by warming the solution, and that " silver and copper in the form of very thin plates or wires likewise dissolve" [in cyanide solutions].

The paper concludes as follows: " My researches lead me to suppose that hydrocyanic acid at the moment of formation also possesses this property [of dissolving gold]; but it is certain that in future cyanides of potassium must be reckoned among the number of the solvents of gold, and that one must beware of using gold or silver vessels for operations which require the use of these salts." He noticed that ferrocyanide of potassium is also a solvent, though in a very much less degree.

Glassford and Napier on Gold and Silver Cyanides. — In 1844, a paper was published by C. F. O. Glassford and J. Napier,[1] " On the Cyanides of the Metals and Their Combinations with Cyanide of Potassium," giving considerable information about the preparation and properties of the cyanides of gold and silver, and the double cyanides which these form, respectively, with cyanide of potassium; but, strangely enough, the solubility of the metals without the aid of the electric current is not referred to, although the fact is mentioned that when two plates of gold dipping in cyanide solution are connected with an electric battery, the gold dissolves much more rapidly from one plate than it is deposited on the other. This paper contains I believe, the earliest reference to the volumetric method of estimating cyanides by means of silver nitrate solution, now universally used in testing cyanide solutions (*loc. cit.*, p. 70).

Elsner: Influence of Oxygen on the Solution of Metals in Cyanide. — In 1846, L. Elsner,[2] in a paper called " Observations on

---

[1] Glassford and Napier, "Philosophical Magazine," (3) XXV, 56–71 (paper read before Chemical Society, Feb. 19 and March 4, 1844).

[2] Elsner, "Journ. für prakt. Chem.," XXXVII, 441.

the Behavior of Reguline Metals in an Aqueous Solution of Potassium Cyanide," showed that the oxygen of the air played an essential part in the solution of gold, silver, and some other metals in cyanide solutions. In this paper he remarks:

"It is a fact known for some years past, that metallic gold, silver, copper, iron dissolve in aqueous solutions of potassium cyanide, even without the assistance of the galvanic current, at the ordinary temperature of the air (about 12–15° C)" . . . "It was very probable that the solution of the last-named metals (Au, Ag, Cd), as well as that of Cu, Fe, Zn, and Ni, had only taken place through the agency of oxygen, so that if this supposition were correct, oxygen must have been removed from the layer of atmospheric air still present above the column of liquid."

Elsner describes experiments showing that the amount of oxygen in tubes containing gold and silver, standing over cyanide solution, was gradually reduced, and that the air which remained extinguished a lighted match when the latter was slowly introduced. When such tubes were opened under water or mercury, these liquids rose in them to a certain height. Mercury gradually rose in a tube containing cyanide, silver, and air, inverted over it. Finely divided metallic silver placed in a well-stoppered tube filled with cyanide solution showed very slight action, only a little silver cyanide being obtained on acidulating with HCl; but after filtering, the liquid passing through gave a considerable precipitate with this acid, showing that further action had taken place on exposure to air.

Elsner formulated no equation to explain the combined action of cyanides and oxygen on gold or silver, although the usually accepted expression of this reaction goes by his name. He classifies the metals according to their action on cyanide, as follows:

1. Insoluble: Pt, Hg, Sn.

2. Soluble with decomposition of water: Fe, Cu, Zn, Ni.

3. Soluble with aid of atmospheric air, without decomposition of water: Au, Ag, Cd.[1]

Early Cyanide Extraction Test. — In 1849 there appeared an account of what seems to have been an early attempt to utilize the solubility of gold in cyanide compounds for metallurgical purposes. This is a description of experiments made in 1848 by Dr. Duflos,[2] entitled, "Entgoldungsversuche der Reichensteiner Arsenik-

---

[1] "An error, see p. 102."

[2] "Journ. für prakt. Chem.," XLVIII, 65–70 (1849).

abbrände." In consequence of the statements of Bagration (quoted above), a quantity of the material was extracted with a very dilute solution of yellow prussiate of potash. The liquid leaching through was digested with hydrated sulphide of iron for the purpose of converting any cyanide of gold and potassium which might be present into sulphide of gold and ferrocyanide of potassium. The residue was burnt, after washing and drying, but the ash was found to contain no gold. Tests made with a mixture of chloride of lime solution and yellow prussiate gave no better result. "Liebig's potassium cyanide was not tried, not so much on account of its high price, as because the considerable amount of ferrous oxide present in the 'Abbrände,' which would form yellow prussiate with the potassium cyanide, already indicated that an unsatisfactory result was to be expected à priori." If this test had not been omitted, the experimenter might very possibly have forestalled Messrs. MacArthur and Forrest by nearly 40 years!

Faraday's Researches on Gold Leaf. — In 1856, Faraday[1] delivered a lecture before the Royal Society entitled, "Experimental Relations of Gold and Other Metals to Light," in which he showed that extremely thin films of gold, capable of transmitting light, could be produced by acting upon gold-leaf, adhering to a plate of glass, with a weak solution of cyanide of potassium. The method of obtaining such films was as follows: "If a clean plate of glass be breathed upon and then brought carefully upon a leaf of gold, the latter will adhere to it; if distilled water be immediately applied to the edge of the leaf, it will pass between the glass and the gold and the latter will be perfectly stretched; if the water be then drained out, the gold-leaf will be left well extended, smooth, and adhering to the glass. If, after the water is poured off, a weak solution of cyanide be introduced beneath the gold, the latter will gradually become thinner and thinner; but at any moment the process may be stopped, the cyanide washed away by water, and the attenuated gold film left on the glass."

Faraday also emphasized the importance of air in this reaction, and gives the following explanation of it (*loc. cit.*, p. 147):

"Air-voltaic circles are formed in these cases and the gold is dissolved almost exclusively under their influence. When one piece of gold-leaf was placed on the surface of a solution of cyanide of potassium, and another, moistened on both sides, was placed

[1] Faraday, "Philosophical Transactions," CXLVII, 145–181 (1857).

under the surface, both dissolved; but twelve minutes sufficed for the solution of the first, whilst above twelve hours were required for the submerged piece. In weaker solutions, and with silver also, the same results were obtained; from sixty to a hundred-fold as much time being required for the disappearance of the submerged metal as for that which, floating, was in contact both with the air and the solvent."

Again (*loc. cit.*, p. 173): "A piece of ruby paper" (colored by finely divided gold) "immersed in a strong solution of cyanide of potassium suffered a very slow action, if any, and remained unaltered in color; being brought out into the air, the gold very gradually dissolved, becoming first blue."

Use of Cyanide in Amalgamation. — The extraction of gold by means of mercury had long been known. Rose[1] states that it has been in use for the last 2000 years; but it had frequently been observed that the surface of the mercury became coated with films of various nature, which impaired its effectiveness for amalgamation. It was found that these films, consisting chiefly of grease or of oxides and carbonates of copper and other base metals, could be removed, and the efficiency of the mercury restored, by treatment with various chemicals, such as nitric acid, caustic soda, sal-ammoniac, etc. Among others, cyanide of potassium was found to be very effective for this purpose, and appears to have been already in frequent use as an aid to amalgamation in America during the 60's of the last century. This practice was, of course, carried out in ignorance of the fact that cyanide was a solvent of gold, and thus cannot in any sense be considered to be an anticipation of the cyanide process. One or two writers, however, drew attention to the solubility of gold in cyanide, only mentioning the fact as a drawback to its use as an aid to amalgamation. Thus, Prof. Henry Wurtz, in a memoir read to the American Association for the Advancement of Science, at Buffalo, Aug. 1, 1866,[2] says: "Early in my investigations I found that no supposition of extraneous films or coatings of air, oxide of iron, talc, or other foreign matter (though undeniably often present and even occasionally active in this way), was available in the general explanation of the phenomena; for it was quickly found that the most powerful

---

[1] "Metallurgy of Gold," 4th edition, p. 91.

[2] See "American Journ. of Mining," IV 354 (Dec. 7, 1867): "On a Theory of Gold Genesis."

chemical agents, save those which act by dissolving and removing a superficial film of the gold itself, gave no aid to quicksilver." The following note is added:

" To this category belongs *cyanide of potassium*, as I need not inform chemists. Despite the highly deleterious, and in fact dangerous, nature (in inexperienced hands) of this agent, it would appear that attempts are now being made to palm it off upon mining men, especially in California, as a new and highly valuable adjunct in amalgamation. As a writer upon this subject, I feel therefore called upon to take this opportunity to say that many experiments were made by myself three or four years since with the cyanide, in the preliminary amalgamation of copper plates, with the idea that it might be of value, *in connection with the amalgam of sodium*, for this purpose; and it is possible that it might be, were it not for one vital obstacle, namely, that the coating thus formed, unlike that produced by the aid of sodium alone, *tarnishes* with great readiness. Alone, the cyanide is very feeble in its action in this way, and much inferior to nitric acid, the common agent used. I find, in fact, that ammonia, mixed with sal-ammoniac (indeed almost anything which will dissolve copper or its oxides), is equally available. The cyanide has been much tested in Colorado, but mostly abandoned. The new project, apparently started in California, of using it in the battery and the pan, is one which could scarcely interest any but those occupied in the vending of miners' nostrums; as the known solubility of both gold and mercury in it would alone be fatal in the view of an expert. In the pan, particularly, another destructive agency would be set up, namely, a solvent action upon the sulphides of the baser metals, and a secondary precipitation of the latter upon the quicksilver. The subject, however, is not worth the space that would be required for its discussion here."

Rae's Patent: Cyanide in Conjunction with Electric Current. — In the following year we come to the patent of Julio H. Rae (U. S. Patent, No. 61,866; Feb. 5, 1867) for an "Improved Mode of Treating Auriferous and Argentiferous Ores." This is the first distinct attempt to apply cyanide as a solvent of gold in ores, though it may perhaps be questioned whether Rae was aware of the solubility of gold in cyanide of potassium without the aid of an electric current. He gives a figure of an agitation apparatus, the object being to dissolve the gold and silver, and precipitate them again on a cathode of copper or suitable material. The process is thus described:

"This invention·consists in treating auriferous and argentiferous ores with a current of electricity or galvanism for the purpose of separating the precious metals from the gangue. In connection with the electric current suitable liquids or chemical preparations, such, for instance, as cyanide of potassium, are used, in such a manner that by the combined action of the electricity and of the chemicals, the metal contained in the ore is first reduced to a state of solution, and afterward collected and deposited in a pure state, and that the precious metals can be extracted from the disintegrated rock or ore at a very small expense and with little trouble or loss of time."

The agitation apparatus consisted of a vertical shaft, to which was attached a metallic cage, the bottom of the shaft resting on a plate of metal (by preference platina) at the bottom of a jar of glass or other suitable material. The shaft was to be connected with the positive pole of a battery, so that shaft, cage, and plate together formed the anode, while the other pole of the battery was to be connected with a thin slip or coil of copper or suitable material (suspended in the liquid above the ore) and "forming a base on which the precious metals are deposited. By the action of the electric current the action of the chemicals on the metals contained in the rock is materially facilitated and a perfect solution thereof is effected."

The use of the expression "materially facilitated" would seem to imply that Rae supposed cyanide of potassium (the only chemical mentioned in the specification) would have some action, even without the electric current.

Cyanide as a Solvent of Metallic Sulphides. — The use of cyanide of potassium as a solvent in the treatment of ores is again mentioned in 1870 by H. C. Hahn[1] ("Schwefel-metalle gegen Cyankalium"), who remarks that "Cyanide of potassium is not only a good solvent for Ag, Cu, etc. (Cl, I, Br, O), but also for several sulphides; thus, CuS (as already known), AuS and AgS dissolve easily; ZnS and FeS after some digestion in strong cyanide solution." He also notes that these sulphides dissolve both in the pure condition and in ores. He describes experiments showing the solubility of silver sulphide in cyanide, and notes that the separation of silver from copper, by means of sulphureted hydrogen, from solutions in which the metals are present as soluble double cyanides, is imperfect, owing to the incomplete precipitation of the silver sulphide. He refers to the application of cyanide in ore treatment as follows: "With regard to AuS, I will mention that its solubility in KCy has already found practical application. The owner of the Union Mine in Amador County, California, treats the concentrates, which contain 200 to 6000 dollars gold per ton, with quicksilver and a strong solution of cyanide of potassium,

[1] Hahn, "Berg- und Hüttenm. Zeit.," XXIX, 66 (Feb. 25, 1870).

and extracts 23 per cent. more of the total gold contents than formerly by ordinary amalgamation."

Hahn seems to have ascribed the action of cyanide to the solubility of gold sulphide rather than of metallic gold; and it is evident also that the practical application referred to was one in which the object was to aid amalgamation, and not to dissolve the gold in cyanide.

Skey on Relative Solubility of Metals in Cyanide. — The objections to the use of cyanide as an aid to amalgamation are again referred to by W. Skey,[1] in an article, read Jan. 29, 1876, " On the Electromotive Order of Certain Metals in Cyanide of Potassium, with Reference to the Use of this Salt in Milling Gold." He noted that cyanide had been used with marked effect in preventing the "flouring" of mercury during the treatment of blanketings at the Thames Gold-Field, New Zealand, and ascribes the action to the solvent effect of cyanide on a film of mercurial salts produced by the action of soluble ferric salts, which "either oxidize or chlorodize the surface of any mercury which they may be in contact with."

To produce this effect, the cyanide must dissolve a portion of the mercury. "Moreover, in thus contemplating the contingencies entailed or risked by the use of any alkaline cyanide in such milling operations, it must be remembered that both gold and silver are not absolutely insoluble in these cyanides. Now the loss of mercury in this way may not be serious, but if *gold* or even *silver* be thus lost (that is, by its solution), even in much less quantity than mercury well could be, the loss then may be serious. Whether the loss of metal certain to be entailed by the use of cyanide of potassium falls upon the mercury, or upon the gold or silver of these blanketings, conjointly or separately, depends entirely upon this relative affinity of these metals for this salt, or in other words it depends upon their electromotive order therein." He shows the commonly accepted view, that mercury is positive to gold and silver in cyanide solutions, to be erroneous, and that mercury is in reality very decidedly negative; thus, metallic gold in contact with a solution of mercuric cyanide would rapidly dissolve and mercury be reduced. He gives the following as the electromotive order of metals in potassic cyanide, from negative to positive:

[1] Skey, "Trans. Proc. N. Z. Inst.,' VIII, 334–337.

− Carbon	Lead
Platinum	Gold
Iron	Silver
Arsenic	Tin
Antimony	Copper
Mercury	+ Zinc.

"Most, if not all, the sulphides or other ores occurring in nature are negative to the whole series. Any of these metals will generally precipitate the ones named above it from its cyanide solution"; thus, gold precipitates mercury and silver gold, taking its place in the liquid. "Thus, it appears, a loss of gold by solution of it must frequently happen whenever cyanide of potassium is employed to assist in the amalgamation of blanketings, or other auriferous stuff. In fact, all that loss of metal occasioned by its solution, and most of which is, as we have seen, a necessity involved in the working of the process itself, falls upon the gold and silver present, the mercury being positively protected from the action of this salt by these more valuable metals." Skey also thinks the loss would be greater if the operation were conducted in iron pans, and also greater with a strong cyanide solution than a weak one, and suggests a method of minimizing the loss by passing the waste liquor in a thin stream over copper plates (presumably with a view to precipitating the dissolved gold and silver).

Dixon on Treatment of Pyritic Ores with Cyanogen Compounds and Oxidizers. — In 1877, W. A. Dixon[1] published a paper "On a Method of Extracting Gold, Silver and Other Metals from pyrites," which was read before the Royal Society of New South Wales, Aug 1, 1877. After giving an account of various methods tried for the treatment of ores containing sulphides of Cu, Pb, Sb, Fe, As, etc., and rich in gold and silver, and noting that amalgamation and chlorination gave unsatisfactory results, he describes a number of experiments made with cyanides and ferrocyanides as solvents, both alone and in conjunction with various oxidizing agents. With regard to cyanide, he says:

"Prince Bagration and Elsner [Watt's "Dict": "Cyanides of Gold"] have observed that precipitated gold is soluble in cyanide of potassium if exposed to the air; and the latter says also in ferrocyanide of potassium. A patent was applied for in America in 1868

---

[1] Dixon, "Proc. Royal Soc. N. S. W," XI, 93–111.

for the use of cyanide of potassium for the extraction of gold from its ores, but I have no particulars of the process." (This possibly refers to Rae's patent, mentioned above.) "It seemed to me, however, that the high price of this salt, its instability when exposed to the air and in solution, and its extremely poisonous properties, precluded its use for this purpose. On trying the reaction between precipitated gold and cyanide of potassium, I found that it was extremely slow if the gold was at all dense. In presence of alkaline oxidizing agents, however, I found that the solution of gold was sufficiently rapid. Thus, on standing overnight, the quantity of gold and cyanide of potassium solutions being similar in each case, with the cyanide alone, traces only of gold were dissolved, but with the addition of calcium hypochlorite, ferrocyanide (ferricyanide?) of potassium, or binoxide of manganese, all the gold was dissolved; with chromate of potassium, a small quantity; with permanganate of potassium, none. With ferrocyanide of potassium alone I did not obtain any gold in solution after standing some days, but I thought that with suitable oxidizing agents it might be obtained in solution according to the equation

$$4Au + 2K_4FeCy_6 + 7O + 4H_2O = 4AuCy_3 + Fe_2O_3 + 8KHO.$$

In the cold, however, with the exception of ferricyanide of potassium, none of the above oxidizing agents had any effect, but heated to 212° F., the reaction with all of them was sufficiently rapid, and I found that this was also the case with permanganate."

Precipitation on Silver and Copper. — For precipitation, he suggests first filtering the hot solution through finely divided metallic silver to recover the gold, and then precipitating the silver as sulphide; he found, however, "that copper in any form precipitated both gold and silver from the solution, or at all events that these metals (*i.e.*, gold and silver) were not dissolved until the copper had all gone into solution; also, that if the copper was present as sulphide, the silver was transformed into sulphide, which is insoluble."

To remove dissolved copper, he suggests digesting the alkaline solution with ferrous hydrate, which would at the same time convert any cyanide present into ferrocyanide, "which has the important advantages of being exceedingly stable and non-poisonous."

Extraction Tests on Ores. — Various experiments are then detailed, on roasted arsenical pyrites, and other pyritic ores, using mixtures of ferrocyanide with different oxidizers in hot solutions. " A portion of roasted arsenical pyrites was digested at 212° for twelve hours with ¼ oz. ferrocyanide of potassium, 32 gr. oxide of manganese (20 lb. per ton), and sufficient water, made alkaline with soda, to make a cream: the solution yielded 9 oz., 8 dwt., 19 gr. gold per ton, leaving 1 oz., 9 dwt., 15 gr. This was the best result obtained with this pyrites, the yield with other oxidizing agents and by more prolonged digestion being all somewhat lower."

In another case the material was previously roasted with salt, and extracted with acid to remove copper. The material then contained

Gold, 8 oz., 0 dwt., 19 gr.;
Silver, 49 " 11 " 5 "

and yielded from 3 oz. 12 dwt. to 5 oz. 1 dwt. of gold and from 46 oz. to 46 oz. 3 dwt. of silver per ton. "This showed that all the silver which had been converted into chloride during the roasting was obtained in solution as cyanide. With the gold, on the other hand, all the results showed that with complex pyrites a portion only could be obtained in solution, either in mercury or in water as cyanide or chloride, whilst none could be obtained as sulphide." These results did not appear to Dixon to be sufficiently satisfactory to warrant further investigation, and the remainder of the article is devoted to the discussion of smelting processes, etc. It would seem, however, that additional tests on the lines of the experiments detailed might probably have led to the discovery of a workable cyanide process.

Solubility of Silver in Cyanide. — Some interesting remarks on the solubility of silver in cyanide solutions are to be found in Percy's "Metallurgy"[1] (published in 1880), although the explanation he gives of the reaction would seem to be quite erroneous. He states that "metallic silver dissolves in a hot aqueous solution of cyanide of potassium, but slowly and sparingly, as shown by the following experiment by R. Smith. A single silver leaf was heated during several hours in a concentrated aqueous solution of cyanide of potassium, and it was found that only a small quantity of silver had been dissolved. Christomanos, on the other hand, asserts that pure silver dissolves readily in a hot solution of this salt (Fresenius, Zeit. für Anal. Chem., VII, 301; 1868). In this case, hydrogen must be evolved according to the following equation,

<hr />

[1] J. Percy, "Metallurgy: Silver and Gold," Part I, p. 115.

unless the silver is oxidized by oxygen derived from any cyanate of potash that might be present; though in the experiment by R. Smith, above recorded, no evolution of hydrogen was observed:

$$4KCy + 2Ag + H_2O = 2KCyAgCy + K_2O + 2H."$$

Cyanide as an Aid to Amalgamation. — In 1880–81 three patents were taken out in the United States, all of them involving the use of cyanide or some cyanogen compound in ore treatment; but the chemical being added not as a solvent but as an aid to amalgamation, these processes have no direct bearing on our subject, though they are of some interest as showing the state of metallurgical knowledge at the time, shortly before the introduction of the cyanide process. So far from utilizing the solvent action of cyanide, the latter would be actually a drawback in any attempted practical application of these patents. The specifications referred to are:

(1) U. S. patent, No. 229,586: Thos. C. Clark:
    Filed Dec. 27, 1879; issued July 6, 1880:
    "Extracting Precious Metals from Ores."
(2) U. S. patent, No. 236,424: Hiram W. Faucett:
    Filed July 13, 1880; issued Jan. 11, 1881:
    "Process of Treating Ore."
(3) U. S. patent, No. 244,080: John F. Sanders:
    Filed April 16, 1881; issued July 12, 1881:
    "Composition for Dissolving the Coating of Gold in Ore."

Clark's Process. — In Clark's process the ore is roasted and thrown while hot into a bath formed of a solution of salt, prussiate of potash, and caustic soda or caustic potash, the object being to effect desulphurization and disintegration, so as to bring the precious metals into a suitable form for amalgamation by freeing them from the union and influence of baser metals.

Faucett Process. — In Faucett's process the crushed ore is to be heated and subjected, under pressure, to the action of "disintegrating chemicals" in solution in a closed vessel, after which it is to be further pulverized and amalgamated. The pressure is effected by the steam generated by contact of the hot ore with the chemical solution. The chemicals contained in the solution are chloride of sodium, nitrate of potash, cyanide of sodium, sulphate of protoxide of iron, and sulphate of copper, with or without admixture of hydrofluoric acid, fluoride of potassium, or fluoride of sodium.

Sanders' Process. — In Sanders' process, a mixture of cyanide of potassium and phosphoric acid is used to remove "the coatings that envelop gold in the ore and that consist usually of various metallic oxides and of silver. " . . . " After agitation the mixture above mentioned will be found "to have dissolved the oxides and the sulphurous coatings of the ore, and the agitation of the barrel or vessel removes the dissolved impurities, leaving the gold free and exposed, and permitting it to be amalgamated by the addition of quicksilver in the usual manner." He also makes the following disclaimer: "I am aware that cyanides have already been used in the extraction of gold; also that gold-bearing ores have been disintegrated in the presence of heat by various chemicals. This I do not claim." The mixture claimed consists of about 16 parts of cyanide of potassium to 1 part of glacial phosphoric acid, with sufficient water to form a thick pulp with the raw gravel. This would undoubtedly dissolve in some cases considerable amounts of gold, but that circumstance seems to have entirely escaped the inventor's observation.

Simpson's Patent: Cyanide with other Chemicals as Gold Solvent. — A patent which has much more direct bearing on the subject and which may legitimately be regarded as in some respects an anticipation of the process of MacArthur and Forrest is that of

Jerome W. Simpson: U. S. patent, No. 323,222:
Filed Oct. 20, 1884; issued July 28, 1885:
"Process of Extracting Gold, Silver, and Copper from their Ores."

The crushed ore is to be treated with certain salts in solution adapted to combine chemically with the metal and form therewith a soluble salt. After thorough agitation, the solid matter is allowed to settle and a piece or plate of zinc is to be suspended in the clear liquid, "which causes the metal dissolved in the salt solution to be precipitated thereon, from which it can be removed by scraping or by dissolving the zinc in sulphuric or hydrochloric acid. The precipitated metal may then be melted into a button."

The salt solution referred to consists of "one pound of cyanide of potassium, one ounce carbonate of ammonia, one-half ounce chloride of sodium, and 16 quarts of water, or other quantities in about the same proportions." It would appear that the carbonate of ammonia was added as a solvent for copper, and the chloride of sodium as a solvent for silver; for in the case of ores containing only gold and copper he omits the chloride of sodium, and for ores rich in silver he employs a proportionately larger quantity of this salt. He also disclaims the exclusive use of cyanide as a solvent, as follows:

"I am aware that cyanide of potassium, when used in connection with an electric current, has been used for dissolving metal, and also that zinc has been employed as a precipitant, and the use of these I do not wish to be understood as claiming broadly." He also disclaims the exclusive use of carbonate of ammonia but claims:

"1. The process of separating gold and silver from their ores which consists in subjecting the ore to the action of a solution of cyanide of potassium and carbonate of ammonia, and subsequently precipitating the dissolved metal, substantially as set forth.

"2. The process of separating metals from their ores, to wit: subjecting the ore to the action of a solution of cyanide of potassium, carbonate of ammonia, and chloride of sodium, and subsequently precipitating the dissolved metals."

The addition of these small quantities of other chemicals almost suggests that they were introduced because Simpson doubted the validity of a patent based on the use of cyanide alone. With regard to this process it may be remarked that the solution specified is inconveniently strong in cyanide, and that the precipitation process, if carried out as described, would be very inefficient. It is nevertheless possible that if the process had been introduced on a working scale it might eventually have been so modified as to become a practical success.

Further Anticipations of the Cyanide Process. — Caveats covering the use of cyanides for extracting precious metals are stated to have been taken out by Endlich and Muhlenberger, in 1885, and by Louis Janin, Jr., in 1886, but no definite particulars are available with regard to these. It is rather curious to note that both Endlich and Janin write in a disparaging way about the prospects of cyanide as a commercially successful solvent of gold and silver (see *Eng. and Min. Journ.*, L, 685 and LI, 86). That the practical difficulties in the way of the application of cyanide as a solvent had not been overcome at the time of the MacArthur-Forrest patents, may be judged from the following remarks of Janin (*ibid.*, L, 685), from a letter dated Nov. 24, 1890:

"To sum up the disadvantages of the cyanide process in a nutshell, I will say that to extract a reasonable percentage of the gold and silver, so great an excess of cyanide is required that the extraction is no longer economical or profitable. Moreover, if this excess is used, it is impossible to completely precipitate by the means advocated by the MacArthur-Forrest people. In fact, as a means of extracting gold I can more conscientiously recommend aqua regia than potassium cyanide."

(C) Introduction of the MacArthur-Forrest Process

Researches on Gold Solvents by the MacArthur-Forrest Syndicate.
— We have now traced the history of cyanide in its application to the treatment of gold and silver as far as the year 1886. In this year experiments were being made in Glasgow by a research syndicate consisting of J. S. MacArthur, R. W. Forrest, W. Forrest, and G. Morton, for the purpose of investigating, and if possible developing into a commercial success, a method of ore treatment known as the "Cassel Gold-Extracting Process." An interesting account of the steps which led to the successful application of cyanide in the treatment of ores as a result of these experiments is given by J. S. MacArthur,[1] in a paper read before the Scottish Section of the Society of Chemical Industry, March 7, 1905, from which a few extracts may here be given. The process under investigation depended on the solvent action of chlorine generated electrolytically in an alkaline solution. It was found, however, that such solvents attacked the base metals in the ore in preference to the gold. Attempts were made to dissolve the gold and limit the action on base metals by adding chlorine or bromine, without the electric current, and introducing a salt which would not absorb Cl or Br, nor precipitate gold, but which would precipitate base metal compounds. Bleaching-powder, borax, and bicarbonate of soda were found to partially answer this purpose, but even with this addition the action of the halogen was directed rather against the base metal than the gold. The efforts of the syndicate were therefore directed to finding a gold solvent that would not be a base-metal solvent. This condition naturally pointed to some alkaline or neutral solution. In making these tests it was decided not to rely on the method of taking the difference between original assay of ore and final assay of residue as indicating gold extracted, but in all cases the gold was to be precipitated from its solution and

[1] "Journ. Soc. Chem. Ind.," XXIV, 311: "Gold Extraction by Cyanide: a Retrospect."

recovered in a visible form, so that it could be actually handled and weighed, sulphureted hydrogen being used as the precipitant.

Cyanide Extraction Trials. — " Among the various solvents on our program for trial, we had included potassium cyanide, and in November, 1886, we tried the effect of it on the tailings of one of the Indian gold mines, and, as usual, treated the solution with H_2S for recovery of the gold; and getting none, we passed on to our next solvent, meanwhile observing our rule of labeling the residue and laying it aside. We had neglected to notice that H_2S did not precipitate gold from its solution in cyanide, and thus our experiment was for the time literally relegated to the shelf. About eleven months after, I had occasion to devise a rapid method of gauging approximately the gold contents in weak solutions of gold chloride, and used for the purpose tin chloride to produce the well-known 'purple of Cassius.' One solution that I had to test contained mercury, and using potassium cyanide to separate the mercury and the gold, I was apprised of the fact that H_2S did not precipitate gold from its cyanide solution. Instantly my mind reverted to the experiment carried out nearly a year before, and I saw that it might have been successful without the success being recognized. Immediately a sample of rich concentrates from a Californian mine was treated, and on this occasion we examined the residue rather than the solution, and found a high percentage of extraction. A sample of poor concentrates from India was now treated, and again a high extraction was obtained. The results were startling. We unearthed the residues from the old experiments (all our work was done in duplicate), and to our intense satisfaction we found that they too had transferred their gold to the cyanide solution. There was now no doubt about the importance of the discovery, and at once a provisional specification was drafted and lodged."

MacArthur-Forrest First Patent. — The provisional specification of the English patent, No. 14,174, was lodged Oct. 19, 1887, by J. S. MacArthur, R. W. Forrest and W. Forrest, and the complete specification taken out July 16, 1888, entitled: "Improvements in Obtaining Gold and Silver from Ores and other Compounds." The essential points in which this patent differs from those of previous inventors above alluded to, are:

1. The solution of the gold and silver is to be effected by means of a liquid to which a cyanide alone is added, without the aid of an electric current or of other chemicals.

2. The patentees emphasize the use of dilute in preference to strong solutions for accomplishing their purpose.

3. Certain definite relations are specified between the quantities of ore, values in gold and silver, and strength and quantity of solution, which, however, were not adhered to in practice and cannot be regarded as being essential to the process.

4. Cyanogen gas is mentioned (probably in error) as one of the solvents claimed.

The following extracts from the specification will illustrate these points:

(1) *Nature of the Solvent.* "In carrying out the invention, the ore or other compound in a powdered state is treated with a solution containing cyanogen or cyanide (such as cyanide of potassium, or of sodium, or of calcium), or other substance or compound containing or yielding cyanogen."

(2) *Use of dilute solutions.* "In practice, we find the best results are obtained with a very dilute solution, or a solution containing or yielding an extremely small quantity of cyanogen or a cyanide, such dilute solution having a selective action such as to dissolve the gold or silver in preference to the baser metals."

(3) *Relation between quantities of ore, solution, etc.* "In preparing the solution we proportion the cyanogen to the quantity of gold or silver, or gold and silver, estimated by assay or otherwise to be in the ore or compound under treatment, the quantity of a cyanide or cyanogen-yielding substance or compound being reckoned according to its cyanogen. . . . In dealing with ores or compounds containing, per ton, 20 ounces or less of gold or silver, or gold and silver, we generally use a quantity of cyanide the cyanogen of which is equal in weight to from one to four parts in every thousand parts of the ore or compound, and we dissolve the cyanide in a quantity of water of about half the weight of the ore. In the case of richer ores or compounds, whilst increasing the quantity of cyanide to suit the greater quantity of gold or silver, we also increase the quantity of water so as to keep the solution dilute."

(4) *Use of cyanogen gas.* "In using free cyanogen, the cyanogen obtained as a gas in any well-known way is led into water to form the solution to be used in our process; or any suitable known mode of setting cyanogen free in solution may be employed."

Reference is made in the specification to the use of agitation, pressure, and increased temperature as aids to extraction, but no claims are made in respect to these.

No special means of precipitation is described in this patent, but the following statement is made: "When all or nearly all the gold or silver is dissolved, the solution is drawn off from the ore or undissolved residue, and is treated in any suitable known way, as, for example, with zinc, for recovering gold and silver. The residuary cyanogen compounds may also be treated by known means for regeneration or reconversion into a condition in which they can be used for treating fresh charges of ores or compounds."

The solvents enumerated are: "Any cyanide soluble in water, . . . such as ammonium, barium, calcium, potassium or sodium cyanide, or a mixture of any two or more of 'them. Or any mixture of materials may be taken which will by mutual action form cyanogen or a substance or substances containing or yielding cyanogen."

Original Claim. — The first claim made was: "The process of obtaining gold and silver from ores and other compounds, consisting in dissolving them out by treating the powdered ore or compound with a solution containing cyanogen or a cyanide or cyanogen-yielding substance, substantially as hereinbefore described."

Amended Claim. — In consequence of litigation, this was modified so as to apply only to dilute solutions and the amended claim (Aug. 20, 1895) read as follows: "Having now particularly described and ascertained the nature of our said invention, and in what manner the same is to be performed, we declare that *we do not claim generally the use of solutions of any strength*, but what we claim is:

"1. The process of obtaining gold and silver from ores and other compounds, consisting in dissolving them out by treating the powdered ore or compound with a *dilute* solution containing cyanogen or a cyanide or cyanogen-yielding substance, substantially as hereinbefore described, *and subject to the above disclaiming note.*"

MacArthur-Forrest Second Patent. — On July 14, 1888, another patent was applied for by MacArthur and Forrest: Eng. patent No. 10,223, entitled, "Improvements in Extracting Gold and Silver from Ores and other Compounds."

Use of Alkalis. — After referring to their previous patent, they describe the use of alkalis, such as potash or lime, for neutralizing the ore previous to cyanide treatment; also various methods of treatment by agitation, grinding in pan-mills, and percolation in tanks with permeable false bottoms, are indicated. The method of precipitation is then described as follows:

Zinc Precipitation. — "The separated solution is next made to pass through a mass of metallic zinc in a state of fine division. We find that the best results are obtained in this part of the process when the zinc has been freshly divided by mechanical or other means, so that its surfaces are as purely metallic as possible; and, further, when the quantity or mass of zinc employed is such that the solution has, in passing through it, ample opportunity for being thoroughly acted on.

"The zinc to be used is reduced to the desired state of fine division by any suitable means. The degree of division is, preferably, such as is obtainable by shaving or by cutting thin strips or very small pieces or grains, by means of a turning tool, circular saw or other suitable tool, from cakes or blocks of zinc of convenient size. By another method, the zinc is brought into the desired state of fine division by passing it in a molten state through a fine sieve and allowing it to fall into water. In order to obtain the best results the finely divided zinc should be used as soon as possible after it has been produced."

The use of zinc in this form is also suggested as applicable to other gold- and silver-extracting processes, in which the metals are dissolved, not as cyanides but as chlorides, bromides, thiosulphates or sulphates.

"After separation of the solution the precious metals may be separated from zinc by distillation. Or the larger portion of the precious metals may be separated from zinc by sieving (by preference under water), when, with a suitable sieve, the greater part of the precious metals will pass through, the greater part of the zinc being left on the sieve."

The claims made are: (1) "In processes for extracting gold and silver from ores or other compounds by means of a cyanide or cyanogen compound, the preparatory treatment of the ores or compounds with an alkali or alkaline earth, substantially as and for the purposes hereinbefore described.

(2) "In precipitating gold and silver from cyanide, chloride, bromide, thiosulphate, or other similar solutions by means of zinc,

the employment of the zinc as freshly prepared in a state of fine division, substantially as hereinbefore described.

(3) "The process for extracting and recovering gold and silver from ores and other compounds, consisting in first treating same with an alkali or alkaline earth, then extracting the gold by means of a cyanide or cyanogen compound, and finally precipitating the gold and silver by means of zinc as freshly prepared in a state of fine division, all substantially as hereinbefore described."

It will be noted that no broad claim to the use of zinc as a precipitant is made, which would of course be untenable, but only to the special form of the metal found suitable in this process, the discovery of which is described by MacArthur in the paper which we have already referred to.[1]

Discovery of Precipitation by Zinc Turnings. — "We knew well that zinc precipitated gold from its cyanide solution; but it remained to make this reaction industrially applicable. We used various forms of finely-divided zinc with more or less advantage; but a picture of some fine zinc shavings, bought with other things in a shilling box of chemicals in my boyish days, haunted my mind, and repeatedly I described it to one of the works foremen without effect, until one day, when making a zinc case for packing cyanide, he made a shaving by a sharp tool and came asking me if this was what I wanted. My reply was 'Yes,' and in half an hour he had prepared the first bundle of zinc shavings for gold precipitation — the pioneer bundle of hundreds of tons of this flimsy but useful material."

American Patents: (MacArthur-Forrest). — During the year 1888, patents were taken out in most of the important mining countries of the world, covering the points included in the two just discussed. The American patents are: (1) U. S. Patent, No. 403,202, May 14, 1889; referring to the dissolving process; and (2) U. S. Patent, No. 418,137, Dec. 24, 1889; referring to the precipitation process.

The former of these differs in several respects from the British patent: it specifies more particularly the classes of ore which had not previously been satisfactorily or profitably treated, and to which the cyanide process may be applied with advantage; and it also emphasizes the points in which the process differs from those of Rae, Simpson, and other inventors, as follows:

[1] J. S. MacArthur, "Journ. Soc. Chem. Ind.," XXIV, 313.

" By treating the ores with the dilute and simple solution of a cyanide, the gold or silver is, or the gold and silver are, obtained in solution, while any base metals in the ores are left undissolved, except to a practically inappreciable extent; whereas, when a cyanide is used in conjunction with an electric current, or in conjunction with another chemically-active agent, such as carbonate of ammonium or chloride of sodium or phosphoric acid, or when the solution contains too much cyanide, not only is there a greater expenditure of chemicals in the first instance, but the base metals are dissolved to a large extent along with the gold or silver, and for their subsequent separation involve extra expense, which is saved by our process." Further particulars are given as to the nature of the containing vessel, and as to mechanical aids to extraction.

The proportions of ore, solution, etc., are the same as in the British patent, but a maximum strength is given and the claim is limited to solutions not exceeding this, being as follows: "The process of separating precious metal from ore containing base metal, which process consists in subjecting the powdered ore to the action of a cyanide solution containing cyanogen in the proportion not exceeding eight parts of cyanogen to one thousand parts of water." (This would be equivalent to 2 per cent. KCN.)

Among the methods of recovering gold and silver from the solution, the process of treating it with sodium amalgam is referred to.

Introduction of the Process. — Arrangements were now made in the chief mining countries to introduce the process on a practical basis: the first works on a commercial scale seem to have been those at Karangahake, New Zealand, established in 1889.[1]

In the early part of 1890, the Cassel Company established an experimental plant near the old Salisbury Battery, at the Natal Spruit, near Johannesburg, Transvaal, where tailings and concentrates were treated in agitation vats, the contents of which were afterwards discharged on to suction filters and the liquid drawn off precipitated by zinc shavings. In April, 1890, these works were in regular operation, and the results were so successful that a plant on a commercial scale was established in the latter part of the year at the Robinson Mine, Johannesburg.

[1] Large scale tests were made on ore from the Crown Mines (New Zealand) in 1888.

Importance of the Cyanide Process in Metallurgy. — Since then the process has spread to every gold and silver mining district in the world, with few exceptions, and although not equally successful with every class of ore, it may fairly be claimed that its introduction has revolutionized the metallurgy of the precious metals.   During the ten years from 1886 to 1896, the world's production of gold was doubled, and it is stated that in the period 1896–1906 it again doubled, mainly through the introduction of new metallurgical methods in which the use of cyanide has played a leading part. " In 1889 the world's consumption of cyanide did not exceed 50 tons per annum.   In 1905 the consumption was nearly 10,000 tons per annum, of which the Transvaal gold field took about one-third." [1]

A royalty was at first charged by the African Gold Recovery Company, representing the Cassel Company (the owners of the MacArthur-Forrest patents) in South Africa.   The validity of the patent was, however, disputed, and the patents were finally set aside in February, 1896 after an appeal to the High Court of the Transvaal.   It must be admitted, however, that although the patents may have been technically invalid, the application of the process in such a form as to be commercially successful was chiefly due to the energy and enterprise of MacArthur, Forrest, and their associates.

Further Developments. — The improvements since introduced have chiefly taken the form of applying well-known mechanical devices previously used in other connections, as adjuncts in the working of the process.   Such alterations in the chemical treatment as have been suggested have as a rule found only a partial and temporary application; they have generally taken the form of additions of oxidizers or other reagents to the ore or solution: these so-called improvements will be referred to later in a special section dealing with them.   We may, however, enumerate the following as being, perhaps, the most interesting developments:

1893.   The Electric Precipitation Process.
  "   Use of Lead Salts in Treatment of Argentiferous Sulphide Ores.
1894.   Zinc-dust Precipitation.

[1] Frankland, " Journ. Soc. Chem. Ind.," XXVI, 175.

1895.   The Bromocyanide Process.
  "   Slimes Treatment by Decantation.
  "   Aeration of Ore and Slimes.
1897.   Zinc-lead Precipitation Process.
1898.   Slimes Treatment by Filter Presses.
  "   Roasting Previous to Cyanide Treatment.
  "   Commercial Application of Cyanide to Treatment of
        Silver Ores.
1902.   Lead Smelting Applied to Zinc Precipitate.
  "   Economical Fine Grinding of Ore as a Preliminary to
        Cyanide Treatment.
1903.   Vacuum Filtration of Slime first Successfully Applied.
1906.   Precipitation with Aluminum Dust.
1907.   Application of Air-lift for Agitation of Slime Pulp.
1911.   Attempted Electrolytic Regeneration of Cyanide in
        Solution after Treatment.

SECTION II

OUTLINE OF OPERATIONS IN THE CYANIDE PROCESS

(A) THE DISSOLVING PROCESS

Stages of Cyanide Treatment. — Having now traced the history of cyanide treatment as far as its establishment on a practical basis by MacArthur and Forrest, we shall proceed to give a brief summary of the actual operations involved in the application of the process, leaving the details for discussion in later chapters. The process described in this section has been modified in recent practice, as detailed in the following section, but the description of the older practice is retained here, as there is a distinct tendency in some particulars to revert to these methods. As in all metallurgical processes for the treatment of ores, the material must undergo a preliminary crushing or disintegration to render the metallic particles accessible to the solvent. After the ore has been thus obtained in a suitable condition, the treatment takes place in three distinct stages, which may be described as:

1. The dissolving process;
2. The precipitation process;
3. The smelting process.

Mechanical Difficulties in Treatment of Crushed Ore. — The method of treatment to be adopted in the dissolving process, and the plant and appliances used, are determined largely by the physical nature of the material to be dealt with. In crushing any kind of ore, whatever appliance be used, certain parts will be reduced to a finer state of division than others; the product is never homogeneous, but consists of particles of all sizes, from the largest that will pass the crushing apparatus to the minutest grains.

Sands and Slimes. — At an early stage in the history of the process it was found that the treatment of the crushed material as a single product was often unsatisfactory. The methods adopted, though varying greatly in detail, may be roughly classified under two heads — Agitation and Percolation. In the agitation process it was found that the coarser and heavier particles showed a tendency to settle and remain stationary in corners or other parts

33

of the vessel where the agitation was least effective, and so largely escaped treatment. In the percolation system the finer material showed a tendency to separate in certain regions, forming bands or masses of clay-like material, practically impervious to the solution, and which thus also escaped effective treatment. Hence the suggestion naturally arose that the best method of handling the material would be to separate the coarser and heavier portion of the crushed ore (technically described as "sands") from the finer and lighter portion (known as "slimes"), and to treat the former by percolation and the latter by agitation. This idea has been almost universally carried out until within the last few years, when the alternative suggestion of crushing the whole of the material so fine that it may be treated satisfactorily by agitation has been rapidly gaining ground.

Amalgamation. — In most cases the ore, after crushing and before cyanide treatment, undergoes some form of amalgamation or treatment with mercury, the usual method being to pass the ore as it leaves the stamp-batteries, together with a sufficient quantity of water, in a thin stream over sheets of copper coated with mercury. The latter combines with and arrests the coarser particles of precious metal, extracting in this way a percentage of the values which varies largely with different kinds of ore, and may perhaps average about 60 per cent.

Hydraulic Separation. — The stream leaving the amalgamated plates then passes to some form of hydraulic separator, the most usual being that known as "Spitzlutte," consisting essentially of a pointed box with a central partition, and an outlet at the bottom near which a jet of water is introduced. The pulp fed in at the top on one side passes downward beneath the partition, where it meets an ascending stream of water; the lighter particles are carried away and overflow on the opposite side at the top of the box, while the coarser and heavier particles fall through the outlet at the bottom. The pulp may be passed through a succession of such boxes to give a sufficiently complete separation of sands and slimes.

Collection of Sands for Treatment. — When the sands are to be treated by percolation they are led, usually, into a collecting-vat, where they are evenly distributed by means of a hose or by some mechanical device. As the vat fills, the surplus water, containing a further quantity of slime, is continuously drawn off. When the vat has been thus filled with sand, the contents are frequently prepared for cyanide treatment by the addition of alkali (generally lime), and a wash of water or very weak solution is given in order to dissolve the alkali and distribute it to every part of the charge. In some cases the cyanide treatment is partially carried out in the collecting-vats, but usually the sand is transferred through discharge doors at the bottom of the collecting-vat into filter-tanks placed underneath. These tanks are nearly always circular, and constructed of wood or iron. The bottom is covered by a wooden framework on which rests a filter composed of cocoanut matting or some similar material covered by thick canvas.

Percolation Process. — The solution is led by iron pipes from the storage-tank to the top of the filter-tank and allowed to flow on until the charge of sand is completely covered. It is then left standing for a longer or shorter period, and drawn off by opening a valve in a pipe beneath the filter bottom; the liquid thus drained off passes either direct to the precipitation boxes or, more generally, to a settling-tank, where any suspended matter is allowed to subside, and from whence the clear liquid is drawn off for precipitation. A number of washes in succession are given in this way, the quantity of liquid used, strength of solution, time of contact and other details being varied according to the nature of the material. The general practice may be described somewhat as follows:

1. One or two washes of very weak (alkaline) solution, for neutralizing acid substances in the ore.

2. The strong solution — say one-fifth the weight of charge — usually about .25 per cent. KCy; left standing 6 to 12 hours or more, then drained as completely as possible.

3. A number of washes, successively weaker in cyanide down to .1 per cent. KCy.

4. A final wash of water, or the weakest solution available.

The total amount of solution used would be from 1 to $1\frac{1}{2}$ times the weight of the dry sand in the charge; total time of treatment, 4 to 8 days. These quantities refer to tailings from ordinary siliceous ores after amalgamation. When the ore is treated "direct" (*i.e.*, without amalgamation), or when rich or refractory material is treated, the quantities and strength of solution and the time of contact may vary within very wide limits.

Collection of Slimes for Treatment. — The slimes carried off by the overflow from the Spitzlutte and collecting-vats are carried, generally by means of wooden launders, to a suitable vessel, generally a large pointed box (spitzkasten), where they undergo partial settlement, in order to remove superfluous water. Lime is generally introduced either in the battery itself or into the stream of pulp leaving the battery, which has the effect of causing the fine particles of ore quickly to coagulate and settle when the rapid flow of the liquid is arrested. The clear water from the spitzkasten may be pumped back to the battery for further use, while the thickened pulp is drawn off from the openings at the bottom of the spitzkasten and passes to the agitation tanks, where it is generally settled to remove a further quantity of water, and then treated with dilute cyanide solution, with alternate agitation and settlement.

Agitation Process. — The agitation is produced either by mechanical stirrers mounted on a vertical revolving axis, or by injecting air under pressure, and circulating the pulp by passing it continuously from the bottom of the tank through a centrifugal or other pump, which throws it back into the same tank. The strength of cyanide necessary in treating slimes is in general considerably less than that required for sands; the values, being mostly in minute particles, dissolve rapidly, especially when arrangements are made for effective aeration.

Separation of Dissolved Values. — The chief difficulties are encountered in the mechanical separation of the solution carrying the dissolved gold and silver, from the residue. Two distinct methods have been extensively adopted for this purpose; these are:

1. The decantation process;
2. The filter-press process.

Decantation Process. — In the decantation process, which is applicable chiefly to very low-grade material, such as the Rand battery slimes, the material is treated in very large tanks by alternate agitation and settlement; the clear settled solution is drawn off, by means of a jointed pipe, from the surface down to the level of the slime-pulp. Fresh solution is then added and the operation repeated two or three times. Sometimes the pulp is finally transferred to a large and relatively deep tank, where the pressure of a high column of liquid causes the slime to settle with a reduced percentage of water; after several successive charges have been added, the clear settled solution is drawn off as far as possible, and the residue, carrying say 30–40 per cent. of moisture, drawn off at the conical bottom of the large tank. As it is not possible, from economical considerations, to give more than one or two successive agitations and decantations on the same charge, the percentage of extraction by this method is usually somewhat low, and with richer material the alternative process of filter-pressing has been found more advantageous.

Filter-Press Process. — In the filter-press process, after the slime-pulp has had a preliminary agitation with cyanide solution, it is transferred to a suitable receiver, whence it is forced by means of compressed air into the filter-presses, consisting of a number of chambers enclosed by metal frames and plates, so arranged that when the chambers are filled with the slime-pulp, liquid (either cyanide solution or water) may be forced through them under pressure, thence passing through filter-cloths into channels connected with the outlet, from which it is conveyed to a settling-tank and thence to the precipitation boxes. Various types of press, operating by pressure or by suction, have been introduced, and will be described in more detail in a special section. In general, it may be said that the filter-press process involves more expense for labor, power, etc., per ton of material treated, than the decantation process, but that the percentage of extraction is higher, and the time of treatment less; also less liquid is required, an important consideration in regions where water is scarce. After several washes of solution and water have been thus forced through the press, the latter is opened, and the cakes of slimes, containing a much lower percentage of moisture than the residues of the decantation process, are removed and discharged.

Fine Crushing for Cyanide Treatment. — In the system now coming into general use the ore is crushed rather coarsely in the battery, with water or dilute cyanide solution containing sufficient lime to make it slightly alkaline. The pulp then passes, with or without previous amalgamation, through a "tube" or "flint" mill, consisting of a revolving steel cylinder with a suitable lining of some hard material, and about half filled with more or less rounded flint pebbles. The ore is effectively and economically ground in this apparatus; the cylinder is set in a slightly inclined position, and the pulp passing out through an opening at the lower end goes to the spitzkasten, whence the coarser portion is returned to the upper end of the tube-mill to be reground. The fine portion, consisting almost entirely of particles small enough to pass a sieve of 150 holes to the linear inch, passes, either direct or after amalgamation, to the agitation tanks, the whole of it being treated as slime and passed through filter-presses.

It will be seen that the treatment is thus much simplified; the high initial cost of leaching and decantation plants is avoided, with a higher percentage of extraction and an increased quantity of ore treated per month.

Extraction by Cyanide Process. — The extraction of gold and silver varies, according to the nature of the material, from under 70 to over 95 per cent. of the total value: many ores hitherto considered refractory have been found to yield a high percentage after fine grinding.

Disposal of Residues. — The residues after treatment are discharged either by shoveling into trucks, or by various mechanical devices, or by sluicing.

(B)   THE PRECIPITATION PROCESS

Clarification of Solutions before Precipitation. — The solutions drawn off from the leaching-tanks, or from slime-treatment tanks or filter-presses, usually contain a certain amount of siliceous and other matter in suspension, which it is desirable to remove by a preliminary operation. This is done by settlement in special tanks, or by passing the liquor through small filter-presses, so that only perfectly clear liquid enters the precipitation boxes.

Zinc-Boxes. — These boxes usually consist of a number of compartments so arranged that the solution flows upward in each compartment through a mass of zinc shavings, then over a partition and downward through a narrow space to the bottom of the next compartment, where it again ascends through a mass of zinc shavings. The shavings are supported on perforated removable trays, so that any precipitate which may become detached may fall to the bottom of the box.

Lead-Zinc Couple. — The efficiency of precipitation is frequently increased by previously immersing the shavings in lead acetate, or by allowing a strong solution of lead acetate to drip slowly into the liquid at the head of the boxes. This produces a deposit of finely-divided lead on the zinc and probably sets up an electrical action which brings about the rapid replacement of gold or silver by zinc in the cyanide solution.

Disposal of Precipitated Solution. — The solution leaving the zinc-boxes passes usually to storage-tanks or sumps, where it is made up to the required strength by addition of fresh cyanide, and pumped back as required for the treatment of fresh charges of ore.

Accumulation of Zinc in Solution. — It might be supposed that, as the zinc is continually dissolving in the precipitation boxes, and the same liquid is repeatedly used in the treatment, the liquid would become gradually saturated with salts of zinc. It is generally found, however, owing to reactions which will be discussed later, that the accumulation of zinc ceases after the solutions have been in use for some time, and that the fresh quantities dissolved are compensated by the introduction of fresh water and the removal of old solution in the moisture of the discharged residues.

Alternative Methods of Precipitation. — Other methods of precipitation have frequently been suggested, since the zinc process, particularly in its earlier forms, presented certain obvious drawbacks. Only two of these, however, have found any extensive application, viz.:

1. Electric precipitation;
2. Zinc-dust precipitation.

The former has now been abandoned in favor of improved forms of the zinc process in all but a few localities, and the latter may be said to be still on its trial.

Electric Precipitation. — The electric process was devised by the late Dr. Wernher Siemens, and was introduced in Russia by A. von Gernet about the same time that the MacArthur-Forrest process was introduced in South Africa. Subsequently (in 1893), von Gernet, representing the firm of Siemens & Halske, of Berlin, introduced the process in the Transvaal, and for several years it was extensively used. The method then employed consisted in electrolyzing the solutions with anodes of iron and cathodes of sheet lead, the gold and silver being deposited on the latter, while the anodes were gradually consumed with formation of soluble ferrocyanides, Prussian blue, and other products. The cathodes were removed from time to time and cupeled for recovery of the gold and silver.

Advantages and Disadvantages of the Process. — The process presented the advantage that weaker cyanide solutions could be precipitated than was at that time possible with the zinc process; the dirty and troublesome operation of cleaning up the zinc-boxes was avoided, and the bullion was obtained in a purer and more marketable form. As, however, a very large area was needed for efficient precipitation, the boxes required were much larger, and in general, the initial outlay, and expenses for maintenance and skilled labor, particularly in the cupellation of the bullion, outweighed these advantages. When improved forms of the zinc process, particularly the use of the zinc-lead couple, were introduced, the electric process was no longer able to hold its own, though a modification of it, which will be described later, is still in use in Mexico.

Zinc-Dust Precipitation. — The use of zinc-dust for precipitation was first suggested by Sulman and Teed (1895), in conjunction with their bromocyanide process. It has been used to a considerable extent in America in conjunction with other forms of the dissolving process. The material used is a by-product obtained in the distillation of commercial zinc, and contains a certain amount of oxide, which is sometimes removed by preliminary treatment with ammonia or an ammonium salt. Being already in a very fine state of division, it presents a large surface and hence forms an efficient precipitant. It is generally agitated with the solution in a special tank, the agitation being obtained by forcing in compressed air.

Consumption of Zinc. — The consumption of zinc, when used in the form of shavings, amounts generally to 0.3 lb. per ton of ore treated; when in the form of dust, the consumption is generally less, but occasionally higher. The precipitate obtained by zinc-dust is more uniform in composition and may conveniently be pressed into cakes prior to smelting.

(C) THE SMELTING PROCESS

Nature of Zinc-Gold Precipitate. — The material collected when the zinc-boxes are periodically cleaned up consists of finely divided gold and silver, mixed with, and perhaps to some extent actually alloyed with, metallic zinc, together with greater or smaller quantities of such impurities as lead, copper, iron, and various insoluble metallic salts, such as carbonates, sulphates, complex cyanogen compounds, silica, alumina, etc. This is removed from the undecomposed zinc shavings as much as possible by rubbing on a sieve; the coarse zinc remaining on the sieve is returned to the boxes, and the remainder, usually, is passed over a finer sieve, the products remaining on and passing this sieve being described as "shorts" and "fines" respectively. These are often submitted to separate treatment.

Acid Treatment. — Formerly, the zinc-gold precipitate was usually simply dried, and smelted direct with suitable fluxes. It is now customary, however, to treat it with dilute sulphuric acid, to remove as much as possible of the metallic zinc and other impurities soluble in this acid. This treatment is generally carried out in a wooden vat, and is aided by stirring by hand or mechanically, and sometimes by heating with steam or otherwise. The vat should be provided with a cover and the operation conducted in a good draft, so that the very poisonous fumes evolved may be rapidly carried off. When the action of the acid is complete the precipitate is allowed to settle and the clear liquid decanted. One or two washes are given with hot water to remove zinc sulphate and other soluble salts, and frequently the residue is put through a small filter-press, after which the cakes are partially dried and smelted with fluxes.

Roasting of Zinc-Precipitate. — Another method often adopted instead of, or in conjunction with, acid treatment is to roast the precipitate in a small reverberatory furnace. In some cases the precipitate is intimately mixed with niter or other oxidizing agent previous to roasting. This operation may be carried out either before or after acid treatment. It is, however, supposed to generally involve considerable losses of gold and silver.

Smelting. — The precipitate, after undergoing one or other of these preliminary operations, is then mixed with a flux consisting of borax, soda, some siliceous material, and sometimes fluor-spar and an oxidizing agent. The fusion is generally made in graphite crucibles, but preferably with a clay lining. Various types of reverberatory and other furnaces have been used in place of crucibles; tilting furnaces, operating in the manner of a Bessemer converter, have also been used. The proportion of fluxes is varied according to the nature of the material to be smelted, and is of course determined largely by the preliminary treatment which it has undergone. Where an oxidizer has to be added manganese dioxide has been found to possess some advantages over niter.

Casting the Bullion. — Immediately before pouring, the bullion is stirred and a sample taken. The slag formed consists chiefly of silicates of zinc and other metals still present after the acid or other treatment. Where the fusion has been properly conducted, the bullion contains 900 to 950 parts gold and silver per 1000.

The slag and any other by-products of the smelting process should be tested for gold and silver; they frequently contain sufficient values to justify working over again by special processes.

Lead Smelting of Zinc Precipitate. — Another method of treating the zinc precipitate, which has been adopted with much success in South Africa and other countries, is the process of smelting with lead, introduced by P. S. Tavener. The precipitate (with or without previous acid treatment and filter-pressing) is mixed with litharge, powdered coal, siliceous slag or sand, and a little scrap-iron. On melting in a suitable furnace, the gold and silver are obtained as an alloy with lead; the slag, carrying the zinc, etc., is tapped off, and the lead bullion cast into bars. These are then cupeled on a bone-ash " test," and the litharge formed collected for use in a subsequent operation. The cupeled metal is then broken up and remelted with borax or other flux in clay or graphite crucibles.

The cost of this process is said to be very much less than that of other methods of smelting; practically all by-products are used up, and the comparatively rich slags from the ordinary fusion methods can be utilized as fluxes in the lead-smelting process. The slag produced in the fusion for lead bullion often contains less than 1 oz. gold per ton.

Refining with Oxygen. — The suggestion has also been made to refine zinc precipitate by passing oxygen gas or air through the molten material, as in the refining of gold by chlorine. A number of experiments were made with this object by T. K. Rose,[1] and some attempts have been made to apply the principle in practice, though we are not aware that it has been regularly adopted as yet.

[1] Rose, *Trans.* I. M. M., XIV, 378.

SECTION III

TENDENCIES OF MODERN CYANIDE PRACTICE

As already noted, the progress of the Cyanide Process since its first commercial application has hitherto been mechanical rather than chemical. The modifications of treatment have chiefly depended on the introduction of machines designed to effect greater economy or efficiency: (1) in the crushing of the ore; (2) in the separation of particles of different sizes; and (3) in the extraction of dissolved values from the treated pulp.

Crushing. — For some years a tendency towards more powerful crushing machinery, e.g., heavier stamps and improved types of fine grinding appliances, has been very noticeable, but there has recently been some reaction against the use of heavy stamps, and in fact there are not wanting signs that the stamp mill may shortly be entirely superseded as a crushing appliance in cyanide treatment. A greatly extended use of tube-mills and similar appliances has been noteworthy, and many efforts have been made to pass the ore from coarse crushers direct to these fine grinding machines.

There has latterly been some reaction against the idea of "all-sliming," and it seems to be generally admitted that greater economy is secured by treating a portion of the sandy particles by percolation.

Crushing in cyanide solution has become nearly universal, and, as a consequence, preliminary water-washing or alkali treatment is generally impossible, the necessary alkali being supplied in the form of lime added to the ore entering the mill.

Classification. — Spitzlutte and similar devices are being generally superseded by classifiers of the scraper or screw-conveyor type, such as the Dorr and Akins machines. It is recognized that a separation of slime should be made at the earliest possible stage, as this facilitates the subsequent handling of the sand. It is customary to return the entire pulp leaving the tube mill to a classifier placed at the head of the mill, the coarse product from

43

which is redelivered to the feed end of the tube-mill, the fine product passing to the agitation plant.

Agitation. — Various improved types of agitation apparatus have been introduced, with or without the air-lift principle, but here again we may note a tendency to revert to the old-fashioned mechanical agitator with revolving arms.

Filtration. — The use of vacuum filters is rapidly extending, but there has recently been a reaction in favor of the old decantation process, mainly in consequence of litigation among the owners of patent rights and a desire on the part of operators to avoid the payment of royalties. Continuous systems of agitation and decantation have been perfected, and in some cases the necessity for filtration has been more or less completely eliminated.

Precipitation. — In the precipitation of solutions we have to note a greatly extended use of zinc dust, the difficulties at first encountered in the application of the method having been practically overcome. Latterly aluminum dust has been used, with still better results, at least in the case of silver-bearing solutions. Filter-presses are extensively used both for separation of the precipitated values from the solution and for clarifying the liquid previous to precipitation.

Smelting. — The use of oil-fired furnaces is general in localities where wood or coal fuel is costly. In some instances a modified form of the Tavener process has found a successful application.

Electrolysis. — Electrolytic processes, in spite of various interesting efforts, have made little or no progress, either as a means of precipitation, or for improving the extractive power of solutions by eliminating cyanicides and regenerating cyanide in an available form.

PART II

CHEMISTRY

SECTION I

CYANOGEN AND ITS COMPOUNDS WITH NON–METALS

WE shall here give a summary of the modes of formation, physical properties, and principal reactions of the more important cyanogen compounds, but we shall confine the discussion principally to the inorganic compounds of the cyanogen radical. Only such organic derivatives as may be of some interest and importance in connection with the cyanide process or with the manufacture of cyanide will be referred to.

For convenience, we may subdivide this discussion of cyanogen and its compounds as follows:

Section I (pages 47–61).

    (*a*) Cyanogen.

    (*b*) Hydrocyanic Acid.

    (*c*) Compounds of Cyanogen with Non-metals.

Section II (pages 62–83).

    (*a*) Simple Metallic Cyanides, including easily decomposed Double Cyanides.

Section III (pages 84–101).

    (*a*) Metallic Compounds of Complex Cyanogen Radicals.

    (*b*) Cyanates, Thiocyanates, and Related Compounds.

For more detailed information respecting these compounds the reader is referred to such works as Watt's "Dictionary of Chemistry," revised by Morley and Muir, or to the articles by Joannis in Frémy's "Encyclopædia."

(*A*) CYANOGEN (CN)

(Molecular formula, C_2N_2)

Formation. — 1. By dry distillation of mercuric cyanide:

$$Hg(CN)_2 = Hg + (CN)_2.$$

A portion of the cyanogen polymerizes, forming the brown solid substance paracyanogen $(CN)_x$.

2. By dry distillation of silver cyanide:

$$2AgCN = 2Ag + (CN)_2.$$

3. By dry distillation of mercuric chloride with a cyanide or ferrocyanide:

$$HgCl_2 + MCN = HgCl + MCl + CN.$$

The gas obtained is in some cases mixed with nitrogen.

4. By action of copper salts on cyanides in solution

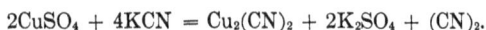

$$2CuSO_4 + 4KCN = Cu_2(CN)_2 + 2K_2SO_4 + (CN)_2.$$

5. Cyanogen cannot be formed by direct union of its elements, but it is a product of certain decompositions of compounds of carbon and nitrogen at a high temperature. Hence it is one of the products obtained by the destructive distillation of coal in the manufacture of illuminating gas.

Physical Properties. — A colorless gas, with a characteristic sharp odor resembling bitter almonds and prussic acid; it is very irritating to the eyes, and poisonous.

Density, 26 (H = 1), or 1.8064 (air = 1).

Easily condensed to a transparent, colorless, very mobile liquid, boiling under ordinary pressure at — 20.7° C., having sp. gr. 0.866 at 17.2° C., and refractive index 1.316.

Liquid cyanogen is a non-conductor of electricity; it has no action on most metallic or non-metallic elements, but dissolves iodine and phosphorus.

Cyanogen gas dissolves to some extent in water and alcohol.

Chemical Characteristics. — Cyanogen in its chemical relationships may be regarded from two different points of view.

(a) In many reactions it plays the part of a *compound radical;* that is to say, it behaves as though the nitrogen and carbon constituted a single element (often represented by the symbol Cy). In this respect it is in many ways analogous to the elements of the halogen group.

Compare	$(CN)_2$	Cl_2	Br_2	I_2
"	KCN	KCl	KBr	KI
"	AgCN	AgCl	AgBr	AgI
"	$Hg(CN)_2$	$HgCl_2$	$HgBr_2$	HgI_2

(b) From another point of view it belongs to the group of organic substances known as *nitriles*, which may be regarded as

derived from organic acids by the substitution of the radical CN
for CO_2H.

Compare Oxalic acid $(CO_2H)_2$ Cyanogen $(CN)_2$
                                           (oxalo-nitrile)

    "   Formic acid $(H \cdot CO_2H)$ Hydrocyanic acid $(H \cdot CN)$
                                           (formo-nitrile)

    "   Acetic acid $(CH_3 \cdot CO_2H)$ Aceto nitrile $(CH_3 \cdot CN)$

Its constitution may be represented by the symbol:

$$N \equiv C - C \equiv N$$

Reactions. — 1. Decomposed by electric arc into carbon and
nitrogen. Not decomposed by heat alone.

2. Burns (in air or oxygen) with a characteristic purplish
(peach-colored) flame, forming carbon-dioxide and liberating
nitrogen:

$$(CN)_2 + 2O_2 = N_2 + 2CO_2.$$

3. Combines with hydrogen when heated with it in a closed
tube or under the influence of the electric current:

$$(CN)_2 + H_2 = 2HCN.$$

Some paracyanogen is also formed.

4. The alkali metals combine slowly with cyanogen at ordinary temperatures, and with incandescence on warming:

$$(CN)_2 + K_2 = 2KCN.$$

Zinc combines slowly on heating. Iron decomposes it into C
and N at a red heat. Hg, Cu, Au, and Pt have no direct action.

5. Certain non-metals (S, P, I) may be vaporized unchanged
in an atmosphere of CN. Perfectly dry Cl has no action on it,
but in presence of moisture a yellowish oil is deposited, apparently a mixture of chlorides of C and N, and also a white,
solid, aromatic substance.

6. A solution of cyanogen in water gradually undergoes
decomposition, turning first yellow, then brown, and depositing
flakes of a brown substance. The first action is probably the'
formation of cyanic and hydrocyanic acids:

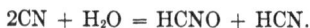

$$2CN + H_2O = HCNO + HCN.$$

Further decomposition then takes place, giving ammonium
oxalate, ammonium carbonate, urea, and azulmic acid (the insol-

uble brown substance),[1] to which the formula $C_4H_5N_5O$ has been assigned.

7. Dry H_2S gas does not combine with cyanogen.   In presence of water two yellow crystalline substances are formed having the composition:

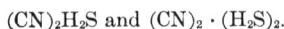

$$(CN)_2H_2S \text{ and } (CN)_2 \cdot (H_2S)_2.$$

The first is formed when CN is in excess; the second when H_2S is in excess.

Solutions of these compounds give precipitates with various metallic salts.

8.   Concentrated aqueous HCl converts cyanogen into oxamide and ammonium oxalate, thus acting as a "hydrating agent":

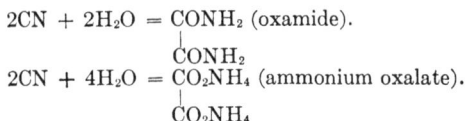

$$2CN + 2H_2O = \underset{\overset{|}{CONH_2}}{CONH_2} \text{ (oxamide).}$$

$$2CN + 4H_2O = \underset{\overset{|}{CO_2NH_4}}{CO_2NH_4} \text{ (ammonium oxalate).}$$

Hydriodic acid gives the same products in the cold, but at a high temperature gives ammonia and ethane, thus:

$$2CN + 12HI = 2NH_3 + C_2H_6 + 6I_2.$$

9.   Alkalis absorb cyanogen and give reactions similar to those of water:

$$Cy_2 + 2KOH = KCy + KCyO + H_2O.$$
[This reaction is analogous to that of chlorine:
$$Cl_2 + 2KOH = KCl + KClO + H_2O.]$$

On treatment of the solution with strong acids, ammonium salts are formed and carbonic acid evolved (by decomposition of the cyanate):

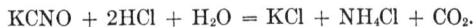

$$KCNO + 2HCl + H_2O = KCl + NH_4Cl + CO_2.$$

A slight decomposition occurs without addition of acids, according to the reaction

$$KCNO + KOH + H_2O = K_2CO_3 + NH_3.$$

When excess of cyanogen is used, the solution becomes brown and deposits an alkaline azulmate.

HgO acts like caustic alkalis: if a current of cyanogen be passed into water containing mercuric oxide in suspension, it

[1] Wöhler, "Pogg. Ann.," III, 177.   Richardson, "Philosophical Magazine," XII, 339.

yields azulmic acid, ammonium carbonate, and mercuric cyanide and oxycyanide.

10. Cyanogen passed over potassium carbonate heated to redness gives a cyanate and cyanide:

$$Cy_2 + K_2CO_3 = KCy + KCyO + CO_2.$$

11. Dry ammonia gas gradually combines, depositing a brown substance (hydrazulmin):

$$2(CN)_2 + 2NH_3 = C_4H_6N_6.$$

Aqueous ammonia gives azulmic acid, oxamide, oxamic acid, and ammonium oxalate.

PARACYANOGEN $(CN)_4$[1]

Preparation. — This polymer of cyanogen is obtained by heating mercuric cyanide in sealed tubes to about 440° C.; also by decomposition of hydrazulmin (see above):

$$C_4H_6N_6 = (CN)_4 + 2NH_3.$$

Properties. — A blackish-brown, light, spongy substance, insoluble in water, soluble in H_2SO_4, and precipitated by water from the solution in this acid.   It absorbs and condenses gases.

(B) HYDROCYANIC ACID (HCy)

(Prussic Acid; Hydric Cyanide)

Preparation. — 1. By heating mercuric cyanide with hydrochloric acid:

$$HgCy_2 + 2HCl = HgCl_2 + 2HCy.$$

The vapor may be rendered anhydrous by passing through tubes containing $CaCO_3$ and $CaCl_2$, which remove HCl and H_2O respectively.   Only about $\frac{2}{3}$ of the theoretical quantity is obtained, owing to $HgCl_2$ forming a compound with HCN.   The yield may be increased by adding NH_4Cl to the mixture.

2. By passing H_2S over $HgCy_2$ or $AgCy$, heated in a tube to 30° − 40° C.

$$H_2S + HgCy_2 = HgS + 2HCy.$$

Lead carbonate is used to absorb excess of H_2S.

3. By action of dilute sulphuric acid on potassium ferrocyanide (the usual method of preparation):

$$K_4FeCy_6 + 3H_2SO_4 = 3KHSO_4 + KFeCy_3 + 3HCy.$$

[1] The formula CN – C = C – CN has been assigned for paracyanogen.
$$\overset{|}{N} = \overset{|}{N}$$

The most suitable proportions are:

Potassium ferrocyanide .... 8 parts
Sulphuric acid (conc.) 9   "
Water 12   "

The water and acid should be mixed before hand, and cooled before adding to the powdered ferrocyanide. The distilling-flask is heated on a sand bath and the distillate collected in a cooled condenser.

4. By the action of acids on almost any metallic cyanide: *e.g.*,

$$KCy + HCl = KCl + HCy.$$

This class of reaction has been utilized for preparing dilute solutions of hydrocyanic acid of known strength for medicinal purposes.

(*a*) $AgCy + HCl = AgCl + HCy$ (Everitt).

(*b*) $PbCy_2 + H_2SO_4 = PbSO_4 + 2HCy$ (Thomson).

(*c*) Tartaric acid acting on KCy yields a nearly insoluble acid potassium tartrate, together with HCy (Clarke).

5. When a series of electric sparks are passed through a mixture of acetylene and nitrogen, there is a gradual combination.

$$\text{Thus: } C_2H_2 + N_2 = 2HCN.$$

Part of the acetylene decomposes with separation of carbon, unless hydrogen be also added.

This reaction is of great interest, as it shows how cyanide compounds may be formed synthetically from their elements: acetylene is produced by direct union of carbon and hydrogen, when electric sparks are passed between carbon-points in an atmosphere of hydrogen.

6. By passing a current of dry ammonia gas over carbon contained in a porcelain tube heated to redness:

$$C + 2NH_3 = NH_4CN + H_2.$$

By passing the vapors through warm dilute sulphuric acid HCy is liberated:

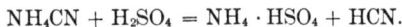

$$NH_4CN + H_2SO_4 = NH_4 \cdot HSO_4 + HCN.$$

A similar reaction takes place when ammonia and carbon monoxide are passed through a red-hot tube:

$$CO + 2NH_3 = NH_4CN + H_2O.$$

7. By dehydration of ammonium formate:

$$HCO_2NH_4 = 2H_2O + HCN.$$

This is a general reaction for forming the "nitriles" of the corresponding organic acids; thus, ammonium oxalate yields cyanogen, and ammonium acetate yields acetonitrile when acted on by a dehydrating agent.

8. By destructive distillation of ammonium oxalate or oxam-ide:

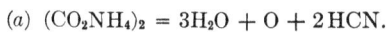

$$(a)\ (CO_2NH_4)_2 = 3H_2O + O + 2HCN.$$
$$(b)\ (CONH_2)_2 = H_2O + O + 2HCN.$$

9. Hydrocyanic acid occurs ready formed in, or is a product of the decomposition of, many vegetable substances. The kernels of many fruits, such as peaches, cherries, almonds, etc., and certain leaves, as those of the cherry-laurel, contain a substance known as amygdalin, which is decomposed by water under the influences of a vegetable ferment (emulsin), the latter undergoing no apparent change in the reaction. This reaction results in the formation of benzaldehyde, glucose, and hydrocyanic acid:

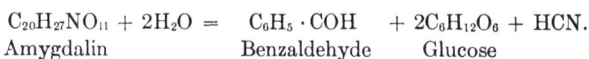

$$C_{20}H_{27}NO_{11} + 2H_2O = \quad C_6H_5 \cdot COH \quad + 2C_6H_{12}O_6 + HCN.$$
$$\text{Amygdalin} \qquad\qquad \text{Benzaldehyde} \qquad \text{Glucose}$$

10. Hydrocyanic acid occurs as one of the products in many reactions of organic bodies; e.g., in the action of ammonia on chloroform:

$$NH_3 + CHCl_3 = 3HCl + HCN;$$

in the combustion of methylamine:

$$CH_3NH_2 + O_2 = 2H_2O + HCN;$$

and in the action of alkalis on nitrobenzene.

The *anhydrous acid* may be prepared by gently heating the aqueous acid, passing the vapors through vessels containing calcium chloride, and condensing the distillate by a freezing mixture.

Physical Properties. — A volatile colorless liquid; extremely poisonous. Boiling-point + 26.1° C.

When allowed to evaporate in the air, a portion solidifies in translucent orthorhombic prisms, having a melting-point − 14° C.

Density of liquid,   0.7058 at 7° C.
                     0.6969 at 18° C.

Density of vapor,   .97 to .90 [air = 1]
                    13.5 [H = 1].

Very soluble in water.   The addition of a small quantity of water to the anhydrous acid raises the boiling-point and lowers the melting-point. The solution of HCN in water is accompanied by condensation and lowering of temperature (an unusual effect).

A mixture of 60 per cent. HCN and 40 per cent. H_2O has the minimum melting-point $(-22.5°$ C.).

Reactions. — 1. Heated in closed vessels at 100° C., it forms a black mass, which when afterward heated in an open tube at 50° C. gives off NH_4CN and $(CN)_2$ leaving a carbonaceous residue. A current of HCN passed through a hot tube gives $(CN)_2$, H, N and C.; this reaction is reversible (see Cyanogen: Reaction No. 3).

2. When the vapor is decomposed by electric sparks it yields acetylene and nitrogen

$$2HCN \rightleftharpoons C_2H_2 + N_2 \text{ (reversible)}.$$

An electric current passed between platinum electrodes, through liquid HCN, gives H at the negative pole, but no gas is evolved at the positive pole, as the cyanogen attacks the platinum.[1]

3. Burns (in air or oxygen) with a white flame, bordered with purple. Explodes in O with great violence, yielding CO_2, H_2O, N, and a little HNO_3.

4. Action of non-metals: Sulphur vapor gives the same products as H_2S on CN (see Cyanogen: Reaction No. 7).

Phosphorus has no action.

Dry Cl and anhydrous HCN give

$$HCN + Cl_2 = HCl + CNCl.$$

Cl and aqueous HCN give only a little CNCl, the chief products being CO_2, CO, HCl and NH_3.

Nascent H gives methylamine:

$$HCN + 2H_2 = CH_3 \cdot NH_2.$$

5. Action of strong acids: Strong HCl and moderately concentrated H_2SO_4 convert it into ammonium formate, which is decomposed on heating, yielding formic acid:

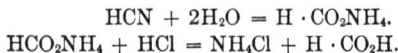

$$HCN + 2H_2O = H \cdot CO_2NH_4.$$
$$HCO_2NH_4 + HCl = NH_4Cl + H \cdot CO_2H.$$

The gaseous acids (HCl, HBr, HI) form addition-compounds with anhydrous HCN, usually of the type $HX \cdot HCN$.

Concentrated H_2SO_4, heated in a closed vessel with HCN to 30–35° C., turns brown; SO_2, CO_2, and N are given off.

6. Action of alkalis: At ordinary temperatures, cyanides are produced:

$$HCN + KOH = KCN + H_2O.$$

---

[1] Davy, "Journ. Science and Arts," I, 288.

On heating, a formate of the metal is produced and ammonia is evolved:

$$KCN + 2H_2O = HCO_2K + NH_3.$$

7. Action of metallic oxides: Most metallic oxides yield cyanides, in some cases with evolution of cyanogen.

8. Action of metallic salts: Hydrocyanic acid is a very weak acid; when absolutely pure it does not redden litmus: it is set free from its compounds by almost every acid, even by CO_2. Alkaline carbonates and most acids are, however, decomposed by HCN, the reaction being to some extent reversible.

Salts of metals which give insoluble cyanides are decomposed by HCN with precipitation of the cyanide (*e.g.*, salts of Ag, Cu, Pb).

Alkaline polysulphides give thiocyanates (MCNS).

9. Spontaneous decomposition: Anhydrous HCN may be preserved in sealed vessels without decomposition. If, however, a trace of ammonia or alkali be present, it decomposes, especially if exposed to light, turning brown and yielding ammonium cyanide and azulmic acid, probably first polymerizing to $(HCN)_3$.

Poisonous Action of Hydrocyanic Acid. — The poisonous action seems to be chiefly paralysis of the nervous centers. Anything which will act as a stimulus to the nerves, such as inhaling chlorine or ammonia, will act as an antidote, although the products of the reaction in these cases are also strong poisons. The symptoms produced by inhaling small quantities of hydrocyanic acid gas are headache, giddiness, a peculiar sensation at the back of the throat, and sometimes nausea. Long continual respiration of small quantities of hydrocyanic acid, as in some operations in the cyanide plant, particularly in cleaning up, may give rise to chronic affections of the heart and throat (palpitations and hoarseness). The poisonous action of soluble cyanides is due to the liberation of HCN by the acids of the stomach. Various antidotes have been proposed, such as salts of cobalt, hypodermic injection of peroxide of hydrogen, etc., but the simplest and most effective seems to be freshly prepared ferrous carbonate.

H. C. Jenkens [1] recommends the following solutions to be prepared and kept handy as an antidote for cases of cyanide poisoning: (*a*) A hermetically sealed bottle containing $7\frac{1}{2}$ grams ferrous sulphate in 30 c.c. of distilled water, previously well boiled. (*b*) A hermetically sealed bottle containing $1\frac{1}{2}$ grams

[1] *Trans.* I. M. M., XIII, 484.

NaOH in 300 c.c. water. (c) A tube containing 2 grams magnesia (MgO). When used, the sealed ends of the two bottles are to be broken off and the solutions mixed, the magnesia added, and the whole administered as rapidly as possible.

Anything which will induce vomiting also has a beneficial effect.

Soluble cyanides introduced into cuts in the skin act as blood poisons, producing sores and eruptions.

(C) Compounds of Cyanogen with Non-Metals

The compounds with sulphur have been already mentioned. The following compounds with the haloid elements are known:

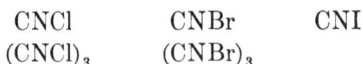

$$CNCl \qquad CNBr \qquad CNI$$
$$(CNCl)_3 \qquad (CNBr)_3$$

The chloride CNCl is gaseous at the ordinary temperature; the others are solid.

Chlorides of Cyanogen

Gaseous Chloride (CNCl)

Prepared: 1. By action of Cl on anhydrous HCN; the excess of Cl is removed by agitation with mercury.

2. By action of Cl on mercuric cyanide (solid or in saturated solution):

$$Hg(CN)_2 + 2Cl_2 = HgCl_2 + 2CNCl.$$

3. By passing Cl through a solution of KCN surrounded by a freezing-mixture: the product is then distilled and the gaseous chloride collected and liquefied in a cooled condenser:

$$KCN + Cl_2 = KCl + CNCl.$$

Physical Properties. — Very volatile liquid; boils at 15.5° C. Density of vapor, 20.2 (H = 1). Solidifies at − 5° C.

Very poisonous; has extremely irritating odor; produces coughing and tears. Soluble in alcohol, ether, and water.

Reactions.— 1. Polymerizes, forming the solid chloride $(CNCl)_3$ in presence of a slight excess of Cl.

2. Decomposed when heated with certain metals, yielding a cyanide or cyanogen:

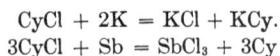

$$CyCl + 2K = KCl + KCy.$$
$$3CyCl + Sb = SbCl_3 + 3Cy$$

3. Decomposed by water in two stages:

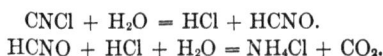

$$CNCl + H_2O = HCl + HCNO.$$
$$HCNO + HCl + H_2O = NH_4Cl + CO_2.$$

4. Alkalis give an analogous reaction:

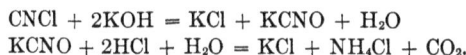

$$CNCl + 2KOH = KCl + KCNO + H_2O$$
$$KCNO + 2HCl + H_2O = KCl + NH_4Cl + CO_2.$$

5. Dry ammonia gas forms ammonium chloride and cyanamide:

$$CNCl + 2NH_3 = NH_4Cl + CN \cdot NH_2.$$

6. Does not precipitate AgCl or AgCN from silver nitrate.

SOLID CHLORIDE $(CNCl)_3$

Preparation. — 1. By exposing to sunlight a mixture of chlorine with excess of $Hg(CN)_2$ in sealed tubes:

$$3Hg(CN)_2 + 6Cl_2 = 3HgCl_2 + 2(CNCl)_3.$$

2. By action of phosphorus pentachloride on cyanuric acid:

$$(CNOH)_3 + 3PCl_5 = 3POCl_3 + 3HCl + (CNCl)_3.$$

3. By slightly heating dry KCNS in a current of dry chlorine:

$$3KCNS + 6Cl_2 = 3KCl + 3SCl_2 + (CNCl)_3.$$

The $(CNCl)_3$ crystallizes from the mixture.

Physical Properties. — Small needles (density 1.32), melting at 140° C. to a transparent liquid which boils at 190° C.

Vapor density, 6.35 (air = 1). Very poisonous. Easily provokes tears. Smells of chlorine, but also has an odor suggestive of mice.

Reactions. — 1. Decomposed by water (more rapidly by alkalis), forming a cyanurate:

$$(CNCl)_3 + 3H_2O = 3HCl + (CNOH)_3.$$

2. Ammonia gives chlorocyanuramide:

$$(CN)_3(NH_2)_2 \cdot Cl.$$

BROMIDES OF CYANOGEN

MONOBROMIDE (CNBr)

Preparation. — 1. By adding a strong solution of KCN gradually to an excess of bromine until the latter is discolored, then distilling and condensing the product

$$KCN + Br_2 = KBr + CNBr.$$

2. By acting upon sodium bromide and bromate with sodium cyanide and sulphuric acid at 70° C. (Göpner.)[1]

$$2NaBr + NaBrO_3 + 3NaCN + 3H_2SO_4 = 3Na_2SO_4 + 3H_2O + 3CNBr.$$

3. By adding bromine drop by drop to a solution of HCN:

$$HCN + Br_2 = HBr + CNBr.$$

When Br is added to an excess of cyanide (reversing the order of procedure in the first method), there is a tendency to formation of the polymer $(CNBr)_3$.

4. One part of Br is added gradually to two parts mercuric cyanide:

$$Hg(CN)_2 + 2Br_2 = HgBr_2 + 2CNBr.$$

The same reaction may be brought about by adding a mixture of mercuric cyanide and hydrochloric acid to bromine in a cooled vessel, and then distilling off the CNBr by placing the vessel in hot water.

Physical Properties. — A transparent crystalline solid, said to melt at 4° − 12° C.; at ordinary pressures, however, it volatizes without melting and is gaseous above 61° C. It is volatile above 15° C. It has a very penetrating odor, attacking the eyes and causing a flow of tears.

Very poisonous, but less so than HCN. Very soluble in water and alcohol. With water it forms a hydrate which is less volatile than the pure bromide of cyanogen. Solution bleaches litmus without reddening it. Crystallizes in long hexagonal needles, which soon change to cubic or tabular crystals.[2]

Reactions. — 1. Gradually decomposed by water with formation of cyanic acid and hydrobromic acid:

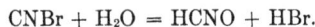

$$CNBr + H_2O = HCNO + HBr.$$

When an aqueous solution of CNBr is evaporated to dryness, ammonium bromide is obtained and carbon dioxide evolved:

$$CNBr + 2H_2O = NH_4Br + CO_2.$$

2. Decomposed rapidly by alkalis, with formation of a bromide and cyanate:

$$CNBr + 2KOH = KBr + KCNO + H_2O.$$

3. Acids (H_2SO_4, HCl, HNO_3) dissolve it without decomposition, but SO_2 is oxidized by CNBr in presence of water:

$$CNBr + SO_2 + 2H_2O = HBr + H_2SO_4 + HCN.$$

[1] See "Journ. Soc. Chem. Ind.," XXV, 1130.
[2] H. L. Sulman, "Proc. Chem. Met. and Min. Soc. of South Africa," I, 114.

4. Soluble cyanides act rapidly, probably evolving cyanogen as follows:

$$KCN + CNBr = KBr + (CN)_2.$$

It is possible, however, that hydrocyanic acid is evolved, as ollows:

$$CNBr + 2KCN + H_2O = KCNO + KBr + 2HCN.$$

The liberation of cyanogen in an active form appears to be the cause of the rapid solvent power on gold of the mixture of KCN and CNBr.

5. *Action on Metals.* — Certain metals are attacked, giving a metallic bromide with evolution of cyanogen:

$$Hg + 2CNBr = HgBr_2 + (CN)_2 \text{ (in solution)}.$$

A similar reaction occurs with the vapor of antimony.

Gold and silver are rapidly dissolved by a mixture of CNBr and an alkali cyanide, forming the soluble double cyanides:

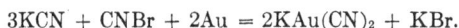

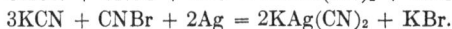

$$3KCN + CNBr + 2Au = 2KAu(CN)_2 + KBr.$$
$$3KCN + CNBr + 2Ag = 2KAg(CN)_2 + KBr.$$

This represents the final effect of the reaction, but it may take place in several stages. Bromide of cyanogen alone does not attack gold or silver.

6. Ammonia decomposes cyanogen bromide, giving bromide and cyanate of ammonium:

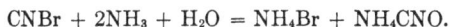

$$CNBr + 2NH_3 + H_2O = NH_4Br + NH_4CNO.$$

POLYMER OF CYANOGEN BROMIDE $(CNBr)_3$

This substance is sometimes obtained during the preparation of ordinary cyanogen bromide. It is formed by heating CNBr for 5 to 6 hours at 130°–140° in sealed tubes, but is partially decomposed with liberation of bromine. It is obtained in a purer condition by heating CNBr with anhydrous ether.

This is a solid body, melting at 300° C., and boiling, with decomposition, at a higher temperature. It is insoluble in alcohol and nearly insoluble in ether.

Exposed to moist air, or heated with water to 100° in sealed tubes, it forms cyanuric acid:

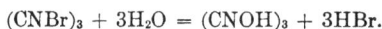

$$(CNBr)_3 + 3H_2O = (CNOH)_3 + 3HBr.$$

IODIDE OF CYANOGEN (CNI)

Preparation. — 1. Not formed by direct union of the elements.

2. Formed by action of free iodine on the cyanides of mercury, silver, or potassium:

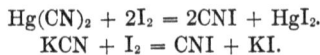

$$Hg(CN)_2 + 2I_2 = 2CNI + HgI_2.$$
$$KCN + I_2 = CNI + KI.$$

The iodide of cyanogen is separated from the mixture by distilling at a temperature below 135° C.

Physical Properties. — Forms long needles, soluble in water, alcohol or ether. Crystallizes from these solutions in four-sided tables. Volatile at ordinary temperatures and boils below 100° C. Has a penetrating odor resembling CN and I. Poisonous. Does not redden litmus.

Reactions. — 1. Decomposed by heat, on passing through a red-hot tube:

$$2CNI = (CN)_2 + I_2.$$

2. Gradually decomposed by water:

$$CNI + H_2O = HCNO + HI.$$

3. Dissolved by alkalis with formation of a cyanide and an iodate, a little cyanate being formed at the same time:

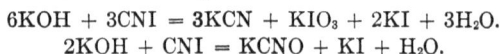

$$6KOH + 3CNI = 3KCN + KIO_3 + 2KI + 3H_2O.$$
$$2KOH + CNI = KCNO + KI + H_2O.$$

(Note difference from CNBr.)

4. *Action of Acids.* — HCl and H_2SO_4 slowly decompose it with liberation of HCN and I at ordinary temperatures.

Sulphurous acid acts as follows:

$$CNI + SO_2 + 2H_2O = HI + H_2SO_4 + HCN.$$

5. Soluble cyanides react as with CNBr, but less vigorously. The action on metals is also similar.

6. *Action of Non–metals.* — Chlorine does not decompose dry CNI.

Phosphorus has a violent action, often accompanied by light; iodide of phosphorus is formed and cyanogen probably liberated. (Phosphorus has only a slight action on CNBr.)

7. Dry H_2S gives:

$$2CNI + H_2S = SI_2 + 2HCN.$$

In presence of water, the reaction is:

$$CNI + H_2S = S + HI + HCN$$

8. Ammonia is slowly absorbed, giving ammonium iodide and cyanamide:

$$CNI + 2NH_3 = NH_4I + CN \cdot NH_2.$$

9. Solution of CNI in water gives no precipitate with solutions of silver salts.

10. On mixing, in the following order, solutions of CNI, KOH, $FeSO_4$ and HCl, a green precipitate is obtained. This reaction does not occur with CNCl or CNBr, unless the $FeSO_4$ be added before the KOH.

SECTION II

SIMPLE METALLIC CYANIDES

(INCLUDING THE EASILY–DECOMPOSED DOUBLE CYANIDES)

WE shall now consider those metallic compounds in which the group CN occurs as the negative radical. This includes all the substances which can correctly be described as metallic cyanides. Compounds such as ferrocyanides, nitroprussides, thiocyanates, etc., have little in common with the true cyanides, and are more conveniently considered as compounds of entirely distinct radicals, $Fe(CN)_6$, $Fe(CN)_5NO$, (SCN), etc.

The true cyanides have the general formula

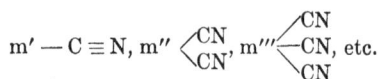

$$m' - C \equiv N, \quad m'' \underset{CN}{\overset{CN}{<}}, \quad m''' \overset{CN}{\underset{CN}{\overset{-CN}{<}}}, \text{ etc.}$$

where m', m'', etc., are positive monad, dyad, etc., elements or radicals. They show, however, a great tendency to combine with each other, forming compounds of the type

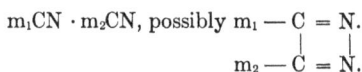

$$m_1CN \cdot m_2CN, \text{ possibly } \begin{array}{c} m_1 - C = N. \\ | \quad | \\ m_2 - C = N. \end{array}$$

where m_1 and m_2 are different positive radicals.

It is also possible that isomeric modifications exist in which the positive radical is united directly with carbon or nitrogen:

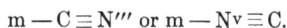

$$m - C \equiv N''' \text{ or } m - N^v \equiv C.$$

corresponding with the organic groups known as nitriles and carbamides.

General Modes of Formation. — 1. By direct action of cyanogen on metals:

$$(CN)_2 + 2K = 2KCN.$$

2. By action of cyanogen on dissolved alkalis with simultaneous formation of a cyanate:

$$(CN)_2 + 2KOH = KCN + KCNO + H_2O.$$

62

3. By action of hydrocyanic acid on metals, oxides, or metallic salts.

4. By double decomposition, with metallic salt and other cyanides:
$$M_1CN + M_2X = M_1X + M_2CN.$$

where M_1, M_2 are two different metallic radicals and X a negative (acid) radical.

5. By calcination of nitrogenous organic matter at a high temperature with alkalis or alkaline carbonates.

6. By passing free nitrogen over a mixture of carbon and an alkaline carbonate heated to redness.

7. By calcining certain organic salts such as acetates and tartrates, with the nitrates or nitrites of the alkalis.

General Properties. — Cyanides of the alkalis and alkaline earths are soluble in water; the simple cyanides of mercury and cadmium are also soluble; other simple cyanides are insoluble. Many of the insoluble cyanides unite with alkaline or other soluble cyanides to form soluble double salts.

Soluble compounds are also formed with ammonium salts, and with certain other metallic salts.

Most soluble cyanides have an alkaline reaction and are very poisonous.

The soluble cyanides are colorless. The insoluble cyanides often have characteristic colors.

General Reactions. — 1. Action of heat: The alkali cyanides, when anhydrous, are not decomposed by ordinary heat. KCN and NaCN may be fused without decomposition. Hydrated NaCN is partially decomposed with evolution of NH_3. Of the cyanides of the alkaline earths, $BaCy_2$ is the most stable and $MgCy_2$ the least. The former may be heated to redness without decomposition. Some cyanides (Zn, Cu) are decomposed by heat into a mixture of C with the metal or a metallic carbide, N being given off; others (Hg, Ag) are decomposed into metal and CN.

2. *Action of Oxygen.* — Alkaline cyanides are converted into cyanates, and by stronger heating into carbonates with evolution of N. The reaction takes place more readily in presence of an oxidizing agent. With other cyanides, CO_2 and N are given off, and an oxide of the metal remains. Cyanides explode in contact with certain oxidizing bodies (nitrates, chlorates).

3. *Action of Chlorine.* — Varies under different conditions. In

general a chloride is formed, with free CN and more or less CNCl. Bromine and iodine act similarly.

4. *Action of Water.* — Decomposes all cyanides at a higher or lower temperature, generally at 100° C. Boiling solutions of alkaline cyanides are converted into alkaline formates with evolution of ammonia:

$$mCN + 2H_2O = HCO_2m + NH_3.$$

Cyanides of the heavy metals, heated in presence of water, decompose, yielding CO_2, CO, HCN, NH_3, the metal remaining as oxide, or in the free state mixed with a little carbon.

Cyanides of Hg and Ag, heated with water in sealed tubes to 280° C., give ammonium carbonate and the free metal.[1]

5. *Action of Acids.* — Dilute mineral acids decompose most metallic cyanides with evolution of HCN. Many organic acids also decompose cyanides. Weak acids, such as CO_2, decompose the cyanides of the alkalis and alkaline earths, and some double cyanides. The cyanides of Hg, Ag, and Pd are not decomposed by dilute HCl, but on the contrary HCN displaces HCl in the chlorides of these metals. Nitric acid decomposes cyanides with liberation of N and CO_2 and formation of oxalic acid. Concentrated H_2SO_4 gives a sulphate of the metal, and of ammonium, and forms CO.

6. Mercuric oxide decomposes most cyanides, when boiled with their solution, giving mercuric cyanide ($HgCy_2$) and the corresponding oxide. This reaction is owing to the great stability of mercuric cyanide.[1]

DETAILS RESPECTING CERTAIN METALLIC CYANIDES

Ammonium Cyanide (NH_4CN): *Preparation.* — 1. By heating ammonium ferrocyanide and condensing the product in a cooled receiver:

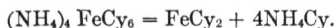

$$(NH_4)_4 FeCy_6 = FeCy_2 + 4NH_4Cy.$$

2. By heating ammonium salts with certain cyanides or ferrocyanides, *e.g.:*

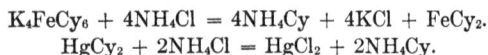

$$K_4FeCy_6 + 4NH_4Cl = 4NH_4Cy + 4KCl + FeCy_2.$$
$$HgCy_2 + 2NH_4Cl = HgCl_2 + 2NH_4Cy.$$

3. By direct union of NH_3 gas and HCN.

---

[1] It is probable that most cyanides are ionized in dilute solution. With mercuric cyanide, however, this is not the case.

4. By passing NH_3 over red-hot carbon:

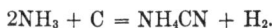

$$2NH_3 + C = NH_4CN + H_2.$$

also

$$2NH_3 + CO = NH_4CN + H_2O.$$

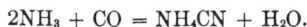

It is also one of the products of combustion of illuminating gas containing ammoniacal vapors, and is formed by action of HNO_3 on some organic substances.

Physical Properties. — Crystallizes in cubes. Very volatile and poisonous. Boils about 36° C., and is easily dissociated into HCN and NH_3. Vapor density at 100° C. 0.79 (air = 1).

Reactions. — 1. Burns in air or oxygen, giving a yellowish flame, depositing ammonium carbonate.

2. Very unstable, particularly at high temperatures. Turns brown and passes into azulmic acid.

3. Decomposed by chlorine or bromine:

$$NH_4Cy + Cl_2 = CyCl + NH_4Cl.$$

CYANIDES OF ALKALI METALS

Potassium Cyanide (**KCN**): *Preparation* (methods mostly applicable also to **sodium cyanide**). — 1. By direct action of hydrocyanic acid on potash (best in alcoholic solution):

$$HCN + KOH = H_2O + KCN.$$

2. By decomposition of ferrocyanide: (*a*) By heating in closed vessels; (*b*) by heating with dry potassium carbonate; (*c*) by action of metallic sodium (giving a mixture of NaCy and KCy). These methods will be discussed in detail under the title "Manufacture of Cyanide."

(*a*) The effect of heating K_4FeCy_6 by itself in a closed vessel is to partially break it up into N and a carbide of iron, mixed with KCN.

(*b*) When heated with K_2CO_3:

$$K_4FeCy_6 + K_2CO_3 = 5KCy + KCyO + CO_2 + Fe.$$

Similar reaction with Prussian blue and K_2CO_3.

(*c*) With metallic sodium:

$$K_4FeCy_6 + Na_2 = 2NaCy + 4KCy + Fe.$$

3. By calcining nitrogenous organic matter with K_2CO_3. A part of the carbon liberated reduces K_2CO_3 to K, which then reacts with N and more C to form KCN.

4. By passing atmospheric nitrogen over a mixture of carbon and potash heated to redness:

$$N_2 + 4C + K_2CO_3 = 2KCN + 3CO.$$

or

$$N_2 + 3C + 2KOH = 2KCN + CO + H_2O.$$

Probably K is first liberated in this case also.

5. By precipitating mercuric cyanide with potassium sulphide:

$$HgCy_2 + K_2S = HgS + 2KCy.$$

Physical Properties. — Anhydrous. Crystallizes in cubic system by fusion and in cubic octahedra from aqueous solutions. Deliquescent; — readily soluble in water; insoluble in absolute alcohol. Strongly alkaline taste and reaction. Very poisonous. Gives off HCN when exposed to air. Melts at red heat and volatilizes at a white heat without decomposition.

Reactions. — 1. When heated in contact with air a little KCNO is always formed. When heated with metallic oxides, such as those of Pb, Sb, Fe, As, Sn, or Mn (MnO_2), larger quantities of cyanate are produced, and the oxide is reduced to metal, or, in the case of MnO_2, to MnO. When heated with oxidizing agents (nitrates or chlorates) it explodes violently.

$KMnO_4$ gives as oxidation products CO_2, HCO_2H, $(CO_2H)_2$, $CO(NH_2)_2$, HNO_2 and NH_3.

Chloride of lime gives calcium cyanate.

Potassium sulphate gives cyanate and sulphide:

$$K_2SO_4 + 4KCN = 4KCNO + K_2S.$$

2. The aqueous solution decomposes even at ordinary temperature, giving potassium formate and ammonia, but much more rapidly on boiling:

$$KCN + 2H_2O = HCO_2K + NH_3.$$

3. On electrolysis it gives CO_2, NH_3 and KOH.

4. When melted with sulphur it forms a thiocyanate:

$$KCN + S = KSCN.$$

Aqueous solution of KCN does not dissolve sulphur, but polysulphides react with KCN in solution as follows:

$$2KCN + K_2S_3 = 2KCNS + K_2S.$$

Selenium reacts in a similar way, but also dissolves in KCN solution, even at ordinary temperature, to form seleno cyanide —

KSeCN (soluble in water to a colorless solution). When fused with tellurium it forms potassium telluride (giving a pink solution):

$$2KCN + Te = K_2Te + (CN)_2.$$

5. Sulphureted hydrogen passed into a solution of KCN gives small yellow needles of "chrysean" and potassium and ammonium sulphides in solution:

$$4KCN + 5H_2S = 2K_2S + NH_4HS + C_4H_5N_3S_2.$$

6. On heating with alkalis, it is transformed first into HCO_2K and NH_3; but on heating to redness H is evolved and an alkaline carbonate remains.

7. Thiosulphates convert it into thiocyanate:

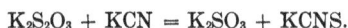

$$K_2S_2O_3 + KCN = K_2SO_3 + KCNS.$$

8. The reactions with Cl, Br and I are given under Section I (c).

9. Certain metals are dissolved in aqueous solution of potassium cyanide:

(a) In presence of oxygen:

$$2Au + 4KCy + O + H_2O = 2KAuCy_2 + 2KOH.$$
$$2Ag + 4KCy + O + H_2O = 2KAgCy_2 + 2KOH.$$

(b) With evolution of hydrogen:

$$Zn + 4KCy + 2H_2O = K_2ZnCy_4 + 2KOH + H_2.$$

(also with Cu, Fe and Ni).

(c) The following are insoluble: Pt, Hg, Sn. (Hg, however, dissolves to some extent, under the conditions of cyanide treatment, from amalgamated plates in the battery.)

10. Many metallic oxides dissolve to form the corresponding cyanide, or double cyanide of the metal and potassium.

Sodium Cyanide (NaCN): *Formation* (see under "Potassium Cyanide," above). — 1. By passing a current of anhydrous HCN into alcoholic solution of caustic soda. Sodium cyanide is precipitated in the anhydrous form, and must be rapidly washed with alcohol and dried.

2. When ammonia is passed over a heated mixture of metallic sodium, or sodium carbonate and carbon, it forms sodium cyanide, with sodium cyanamide as intermediate product:

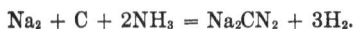

$$Na_2 + C + 2NH_3 = Na_2CN_2 + 3H_2.$$
$$Na_2CN_2 + C = 2NaCN.$$

3. By fusing barium cyanide with sodium carbonate:

$$BaCy_2 + Na_2CO_3 = BaCO_3 + 2NaCy.$$

The sodium cyanide is separated from the insoluble $BaCO_3$ by extracting with water.

4. By fixation of atmospheric nitrogen, using the following series of reactions:

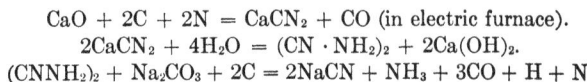

$$CaO + 2C + 2N = CaCN_2 + CO \text{ (in electric furnace).}$$
$$2CaCN_2 + 4H_2O = (CN \cdot NH_2)_2 + 2Ca(OH)_2.$$
$$(CNNH_2)_2 + Na_2CO_3 + 2C = 2NaCN + NH_3 + 3CO + H + N$$

(These processes will be more fully discussed under "Manufacture of Cyanide.")

Properties. — Sodium cyanide is a white crystalline solid, similar in general appearance and properties to potassium cyanide. It is very deliquescent, and forms two hydrates, of the composition, $NaCN \cdot 2H_2O$ and $2NaCN \cdot H_2O$, respectively. The first is white, crystallizing in thin plates, the second yellowish. Both these hydrates are unstable, and give off ammonia, even when protected from the air, at the ordinary temperature.

Reactions. — Analogous to those of potassium cyanide. Certain alleged differences in the behavior of the two cyanides in the treatment of ores will be discussed later.

CYANIDES OF THE ALKALINE EARTH METALS

Barium Cyanide ($Ba(CN)_2$): *Preparation.* — 1. By calcination of $Ba_2Fe(CN)_6$.

2. By passing anhydrous HCN over $Ba(OH)_2$.

3. By passing a current of air over a mixture of BaO and carbon at a high temperature, or, better, by passing air previously freed from oxygen and water vapor over a heated mixture of $BaCO_3$ and charcoal, tar, or other carbonaceous matter.

Barium carbide (BaC_2) is first formed, which combines with nitrogen, giving $Ba(CN)_2$.

Properties. — White, anhydrous, or crystallizing as $BaCy_2 \cdot 2H_2O$. Loses water in vacuo at ordinary temperature, forming $BaCy_2 \cdot H_2O$. Very soluble and deliquescent. Completely dehydrated by heating in a current of dry air at 75° and then at 100° C.

Reactions. — 1. Decomposes on heating:

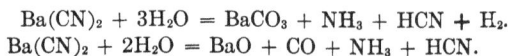

$$Ba(CN)_2 + 3H_2O = BaCO_3 + NH_3 + HCN + H_2.$$
$$Ba(CN)_2 + 2H_2O = BaO + CO + NH_3 + HCN.$$

2. Decomposed by carbon dioxide:

$$Ba(CN)_2 + CO_2 + H_2O = 2HCN + BaCO_3.$$

3. Fused with alkaline carbonates it yields $BaCO_3$ and the corresponding alkaline cyanide.

Strontium Cyanide ($Sr(CN)_2$). — Analogous in preparation and properties to the above. Crystallizes as $SrCy_2 \cdot 4H_2O$. Less stable than $BaCy_2$; more so than $CaCy_2$.

Calcium Cyanide ($Ca(CN)_2$): *Preparation.* — 1. By action of HCN on lime.

2. By passing nitrogen over calcium carbide, the chief product is calcium cyanamide:

$$CaC_2 + N_2 = CaCN_2 + C \text{ (at 800° C.).}$$

When this is heated to a higher temperature (preferably with addition of salt to facilitate the reaction) it is decomposed as follows:
$$CaCN_2 + C = Ca(CN)_2$$

Physical properties. — Said to have been obtained in anhydrous cubes, but very unstable.

Reactions. — 1. Dilute solutions are tolerably stable, but a concentrated solution placed in vacuo gradually decomposes, blackens, and gives the ordinary products of decomposition of HCN. In presence of dehydrating agents and KOH, small crystalline needles of the oxycyanide, $6CaO \cdot CaCy_2 \cdot 15H_2O$, are formed.

2. Concentrated alcohol gives a precipitate of calcium hydrate.

Magnesium Cyanide ($Mg(CN)_2$) [existence doubtful]: *Preparation.* — 1. By dissolving freshly precipitated $Mg(OH)_2$ in HCN (Scheele).

2. By passing CO or CO_2 over heated magnesium nitride:

$$Mg_3N_2.$$

Properties. — Very unstable. Solution decomposes on evaporation, liberating HCN and leaving $Mg(OH)_2$. Decomposed by CO_2 with precipitation of $MgCO_3$.

Magnesium salts decompose alkaline cyanides thus:

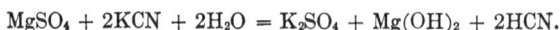

$$MgSO_4 + 2KCN + 2H_2O = K_2SO_4 + Mg(OH)_2 + 2HCN.$$

Cyanides and Double Cyanides of Zinc

Zinc Cyanide ($Zn(CN)_2$): *Preparation.* — 1. By precipitating a salt of zinc with an alkaline cyanide in suitable proportions:

$$2KCy + ZnSO_4 = K_2SO_4 + ZnCy_2.$$

2. By the action of HCN on the oxide or acetate of zinc. These reactions are incomplete and appear to be reversible:

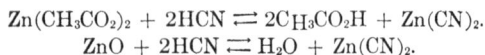

$$Zn(CH_3CO_2)_2 + 2HCN \rightleftharpoons 2CH_3CO_2H + Zn(CN)_2.$$
$$ZnO + 2HCN \rightleftharpoons H_2O + Zn(CN)_2.$$

Properties. — White, amorphous or crystallized in orthorhombic prisms. Insoluble in water. Unalterable when dry.

Reactions. — 1. Heated in a closed vessel, gives a black carbonaceous residue.

2. Dissolves in alkalis, forming a double cyanide and zincate:

$$2ZnCy_2 + 4KOH = K_2ZnCy_4 + Zn(OK)_2 + 2H_2O.$$

probably in two stages:

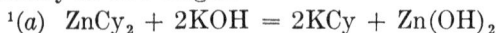

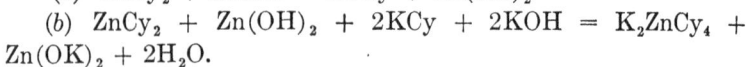

[1](a) $ZnCy_2 + 2KOH = 2KCy + Zn(OH)_2$

(b) $ZnCy_2 + Zn(OH)_2 + 2KCy + 2KOH = K_2ZnCy_4 + Zn(OK)_2 + 2H_2O.$

3. Dissolves in alkaline cyanides, forming double cyanides $(ZnCy_2 \cdot 2mCy)$:

$$ZnCy_2 + 2mCy = m_2ZnCy_4 \text{ [m = K, Na, NH}_4].$$

Zinc-Potassium Cyanide ($K_2Zn(CN)_4$): *Preparation.* — 1. By dissolving $ZnCy_2$ in excess of KCy (see above).

2. By acting on metallic zinc with KCy in excess:

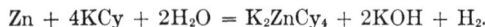

$$Zn + 4KCy + 2H_2O = K_2ZnCy_4 + 2KOH + H_2.$$

3. By dissolving ZnO or $ZnCO_3$ in a mixture of KCy and HCy:

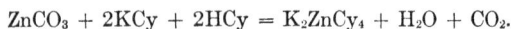

$$ZnCO_3 + 2KCy + 2HCy = K_2ZnCy_4 + H_2O + CO_2.$$

Properties. — Regular transparent colorless octahedra. Can be fused without decomposition. Very soluble in water at all temperatures.

Reactions. — 1. Decomposed by dilute acids, first with precipitation of $ZnCy_2$, which is further decomposed by excess of acid:

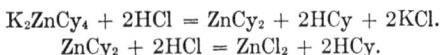

$$K_2ZnCy_4 + 2HCl = ZnCy_2 + 2HCy + 2KCl.$$
$$ZnCy_2 + 2HCl = ZnCl_2 + 2HCy.$$

2. Gives precipitates of insoluble double cyanides of zinc with various solutions of metallic salts (*e.g.*, those of Ba, Ca, Ni, Co, Al, Fe, Hg, Cu, Pb).

3. With silver nitrate it gives first a precipitate of $ZnCy_2$, and on adding excess of $AgNO_3$ this is decomposed, giving AgCy:

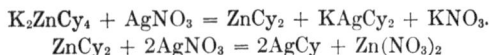

$$K_2ZnCy_4 + AgNO_3 = ZnCy_2 + KAgCy_2 + KNO_3.$$
$$ZnCy_2 + 2AgNO_3 = 2AgCy + Zn(NO_3)_2$$

[1] This reaction is discussed in detail in Section V.

Other Double Cyanides of Zinc. — The double cyanides of zinc, with NH_4, Na, K, and Ca, are soluble in water; the remainder are more of less insoluble.

The double cyanide of Zn and ammonium is unstable, and decomposes in solution:

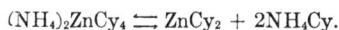

$$(NH_4)_2ZnCy_4 \rightleftharpoons ZnCy_2 + 2NH_4Cy.$$

CYANIDES OF THE IRON GROUP

The simple cyanides of iron, manganese, nickel, and cobalt are unstable; the chromic cyanide is tolerably stable. No simple cyanide of aluminium is known. All these compounds show a great tendency to form double salts by reacting with alkaline cyanides.

The double cyanides of nickel are easily decomposable, and are more or less analogous to those of zinc; they are of the general type $NiCy_2 \cdot 2mCy$, and are decomposed by dilute acids in the same way. Thus:

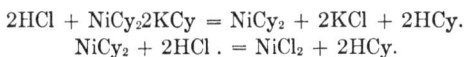

$$2HCl + NiCy_2 2KCy = NiCy_2 + 2KCl + 2HCy.$$
$$NiCy_2 + 2HCl . = NiCl_2 + 2HCy.$$

The remainder form characteristic compounds of complex cyanogen radicals, which are not decomposed by dilute acids, and will be described later.

The characteristic colors of the simple cyanides of this class are as follows:

Iron (ferrous). . . . . .Reddish yellow.
  " (ferric) . . . . .Deep brown(solution only).
Manganese (manganous) . . . . . . .White, turning brown.
Nickel . . . . .Apple green.
Cobalt . . . . .Pink.
Chromium (chromous) . . . . .White.
  " (chromic) . . . . .Grayish blue.

CYANIDES OF THE PRECIOUS METALS

Cyanides of Gold. — Gold forms two simple cyanides: Aurous cyanide (AuCN) and Auric cyanide $(Au(CN)_3)$. It also forms two corresponding classes of double cyanides: Aurous $(mAu(CN)_2)$ and Auric $(mAu(CN)_4)$.

Aurous cyanide (AuCN): *Preparation.* — 1. By decomposing auro-potassic cyanide with acids:

$$KAuCy_2 + HCl = AuCy + HCy + KCl.$$

The mixture remains clear at ordinary temperatures, but becomes turbid and deposits AuCy at 50° C.

2. By evaporating to dryness a mixture of $HgCy_2$ and $AuCl_3$. Only part of the Au is obtained as AuCy, the remainder forming a double cyanide of Au and Hg.

Properties. — A citron-yellow crystalline powder, inodorous, tasteless, unalterable in air; not affected by light if dry, but in presence of moisture takes a greenish tint. Insoluble in water.

Reactions. — 1. Decomposed by heat into Au and CN.

2. Burns, leaving residue of metallic gold.

3. Most acids do not decompose it, even on boiling. When recently precipitated it is soluble in some acids without change. H_2SO_4 and $HNO_3 + HCl$ dissolve it slowly on boiling. H_2S does not decompose it.

4. Ammonia and ammonium sulphide dissolve it; sulphide of gold is precipitated from the latter solution on adding acid.

5. KOH, on boiling, gives Au and $KAuCy_2$.

6. Dissolves in alkaline cyanides:

$$KCy + AuCy = KAuCy_2.$$

7. Dissolves in sodium thiosulphate.

Auric Cyanide $(Au(CN)_3)$: *Preparation.* — 1. By dissolving $KAuCy_2$ in an acid and evaporating.

2. When potassium auricyanide $(KAuCy_4)$ is treated with excess of $AgNO_3$, a precipitate is formed, consisting of $AgAuCy_4$. This, on treatment with HCl, decomposes as follows:

$$AgAuCy_4 + HCl = AgCl + HCy + AuCy_3.$$

The filtrate from AgCl is evaporated in vacuo.

Properties. — Small colorless crystals, melting at 50° C. Easily soluble in water, alcohol, or ether.

Reactions. — 1. Decomposes on gentle heating, first giving the aurous cyanide AuCy, then Au and CN.

2. Mercurous nitrate forms AuCy and $HgCy_2$.

3. Oxalic acid does not precipitate Au from a solution of $AuCy_3$ in water.

Double Aurous Cyanides. — The following are known:

Auropotassic cyanide ($KAuCy_2$).
Aurosodic cyanide ($NaAuCy_2$).
Auroammonic cyanide (NH_4AuCy_2),
Aurobarytic cyanide ($Ba(AuCy_2)_2$).
Aurozincic cyanide ($Zn(AuCy_2)_2$).
Auroargentic cyanide ($AgAuCy_2$).

Auropotassic Cyanide ($KAu(CN)_2$): *Preparation.* — 1. By dissolving metallic gold in potassium cyanide, in presence of air or oxygen: $4 KCy + 2Au + O + H_2O = 2KOH + 2KAuCy_2$.

[This, and the analogous reaction with NaCy, are probably the chief means by which gold is dissolved in cyanide treatment].

2. By dissolving AuCy in KCy (see above).

3. By dissolving oxide of gold or fulminating gold ($AuCNO)_2$ in KCy.

Properties. — Crystallizes in colorless plates composed of orthorhombic prisms with octahedral faces.   Dissolves in 4 parts of cold water and 0.8 parts of boiling water.   Very slightly soluble in alcohol.   Unalterable in air.

Reactions. — 1. Decomposed by heat into Au, Cy, and KCy.

2. Slowly decomposed by acids, giving off HCy and depositing AuCy.

3. Not decomposed by H_2S.

4. Decomposed by mercuric chloride, forming a chlorocyanide of mercury and potassium:

$$HgCl_2 + KAuCy_2 = KCl \cdot HgCy + AuCy + Cl.$$

5. Forms addition-products with the haloid elements, having the general formula

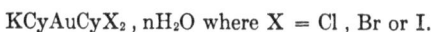

$$KCyAuCyX_2, nH_2O \text{ where } X = Cl, Br \text{ or } I.$$

Iodine decomposes it into AuCy, Cy, and KI.

Aurosodic Cyanide ($NaAu(CN)_2$). — Its preparation and reactions are analogous to those of $KAuCy_2$.   Also prepared by double decomposition, thus:

$$Na_2SO_4 + Ba(AuCy_2)_2 = BaSO_4 + 2NaAuCy_2.$$

Less soluble in water than $KAuCy_2$.   Very slightly soluble in alcohol.   Crystallizes in anhydrous scales which decompose at about 200°.   Also forms addition compounds with Cl, Br, and I.

Auripotassic Cyanide ($KAu(CN)_4$): *Preparation.* — 1. By acting on a mixture of Au and $AuCl_3$ with concentrated solution of KCy, and crystallizing.

Properties. — Colorless crystals. Containing H_2O, which is only completely given off at 200° C.

Reactions. — 1. Melts to a brown liquid, evolving Cy and leaving $KAuCy_2$.

2. Slowly decomposed by acids.

3. Chlorine decomposes it on heating, giving CNCl. Iodine gives $KAuCy_2I_2$.

4. Mercuric chloride precipitates $AuCy_3$.

$$HgCl_2 + KAuCy_4 = AuCy_3 + KCl + HgClCy_2.$$

Analogous compounds: Auriammonic cyanide $(NH_4Au(CN)_4)$ and auriargentic cyanide $(AgAu(CN)_4)$.

Cyanides of Silver. — Silver forms a cyanide (AgCN) and double cyanides of the type $mAg(CN)_2$.

Cyanide of Silver (AgCN): *Preparation.* — 1. By precipitating a silver salt with HCN or a metallic cyanide in suitable proportions. The best method of obtaining it pure is by precipitating the double cyanide of silver with silver nitrate:

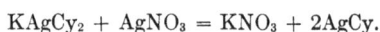

$$KAgCy_2 + AgNO_3 = KNO_3 + 2AgCy.$$

(It is thus obtained free from carbonates, cyanates, chlorides and ferrocyanides, which might be present in commercial KCy or NaCy).

2. Many cyanides, when treated, especially by the aid of heat, with $AgNO_3$, are decomposed, yielding AgCN. Thus,

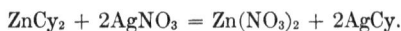

$$ZnCy_2 + 2AgNO_3 = Zn(NO_3)_2 + 2AgCy.$$

3. By adding an acid to a solution of the double cyanide:

$$KAgCy_2 + HNO_3 = KNO_3 + HCy + AgCy.$$

Properties. — White, curdy substance resembling AgCl. Insoluble in water and in cold, moderately concentrated acids. Turns brown on exposure to light. (According to Fresenius, it is not affected by light.)

Reactions, — 1. When heated without access of air, it first melts, then decomposes, swelling considerably, giving up CN and leaving a compound, $Ag_6(CN)_3$, which, on further heating, yields N and carbide of silver (Liebig).

2. Heated in contact with air or oxygen, it leaves only metallic silver.

3. Chlorine forms AgCl and CN. When the cyanide is completely decomposed, the CN unites with excess of Cl, forming CNCl.

4. Sulphur heated with AgCN gives AgSCN.

5. Hot acids decompose it, giving HCN. It is completely dissolved and decomposed by hot 50 per cent. H_2SO_4. [AgCl is not decomposed under these conditions].

6. Heated with water in a sealed tube to 280° C., it is decomposed into Ag and $(NH_4)_2CO_3$.

7. Boiling solution of KCl decomposes it:

$$AgCy + KCl = AgCl + KCy.$$

8. Soluble in ammonia, ammoniacal salts, ferrocyanides, and thiosulphates.

9. Soluble in alkaline cyanides, forming double salts:

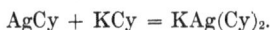

$$AgCy + KCy = KAg(Cy)_2.$$

10. Decomposed by a solution of sulphur in CS_2, forming compounds which appear to be sulphides of cyanogen.

Argento Potassic Cyanide (**$KAg(CN)_2$**): *Preparation.* — 1. By dissolving metallic silver in a solution of potassium cyanide in presence of air or oxygen:

$$4KCy + 2Ag + O + H_2O = 2KOH + 2KAgCy_2.$$

2. By dissolving certain insoluble salts of silver (*e.g.*, AgCN, AgCl, AgI, Ag_4FeCy_6, Ag_3FeCy_6, and AgCNO) in an excess of KCN.

Properties. — Crystallizes in colorless, regular octahedra, with faces often depressed to the center. Inodorous, with metallic taste. Soluble in 8 parts of cold water, in 4 parts water at 20° C., and in 1 part boiling water. Also soluble in alcohol.

Reactions. — 1. Decomposed by acids with formation of AgCN (see above).

2. Decomposed by H_2S or alkaline sulphides, with precipitation of Ag_2S:

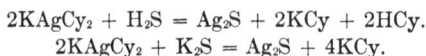

$$2KAgCy_2 + H_2S = Ag_2S + 2KCy + 2HCy.$$
$$2KAgCy_2 + K_2S = Ag_2S + 4KCy.$$

3. Many metallic solutions precipitate insoluble double cyanides of silver; *e.g.*,

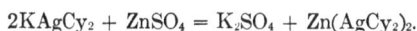

$$2KAgCy_2 + ZnSO_4 = K_2SO_4 + Zn(AgCy_2)_2.$$

4. Certain metals, such as zinc, decompose the double cyanide,

with precipitation of silver, probably as the result of secondary reactions:

$$KAgCy_2 + Zn + H_2O = KOH + H + ZnCy_2 + Ag.$$

Other Double Cyanides of Silver. — Among others, the following are known:

Argentocyanic acid ($HAgCy_2$).

Sodic argentocyanide ($NaAgCy_2$).

Barium argentocyanide ($Ba(AgCy_2)_2 \cdot H_2O$).

Sodio-potassic argentocyanide ($NaK_3(AgCy_2)_4$) = $NaAgCy_2$ + $3KAgCy_2$.

Zinc argentocyanide ($Zn(AgCy_2)_2$).

Similar compounds are formed with Cd, Ni, Co, Mn, Fe, Cu, Pb, and Hg.

Ammonia forms the compound NH_3AgCy.

The compound $AgNO_3 \cdot 2AgCy$ is formed by dissolving AgCy in a concentrated solution of $AgNO_3$.

Cyanides of Other Precious Metals. — Cyanides and various classes of double cyanides are formed with the platinum metals and other less common elements, which need not be considered here.

CYANIDES OF COPPER

Three simple cyanides have been described, *viz.*, cuprous cyanide ($Cu_2(CN)_2$), cupric cyanide ($Cu(CN)_2$), cuproso-cupric cyanide ($Cu_3(CN)_4$). There are also numerous double cyanides of various types.

Cuprous Cyanide ($Cu_2(CN)_2$): *Preparation.* — 1. By the action of HCN on recently precipitated cuprous hydrate, on cuprous chloride dissolved in HCl, or on cupric chloride in presence of SO_2.

2. By decomposition of cupric cyanide in presence of water.

3. By precipitation of cupropotassic cyanide with acids:

$$2KCuCy_2 + H_2SO_4 = K_2SO_4 + Cu_2Cy_2 + 2HCy.$$

Properties. — A white, curdy precipitate, insoluble in water and dilute acids. This is the most stable of the cyanides of copper.

Reactions. — 1. On heating it melts, giving off much water and leaving a swollen brown mass.

2. Decomposed by nitric acid with evolution of N_2O_2.

3. Dissolves without change in concentrated HCl, and is precipitated from this solution by H_2O or KOH.

4. Soluble in ammonia and in many ammonium salts, forming complex cyanides.

5. Soluble in alkaline cyanides:

$$Cu_2Cy_2 + 2KCy = 2KCuCy_2.$$

With excess of KCy, other compounds are formed (see below).

6. Decomposed by ferric chloride in the cold, with evolution of cyanogen:

$$Fe_2Cl_6 + Cu_2Cy_2 = 2FeCl_2 + Cu_2Cl_2 + Cy_2.$$

7. Decomposed by acetic acid in presence of an oxidizing agent, giving cyanogen.

Cupric Cyanide ($CuCy_2$): *Preparation.* — 1. By treating carbonate of copper with HCy in solution:

$$CuCO_3 + 2HCy = CuCy_2 + H_2O + CO_2.$$

2. By acting on acetate of copper with HCy.

3. By adding KCy to an excess of a cupric salt:

$$2KCy + CuSO_4 = CuCy_2 + K_2SO_4.$$

Properties. — A yellowish-brown, unstable substance, only known in hydrated condition.

Reactions. — 1. Changes rapidly and spontaneously at the ordinary temperature, forming greenish or yellowish-gray cuprosocupric cyanide ($?Cu_3Cy_4 \cdot H_2O$), which, on heating, gives $Cu_2Cy_2 + Cy$.

Cuproso-Cupric Cyanides. — There appear to be two of these, viz.:

$$Cu_2Cy_2 \cdot CuCy_2 = Cu_3Cy_4.$$
$$2Cu_2Cy_2 \cdot CuCy_2 = Cu_5Cy_6.$$

These are, perhaps, only mixtures of Cu_2Cy_2 and $CuCy_2$.

Preparation. — 1. By treating cuprous oxide with HCy.

2. By adding HCy to $CuSO_4$.

3. By spontaneous decomposition of cupric cyanide:

$$3CuCy_2 = Cu_3Cy_4 + Cy_2.$$

Properties. — According to mode of formation, it is yellowish-gray, or crystallizes in transparent prisms of a brilliant canary-green color.

Reactions. — 1. Heated to 100° it loses Cy and water, and is converted into Cu_2Cy_2 without change of form.

2. Strong acids liberate HCy, leaving a mixture of cuprous and cupric salts.

3. Alkalis form double salts of copper and the alkali metal, probably $3Cu_2Cy_2 \cdot 4mCy$, and precipitating $Cu(OH)_2$.

4. Alkaline cyanides give (when in excess) the compound $Cu_2Cy_2 \cdot 6mCy$. (m = K, Na, etc.)

5. Ammonia dissolves it to a blue solution. The compound is only slightly soluble, and sometimes deposits green crystals (see below).

6. Ammonium carbonate dissolves it in the cold, other ammonium salts only on heating.

Double Cyanides of Copper. — These are very numerous, the best known being those of ammonium and potassium. The ammonio-cyanides of copper are of two distinct kinds, represented respectively by the formulæ.[1]

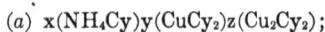

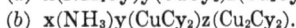

$$(a) \ x(NH_4Cy)y(CuCy_2)z(Cu_2Cy_2);$$
$$(b) \ x(NH_3)y(CuCy_2)z(Cu_2Cy_2).$$

The potassium compounds have the general formula, $x(KCy)$ $y(Cu_2Cy_2)$.

Double Cyanides of Copper and Ammonium. — The compounds of class (a), which are true double cyanides of copper and ammonium, are colorless, highly soluble, and stable in absence of other metallic salts. They act feebly as solvents of gold. They are decomposed by excess of a copper salt, yielding a mixture of cuprous and cupric cyanides.

The compounds of the second class (b), having the formula $x(NH_3)y(CuCy_2)z(Cu_2Cy_2)$, are formed by dissolving $mCuCy_2 + nCu_2Cy_2$ in ammonia. They are generally regarded as cuprammonium cyanides, i.e., as compounds in which the ammonia is intimately united with copper, forming a complex radical, to which Rose[2] gives the formula $Cu(NH_3)_4$ (cuprammonium), and which enters into unstable combinations with cyanogen.

These substances are generally blue or green crystalline bodies, only slightly soluble in water. They yield deep blue solutions and appear to be "ionized" in dilute solutions, with liberation of $\overset{-}{CN}$ and the cuprammonium ion $\overset{+}{Cu}(\overset{+}{NH_3})_4$, to which the blue color of the liquid is due.

[1] Sulman, *Trans.* I. M. M., XIV, 363.
[2] *Trans.* I. M. M., XIV, 369.

Preparation and Properties. — 1. When a salt of copper is precipitated by ammonium cyanide, cyanogen is abundantly evolved and a greenish-blue substance formed having the formula $2NH_3 \cdot CuCy_2 \cdot Cu_2Cy_2 \cdot H_2O$. This is an amorphous powder, slightly soluble in cold water and decomposed at 100° C.

2. By acting on the above with further quantities of ammonia, a blue liquid is obtained which, on evaporation, yields $4NH_3 \cdot CuCy_2 \cdot 2Cu_2Cy_2$, in small, bright-green, needle-like crystals, prismatic, with metallic luster.

The same substance is obtained by passing a current of HCN into oxide of copper suspended in ammonia.

3. When the preceding compound is heated with ammonia and a current of NH_3 gas is passed through the liquid, the latter deposits, on cooling, the substance $6NH_3 \cdot CuCy_2 \cdot Cu_2Cy_2$. This is a crystalline precipitate, in prismatic needles of a fine blue color. Exposed to the air, it loses NH_3 and becomes green.

Reactions. — Solutions of these compounds readily give up cyanogen to any substance capable of combining with it. Hence they act as powerful solvents of gold. This reaction is accompanied by deposition of the slightly soluble dicuproso-cupric cyanide $4NH_3 \cdot CuCy_2 \cdot Cu_2Cy_2$, and takes place without the intervention of oxygen.

Double Cyanides of Copper and Potassium. — The following are known: [1]

$$(a) \quad 4KCy \cdot 3Cu_2Cy_2.$$
$$(b) \quad 2KCy \cdot Cu_2Cy_2.$$
$$(c) \quad 4KCy \cdot Cu_2Cy_2.$$
$$(d) \quad 6KCy \cdot Cu_2Cy_2.$$

Preparation and Properties. — (a) The first of these is obtained by the action of fused cuprous cyanide on potash and is little known.

(b) This compound is formed by the action of KCy on Cu_2Cy_2 in the proportions indicated by the formula of the compound. It is also formed by acting on cupric hydrate with KCN solution and slowly evaporating,[2] or by acting on cuproso-cupric cyanide with KCN.

It forms transparent, colorless clinorhombic prisms, fusible

[1] Sulman, *Trans.* I. M. M., X, 125.
[2] Virgoe, *Ibid.*, X, 107.

without decomposition, but liberating metallic copper on continued fusion.

The compound $2KCy \cdot Cu_2Cy_2$ is sparingly soluble in water, and partially decomposes in solution into Cu_2Cy_2 and $6KCy \cdot Cu_2Cy_2$. It is not a solvent of gold or silver.

(c) This compound, $4KCy \cdot Cu_2Cy_2$, is the form in which copper, according to Virgoe (loc. cit., p. 141), normally exists in cyanide solutions, though Sulman (loc. cit., p. 127) considers that it is more frequently present as sulphocyanide ($Cu_2(SCy)_2$) dissolved in KCy.

It is prepared by agitating cupric hydrate with excess of KCy:

[1] $2Cu(OH)_2 + 8KCy = 4KCy \cdot Cu_2Cy_2 + 4KOH + Cy_2$.

Virgoe considers that an intermediate product exists, to which he gives the formula $2KCy \cdot Cu_2Cy_2 \cdot 4KCy \cdot Cu_2Cy_2$, which might be written $3KCy \cdot Cu_2Cy_2$. This latter is said to be a good solvent of gold and silver (loc. cit., p. 142).

(d) $6KCy \cdot Cu_2Cy_2$. This is obtained by dissolving metallic copper in KCN:

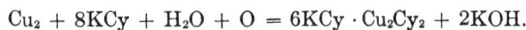

$$Cu_2 + 8KCy + H_2O + O = 6KCy \cdot Cu_2Cy_2 + 2KOH.$$

Colorless transparent prisms, with 6 sides, terminated by 6-sided pyramids, almost unalterable in air, but gradually becoming blue. May be fused with separation of a little copper. [The salts described by Virgoe as existing in solution may possibly be only mixtures of $2KCy \cdot Cu_2Cy_2$ and $6KCy \cdot Cu_2Cy_2$].

Reactions. — 1. The double cyanides of K and Cu are all decomposed by acids, forming Cu_2Cy_2 and HCy.

2. H_2S does not precipitate copper sulphide from their solutions, or only to a very slight extent from dilute solutions. Alkaline sulphides give no precipitate.

3. Ferric salts precipitate Cu_2Cy_2, $Fe(OH)_6$ and liberate HCy.

4. Mercuric salts give Cu_2Cy_2, $HgCy_2$, and the potassium salt of the radical originally combined with the mercury.

5. Excess of copper salts give a pale yellow precipitate which may be a cuproso-cupric cyanide — $xCuCy_2 \cdot yCu_2Cy_2$.

6. Silver nitrate converts a part of the KCy in these compounds

[1] The reaction (in presence of an excess of alkali) which, according to the writer's experiments, seems most in accordance with the facts, is as follows:

$2Cu(OH)_2 + 7KCy = 4KCy \cdot Cu_2Cy_2 + 2KOH + KCyO + H_2O$

This indicates a consumption of 3.6 parts KCy for every part of Cu dissolved.

into the soluble $KAgCy_2$ before any precipitation of AgCy takes place. The usual reaction is (Virgoe, *loc. cit.*, p 139):

$$AgNO_3 + 4KCy \cdot Cu_2Cy_2 = 2KCy \cdot Cu_2Cy_2 + KAgCy_2 + KNO_3.$$

On further addition of $AgNO_3$, there is a precipitation of Cu_2Cy_2 as follows:

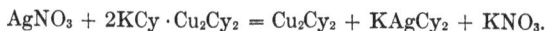

$$AgNO_3 + 2KCy \cdot Cu_2Cy_2 = Cu_2Cy_2 + KAgCy_2 + KNO_3.$$

When KI is added, however, there is an immediate precipitation of AgI, which is not soluble in the cupropotassic cyanides.

Other Double Cyanides of Copper. — Other compounds are described with Na, Ba, Zn, Mn, Co, Ni, Fe, Pb, and other metals.

CYANIDES OF MERCURY

Only one well-defined simple cyanide exists, namely, mercuric cyanide $(Hg(CN)_2)$.

There are two basic cyanides (oxycyanides), namely, $HgO \cdot HgCy_2$, and $HgO \cdot 3HgCy_2$, besides numerous double cyanides with the alkali metals and ammonia.

Many chlorocyanides are also known, of the general formula $HgCy_2 \cdot RCl + nH_2O$.

Mercuric Cyanide ($Hg(CN)_2$): *Preparation.* — 1. By action of HCy on mercuric oxide suspended in water:

$$2HCy + HgO = HgCy_2 + H_2O.$$

2. By double decomposition of mercuric salt and hydrocyanic acid:

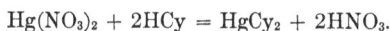

$$Hg(NO_3)_2 + 2HCy = HgCy_2 + 2HNO_3.$$

[With mercurous salts a mixture of $HgCy_2$ and metallic mercury is obtained].

3. By the action of mercuric oxide on cyanides or ferrocyanides, or (with boiling) on Prussian blue.

4. By heating mercuric sulphate with ferrocyanide.

5. By acting on the ferrocyanides of lead or barium with mercuric sulphate, and filtering from the insoluble $PbSO_4$ or $BaSO_4$.

Properties. — Crystallizes in colorless, rectangular prisms with square base, but occurs in other forms. Anhydrous. Soluble in 8 parts cold water; more soluble in hot water. Insoluble in absolute alcohol.

Reactions. — 1. On heating, gives Hg and $(CN)_2$ (see above).

2. Decomposed by concentrated acids, giving HCN and the corresponding mercury salt.

3. Chlorine in the presence of moisture or on exposure to light gives $HgCl_2$ and CNCl.

Bromine and iodine act similarly.

4. Water at 280° C. forms Hg and $(NH_4)_2CO_3$.

5. Decomposed by HCl, HBr, HI, and H_2S, giving the corresponding Hg salt and HCy. Gradually decomposed by alkaline sulphides, forming HgS and a thiocyanate of the alkali.

Basic Cyanides of Mercury. — These are white or yellowish crystalline explosive bodies, formed by acting on $HgCy_2$ with HgO in various proportions.

Double Cyanide of Mercury and Potassium $(K_2Hg(Cy)_4)$: *Preparation.* — 1. By dissolving $HgCy_2$ in KCy and evaporating till the salt crystallizes:

$$HgCy_2 + 2KCy = K_2HgCy_4.$$

2. By treating $HgCy_2$ with HCy and K_2CO_3:

$$HgCy_2 + 2HCy + K_2CO_3 = K_2HgCy_4 + H_2O + CO_2.$$

3. By dissolving HgO in KCy:

$$HgO + 4KCy + H_2O = K_2HgCy_4 + 2KOH.$$

4. By dissolving $HgCl_2$ in KCy:

$$HgCl_2 + 4KCy = K_2HgCy_4 + 2KCl.$$

Properties. — Crystallizes in regular octahedra. Transparent. Unalterable in air. Dissolves in 4 parts cold water. Solution smells slightly of HCy.

Reactions. — 1. Decrepitates violently on heating, and melts to a brown liquid which gives off Cy_2 and Hg.

2. Decomposed by acids, either completely or with formation of $HgCy_2$ and HCy:

$$4HCl + K_2HgCy_4 = 2KCl + HgCl_2 + 4HCy.$$
$$2HNO_3 + K_2HgCy_4 = 2KNO_3 + HgCy_2 + 2HCy.$$

3. Decomposed by alkaline sulphides or by H_2S with precipitation of HgS: $\quad K_2S + K_2HgCy_4 = HgS + 4KCy.$

4. Acts as a solvent of gold and silver, with deposition of mercury and without intervention of oxygen:

$$K_2HgCy_4 + 2Au = Hg + 2KAuCy_2.$$
$$K_2HgCy_4 + 2Ag = Hg + 2KAgCy_2.$$

5. Decomposed by other metals (zinc, for example) with deposition of mercury:

$$K_2HgCy_4 + Zn = K_2ZnCy_4 + Hg.$$

[A similar double cyanide is formed with sodium, namely, Na_2HgCy_4. Double cyanides of mercury with Zn and Pb are also known.

Many double salts are formed by combination of $HgCy_2$ with other metallic salts, *e.g.*, nitrates, thiosulphates, thiocyanates, ferrocyanides, etc.

CYANIDES OF LEAD

No simple cyanide of lead of the formula $Pb(CN)_2$ is known. The precipitate obtained by acting on lead salts with an alkaline cyanide appears to have the composition $2PbO \cdot Pb(CN)_2 \cdot H_2O$. It is best prepared by acting on neutral lead acetate with KCN in presence of ammonia.

It is a white precipitate, insoluble in water, but decomposed by dilute acids with evolution of HCy.

Double Cyanide of Zinc and Lead. — When a salt of lead is precipitated by K_2ZnCy_4, a white precipitate is obtained having the composition $2ZnCy_2 \cdot PbCy_2$.

SECTION III

METALLIC COMPOUNDS OF COMPLEX CYANOGEN RADICALS

(A) METALLIC COMPOUNDS OF THE RADICAL RC_6N_6

THE compounds we now have to consider cannot legitimately be described as cyanides. Although under certain conditions they may be obtained from, or may yield, cyanogen or cyanides of the type previously discussed, they differ widely from the cyanides in their general properties and reactions, and behave as though containing a negative radical into the composition of which a metal (usually of the iron type) enters as a constituent part. This radical goes in and out of combination unchanged, and may be replaced by different non-metallic elements or radicals.

The general formula for this radical is

$$RC_6N_6,$$

where R is an atom of a metal such as Fe, Mn, Cr, Co, or of the platinum group, such as Pt, Ir, Pd, Os, Rh, Ru.

A closely related class of substances, known as the nitroprussides, contain the radical

$$RC_5N_6O, \text{ probably } RC_5N_5(NO).$$

The radical RC_6N_6 is capable of acting as a dyad, triad, or tetrad, forming compounds of the types —

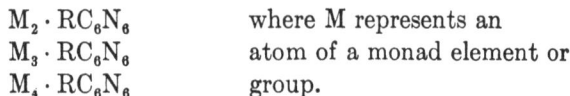

$M_2 \cdot RC_6N_6$	where M represents an
$M_3 \cdot RC_6N_6$	atom of a monad element or
$M_4 \cdot RC_6N_6$	group.

Compounds of the first type are very unstable; those of the second may perhaps have the constitution

$$M_3 \equiv (RC_6N_6)$$

$M_3 \equiv (RC_6N_6)$ on the assumption that the group RC_6N_6 is normally tetrad.

84

COMPOUNDS IN WHICH IRON ENTERS INTO THE NEGATIVE
RADICAL

These compounds are as follows:

(a) Ferrocyanides, $M_4FeC_6N_6$.
(b) Ferricyanides, $M_3FeC_6N_6$.
(c) Perferricyanides, $M_2FeC_6N_6$. (?)
(d) Nitroprussides, $M_2Fe(CN)_5 \cdot (NO)$.

For simplicity we may use the symbol $FeCy_6$ to indicate the radical FeC_6N_6 in the first three of these groups of compounds.

(a) FERROCYANIDES (M_4FeCy_6).

General Characteristics.— 1. Most ferrocyanides are colored; the characteristic colors of these compounds are of value in analysis.

2. The soluble alkali ferrocyanides are non-poisonous.

3. They do not evolve HCN when treated with very dilute acids, but generally yield hydroferrocyanic acid (H_4FeCy_6).

4. Those ferrocyanides which are completely dehydrated by heat without decomposition yield on further heating N and carbide of iron. In addition to these there may be formed a cyanide or carbide of the positive metal, or the metal may be liberated with evolution of $(CN)_2$.

5. Those ferrocyanides which cannot be completely dehydrated without decomposition yield HCN, NH_3, CO_2 and a mixture or compound of each of the metals with carbon.

6. When aqueous solutions of alkali ferrocyanides are electrolyzed, the alkali separates at the $-$ pole, and HCN and Prussian blue at the $+$ pole; in some cases a cyanide of the metal forming the electrode is produced.

7. When heated with concentrated H_2SO_4, sulphates of Fe, NH_4, and the metal forming the $+$ radical are produced, and SO_2, CO, CO_2, and N are evolved.

8. H_2S has no action on alkaline ferrocyanides. Some others are decomposed, giving a sulphide of the $+$ radical, and H_4FeCy_6.

9. Ferrocyanides of heavy metals are generally decomposed by aqueous alkalis, giving a ferrocyanide of the alkali and hydrate of the heavy metal.

10. Decomposed by boiling with HgO and water, forming $HgCy_2$ and oxides or hydrates of the metals.

11. Alkaline ferrocyanides are converted into the corresponding ferricyanides by oxidizing agents.

Potassium Ferrocyanide $(K_4FeCy_6 \cdot 3H_2O)$:   *Preparation.* —
1. By boiling Prussian blue with potash:

$$Fe_4(FeCy_6)_3 + 12KOH = 3K_4FeCy_6 + 4Fe(OH)_3.$$

2. By fusing nitrogenous animal matter with K_2CO_3 and scrap iron, lixiviating with water, filtering, and crystallizing. KCy is first formed, which reacts on the iron compounds during the extraction with water, yielding K_4FeCy_6. [This process will be discussed more fully under the heading "Manufacture of Cyanide."]

3. By heating ammonium thiocyanate (NH_4SCN) with scrap iron to dull redness and extracting with water.

4. By action of KCN on ferrous compounds:

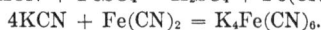

$$2KCN + FeSO_4 = K_2SO_4 + Fe(CN)_2.$$
$$4KCN + Fe(CN)_2 = K_4Fe(CN)_6.$$

Similarly with FeS, $FeCO_3$, $Fe(OH)_2$, etc.

5. By acting upon various metallic ferrocyanides with KOH.

Properties. — Crystallizes with 3 molecules H_2O as yellow quadratic pyramids. Non-poisonous. Loses water at 60° to 80° C. Unchanged at ordinary temperatures. Soluble in 4 parts cold and 2 parts boiling water. Insoluble in alcohol.

Reactions. — 1. Heated in a closed vessel, it melts a little above red heat, evolves N, and leaves a mixture of Fe carbide and KCy. When the hydrated salt is used, CO_2, NH_3, HCN, and N are evolved.

2. Aqueous solution is gradually decomposed on exposure to light, forming Prussian blue and evolving HCN.

3. When heated to redness in air, or with reducible metallic oxides, it gives KCyO. Ozone changes it slowly into K_3FeCy_6. Oxidizers such as $KMnO_4$, MnO_2, PbO_2, etc., also form K_3FeCy_6:

$$2K_4FeCy_6 + RO_2 = RO + K_2O + 2K_3FeCy_6.$$

Chlorine and bromine act as follows:

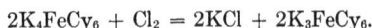

$$2K_4FeCy_6 + Cl_2 = 2KCl + 2K_3FeCy_6.$$

Iodine gives a double compound of KI and K_3FeCy_6.

4. [1] On electrolysis it gives K_3FeCy_6 at the $+$ pole and KOH and H at the $-$ pole (Joannis).

[1] According to Morley and Muir (Watts, "Dict. of Chemistry," II, 333), when aqueous solutions of alkali ferrocyanides are electrolyzed, alkali separates at the $-$ pole and HCy and Prussian blue at the $+$ pole.

5. Dilute cold sulphuric acid gives H_4FeCy_6:

$$K_4FeCy_6 + 2H_2SO_4 = H_4FeCy_6 + 2K_2SO_4.$$

Dilute warm sulphuric acid gives

$$2K_4FeCy_6 + 3H_2SO_4 = 3K_2SO_4 + K_2Fe \cdot FeCy_6 + 6HCy.$$

Concentrated sulphuric acid gives

$$K_4FeCy_6 + 6H_2SO_4 + 6H_2O = 2K_2SO_4 + 3(NH_4)_2SO_4 + FeSO_4 + 6CO.$$

6. Fairly concentrated HNO_3 (say 50 per cent. HNO_3 by volume) forms nitroprussic acid ($H_2FeCy_5 \cdot NO \cdot H_2O$).

Very concentrated HNO_3 decomposes K_4FeCy_6 into N, CN, NO, CO_2, KNO_3 and Fe_2O_3.

7. Boiled with mercuric oxide:

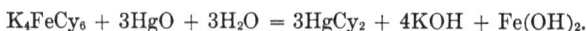

$$K_4FeCy_6 + 3HgO + 3H_2O = 3HgCy_2 + 4KOH + Fe(OH)_2.$$

8. Boiled with NH_4Cl, it gives NH_4Cy, which volatilizes:

$$K_4FeCy_6 + 6NH_4Cl = 6NH_4Cy + 4KCl + FeCl_2.$$

9. Heated with ammoniacal silver nitrate, it forms $Fe(OH)_3$ and $KAgCy_2$.

10. Most metallic salts give precipitates of insoluble ferro-cyanides.

CHARACTERISTIC INSOLUBLE FERROCYANIDES

The more important of these are as follows:

Barium	$BaFeCy_6 \cdot 6H_2O$	White, slightly soluble.
Cobalt	$Co_2FeCy_6 \cdot 7H_2O$	Blue, changing to red.
Copper	$Cu_2FeCy_6 \cdot xH_2O$ (cupric)	Reddish brown.
Iron	$Fe_2(FeCy_6)$ (ferrous)	White.
"	$Fe_4(FeCy_6)_3$ (ferric)	Blue.
Lead	$Pb_2FeCy_6 \cdot 3H_2O$	White.
Manganese	$Mn_2FeCy_6 \cdot 7H_2O$	White.
Mercury	composition doubtful	White.
Nickel	$Ni_2FeCy_6 \cdot$ (7 to 11) H_2O	Greenish white.
Tin	$Sn_2FeCy_6 \cdot 4H_2O$ (stannous)	White.
"	$SnFeCy_6 \cdot 4H_2O$ (stannic)	Brownish.
Zinc	$Zn_2FeCy_6 \cdot 3H_2O$	White.
Silver	Ag_2FeCy_6	White.

The more important soluble ferrocyanides are those of NH_4, Ca, H, Mg, Na, K.

There are also numerous compounds in which more than one metal forms part of the positive radical, for example, $K_2ZnFeCy_6$.

Hydrogen Ferrocyanide:

Hydroferrocyanic Acid (H_4FeCy_6). — This is the acid corresponding to the metallic ferrocyanides. It may be prepared:

1. By decomposing Ba_2FeCy_6 with an equivalent of H_2SO_4, and filtering from $BaSO_4$.

2. By the action of cold concentrated HCl on an air-free solution of K_4FeCy_6, and washing the precipitate with alcohol and ether.

Properties. — A white, crystalline powder, becoming blue in moist air with evolution of HCy. Soluble in water. Strongly acid to litmus.

Reactions. — Solution decomposes carbonates, acetates, oxalates, and tartrates. When boiled with water it decomposes:

$$2H_4FeCy_6 = FeH_2(FeCy_6) + 6HCy.$$

Ferrocyanides of Copper:

Cuprous Ferrocyanide (Cu_4FeCy_6) is said to be formed by adding K_4FeCy_6 to Cu_2Cl_2 dissolved in HCl.

Cupric Ferrocyanide ($Cu_2(FeCy_6) \cdot xH_2O$) is formed:

1. By adding K_4FeCy_6 to a cupric salt. By this means the precipitate formed is mixed with double ferrocyanides $CuK_2(FeCy_6)$, and $Cu_3K_2(FeCy_6)_2$.

2. By adding H_4FeCy_6 to a cupric salt. A reddish-brown precipitate, insoluble in dilute acids; soluble in ammonia and alkalis.

Ferrocyanides of Iron:

Ferrous Ferrocyanide $Fe_2(FeCy_6)$. — Obtained by adding ferrous salts to H_4FeCy_6, both compounds being freed from air and ferric compounds. A white amorphous precipitate, soon oxidizing, with formation of blue compounds:

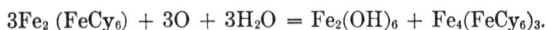

$$3Fe_2 (FeCy_6) + 3O + 3H_2O = Fe_2(OH)_6 + Fe_4(FeCy_6)_3.$$

Potassium Ferrous Ferrocyanide ($K_2 \cdot Fe \cdot FeCy_6$) (Everitt's salt). — Obtained by decomposing K_4FeCy_6 with hot dilute sulphuric acid. (See under "Potassium Ferrocyanide.")

White, microscopic quadratic crystals, becoming blue in air. Converted by oxidizers into potassium-ferrous ferricyanide ($FeK \cdot FeCy_6$).

Ferric Ferrocyanide (Prussian Blue: $Fe_4(FeCy_6)_3$): *Preparation.* — 1. By acting on a ferric salt with potassium ferrocyanide, keeping the ferric salt in excess:

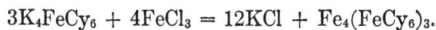

$$3K_4FeCy_6 + 4FeCl_3 = 12KCl + Fe_4(FeCy_6)_3.$$

2. By acting on a soluble cyanide with a ferrous and ferric salt, adding excess of KOH or NaOH, then excess of acid:

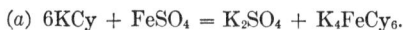

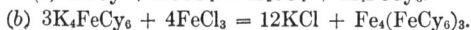

$$(a)\ 6KCy + FeSO_4 = K_2SO_4 + K_4FeCy_6.$$
$$(b)\ 3K_4FeCy_6 + 4FeCl_3 = 12KCl + Fe_4(FeCy_6)_3.$$

3. Prussian blue is obtained commercially by mixing K_4FeCy_6 with partially oxidized ferrous sulphate, and heating the light-blue precipitate so obtained with various oxidizing agents. In addition to ferric ferrocyanide ($Fe_4(FeCy_6)_3$) it usually contains also — Ferrous ferrocyanide ($Fe_2 \cdot FeCy_6$), Ferrous ferricyanide ($Fe_3(FeCy_6)_2$,) and sometimes Everitt's salt ($K_2Fe \cdot FeCy_6$).

Properties. — A dark-blue amorphous solid with luster resembling that of copper. Retains water up to 250° C., at which temperature it begins to decompose. Insoluble in water, alcohol, ether, and dilute acids.

Reactions. — 1. When strongly heated it ignites, leaving Fe_2O_3. Heated in a partially closed vessel, it first gives off NH_4Cy and $(NH_4)_2CO_3$, and finally HCN, CO, and CO_2.

2. Alkalis (including MgO) decompose it, yielding ferric hydrate and a ferrocyanide of the alkali.

Ammonia gives $Fe(OH)_3$ and $(NH_4)_4FeCy_6$.

Alkaline carbonates give similar reactions.

3. Boiled with HgO and water, it gives Fe_2O_3 and $HgCy_2$.

4. Boiled with lead oxide, it gives lead ferrocyanide, a ferricyanide, and Fe_2O_3.

5. Reduced by H_2S or by Fe or Zn, it gives white ferrous ferrocyanide ($Fe_2 \cdot FeCy_6$).

Sodium Ferrocyanide ($Na_4FeCy_6 \cdot 12H_2O$). — Obtained by boiling Prussian blue with NaOH, filtering, and cooling. Pale yellow monoclinic crystals, which effloresce in air. Reactions similar to those of K_4FeCy_6.

Zinc Ferrocyanide ($Zn_2FeCy_6 \cdot 3H_2O$). — Obtained pure by adding excess of $ZnSO_4$ to H_4FeCy_6. The precipitates formed by $ZnSO_4$, or other zinc salts, with ferrocyanides contain more or less of double ferrocyanides of zinc and the alkali metal.

White precipitate, insoluble in dilute acids, soluble in alkalis and ammonia. Double salts are also formed with ammonia, one of which is $Zn_2FeCy_6 \cdot 3NH_3 \cdot H_2O$.

The white precipitate found in the zinc-boxes used in precipitation of gold-cyanide solutions, and formed when insufficient alkali is present, appears to be a mixture of $ZnCy_2$, $Zn(OH)_2$ and double ferrocyanides, to which the formulæ $K_2Zn(FeCy_6)$ and $K_2Zn_3(FeCy_6)_2$ have been assigned.[1] (See Section V: "Chemistry of Precipitation.")

(b) FERRICYANIDES (M_3FeCy_6)

These substances are formed generally by the action of oxidizing agent on ferrocyanides:

$$2M_4FeCy_6 + RO = M_2O + R + 2M_3FeCy_6;$$

where RO represents any substance containing "available" oxygen. The alkali ferricyanides are soluble in water: most others are insoluble. Many have characteristic colors. They act as oxidizing agents by reaction the converse of that given above.

CHARACTERISTIC FERRICYANIDES

Metal	Color	Solubility, etc.
NH₄	Ruby-red	Soluble
Ca	Red	Deliquescent
Co	Red-brown	Insoluble
Cu { Cu′	Brown-red	Insoluble (cuprous)
Cu″	Yellowish	Insoluble (cupric)
Fe″	Blue	Insoluble (ferrous)
Fe″Fe‴	Green	Insoluble (ferrosoferric)
H	Brownish green	Very Soluble
Pb	Dark red	Insoluble
Mg	Red-brown	Soluble
Mn	Brown	Insoluble
Ni	Reddish brown	Insoluble
K	Red	Soluble
Ag	Orange-yellow	Insoluble
Na	Ruby-red	Deliquescent
Sn	White	Gelatinous

The insoluble ferricyanides are soluble in alkalis and ammonia.
Potassium Ferricyanide (K_3FeCy_6): *Preparation.* — 1. By act-

[1] "Proc. Chem., Met. and Min. Soc. of South Africa," V. 56, 75.

ing upon K_4FeCy_6 with Cl, Br, ozonized oxygen, PbO_2, $KMnO_4$, or bleaching powder. In some cases the action requires the presence of an acid, thus:

$$5K_4FeCy_6 + 4H_2SO_4 + KMnO_4 = 5K_3FeCy_6 + 3K_2SO_4 + MnSO_4 + 4H_2O.$$

When chlorine is used, it is passed into cold solution of K_4FeCy_6, with constant agitation, till a few drops of the liquid give a brown-red color with $FeCl_3$, instead of a blue color.

K_3FeCy_6 is also one of the products of electrolysis of K_4FeCy_6.

Properties. — Large red prismatic monoclinic crystals. More soluble in hot than in cold water. Nearly insoluble in alcohol.

Reactions. — 1. Heated in a closed vessel, it decrepitates, evolves $(CN)_2$ and a little N, leaving a residue of KCy, K_4FeCy_6, $Fe_4(FeCy_6)_3$, C, Fe, and probably paracyanogen $(CN)x$.

2. Heated in air $(CN)_2$ is evolved, and Fe_2O_3 and KCy remain.

3. Reduced to K_4FeCy_6 by many reagents, such as H_2S, Na_2S, $Na_2S_2O_3$; also by finely divided metals (Ag, Fe, Zn, *etc.*), by SO_2, by hot ferrous salts, phosphites, *etc.*, especially in presence of alkalis. The general type of these reactions may be represented thus:

$$K_3FeCy_6 + KOH + R = K_4FeCy_6 + R \cdot OH.$$

R being an oxidizable substance.

4. Certain metallic oxides (*e.g.*, PbO, Cr_2O_3, MnO, SnO, etc.), when boiled with K_3FeCy_6 in presence of KOH, pass to a higher state of oxidation and form K_4FeCy_6, thus:

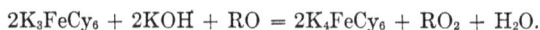

$$2K_3FeCy_6 + 2KOH + RO = 2K_4FeCy_6 + RO_2 + H_2O.$$

5. Alkaline peroxides are decomposed with evolution of oxygen, thus:

$$K_3FeCy_6 + K_2O_2 + H_2O = K_4FeCy_6 + KOH + O_2.$$

6. When alkalis are boiled with concentrated ferricyanide solutions, Fe_2O_3 is precipitated and ferrocyanide, cyanide, and $(CN)_2$ produced.

Ammonia forms ferrocyanides of K and NH_4, and evolves N. Mercuric oxide gives $HgCy_2$ and Fe_2O_3:

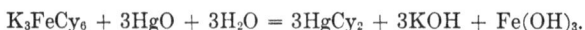

$$K_3FeCy_6 + 3HgO + 3H_2O = 3HgCy_2 + 3KOH + Fe(OH)_3.$$

7. A mixture of KCy and K_3FeCy_6 acts as a solvent of gold or silver without intervention of oxygen:

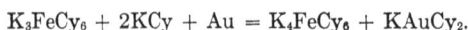

$$K_3FeCy_6 + 2KCy + Au = K_4FeCy_6 + KAuCy_2.$$

8. Decomposed by chlorine or bromine, forming first the haloid

cyanogen compounds (CNCl and CNBr) and HCy; on continued boiling various ferrosoferric ferricy anides are formed.

9. Potassium iodide forms an unstable compound which readily breaks up into I and K_4FeCy_6.

10. Boiling HCl forms KCl, $FeCl_3$ and "Turnbull's blue," $Fe_3 (FeCy_6)_2$.

11. Evaporation with nitric acid yields potassium nitroprusside (K_2FeCy_5NO) and KNO_3.

Hydrogen Ferricyanide (H_3FeCy_6):

Hydroferricyanic Acid. — Prepared by action of H_2SO_4 on lead ferricyanide, or by adding cold concentrated K_3FeCy_6 to 2 to 3 times its volume of concentrated HCl and filtering. Lustrous brownish-green needles. Very soluble in water or alcohol. Decomposes in air, giving HCy and a blue residue.

Ferricyanides of Copper:

Cuprous Ferricyanide ($Cu_3(FeCy_6)$ or $Cu_6(FeCy_6)_2$). —Brownish-red precipitate, obtained by adding a solution of Cu_2Cl_2 in HCl to K_3FeCy_6. Soluble in NH_3 but not in NH_4 salts.

Cupric Ferricyanide ($Cu_3(FeCy_6)_2$). — Yellowish precipitate, obtained by adding a cupric salt to K_3FeCy_6. Probably contains K. Soluble in NH_3 and in NH_4 salts.

Ferricyanides of Iron:

Ferrous Ferricyanide (Turnbull's Blue: $Fe_3(FeCy_6)_2$). — Prepared by adding K_3FeCy_6 or, better, H_3FeCy_6 to a ferrous salt, and, after digesting for some time, washing the precipitate with hot water. Also prepared by partial oxidation of ferrous ferrocyanide, Fe_2FeCy_6:

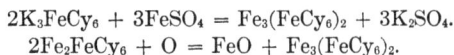

$$2K_3FeCy_6 + 3FeSO_4 = Fe_3(FeCy_6)_2 + 3K_2SO_4.$$
$$2Fe_2FeCy_6 + O = FeO + Fe_3(FeCy_6)_2.$$

A deep blue powder, with a tinge of copper-red. Insoluble in water, alcohol, and dilute acids. Soluble in oxalic acid. Cannot be dehydrated without decomposition. Exposed to moist air, it yields Prussian blue and ferric oxide:

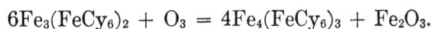

$$6Fe_3(FeCy_6)_2 + O_3 = 4Fe_4(FeCy_6)_3 + Fe_2O_3.$$

Decomposed by alkalis, alkaline carbonates, and ammonia, giving a ferricyanide and Fe_3O_4.

Ferrosoferric Ferricyanide (Prussian Green; Pelouze's Green: $Fe_7(FeCy_6)_6$). — A green precipitate, obtained by passing Cl into K_3FeCy_6 and boiling, then washing the residue with concentrated boiling HCl.   Changes to Prussian blue on contact with air.

A black ferrosoferric ferricyanide $Fe_5 (FeCy_6)_4$? is formed by acting on K_3FeCy_6 with Br.

Potassium Ferrous Ferricyanide (Soluble Prussian Blue: FeK·FeCy$_6$).—Obtained by mixing solutions of $FeCl_3$ and K_4FeCy_6 in equivalent proportions, stirring, and at once washing with cold water.   Also by action of pure $FeSO_4$ on K_3FeCy_6 in absence of air. When dried in vacuo it has the composition $4 (FeK·FeCy_6) 7H_2O$.

A blue solid, soluble in cold water.   Solution decomposed by boiling, giving a yellowish precipitate; on adding acids or metallic salts, a blue precipitate.   Alkalis and ammonia give $Fe(OH)_3$ and a ferrocyanide.   Ferric salts give Prussian blue.   Ferrous salts give Turnbull's blue.

Ferricyanide of Silver (Ag$_3$FeCy$_6$). — A reddish-yellow precipitate obtained by adding K_3FeCy_6 to $AgNO_3$.   Acted on by ammonia, forming a reddish double salt, which is soluble in excess of ammonia.

(c)   Perferricyanides (M_2FeCy_6)

When K_4FeCy_6 solution is heated with iodine a greenish-brown liquid is formed, from which alcohol precipitates a crystalline salt, which appears to be K_2FeCy_6.   It can be better prepared by acting on K_4FeCy_6 with HCl and $KClO_3$, neutralizing, and precipitating with alcohol.   Nearly black.   Dissolves to a deep violet solution.   Gives green precipitates with many metallic salts.   Very unstable.   Acts as a powerful oxidizer.

(d)   Nitroprussides ($M_2FeCy_5(NO)$)

Formation. — These compounds are formed by acting on alkali, ferrocyanides, or ferricyanides with HNO_3.   Two parts powdered K_4FeCy_6, 5 parts HNO_3, and 5 parts water are warmed on a water-bath.   A coffee-colored liquid is formed which gives off HCy, Cy, N, and CO_2.   When the liquid gives a dark-green or slate-colored precipitate with a ferrous salt (instead of blue), the reaction is complete.   After cooling and filtering, the liquid is neutralized with an alkali, again filtered, and evaporated till the alkaline nitroprusside formed crystallizes.

They are formed also by the action of nitrites on ferrocyanides, or on a boiling mixture of $FeCl_3$ and KCy.

General Properties. — The alkaline nitroprussides are soluble; also those of the alkaline earths. Most others are insoluble. Generally colored.

CHARACTERISTIC NITROPRUSSIDES

Metal	Color of Salt	Remarks
NH_4	....	Very Unstable
Ba....................	Dark red	Very Soluble
Ca	....	Very Soluble
H....................	Dark red ·	Deliquescent
K....................	Dark red	Deliquescent
Na....................	Ruby-red	Non-deliquescent
Cu	Greenish	Insoluble
Fe″	Yellowish pink	Insoluble
Ag....................	Flesh-colored	Insoluble
Zn	Yellow-rose	Insoluble

General Reactions. — 1. Solutions of alkali nitroprussides give a deep, brilliant purple color with alkali sulphides. The color soon fades,[1] but is often used as a characteristic test for sulphides in solution.

2. Aqueous solutions decompose on heating or exposure to light, giving a blue precipitate.

3. When boiled with alkalis, $Fe(OH)_3$ is precipitated, N evolved, and a nitrite and ferrocyanide formed.

4. H_2S precipitates Prussian blue, leaving a ferrocyanide in solution.

5. Soluble nitroprussides act as powerful oxidizing agents.

6. $KMnO_4$ forms $NaNO_3$ and a ferricyanide.

COMPOUNDS CONTAINING A METAL ANALOGOUS TO IRON, FORMING A COMPLEX RADICAL WITH CYANOGEN

The chief classes of these compounds are:

Chromocyanides (M_4CrCy_6)
Chromicyanides (M_3CrCy_6)
Cobaltocyanides $(M_4CoCy_6$: unstable)
Cobalticyanides (M_3CoCy_6)
Manganocyanides (M_4MnCy_6)
Manganicyanides (M_3MnCy_6).

[1] The products of decomposition are ferrocyanide, thiocyanate, nitrite, sulphur, and ferric oxide.

The chromo-, cobalto-, and mangano-cyanides (corresponding to ferrocyanides) are obtained by acting on the salts of the lower oxide of the respective metals with strong solutions of KCy.

The chromi-, cobalti-, and mangani-cyanides are obtained by oxidizing the above compounds, or by acting on the higher oxides or salt of the respective metals with KCy.

Some of the acids belonging to this series, namely, H_4CrCy_6, H_4CoCy_6, H_3CoCy_6, and H_4MnCy_6, have been prepared.

It has been suggested that mixtures of KCy with cobalti-cyanides or manganicyanides might be used as gold solvents.

(B) Metallic Compounds of Radicals in which Cyanogen is Combined with Oxygen or a Similar Element.

A number of substances are known in which cyanogen appears to form part of a negative radical, in which it is intimately united with oxygen or another element (S, Se) of the oxygen group.

These compounds have the general formula

$$(MR''CN)x,$$

where M is a + metal or group, and R'' an element (O, S, or Se) forming part of a negative radical containing C and N in the proportions to form cyanogen.

Since we may suppose the element R'' to be directly united either with the C or with the N of the cyanogen, we have two classes of isomeric compounds, corresponding with the formulæ

$$^1M - R'' - C \equiv N''' \text{ and } M - R'' - N^v \equiv C, \text{ respectively.}$$

The latter are distinguished by the prefix "iso."

Compounds of the type $(MRCN)x$ show a great tendency to polymerize, and different classes of bodies are known in which $x = 1, 2,$ and 3, respectively.

Different theories have been advanced to explain their constitution, which need not concern us here, and innumerable organic derivatives have been described, often of great complexity, in which M is a positive radical of the "alkyl" type (*e.g.*, CH_3, C_2H_5, C_6H_5, etc).

The most important substances for our purpose fall under the following groups:

[1] Other arrangements, *e.g.*, $M - C \equiv N = R''$ and $M - N = C = R''$, are also possible, and may occur in certain compounds.

Cyanates, MOCN. Isocyanates, MONC.
Fulminates, (MOCN)$_2$. Isodicyanates, (MONC)$_2$.
Cyanurates, (MOCN)$_3$. Isocyanurates, (MONC)$_3$.
Thiocyanates, MSCN.
Dithiocyanates, (MSCN)$_2$.
Thiocyanurates, (MSCN)$_3$.
Selenocyanates, MSeCN.

CYANATES (MOCN)

Cyanic Acid (HOCN). — This acid is only formed in small quantity by decomposing cyanates by stronger acids, as it immediately undergoes further decomposition. It may be made by the action of heat on its polymer, cyanuric acid (HOCN)$_3$, or by dehydrating urea (with phosphorus pentoxide), or by passing dry HCl over dry AgCNO.

It is a thin, colorless liquid which reddens litmus and has a pungent odor resembling acetic acid. It excites tears, and raises blisters on the skin. Soluble without decomposition in ice-cold water. Readily polymerizes. It is decomposed by water as follows:
$$HOCN + H_2O = NH_3 + CO_2.$$

Metallic Cyanates: *General Methods of Formation.* — 1. By fusing alkaline cyanides or ferrocyanides with easily-reduced oxides, such as PbO_2, or with oxidizing agents (nitrates, bichromates, etc.):
$$MCN + O = MOCN.$$

2. By passing cyanogen into a solution of an alkali or alkaline earth; *e.g.*,
$$(CN)_2 + 2KOH = KCN + KOCN + H_2O.$$

3. By heating alkaline carbonates to low redness in an atmosphere of cyanogen, or with mercuric cyanide:
$$M_2CO_3 + (CN)_2 = 2MOCN + CO.$$

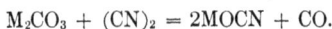

4. By electrolysis of soluble cyanides.

General Properties and Reactions. — 1. Most metallic cyanates are soluble in water; those of Ag, Cu, Pb, and Hg are only slightly soluble.

2. Alkali cyanates are not decomposed by heating to dull redness in dry air. In moist air they give carbonates of ammonium and the alkali:
$$2MOCN + 4H_2O = M_2CO_3 + (NH_4)_2CO_3.$$

Cyanates of the heavy metals are decomposed by heat into CO_2 and a cyanide of the metal:

$$2M''(OCN)_2 = M''(CN)_2 + M'' + 2CO_2 + N_2$$

3. Easily decomposed in solution by water into carbonates and ammonium salt (see above); or by dilute acids into CO_2, an ammonium salt of the acid, and a metallic salt of the acid; e.g.,[1]

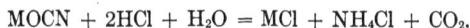

$$MOCN + 2HCl + H_2O = MCl + NH_4Cl + CO_2.$$

4. Reduced to cyanides by heating with certain reducing agents — H, C, K, or Fe.

5. Melted with sulphur, they give sulphides, sulphates, and thiocyanates.

6. Sodium amalgam reacts with neutral cyanates to produce formamide ($HCONH_2$).

7. Certain metallic cyanates are obtained by acting on salts of the required metal with alkaline cyanates, and crystallizing from alcohol.

8. *Silver Cyanate*: Silver salts give a white precipitate (perhaps AgONC), insoluble in cold water, soluble in boiling water; dissolved and decomposed by dilute nitric acid:

$$AgOCN + 2HNO_3 + H_2O = AgNO_3 + CO_2 + NH_4NO_3.$$

Silver cyanate is decomposed by heat into Ag mixed with C and some N and CO_2. It is soluble in NH_3, giving a double compound which loses NH_3 in the air.

Potassium Cyanate (KOCN): *Preparation.* — By the general methods given above; best as follows: 4 parts K_4FeCy_6 and 3 parts $K_2Cr_2O_7$, both dry and pulverized, are mixed, and the mixture heated, a little at a time, till it blackens. After cooling it is extracted with boiling alcohol, and KCNO crystallized out and dried in vacuo.

Properties. — Small, colorless, odorless laminæ resembling $KClO_3$. Soluble in water; fairly soluble in boiling hydrated alcohol. Insoluble in absolute alcohol.

1. According to some writers,[2] the salt present in commercial cyanide is an isocyanate (presumably KONC or KNCO), which is gradually transformed in solution into the normal cyanate

[1] Concentrated HCl or dilute H_2SO_4 give a certain amount of HOCN.

[2] Bettel and Feldtmann, " Proc. Chem., Met. and Min. Soc. of South Africa," I, 273.

(KOCN), which latter has no action on silver nitrate. This, however, is denied by other authorities,[1] who maintain that the decomposition taking place is merely the gradual formation of ammonium and potassium carbonate, referred to above:

$$2KOCN + 4H_2O = K_2CO_3 + (NH_4)_2CO_3.$$

2. When $BaCl_2$ is added to a solution of KOCN, it gives at first no precipitate; but on boiling, barium carbonate is thrown down:

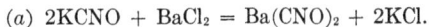

(a) $2KCNO + BaCl_2 = Ba(CNO)_2 + 2KCl.$
(b) $Ba(CNO)_2 + 4H_2O = BaCO_3 + (NH_4)_2CO_3.$

3. A solution of cobalt acetate gives, with potassium cyanate, dark blue quadratic crystals of a double cyanate $(CoK_2(CNO)_4)$. This reaction has been proposed as a test for cyanates.

FULMINATES $(MOCN)_2$

These are unstable, often explosive substances, formed by the action of alcohol on metallic nitrates. The acid itself has not been isolated. The most important are those of silver, mercury, and gold.

Mercuric fulminate $(Hg(OCN)_2)$. — Used in the manufacture of percussion caps. It is prepared by dissolving mercury in cold nitric acid, and pouring the solution into 90 to 92 per cent. alcohol. The salt deposits and is recrystallized from water. Crystallizes from alcohol in octahedra; from water in needles. Very slightly soluble in cold water, more soluble in hot water.

Explodes by heat, friction, percussion, or treatment with H_2SO_4, yielding Hg, N, and CO. When heated with water it polymerizes to fulminurate. Forms double salts with cyanides.

CYANURATES $(MOCN)_3$

Cyanuric acid is a product of the decomposition of urea by heat:

$$3CO(NH_2)_2 = (HOCN)_3 + 3NH_3.$$

It is precipitated, on dissolving the residue and diluting with water, in colorless, oblique, rhombic prisms (with $2H_2O$). Effloresces. Converted into anhydrous octahedra at $100°–120°$ C. Slightly soluble in cold water, more soluble in hot water. Soluble in hot acids without decomposition. Heated in a closed tube, it gives cyanic acid:

$$(HCNO)_3 = 3HOCN$$

[1] Victor, in "Zeit. für anal. Chem.," XL, 462–465.

When dissolved in NH_3 it gives a fine pink precipitate with ammonio-sulphate of copper.

PCl_5 forms $(CNCl)_3$.

Cyanurates are precipitated by Ba salts but not by Ca salts. They may occur sometimes in working cyanide solutions. The following formula has been suggested as representing their constitution:

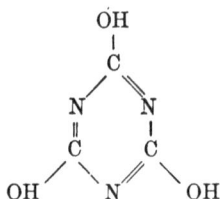

$$\begin{array}{c} \overset{\displaystyle OH}{\underset{\displaystyle C}{|}} \\ N \diagup \quad \diagdown\!\!\!\diagdown N \\ C \qquad C \\ \diagup \quad \diagdown N \diagup\!\!\!\diagup \quad \diagdown \\ OH \qquad OH \end{array}$$

Only organic derivatives of the isodicyanates and isocyanurates are known.

THIOCYANATES (MSCN)

General Methods of Formation. — 1. By fusion of cyanides or ferrocyanides with sulphur or with alkaline sulphides or polysulphides, or thiosulphates.

2. By the action of alkaline polysulphides on cyanides in aqueous solution.

Properties. — Soluble, often deliquescent salts. The thiocyanates of copper (cuprous), lead, mercury, and silver are white or yellowish insoluble salts.

General Reactions. — 1. Solutions give a deep-red color with ferric salts:
$$3KCNS + FeCl_3 = 3KCl + Fe(SCN)_3.$$

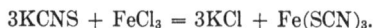

2. Alkali thiocyanates are decomposed by heat in presence of air, giving a sulphate and cyanate of the metal, and evolving SO_2. Thiocyanates of the heavy metals give a sulphide, S, CS_2, and finally $(CN)_2$ and N.

3. Oxidizing agents form sulphates: *e.g.*,
$$5KSCN + 6KMnO_4 + 12H_2SO_4 = 11KHSO_4 + 6MnSO_4 + 5HCN + 4H_2O.$$

4. Copper salts in presence of a reducing agent precipitate white cuprous thiocyanate, insoluble in water and dilute acids:
$$2KCNS + 2CuSO_4 + Na_2SO_3 + H_2O = Cu_2(SCN)_2 + K_2SO_4 + 2NaHSO_4,$$

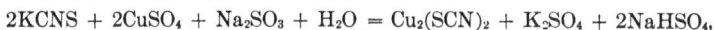

5. Silver salts give a white curdy precipitate of silver thiocyanate:
$$KCNS + AgNO_3 = AgSCN + KNO_3.$$

AgSCN is insoluble in water and dilute acids, soluble in ammonia, alkaline cyanides, and thiocyanates, and also in mercurous nitrate. Forms soluble double thiocyanates with KSCN. $(NH_4)SCN$, etc.

6. Salts of mercury give white precipitates; the mercurous salt is unstable and readily decomposes with separation of Hg. Mercuric thiocyanate swells greatly on heating, giving off Hg, N, and CS_2.

Ammonium Thiocyanate ($NH_4 \cdot SCN$): *Preparation.* — 1. By digesting HCN with ammonium polysulphide (S dissolved in $(NH_4)_2S$). On boiling, S separates, and the filtrate contains NH_4SCN.

2. By decomposing cupric thiocyanate with $NH_4 \cdot HS$, filtering and evaporating.

3. By evaporating a mixture of ammonia with alcoholic solution of CS_2.

4. By heating ammonium sulphate with carbon and sulphur:

$$(NH_4)_2SO_4 + C + S = NH_4SCN + SO_2 + 2H_2O.$$

Properties. — Large, white, deliquescent plates, very soluble in water and alcohol. Melts at 159° C. Dissolves in water with great absorption of heat.

Reactions. — 1. Heated to 180°–190° C., it gives off CS_2, H_2S, and NH_3.

2. Heated for some time nearly to melting-point, it is converted into its isomer, thiourea $(CS(NH_2)_2)$.

3. Forms double salts with $HgCy_2$ and with various thiocyanates.

Potassium Thiocyanate (KSCN): *Formation.* — 1. By fusing potassium sulphide with ferrocyanide, $K_2S_2O_3$ is first formed; at a higher temperature this reacts to form KSCN.

2. By acting on ammonium thiocyanate with potassium sulphide or potash at high temperature:

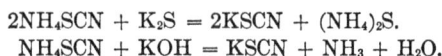

$$2NH_4SCN + K_2S = 2KSCN + (NH_4)_2S.$$
$$NH_4SCN + KOH = KSCN + NH_3 + H_2O.$$

Properties. — Long, white, striated prisms resembling niter. Very soluble in water, with great absorption of heat. Melts at 161.2° C. Non-poisonous.

Reactions. — 1. On melting, the salt turns green, then indigo, but becomes white again on cooling. Heated in moist air, it gives CO_2, NH_3, and K_2S.

2. Aqueous solution slowly decomposes (rapidly on boiling), evolving ammonia:

$$2KSCN + 4H_2O = K_2S + (NH_4)_2S + 2CO_2.$$

3. Heated with iron, it forms FeS, K_2S, and K_4FeCy_6.

4. On electrolysis, it gives H_2SO_4, SO_2, HCN, and S.

5. HCl gas reacts violently with molten KSCN, forming HCN, CS_2, and NH_4Cl.

6. Concentrated nitric acid forms "pseudo-sulphocyanogen" ($C_3N_3HS_3$). The same compound is formed by action of Cl on concentrated solution of KSCN. With excess of Cl the products are NH_4Cl, $(NH_4)_2SO_4$, and CO_2. Cl, passed into molten KSCN, gives S_2Cl_2 and $(CNCl)_3$.

7. Heated gently with PCl_5, it gives KCl, CNCl, and $PSCl_3$.

8. Forms double salts with $HgCy_2$ and HgI_2.

Thiocyanates of Gold and Potassium:

Auro potassic Thiocyanate (KAu(SCN)$_2$). — When auric chloride ($AuCl_3$) is added drop by drop to KSCN at 80° as long as the precipitate dissolves, and the liquid is evaporated and crystallized, straw-yellow prisms are obtained, melting at 100° C. Decomposed by heat into Au, S, CS_2, and KSCN. Solution blackens in light. Gives precipitates with various salt solutions. NH_3 precipitates $NH_3 \cdot AuSCN$.

Auripotassic Thiocyanate (KAu(SCN)$_4$). — When $AuCl_3$ is added to excess of KSCN in the cold, this substance is formed, which crystallizes from warm water in orange-red needles. Soluble in water, alcohol, and ether. Forms double compounds.

An analogous **aurisodic thiocyanate**, $NaAu(SCN)_4$, is also known.

SECTION IV

CHEMISTRY OF THE DISSOLVING PROCESS

Necessity of Oxygen in Ordinary Cyanide Process. — The researches of Bagration, Elsner, Faraday, and latterly of J. S. Maclaurin,[1] have demonstrated that under ordinary circumstances the presence of oxygen is necessary for the solution of metallic gold and silver in cyanide solutions. The reaction may in fact be expressed in the form frequently quoted as " Elsner's equation," though not originally due to him:

$$2Au + 4KCy + O + H_2O = 2KAuCy_2 + 2KOH.$$

Intermediate Products of Reaction. — The correctness of this equation as representing the final result of the reaction has been abundantly verified by Maclaurin and other investigators, but it is quite possible that intermediate products are formed, so that the actual mechanism of the reaction is not so simple as would appear from the above. Bodländer[2] states that hydrogen peroxide is first produced, and that a portion of this is further decomposed, an additional amount of gold being dissolved. The reactions he gives are as follows:

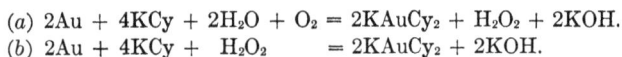

(a) $2Au + 4KCy + 2H_2O + O_2 = 2KAuCy_2 + H_2O_2 + 2KOH.$
(b) $2Au + 4KCy + \quad H_2O_2 \quad = 2KAuCy_2 + 2KOH.$

It is stated that H_2O_2, or some substance having similar reactions, can be detected in the solution. Bettel, however, states that the intermediate product is potassium auricyanide, and proposes the following:

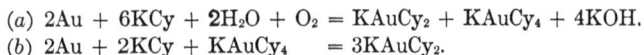

(a) $2Au + 6KCy + 2H_2O + O_2 = KAuCy_2 + KAuCy_4 + 4KOH.$
(b) $2Au + 2KCy + KAuCy_4 \quad = 3KAuCy_2.$

The reactions in the case of silver appear to be precisely analogous.

---

[1] Maclaurin, "Proc. Chem. Soc.," CXII, p. 81; CXLVI, pp. 7–8; CLXVIII, p. 149.

[2] Bödlander, " Zeit. für angew. Chem.," 1896, pp. 583–587.

Use of Artificial Oxidizers. — The necessity for oxygen being thus recognized, it seemed natural to suppose that the reaction would be assisted and hastened by the addition of substances which would readily yield oxygen under the conditions of cyanide treatment. The advantage, however, was only apparent in a few cases, namely, in those of alkaline peroxides, ferricyanides, and permanganates. In the case of peroxides the decomposition takes place rapidly in presence of water:

$$Na_2O_2 + H_2O = 2NaOH + O;$$

and consequently they are of little use unless added to dry ore previous to cyanide treatment. In the case of ferricyanides, the reaction is very probably quite independent of oxygen, as already pointed out. The reaction with permanganate is somewhat complicated; formates, oxalates, nitrites, urea, and ammonia being among the products (in alkaline solutions), and it does not appear to be known with any certainty how permanganates act in aiding the solution of gold. If manganicyanides are formed, the reaction is probably analogous to that of ferricyanides. Solutions completely deprived of oxygen were found to have no solvent effect whatever; and the addition to such solutions of chlorates, perchlorates, chromates, bichromates, nitrates, nitrites, or bleaching powder had no effect, and did not bring about solution of the gold.[1]

Solution of Metals in Cyanide Independently of Oxygen. — We have already noted several types of reaction in which gold is dissolved (usually by the interaction of two different cyanogen compounds, one of them being a simple alkali cyanide), in which oxygen plays no part. We may here briefly enumerate the following cases, some of which have been already referred to:

(1) The haloid compounds of cyanogen; *e.g.*,

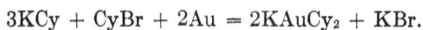

$$3KCy + CyBr + 2Au = 2KAuCy_2 + KBr.$$

(2) Double cyanides of copper and alkali metal; *e.g.*,

$$Cu_2Cy_2 \cdot 4KCy.$$

(3) Cuprammonium cyanides of the type

$$x(NH_3) \cdot y(CuCy_2) \cdot z(Cu_2Cy_2).$$

These readily decompose with formation of the insoluble salt

[1] Bettel and Marais; quoted by T. K. Rose, "Metallurgy of Gold," 4th ed., p. 380.

$4NH_3 \cdot CuCy_2 \cdot 2Cu_2Cy_2$ and liberation of cyanogen, which at the moment of formation attacks gold in presence of KCy:

$$KCy + Cy + Au = KAuCy_2.$$

(4) Double cyanides of mercury and alkalis:

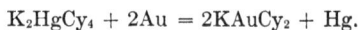

$$K_2HgCy_4 + 2Au = 2KAuCy_2 + Hg.$$

(5) Double cyanides of zinc and alkalis. [These are stated to be solvents of gold by themselves, but it is evident that no reaction analogous to No. 4, above, can take place. It seems more probable that the molecule breaks up in solution:

$$K_2ZnCy_4 = 2KCy + ZnCy_2.$$

and that the dissolving of gold is due to the ordinary reaction of KCy in presence of O and H_2O.]

(6) Ferricyanides and analogous compounds:

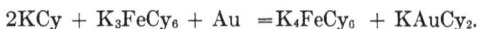

$$2KCy + K_3FeCy_6 + Au = K_4FeCy_6 + KAuCy_2.$$
$$2KCy + K_3MnCy_6 + Au = K_4MnCy_6 + KAuCy_2.$$

In practice, these aids to solution have had only a limited application, but arrangements for bringing air into intimate contact with the material undergoing treatment, and with the cyanide solution, have sometimes proved of great value. In such cases it may be doubted whether the oxygen directly aids the solution of gold, or whether the beneficial action is not rather due to the destruction of cyanicides, or of precipitants of gold, present in the material to be treated. According to H. F. Julian,[1] free oxygen plays no primary part in the dissolution of gold by cyanide, but, on the contrary, exerts a retarding influence. He ascribes the solvent action of cyanide (as did Faraday) to the formation of local voltaic circuits. "These, in the first instance, deposit hydrogen and oxygen, which it may be assumed become occluded at their respective electrodes until the systems are in equilibrium." When this condition is reached, less expenditure of energy is required to remove Cy from the solution than to occlude a further quantity of O, and accordingly AuCy is formed and is deposited in films. This compound has a high potential, and acts as an electrode, forming a couple with the uncombined gold. The AuCy then dissolves in the excess of KCy, "as one salt dissolves

[1] Julian, "Brit. Assoc. Reports," South Africa, 1905, p. 369.

in the solution of another." When the solution contains dissolved oxygen, this aids in a secondary manner by oxidizing the occluded hydrogen produced by the local voltaic currents, and thus upsetting the equilibrium.

Factors that Influence Solubility of Gold and Silver. — Whatever may be the correct theoretical explanation of the reaction, it has been observed in practice that certain factors exert an influence on the rate of solution of the precious metals in cyanide. The most important of these are:

(1) Degree of concentration of the solution.

(2) Temperature.

(3) Presence or absence of "Cyanicides," *i.e.*, of substances other than the precious metals, capable of decomposing or combining with cyanides.

(4) Amount of "available oxygen," either free or in some easily decomposed compound.

(5) Physical condition of the surfaces of metallic gold and silver in contact with the solution.

(6) Area of metal exposed in proportion to weight.

(7) Relative masses of solvent, and matter undergoing treatment in contact with it.

Influence of Strength of Solution. — It has been found by Maclaurin [1] that, other things being equal, both gold and silver dissolve most rapidly in solutions of a strength corresponding to 0.25 per cent. KCy (*i.e.*, 0.10 per cent. Cy). The rate of solution falls off gradually above or below this strength, but is nearly constant between 0.1 per cent., and 0.25 per cent. KCy. His experiments show that the rate of solution of silver is about two-thirds that of gold, but otherwise the phenomena are similar. L. Janin's experiments,[2] however, show, for cement silver, a maximum solubility at 1 per cent. KCy. The reduced efficiency of stronger solutions has been ascribed to the diminished solubility of oxygen in strong KCy. It would seem, however, that the dissociation of cyanide into its ions, $\overset{+}{K}$ (or $\overset{+}{Na}$) and $\overset{-}{CN}$, is an essential condition to the solution of gold or any metal therein. Since this dissociation takes place the more completely, the more dilute the solution, there must be some point of maximum effi-

---

[1] J. S. Maclaurin, "Journ. Chem. Soc.," LXIII, 731; LXVII, 199.

[2] L. Janin, "Eng. and Min. Journ.," Dec. 29, 1888.

ciency at which the solution contains the greatest possible number of $\overline{CN}$ ions *per unit volume.*

Influence of Temperature. — It has long been known that
hot solutions dissolve the metals more rapidly than cold ones,
and the greater efficiency of hot cyanide solution for dissolving
gold was pointed out by Bagration in 1843. Experimental trials
with solutions heated to about 130° F. (55° C.) have generally
shown a somewhat higher extraction than at the ordinary temperature. In the practical application of this observation, however, there are economic difficulties, depending on the cost of
raising large volumes of liquid to a high temperature; and moreover, there is in hot solutions a somewhat rapid decomposition
of cyanide, as described below.

*Influence of other Soluble Substances: Electrochemical Order of
Metals in Cyanide Solution:* The relative solubilities of metals
in cyanide (or other solvent) depend upon their electrochemical
order in that solvent. Determinations of the electrochemical
order of metals in potassium cyanide solution, under various
conditions of dilution and temperature, have been published
by Gore, Skey, Christy, and others. Generally speaking, the
order, from positive to negative, of the ordinary metals, is as
follows:

+ Mg	Ag
Al	Hg
Zn	Pb
Cu	Fe
Au	− Pt

Carbon is − to Pt.

From this table it may be inferred that any metal in the list
would tend to dissolve in cyanide more readily than the metal
below, and would displace that metal from solution and precipitate
it; for example, zinc will precipitate Cu, Au, or Ag; copper will
precipitate Au or Ag; gold or silver will precipitate mercury;
magnesium or aluminium will precipitate gold or silver more
readily than will zinc.

The solubility, and in some cases the order, is affected by
differences of concentration and temperature.

Experiments on the electromotive force of minerals in cyanide solutions[1] show that most minerals are electronegative to metallic gold, so that gold tends to dissolve in preference to such minerals.

In ore treatment, it is generally found that the percentage of gold extracted is higher than that of silver, but this may easily be accounted for when it is considered that the actual weight of silver present in an ore is generally much greater than that of gold, and that the silver is largely present, not as metal, but as a compound such as Ag_2S, only slightly soluble in KCy.

Selective Action of Dilute Solutions. — It is claimed that dilute cyanide solutions have a "selective action" for gold and silver in preference to base metals. In the case of the majority of ores, the actual experimental facts may be stated as follows:

(1) It is not usually the case that the actual amount of base metal dissolved is less than that of precious metal; but whether the solution be strong or weak, the amount of base metal dissolved relatively to the total quantity of base metal present is less than the quantity of precious metal dissolved relatively to the total quantity of precious metal present.

(2) The weight of precious metal dissolved per unit of base metal or compound dissolved in a given case is the greater the more dilute the solution. In all actual cases, the precious and base metals are dissolved simultaneously, but the rate of solution of the precious metals is less diminished by dilution than that of the base metals.

These phenomena are not by any means confined to cyanide solutions: the same laws are observed in the action of thiosulphates on silver chloride, in ores which have undergone a chloridizing roast, and in the action of acids on powdered rock-mixtures containing different percentages of almost any two or more ingredients soluble in the acid. In general terms, it may be said that the ingredient which is present in the smaller percentage is dissolved in the greater relative amount; the difference becoming more marked the more dilute the solution.

Since the formation of base-metal compounds is the main source of consumption of cyanide in the treatment of most ores, it follows that the consumption of cyanide may be taken, roughly, as an index of the amount of base metal dissolved in any given case. It is nearly always found that the consumption of cyanide increases regularly as the strength of the solution is increased,

[1] Christy, *Trans.* A. I. M. E. (Sept., 1899), XXX, 1–83.

but that the amount of gold or silver dissolved only increases slightly, if at all, with increasing strength of cyanide, provided a reasonably sufficient time be given for the action. The experiments by MacArthur, quoted by Rose,[1] illustrate this point, as do also the following tests by Caldecott;[2] in which charges of 34 tons each, assaying originally 100 grains per ton, were treated for three days with various strengths of solution, and subsequently water-washed.

Strength of Cyanide Used. KCy, Per Cent.	Strength after Treatment. KCy, Per Cent.	KCy Consumed. Per Cent.	Gold Extracted. Per Cent.
.041	.010	.017	83
.110	.068	.024	84
.373	.277	.055	83
1.023	.860	.095	85
3.333	3.020	.180	83

Many other experiments might be quoted from which it would appear that while the extraction is practically constant, the consumption is higher the greater the cyanide strength.

Decomposition of Cyanide by External Influences. — In its early days, objections were frequently raised against the practical application of the cyanide process on account of the instability of potassium cyanide when exposed to atmospheric influences. It was urged that this substance undergoes spontaneous decomposition, and that it is rapidly affected by atmospheric oxygen and carbonic acid. A dilute solution of KCN, as already pointed out, is probably to a large extent dissociated into the ions $\overset{+}{K}$ and $\overset{-}{CN}$.

Hydrolysis. — In addition to this, but to a smaller extent, a decomposition known as "hydrolysis" occurs, as follows:

$$KCy + H_2O = KOH + HCy.$$

The HCy thus formed is continually volatilized, and the smell of it is always noticeable over vessels containing a solution of any alkali cyanide.

The alkali cyanides in the solid state are not stable unless

---

[1] "Metallurgy of Gold," 4th edition, p. 392.
[2] "Proc. Chem., Met. and Min. Soc. of South Africa," I, 294.

perfectly dry; sodium cyanide is especially liable to decomposition when existing as a hydrate, and reacts with water evolving ammonia; thus: $NaCN + 2H_2O = HCO_2Na + NH_3$.

A similar decomposition occurs with all soluble cyanides of the alkalis and alkaline earths, but much more rapidly in hot solutions, producing a formate of the metal and evolving ammonia gas. The hydrolytic decomposition into formates and ammonia is also promoted by the presence of large amounts of alkali.

It has been stated that sodium cyanide is less stable in solution than potassium cyanide of corresponding cyanogen strength, or than the mixed cyanides frequently obtained as a commercial product.

Oxidation. — The cyanides also, both in the solid state and in solution, gradually absorb oxygen and carbonic acid from the air; the solid cyanides are deliquescent and are gradually acted upon by the moisture they absorb from the air. They are ultimately converted into carbonates, possibly in some cases with formation of cyanates as intermediate products, the ultimate reaction being:

$$2KCN + O_2 + 4H_2O = K_2CO_3 + (NH_4)_2CO_3.$$

The oxidation in solution is, however, very slow under ordinary circumstances, and air or oxygen may be injected under pressure into a cyanide solution for a long time without any appreciable oxidation of cyanide taking place.

Action of Carbonic Acid. — In the absence of free alkali, carbonic acid decomposes cyanide; thus:

$$KCN + CO_2 + H_2O = HCN + KHCO_3.$$

This does not take place, however, if sufficient caustic alkali ($NaOH$, KOH, $Ca(OH)_2$, $Ba(OH)_2$) or ammonia, or alkaline carbonate (K_2CO_3, Na_2CO_3), be present to neutralize the CO_2 according to the reactions:

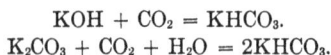

$$KOH + CO_2 = KHCO_3.$$
$$K_2CO_3 + CO_2 + H_2O = 2KHCO_3.$$

Alkali bicarbonates afford no protection against CO_2; as soon as all hydrates and carbonates have been converted into bicarbonates, any further quantities of CO_2 entering the solution at once decompose the cyanide. These decompositions are, however, of less importance in dilute solutions than might be supposed, and the addition of a slight excess of alkali prevents the rapid evolution of HCy, even when solutions are exposed to the air in open vessels.

Action of Base Metal Compounds: Iron Compounds. — In the treatment of ores a certain amount of cyanide is always consumed in attacking the compounds of base metals contained therein. Of these metals the commonest is *iron*, which occurs in a variety of conditions, some of which are unaffected by cyanide, while others are rapidly attacked.

Iron in the form of ferric oxide is quite unaffected by cyanides; the same is also probably true of anhydrous ferrous oxide and magnetic oxide. Ferrous hydrate, however, is rapidly acted upon with formation of ferrocyanide:

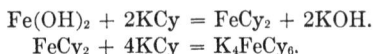

$$Fe(OH)_2 + 2KCy = FeCy_2 + 2KOH.$$
$$FeCy_2 + 4KCy = K_4FeCy_6.$$

Similarly with ferrous carbonate:

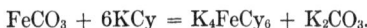

$$FeCO_3 + 6KCy = K_4FeCy_6 + K_2CO_3.$$

which reaction explains the efficiency of ferrous carbonate as an antidote for cyanide poisoning.

Iron pyrites exist in ores in two different crystalline conditions, known as "pyrite" and "marcasite." Pyrite, which occurs in dense yellowish cubical crystals, is only slowly acted on by cyanide, whereas marcasite, which is softer, is much more readily attacked, perhaps directly as follows:

$$FeS_2 + KCy = FeS + KSCy.$$

In any case, under ordinary conditions of treatment, pyrites rapidly undergoes oxidation, forming sulphates and basic sulphates of iron, which act readily as cyanicides. Caldecott[1] gives the following as the main stages in the decomposition of pyrite or marcasite:

(1) FeS_2 ....Iron pyrites.
(2) $FeS + S$ ....Ferrous sulphide, sulphur.
(3) $FeSO_4 + H_2SO_4$ ....Ferrous sulphate, sulphuric acid.
(4) $Fe_2(SO_4)_3$ ....Ferric sulphate.
(5) $2Fe_2O_3 \cdot SO_3$ ......Insoluble basic ferric sulphate.
(6) Fe_2O_3 ....Ferric oxide.

Ferrous sulphide is injurious owing to its power of absorbing

[1] "Proc. Chem., Met. and Min. Soc. of South Africa," II, 98.

oxygen, and may perhaps actually decompose $KAuCy_2$ so as to precipitate gold that has been previously dissolved. Thiocyanates are almost invariably present in cyanide solutions after use in extraction of ores. As sulphur does not dissolve directly in cyanide, they may be due to the action on pyrites, as stated above, or to action on ferrous sulphide:

$$FeS + 7KCy + H_2O + O = KCyS + K_4FeCy_6 + 2KOH.$$

All the iron in working cyanide solutions probably ultimately takes the form of soluble alkali ferrocyanides (K_4FeCy_6 or Na_4FeCy_6), since ferrous thiocyanate, if formed at all, is unstable and would react with KCy:

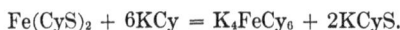

$$Fe(CyS)_2 + 6KCy = K_4FeCy_6 + 2KCyS.$$

Ferrous and ferric sulphates act directly as cyanicides, ultimately giving rise to Prussian blue and similar compounds, or to hydrocyanic acid.

In the case of ferrous sulphate, some such series of reactions as the following may be supposed to take place:

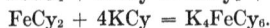

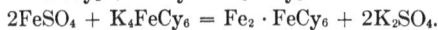

$$FeSO_4 + 2KCy = FeCy_2 + K_2SO_4.$$
$$FeCy_2 + 4KCy = K_4FeCy_6.$$
$$2FeSO_4 + K_4FeCy_6 = Fe_2 \cdot FeCy_6 + 2K_2SO_4.$$
$$3Fe_2FeCy_6 + 3O + 3H_2O = Fe_4(FeCy_6)_3 + 2Fe(OH)_3.$$

In the case of ferric sulphate:

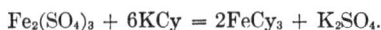

$$Fe_2(SO_4)_3 + 6KCy = 2FeCy_3 + K_2SO_4.$$
$$FeCy_3 + 3H_2O = Fe(OH)_3 + 3HCy.$$

In presence of sufficient alkali the soluble ferric salts are precipitated as ferric hydrate, which is unacted upon by cyanide;[1] ferrous salts give ferrous hydrate, which readily forms ferrocyanide, as above described.

The basic sulphates, although insoluble in water, are attacked by cyanide, giving a similar series of reactions.

Action of Copper Compounds on Cyanide. — Copper exists in ores most frequently as sulphide, often combined with sulphides of other metals, copper pyrites being a compound, or intimate mixture, of the sulphides of iron and copper. It also occurs as native copper, and in oxidized ores as carbonate, oxide, silicate, and other forms.

[1] According to Julian and Smart ("Cyaniding Gold and Silver Ores," 2d edition, p. 112), "ferric hydrate" is an indefinite compound, $Fe_2O_3 \cdot xH_2O$, which, under certain conditions, may dissolve in cyanide, forming a ferrocyanide.

Action of Cuprous Sulphide. — The natural sulphides of copper are much less readily acted on by cyanide than are the carbonates, or than artificially prepared sulphide, but they gradually dissolve, probably with formation, finally, of cupric thiocyanate, by some such series of reactions as the following:

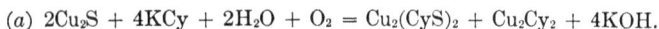

(a) $2Cu_2S + 4KCy + 2H_2O + O_2 = Cu_2(CyS)_2 + Cu_2Cy_2 + 4KOH.$

The insoluble cuprous thiocyanate and cyanide readily dissolve in excess of KCy:

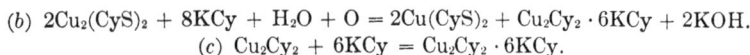

(b) $2Cu_2(CyS)_2 + 8KCy + H_2O + O = 2Cu(CyS)_2 + Cu_2Cy_2 \cdot 6KCy + 2KOH.$

(c) $Cu_2Cy_2 + 6KCy = Cu_2Cy_2 \cdot 6KCy.$

Since the solutions almost invariably contain an excess of alkali thiocyanate, the further reaction is probably as follows:

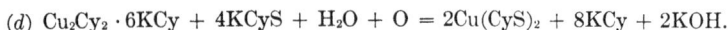

(d) $Cu_2Cy_2 \cdot 6KCy + 4KCyS + H_2O + O = 2Cu(CyS)_2 + 8KCy + 2KOH.$

Action of Carbonate of Copper. — Carbonate of copper, both native and artificial, is attacked by cyanide with extreme rapidity, and is one of the most troublesome cyanicides. The action, when KCy is in excess, appears to be as follows:

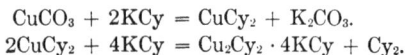

$$CuCO_3 + 2KCy = CuCy_2 + K_2CO_3.$$
$$2CuCy_2 + 4KCy = Cu_2Cy_2 \cdot 4KCy + Cy_2.$$

When sufficient alkali is also present:

$$Cy_2 + 2KOH = KCy + KCyO + H_2O;$$

so that the entire reaction becomes:

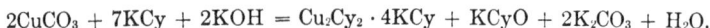

$$2CuCO_3 + 7KCy + 2KOH = Cu_2Cy_2 \cdot 4KCy + KCyO + 2K_2CO_3 + H_2O.$$

Soluble cupric salts ($CuSO_4$, etc.) act in an analogous way, evolving cyanogen in unprotected solutions, and forming double cyanides probably of the type $Cu_2Cy_2 \cdot nKCy.$

Removal of Copper Before or During Treatment. — Copper, when present in the form of carbonate, oxide, hydrate, etc., may be largely removed from the ore:

(1) By preliminary acid treatment, with H_2SO_4 or H_2SO_3; any acid remaining in the charge after this operation may be neutralized with alkali previous to cyanide treatment.

(2) By preliminary treatment with ammonia, which dissolves many copper compounds that are insoluble in water.

(3) In the course of cyanide treatment, by the use of a mixture of cyanide with ammonia or ammonium salts. Such mixtures

are largely dissociated in dilute solutions into the ions $\overset{+}{NH_4}$, $\overset{+}{K}$, $\overset{+}{Na}$, and $\overset{-}{CN}$, $\overset{-}{Cl}$, $\overset{-}{OH}$, etc.; they, therefore, act as powerful solvents, and appear to attack gold under certain conditions in preference to copper. The copper forms "cuprammonium" compounds of the type $xNH_3 \cdot yCuCy_2 \cdot zCu_2Cy_2$, which are only slightly soluble in water, forming blue solutions from which green, insoluble, needle-like crystals of ammonium dicuprosocupric cyanide ($4NH_3 \cdot CuCy_2 \cdot 2Cu_2Cy_2$) are deposited. At the same time cyanogen is liberated, which at the moment of formation readily attacks gold in presence of excess of KCy:

$$KCy + Cy + Au = KAuCy_2.$$

Gold is thus dissolved and copper precipitated at the same time.

Copper may also be precipitated from solutions by the addition of acids; thus:

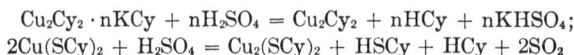

$$Cu_2Cy_2 \cdot nKCy + nH_2SO_4 = Cu_2Cy_2 + nHCy + nKHSO_4;$$
$$2Cu(SCy)_2 + H_2SO_4 = Cu_2(SCy)_2 + HSCy + HCy + 2SO_2$$

forming the insoluble cuprous cyanide and thiocyanate.

General Action of Metallic Sulphides. — Many metallic sulphides are more or less soluble in alkalis, with formation of alkaline sulphides:
$$2KOH + R''S = R''(OH)_2 + K_2S.$$

Since the cyanide solutions almost invariably contain an excess of caustic alkali, this reaction takes place to some extent in the treatment of ores containing such sulphides. The soluble alkaline sulphide locally formed may sometimes act as an injurious factor in the process, owing to its powerful reducing action, which removes the oxygen necessary for the solution of the gold. It is, however, probably soon converted into thiocyanate by the reaction
$$K_2S + KCy + H_2O + O = KCyS + 2KOH.$$

and also thrown out of solution by the reaction

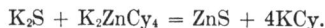

$$K_2S + K_2ZnCy_4 = ZnS + 4KCy.$$

It is partly owing to this latter reaction that the zinc, dissolved as double cyanide in the precipitation process, is removed from the solutions before it has accumulated to an injurious extent.

As a general rule, clean unoxidized minerals have only a slight action on cyanide solutions. Christy's experiments[1] show that nearly all such minerals are electronegative to gold in cyanide solutions, so that the latter would dissolve in preference. Nevertheless, a large consumption of cyanide is sometimes observed in the treatment of ores containing, for example, mispickel, realgar, stibnite, cinnabar, galena, blende, fahl ore, sulphides of nickel and cobalt, etc., especially when the sulphides have undergone partial oxidation.

In some cases, as with the sulphides of arsenic, antimony, and tin, it is possible to extract the cyanicide by a preliminary operation, in which the ore is treated with a (preferably hot) solution of an alkali or alkaline sulphide. Zinc-blende is not much attacked by cyanide, but calamine is a rapid cyanicide; thus:

$$ZnCO_3 + 4KCy = K_2ZnCy_4 + K_2CO_3.$$

Action of Selenium and Tellurium. — Metallic selenides, especially when finely crushed, are dissolved in cyanide without much difficulty, the element passing into solution as a selenocyanide; *e.g.*, KSeCy.

Tellurium dissolves with more difficulty, perhaps first forming an alkaline telluride, K_2Te, which is rapidly converted into a tellurite, K_2TeO_3.

Ores containing considerable quantities of refractory minerals, especially those of arsenic and tellurium, are perhaps best treated by preliminary roasting, to remove or oxidize the greater part of these elements; but in some cases they have been successfully treated by very fine crushing, followed by bromocyanide.

When an ore has been partially oxidized, either by natural processes or by roasting, it is very necessary to extract the soluble matter by water-washing previous to cyanide treatment. Such partially oxidized material may contain large amounts of soluble sulphates, which may cause a very much larger consumption of cyanide than the raw unoxidized ore. Addition of alkalis, without previous water-washing, is not always a satisfactory remedy, as metallic hydrates, such as $Fe(OH)_2$, may be precipitated, which are themselves cyanicides.

Oxides of manganese sometimes appear to form very easily decomposable double cyanides, which readily deposit the brown hydrated oxide.

[1] *Trans.* A. I. M. E. (Sept., 1899), XXX, 33.

Regeneration of Solutions. — Since the main cause of cyanide consumption is the formation of soluble double cyanides or complex cyanogen compounds of the base metals, and since solutions highly charged with such compounds are more or less inefficient as solvents of gold and silver, it has been suggested that the cyanogen in such liquors might be recovered in the form of simple alkali cyanides, by treatment with suitable chemicals. This has been carried out in practice in some cases, but generally speaking the cost of chemicals, power, labor, etc., required for such treatment outweighs the advantage gained by it. The following methods have been suggested:

(1) The solution is acidulated, and the cyanogen converted wholly or partially into hydrocyanic acid, in some cases with precipitation of a portion as insoluble cyanides; thus:

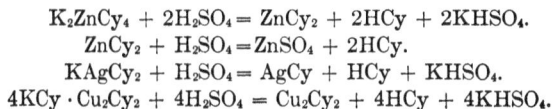

$$K_2ZnCy_4 + 2H_2SO_4 = ZnCy_2 + 2HCy + 2KHSO_4.$$
$$ZnCy_2 + H_2SO_4 = ZnSO_4 + 2HCy.$$
$$KAgCy_2 + H_2SO_4 = AgCy + HCy + KHSO_4.$$
$$4KCy \cdot Cu_2Cy_2 + 4H_2SO_4 = Cu_2Cy_2 + 4HCy + 4KHSO_4.$$

(2) Ferric salts may be added, to precipitate ferrocyanides as Prussian blue (for reactions see above). After drying, the Prussian blue is converted into an alkali cyanide by fusion in a closed vessel with alkali or alkaline carbonate.

(3) Alkaline sulphides decompose the double cyanides of silver, mercury, zinc, cadmium and nickel, precipitating these metals more or less completely as sulphides and forming a soluble alkali cyanide; *e.g.*:

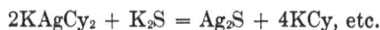

$$2KAgCy_2 + K_2S = Ag_2S + 4KCy, \text{ etc.}$$

The reaction, however, is never complete, as the sulphides formed are not absolutely insoluble in the excess of cyanide. Moreover, any excess of sulphide must be removed and the solution thoroughly aerated before it can be successfully used in treating fresh charges of ore.

Reprecipitation of Gold and Silver during Dissolving Process. — Under certain conditions, a premature precipitation of gold and silver may occur in the leaching tanks or other parts of the dissolving plant, and may lead to losses. It is well known that charcoal is a precipitant of gold from cyanide solutions; an admixture of carbonaceous matter may therefore cause local precipitation of values previously dissolved. This was a frequent cause of trouble with the black cyanide formerly in use (containing carbon and carbides of iron). When this cyanide was dissolved, it left a layer of black sediment, which sometimes contained considerable values after being some time in contact with the auriferous solutions. Decaying organic matter, such as grass roots, etc., may likewise cause losses in this way.

Local acidity of the ore may cause precipitation of insoluble aurous cyanide:

$$KAuCy_2 + H_2SO_4 = AuCy + HCy + KHSO_4.$$

Under some circumstances gold and silver may be thrown down as insoluble salts, by the action of soluble base metal salts; thus:
$$2KAgCy_2 + ZnSO_4 = K_2SO_4 + ZnAg_2Cy_4.$$

The action of alkaline sulphides on silver double cyanide has already been noted; gold is not precipitated by these reagents. Some metallic sulphides, such as Cu_2S, seem to have the power of throwing down gold from a cyanide solution.

Action of Cyanide on Ferricyanide. — Morris Green [1] suggests the following equation as representing the reaction between cyanides and ferricyanides in aqueous solution

$$2K_3Fe(CN)_6 + 3KCN + H_2O =$$
$$2K_4Fe(CN)_6 + KCNO + 2HCN.$$

As the ferrocyanide is detrimental and the cyanate inert, this reaction does not explain the efficiency of ferricyanides as an aid in gold extraction.

Dr. J. Moir, discussing Green's suggestion [*loc. cit.*, March, 1913, p. 422] thinks the activity of ferricyanide for this purpose is due to the momentary liberation of nascent cyanogen:

$$KCN + K_3Fe(CN)_6 = K_4Fe(CN)_6 + (CN).$$

The cyanogen is only active at the moment of formation, as $(CN)_2$ or cyanogen gas is quite inert.

A secondary reaction appears to take place between cyanic acid and cyanides, liberating hydrocyanic acid; as this is a "balanced" or incomplete reaction, the decomposition shown in Green's equation is never complete, as in fact was found by experiment.

[1] "J. Chem. Met. and Min. Soc. of S. Afr.," Feb. 1913, p. 358.

The entire series of reactions, taking place, according to Dr. Moir is as follows:

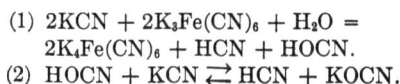

(1) $2KCN + 2K_3Fe(CN)_6 + H_2O =$
$2K_4Fe(CN)_6 + HCN + HOCN.$

(2) $HOCN + KCN \rightleftarrows HCN + KOCN.$

In the presence of sufficient free alkali the entire reaction may be represented thus:

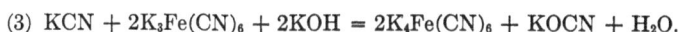

(3) $KCN + 2K_3Fe(CN)_6 + 2KOH = 2K_4Fe(CN)_6 + KOCN + H_2O.$

On standing, or on boiling, the cyanate is decomposed by hydrolysis:

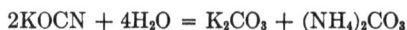

$$2KOCN + 4H_2O = K_2CO_3 + (NH_4)_2CO_3$$

the solution becoming alkaline.

Action of Cyanide on Silver Sulphide. — The reaction $Ag_2S + 4NaCN = Na_2S + 2NaAg(CN)_2$ is more or less reversible, since alkaline sulphides precipitate silver from solutions of the double cyanide.

In practice it is found that no sulphur remains in the solution as a soluble alkaline sulphide, but that the whole of it takes the form of thiocyanate, by the reaction

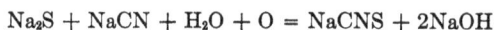

$$Na_2S + NaCN + H_2O + O = NaCNS + 2NaOH$$

so that the equation for the reaction between silver sulphide and cyanide would perhaps be better expressed as follows:

$$Ag_2S + 5NaCN + H_2O + O = 2NaAg(CN)_2 + NaCNS + 2NaOH.$$

Noel Cunningham (Min. Sci. Press, Nov. 16, 1912, p. 634) states that in the treatment of concentrates the amount of silver dissolved could be approximately ascertained by titrating the thiocyanate in the solution. Until the extraction of silver was complete, only the sulphur originally present as silver sulphide entered the solution as thiocyanate. When treatment was continued beyond this point, a considerable amount of sulphur originally in other combinations passed into solution.

As noted elsewhere, the addition of lead salts in some cases improves the extraction of silver from ores in which that metal is present as sulphide.

Influence of Lead Salts on Silver Extraction. — In treating ores containing sulphide of silver it is customary to add a small quantity of lead acetate or litharge at some stage of the treatment. The efficiency of these compounds is commonly explained by supposing that the lead precipitates any sulphur present as alkaline sulphide and thus prevents the absorption of oxygen by these sulphides, the reactions being:

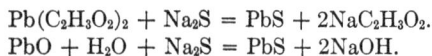

$$Pb(C_2H_3O_2)_2 + Na_2S = PbS + 2NaC_2H_3O_2.$$
$$PbO + H_2O + Na_2S = PbS + 2NaOH.$$

This explanation is not altogether satisfactory, as alkaline sulphides hardly ever occur in the solutions, even when lead salts are absent. It is preferable to consider that the lead salt aids the cyanide in attacking the silver mineral in the ore by some such reaction as the following:

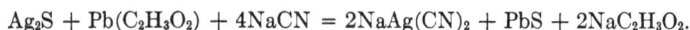

$$Ag_2S + Pb(C_2H_3O_2) + 4NaCN = 2NaAg(CN)_2 + PbS + 2NaC_2H_3O_2.$$

Tests made by the writer and others have shown that practically all lead salts, soluble or insoluble, are effective in increasing the silver extraction with certain ores, the efficiency being generally equal for quantities of these different compounds containing equal amounts of lead, the only noteworthy exception observed being in the case of lead tartrate, where the organic radicle perhaps exerts an injurious effect. Even finely powdered galena was found to have an effect in increasing the extraction.

The amount of lead compound which must be·added to produce the most beneficial effect is by no means a matter of indifference. In every case there is a maximum, beyond which further additions of lead are injurious. In tests on a particular Mexican silver ore the best effects were obtained by the addition of about 1.5 kilos of lead acetate or 1 kilo of litharge per ton of ore.

In some cases the lead in the ore itself may affect the silver extraction, and may dissolve in the cyanide solution in sufficient quantity to exert injurious effects. Native carbonate of lead is soluble to an appreciable extent in cyanide solutions, and the lead thus introduced consumes cyanide and precipitates on zinc, producing base bullion.

Lead may be precipitated from the solution by the use of sodium carbonate or bicarbonate which throw down lead and calcium without precipitating gold or silver.

The only lead compounds hitherto used in practice are the acetate, oxide (litharge) and nitrate, the latter being preferred to the acetate at the Homestake mine, on account of lesser cost per unit of lead.

Precipitating Effect of Carbon and Carbonaceous Matter. — Many observations have been made showing that the presence of coal, shale, charcoal, wood-ashes, decaying wood or other vegetable matter, in the material undergoing cyanide treatment has a prejudicial effect on the extraction of gold and silver. This effect may be due to the removal of oxygen by the direct action of these substances, or to the precipitation of gold and silver already dissolved. The action has generally been ascribed to the decomposition of the double cyanides by the gases occluded in the pores of the wood or charcoal.

Most observers are agreed that pure graphite has no precipitating action whatever.

A. J. Clark and W. J. Sharwood in Proc. Chem. Met. and Min. Soc. of South Africa, Jan. 1910, p. 234, give the results of a number of experiments with different materials, showing that anthracite has no action, bituminous coal a slight action, coke a more marked effect, and charcoal more still. It was found possible to precipitate as much as 99 per cent. of the gold in a cyanide solution by passing it slowly through a relatively large amount of bituminous coal.

In the case of tailings which normally gave an extraction of 70 to 75 per cent., it was found that the addition of 8 per cent. of wood-ashes reduced the extraction to 3.5 per cent. There was no distinct evidence that undecayed wood, or other material such as canvas or cocoa-nut matting, consisting chiefly of cellulose, have any precipitating effect; the values obtained by assaying such material after soaking with gold-bearing solution being little more than could be accounted for by simple absorption. Decaying wood was found to be an active precipitant. In several cases where ore from a mine has been mixed with material from decayed or charred timbers, much difficulty has been experienced in getting a satisfactory extraction. When wood is well painted with a suitable medium, little more than traces of precious metal are absorbed. Gold is precipitated by charcoal in the metallic form, as may be seen by microscopic examination.

E. H. Croghan (Journ. Chem. Met. and Min. Soc. of South Africa, May, 1910, p. 391) describes a number of experiments on carbon (as half-burnt coal) separated from lime used in cyaniding. The carbon was extracted by dissolving most of the lime in cold hydrochloric acid, obtaining a product containing 60 per cent. of carbon. Different quantities of this material were agitated with measured amounts of a cyanide and gold solution, and the amount of gold precipitated after 18 hours' contact was determined by filtering, drying the carbon residue, and scorifying with lead. The result showed for quantities ranging from 0.5 to 0.1 per cent. carbon in lime, a precipitation of .01 mgr. of gold; for smaller quantities the gold precipitated was unweighable. The method of preparing the carbon for the test was, however, criticised by C. Toombs and Jas. Gray (Journ. Chem. Met. and Min. Soc. of South Africa, July, 1910, p. 11) on the ground that the hydrochloric acid used would liberate occluded gases, and break up sulphides which might be associated with the carbon, and tests were made showing that carbon present in lime, when not treated with acid as above, may have a greater reducing action on gold in cyanide solution than carbon which has been separated with hydrochloric acid.

R. K. Cowes (Proc. Aus. Ins. Min. Eng., Jan., 1911, quoted in Journ. Chem. Met. and Min. Soc. of South Africa, May, 1911, 575) found that precipitation of gold and silver was caused by the presence of carbon in tailings treated at the Waihi-Paeroa Gold Extraction Co.'s plant in New Zealand. During treatment the silver continued to dissolve for some time, while the gold residues increased until practically all the gold previously dissolved was precipitated, after which the silver residue also began to increase, showing that carbon has no effect on the silver as long as any gold is present to be precipitated. It was found that fresh carbon precipitates gold and silver more readily than carbon which has become saturated with water. This may easily be the case if the precipitation is due to occluded gases.

Morris Green (Journ. Chem. Met. and Min. Soc. of South Africa, Sept., 1912, and Bull. I. M. M. No. 109, Oct. 9, 1913) describes experiments on the cause of the precipitation of gold from cyanide solutions by means of charcoal. It was proved (1) that carbon per se has no precipitating effect. Samples of graphite, whether in lump or powder, precipitated no gold. When gold cornets were agitated with powdered graphite and cyanide, the rate of solution of the gold was accelerated. It was also found that carbon, prepared by the action of sulphuric acid on sugar, has no precipitating action on gold. (2) It has been supposed that the precipitation of gold by charcoal is due to an osmotic effect whereby the dissolved substance is separated by some action similar to that of a semi-permeable membrane. This theory was likewise disproved. A gold cyanide solution separated from water by charcoal walls $\frac{1}{4}$ to $\frac{1}{2}$ in. thick showed no osmotic pressure. The cyanide readily diffused through the charcoal into the water. Cyanide does not pass to any considerable extent through a membrane of copper ferrocyanide precipitated within the walls of an unglazed porcelain vessel. (3) The nature of the occluded gases was examined. It was shown that the precipitating power of charcoal varied according to the method of preparation and the length of time to which it had been exposed to the air. The following table shows the percentage of gold precipitated under the same conditions by equal weights of different forms of charcoal:

	Gold precipitated per cent.
Freshly ignited wood charcoal	90.3
Wood charcoal exposed in lumps for about 6 months	50.0
The same, re-heated to incipient redness	84.2
Sugar charcoal, made by heating sugar in closed retorts	62.1
Sugar charcoal, by partially burning sugar in contact with air . .	5.6

It is noteworthy that charcoal gradually loses its precipitating power on exposure, but that this power can be partially restored by reheating.

Experiments were made with various kinds of charcoal, exhausted with a gas burette at ordinary temperature and also at temperatures of 70° C. and 200° C. The gases evolved consisted chiefly of oxygen, nitrogen and hydrogen, and the precipitating power of the charcoal was little if at all impaired.

A sample of the charcoal was then heated to 500° C. in a silica tube, and the evolved gases extracted by a powerful vacuum pump. They had the following composition:

	Volumes per cent.
Carbon dioxide	12.4
Oxygen	0.7
Carbon monoxide	26.2
Hydrogen	2.5
Nitrogen	58.2

Green considers that the gases as actually present in the charcoal probably contained more CO and less CO_2 than in the above, as some CO_2 would be formed by dissociation according to the reaction $2CO = CO_2 + C$.

Graphite similarly treated evolved no CO.

The charcoal, after exhausting at 500° C. as above, showed a greatly diminished precipitating power. In the case of freshly prepared charcoal this was reduced from 93.3 to 17 per cent., and in the case of commercial wood charcoal previously exposed to the atmosphere for about a year, from 41.2 to 14.8 per cent.

The general conclusion drawn is that precipitation is mainly due to occluded carbon monoxide, in a condensed and, therefore, more active condition. Hydrogen and electrical effects may also play a minor part in the phenomenon.

SECTION V

CHEMISTRY OF PRECIPITATION AND SMELTING PROCESSES

It has already been pointed out that gold and silver are electronegative to zinc in cyanide solutions; hence we should expect that zinc would act as a precipitant of the precious metals. This effect, however, is obviously complicated by various secondary reactions. The following points are always noticed in this connection:

(a) Precipitation only takes place in presence of sufficient free cyanide.

(b) It is always accompanied by increase in the alkalinity of the solution.

(c) It is always accompanied by evolution of hydrogen.

The entire effect, as regards precipitation of gold, is probably expressed by the equation:

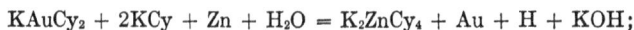

$$KAuCy_2 + 2KCy + Zn + H_2O = K_2ZnCy_4 + Au + H + KOH;$$

but the solution of zinc goes on to a large extent quite independently of the precipitation of gold, and bears no necessary proportion to it. The probable reaction is

$$Zn + 4KCy + 2H_2O = K_2ZnCy_4 + 2KOH + H_2,$$

though some writers have supposed that an alkaline zincate, such as $Zn(OK)_2$, may be formed.

It would appear that the electric couples $Zn:Pb$ and $Zn:Au$, etc., give rise to local currents which electrolyze water, the oxygen at the moment of formation attacking zinc to form hydroxide:

$$Zn + O + H_2O = Zn(OH)_2.$$

Under the usual working conditions this redissolves in the excess of cyanide; thus:

$$Zn(OH)_2 + 4KCy = K_2ZnCy_4 + 2KOH;$$

but an insufficiency of free cyanide may lead to its deposition in the form of a white precipitate on the surface of the shavings.

·123

As before pointed out, the double cyanides of zinc and alkali metals may be assumed partially to dissociate in dilute solutions, so that we have KCy, NaCy, and $ZnCy_2$ actually in solution, and these are, perhaps, still further split up. In the absence of sufficient free alkali or alkaline cyanide, this $ZnCy_2$ may also be deposited.

Sharwood[1] has investigated the conditions under which zinc oxide dissolves in cyanide, and zinc cyanide in alkaline hydroxides. He finds that ZnO is soluble in .65 per cent. KCy with formation of a double cyanide, and of a compound of zincate with excess of alkaline hydrate:

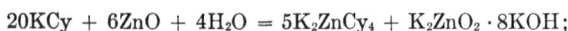

$$20KCy + 6ZnO + 4H_2O = 5K_2ZnCy_4 + K_2ZnO_2 \cdot 8KOH;$$

the latter being the result of a secondary reaction; thus:

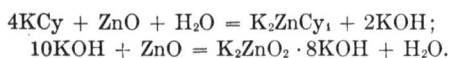

$$4KCy + ZnO + H_2O = K_2ZnCy_4 + 2KOH;$$
$$10KOH + ZnO = K_2ZnO_2 \cdot 8KOH + H_2O.$$

The action of KOH on $ZnCy_2$ is as follows:

(a) When the $ZnCy_2$ is in less proportion than $ZnCy_2$: 2KOH it dissolves completely and permanently:

$$4KOH + 2ZnCy_2 = K_2ZnCy_4 + K_2ZnO_2 + 2H_2O.$$

On heating, the zincate decomposes:

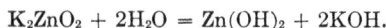

$$K_2ZnO_2 + 2H_2O = Zn(OH)_2 + 2KOH.$$

(b) When $ZnCy_2$ is in the proportion $ZnCy_2$: KOH, it is temporarily dissolved, but soon begins to precipitate $Zn(OH)_2$, as follows:

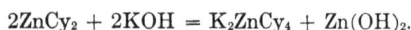

$$2ZnCy_2 + 2KOH = K_2ZnCy_4 + Zn(OH)_2.$$

(c) With larger proportions of $ZnCy_2$ any excess remains unaffected by KOH.

NaOH acts in a similar way.

It seems probable that any excess of KCy would convert the zincate into the double cyanide, but as increasing the alkalinity undoubtedly increases the solvent efficiency of the solution, it may be that we have here a reversible reaction:

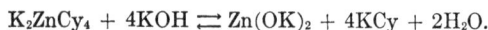

$$K_2ZnCy_4 + 4KOH \rightleftharpoons Zn(OK)_2 + 4KCy + 2H_2O.$$

Various investigators have made more or less complete analyses of the white precipitate above referred to; somewhat conflicting results have been obtained, and the substance no doubt varies according to the conditions of formation. Experience has shown

[1] "Bull. Univ. Cal.," April 27, 1904.

that the formation of the precipitate is greatly promoted by exposure of the zinc shavings to the air. It occurs chiefly as a flocculent, whitish deposit in the upper compartments of the boxes, and often interferes seriously with effective precipitation of the gold and silver. The general conclusion is that it consists principally of zinc hydrate, with smaller quantities of cyanide and ferrocyanide of zinc, with small amounts of the sulphates and carbonates of calcium, lead, etc.; also alumina, silica, and other ingredients of less importance. The following analyses give some idea of its composition:

(1) By A. Whitby[1] (the metals being calculated as their equivalent oxides):

ZnO	54.00	per cent.
SiO_2 and insoluble	2.10	"
$Fe_2O_3 + Al_2O_3$	.75	"
PbO	.73	"
CaO	.50	"
CN	2.50	"
K_2O	.70	"
Loss on ignition	36.80	"

(2) By Bay and Prister [2]:

Zinc hydrate, $Zn(OH)_2$	54.79	per cent.
Zinc cyanide, $ZnCy_2$	22.73	"
Zinc-potassium ferrocyanide, $K_2Zn_3(FeCy_6)_2$	10.45	"
Silica, SiO_2	1.03	"
Ferric oxide, Fe_2O_3	1.00	"
Cupric oxide, CuO	.40	"

The zinc hydrate and cyanide are readily soluble in excess of KCy or of caustic alkali (also in dilute acids), but the ferrocyanide is not easily soluble, and if once formed generally remains in the boxes until the clean-up.

There are grounds for supposing that gold and silver are precipitated on zinc, either in the metallic condition or as an alloy, according to the concentration of the liquors undergoing precipitation in metal and cyanide. Argall states that when the solutions pass 1.5 ounces gold per ton, the gold on the shavings assumes a yellow color, and passing 2 ounces and upward is usually golden yellow in the first compartment, shading off in

---

[1] " Proc. Chem., Met. and Min. Soc. of South Africa," V, 55.
[2] Ibid., V, 77.

the following compartments to the usual black-colored deposit obtained from poor solutions.

The alloys so formed are more energetically attacked than pure zinc by the cyanide solutions, and hence act more efficiently as precipitants. Silver also presents a totally different appearance, according to the conditions of precipitation.

Action of Copper in Precipitation with Zinc. — Solutions containing copper deposit this metal in the upper compartments of the zinc-box; the copper forms a closely-adherent metallic film, which coats the zinc shavings in such a way that all further action ceases and the precipitation of gold and silver is prevented. In strong solutions, copper is precipitated in preference to gold and silver. When copper has accumulated to a certain extent in the solution, it becomes practically impossible to use the liquor either for extraction of gold and silver or for precipitation, and the copper must be removed by some means or the liquor run to waste.

The zinc-lead couple has been sometimes used with advantage for cupriferous solutions; certain compartments filled with zinc shavings previously dipped in lead acetate are found to collect most of the copper. Barker,[1] however, says that although "with a good spongy coating of lead on the zinc the precipitation of the copper is at first good, the copper coming down in a loose spongy form," the action is not continuous. He therefore suggests precipitating as cuprous cyanide by the addition of sulphuric acid. When the copper exists in the solutions as a double cyanide (*e.g.*, $Cu_2Cy_2 \cdot 4KCy$, or $Cu_2Cy_2 \cdot 6KCy$), the precipitation of copper is accompanied by regeneration of cyanide, as shown by the following reactions:

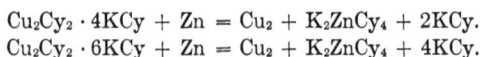

$$Cu_2Cy_2 \cdot 4KCy + Zn = Cu_2 + K_2ZnCy_4 + 2KCy.$$
$$Cu_2Cy_2 \cdot 6KCy + Zn = Cu_2 + K_2ZnCy_4 + 4KCy.$$

Small quantities of mercury are almost invariably present in the solution before precipitation. Mercury is carried off mechanically with the tailings from the battery and gradually dissolves to form K_2HgCy_4. In some cases it may also be partly derived from mercury minerals contained in the ore itself. The compound is directly decomposed by zinc, as follows:

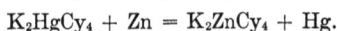

$$K_2HgCy_4 + Zn = K_2ZnCy_4 + Hg.$$

[1] *Trans.* I. M. M., XII, 399.

As already pointed out, the double cyanides of mercury are gold solvents, and also act by decomposing soluble sulphides, so that the presence of this metal in small quantities is a distinct advantage.

The precipitate formed in the zinc-boxes occasionally contains other elements; such as arsenic, antimony, selenium, tellurium, etc., derived from the material treated. The removal of these in the subsequent treatment may involve special operations, as noted below. Whenever the zinc shavings have been treated with lead acetate, the precipitate will contain a certain amount of metallic lead.

Electrolytic Precipitation. — Many methods have been tried or suggested for precipitating gold and silver electrolytically from cyanide solutions. The best known is that of Siemens and Halske, which, in the form originally adopted in South Africa, used anodes of iron and cathodes of lead. The cathodes were not attacked under ordinary circumstances, but an adherent deposit of gold was formed upon them. Under the influence of the electric current the aurocyanide is ionized into

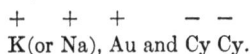

$$\overset{+}{K}(\text{or Na}), \overset{+}{\text{Au}} \text{ and } \overset{-}{\text{Cy}} \overset{-}{\text{Cy}}.$$

The K, of course, goes to form KOH and liberates H at the cathode. The Cy attacks the iron anode, ultimately forming ferrocyanide and a certain amount of Prussian blue, ferric hydrate, and perhaps other by-products.

Afterward, the cathodes and anodes in use were both of lead, but the lead anodes were found to become brittle after a time and were coated superficially with a white deposit, probably oxycyanide of lead.

Tinned iron cathodes with lead anodes were also used at one time; the current density was so regulated that the gold was precipitated, not in an adherent form, but as a loose powder which continually fell off the cathodes and settled at the bottom of the box, whence it could be easily collected in cleaning up. In the latest form of the process, adopted at the San Sebastian Mine, Salvador, Central America,[1] the cathodes are rolled lead plates, and the anodes are similar lead plates which have been previously coated with a layer of peroxide of lead by immersing

[1] C. P. Richmond, "Eng. and Min. Journ.," LXXX, 512.

in a solution of permanganate. The difficulty with regard to copper is got over in the plant referred to by using electric precipitation for recovering part of the copper and gold and then passing the solution, comparatively free from copper, through ordinary zinc-boxes to recover the residual gold.

Other electrolytic methods, using amalgamated zinc or copper as electrodes, have also been tried, but with little practical success. Aluminium has been suggested as a material for the electrodes, but although effective as a precipitant, it causes much difficulty by forming infusible compounds in the subsequent smelting operations.

Other Precipitation Processes. — Aluminium in the form of shavings has also been suggested, and occasionally tried on a working scale, as a substitute for zinc, without the aid of electricity. It is even more electro-positive than zinc, but a deposit of alumina is formed on the shavings and causes much trouble in the treatment of the precipitate.

Sodium amalgam has been used, either directly or in conjunction with an electric current. Theoretically, the action is very simple:
$$KAuCy_2 + Na = Au + KCy + NaCy;$$

but in practice it is found that the large surface required precludes the economic application of the process.

In the Gilmour-Young process the ore is subjected to combined pan-amalgamation, and cyanide treatment. After grinding in pans with mercury and cyanide solution, copper and zinc amalgam are added, which precipitate the gold from the solution. The gold amalgamates and can be recovered by retorting, like ordinary battery gold.

Other methods involving the use of amalgams are the Pelatan-Clerici and Riecken processes.

Charcoal was at one time extensively used as a precipitant in Victoria, Australia. The chemical action by which it precipitates gold is somewhat obscure, but is supposed to depend on the occlusion of certain gases in the pores of the charcoal which act on and decompose the aurocyanide in the percolating solutions. An enormous volume is required for effective precipitation.

Other suggested methods of precipitation involve the use of chemicals which destroy, or at least enter into combination with, the whole of the cyanogen present in the solution as simple cyanides. These cannot be considered to have passed the experimental stage. S. B. Christy and P. de Wilde have observed independently that soluble cuprous salts, added to a cyanide solution, produce a precipitate of white cuprous cyanide, which carries down with it practically the whole of the gold and silver; but this reaction only takes place in acid solutions, and would involve the neutralization of enormous quantities of alkali.

Soluble silver salts give a precipitate of silver cyanide when added in sufficient excess, or to an acidulated solution; practically the whole of the gold is carried down; but this, of course, is only a laboratory method.

Final Operations. — In the clean-up, apart from mechanical operations, we have to consider three processes involving chemical reactions: (1) Acid treatment; (2) roasting; (3) smelting.

Acid Treatment. — The precipitate as obtained from the zinc-boxes invariably contains a considerable amount of metallic zinc in fine shreds or particles which pass through the sieves. This can be largely removed by treatment with sulphuric or hydrochloric acid, the zinc dissolving as sulphate or chloride. In some cases sodium bisulphate ($NaHSO_4$), which is a by-product in the manufacture of nitric acid and may be obtained at a low cost, has been used with advantage as a substitute for H_2SO_4, the reaction being:

$$2NaHSO_4 + Zn = Na_2SO_4 + ZnSO_4 + H_2.$$

In the case of precipitates containing As, Sb, Se, and Te, these elements are partially evolved as the hydrides AsH_3, SbH_3, SeH_2, and TeH_2, on treatment with sulphuric or hydrochloric acids; these gases are very offensive and poisonous, and in the case of AsH_3 have several times been the cause of fatal accidents. In all cases, considerable amounts of HCy are evolved from the decomposition of the cyanogen compounds contained in the precipitate. Preliminary treatment with nitric acid has therefore been recommended; this converts cyanides into cyanates, arsenic into arsenic acid, selenium and tellurium into selenious and tellurous acids. These compounds, being non-volatile, give rise to no noxious fumes, and are carried off in solution after the treatment. There may, however, be losses of silver and possibly also of gold.

When zinc dust is used as a precipitant, some difficulty has been experienced (as at Mercur, Utah), in treating the precipitate obtained, when sulphuric acid alone has been used. It packs very hard, and the particles of zinc appear in some cases to become coated with a layer of gold which protects them from further action. A mixture of sulphuric and nitric acids was found to be a more efficient solvent.

Objection has been raised to the use of hydrochloric acid on the ground that it may liberate chlorine under certain circumstances, thus involving loss of gold. It is, moreover, considerably more costly than sulphuric acid in most places.

Whatever treatment be adopted, the solution of the zinc is always incomplete. It is possible that among the various alloys which may be formed there are some that are not readily attacked by acids; also in some cases protective coatings of $PbSO_4$ or $CaSO_4$ may be formed.

Roasting. — The main object of this operation is to oxidize the metallic zinc to ZnO, which in the subsequent smelting passes into the slag as silicate of zinc, whereas metallic zinc would pass into the bullion. Any other oxidizable metals are also oxidized, and cyanogen compounds either completely destroyed or converted into cyanates. Arsenic, antimony, selenium and tellurium are partially volatilized as oxides, but it is not possible to completely remove them in this way, and the fumes may mechanically carry off considerable amounts of gold and silver.

A partial oxidation of the zinc is sometimes brought about in a preliminary operation by mixing the precipitate with niter and drying in shallow pans at a low heat.

The roasting may be done in pans over an open fire, in a muffle, or in a reverberatory furnace.

Smelting. — This operation aims at producing bullion as free as possible from base metals and other impurities. As the precipitate, after acid treatment and roasting, still contains silica, insoluble calcium salts, and considerable amounts of zinc, lead, and other base metals, the flux required must be varied according to its composition. The chief ingredients used as fluxes are:

1. *Borax,* to produce fusible borates of iron, aluminium, and other metals.

2. *Silica* (if not already present in sufficient quantity), to form silicate of zinc.

3. *Carbonate of soda*, to form a more fusible silicate, to flux any excess of silica, and in some cases to form fusible double sulphides.

4. *Fluorspar*, to flux infusible calcium sulphate, and generally to increase the fluidity of the slag.

5. *Oxidizing agents* (niter, MnO_2, etc.), to oxidize any base-metal sulphides, selenides, tellurides, carbonaceous matter, etc.

Metallic iron is also sometimes added, to reduce base-metal sulphides, etc. In some cases a matte is formed, which requires special treatment for reduction to bullion. The slags are sometimes remelted with addition of fresh flux, to which litharge is added, whereby a large part of the value is collected as lead bullion, and the resulting low-grade slag may safely be rejected.

Lead Smelting of Zinc Precipitate. — In the Tavener process the zinc precipitate is smelted direct with litharge, siliceous material, and a reducing agent, the object being to form a lead alloy of the gold and silver and to flux off the zinc, etc., as fusible silicates in one operation. The litharge is reduced as follows:

$$2PbO + C = CO_2 + 2Pb.$$

The lead bullion is then cupeled, and the litharge formed in this operation recovered for use in future smelting charges.

Bullion is often further refined by remelting with fresh flux, and in some cases by oxidizing by means of a blast of air directed on the surface of the molten metal. Attempts have also been made to refine by injecting air or oxygen into the crucible containing the molten bullion.

According to T. K. Rose,[1] the following is the order in which metals are oxidized under such conditions; the table given also

Metal	Oxide	Heat of Combination with O
Zinc	ZnO	827
Iron	Fe_2O_3	637
Lead	PbO	503
Nickel	Ni_2O_3	401
Copper	Cu_2O	372
Platinum	PtO	179
Silver	Ag_2O	59
Gold	AuO	— 44

[1] *Trans.* I. M. M., XIV, p. 384.

shows the heat of combination, the number given being the number of grams of water which would be raised from 0° to 100° C., by the combination of 16 grams of oxygen with the metal in question.

SECTION VI

MANUFACTURE OF CYANIDE

THE various processes for the manufacture of cyanide may be classified according to the source from which the nitrogen is derived. The principal methods in use are:

(a) Those in which refuse animal matter is used as the nitrogenous raw material, ferrocyanide being generally produced as an intermediate product.

(b) Those in which atmospheric nitrogen is employed to form cyanide compounds, directly or indirectly.

(c) Those in which ammonia or ammonium compounds form the nitrogenous raw material, including methods which utilize residues from gas-works.

(a) PRODUCTION OF CYANIDES FROM REFUSE ANIMAL MATTER

Until about the year 1890, this was the method almost universally used. The raw materials required are: (1) Nitrogenous animal matter, such as horns, hoofs, dried blood, wool, woollen rags, hair, feathers, leather-clippings, etc. (2) An alkaline carbonate, such as pearl-ash, soda-ash, etc. (3) Iron filings or borings.

The alkali and the iron are first fused together at a moderate heat in an iron pan, or other suitable vessel, contained in a reverberatory furnace. The well-dried animal matter is then introduced in small quantities at a time and stirred in. The heat is then raised and the furnace closed so as to maintain a reducing atmosphere. The hard black mass which forms is then taken out and lixiviated with nearly boiling water. The crude ferrocyanide containing sulphides, sulphates, carbonates and thiocyanates, is crystallized out and purified by recrystallization. The ferrocyanide was formerly converted into cyanide by first dehydrating and then fusing, either alone or with alkaline carbonate:

$$K_4Fe(CN)_6 + K_2CO_3 = 5KCN + KCNO + Fe + CO_2$$

The cyanide so formed is always contaminated with cyanates and carbonates, and generally with small amounts of other salts (sulphides, chlorides, thiocyanates, etc.).

The procedure frequently adopted at present is to fuse with metallic sodium:

$$K_4FeCy_6 + 2Na = 4KCy + 2NaCy + Fe;$$

thus yielding a mixture of potassium and sodium cyanides free from cyanates, etc. The presence of sulphides and thiocyanates in the product is due chiefly to the sulphur contained in the organic matter. These compounds are partially decomposed and removed by metallic iron during the fusion. When the cyanide is made by direct fusion of ferrocyanide, the product contains carbide of iron, some nitrogen being given off in the process. Most of the volatile organic nitrogen is lost in the form of ammonia or nitrogen gas during the fusion for ferrocyanides, in the first stage of the process.

(b) Production of Cyanides from Atmospheric Nitrogen

It was observed by Scheele that when nitrogen is passed over a mixture of K_2CO_3 and charcoal heated to redness, a cyanide of potassium is formed. Many attempts were made throughout the nineteenth century to utilize this reaction for industrial purposes. One of the earliest was that of Possoz and Boissière, who used a mixture of charcoal with 30 per cent. of potassium carbonate. This was kept at a red heat in fire-clay cylinders, through which a mixture of N and CO, produced by passing air over red-hot alkalized carbon, was allowed to pass for about 10 hours. The product was then lixiviated with water in presence of ferrous carbonate (spathic iron ore) to give a ferrocyanide.

It was also observed at an early date (by Clark, in 1837), that cyanides are formed as an efflorescence in blast-furnaces, and that the gases of these furnaces contain cyanogen. Bunsen proposed a special blast-furnace for the production of cyanide, in which coke and potash in alternate layers were to be heated by a strong blast, the fused cyanide running off at the bottom of the furnace. It was found, however, that a very high temperature was necessary, as the potassium compound must be reduced to metallic potassium before combination with atmospheric nitrogen takes place.

Better results were obtained by substituting barium carbonate for K_2CO_3, as in the process of Margueritte and DeSourdeval (1861). Air (deoxygenated by hot carbon) was passed over a previously ignited mixture of $BaCO_3$, iron-filings, coal tar, and sawdust, whereby barium cyanide is produced. This is converted into sodium cyanide by fusion with sodium carbonate. The $BaCO_3$ is first reduced to barium carbide (BaC_2):

$$BaCO_3 + 4C = BaC_2 + 3CO.$$

This then combines with nitrogen to form $Ba(CN)_2$. It has been found, however, that only about 30 per cent. of the barium is converted to cyanide, the remainder forming barium cyanamide by a secondary reaction:

$$Ba(CN)_2 = C + BaCN_2.$$

When calcium is substituted for barium in this process, practically the whole is converted into calcium cyanamide:

$$CaC_2 + N_2 = C + CaCN_2.$$

Calcium cyanamide is also formed in the electric resistance furnace by passing nitrogen over a mixture of lime and charcoal:

$$CaO + 2C + N_2 = CaCN_2 + CO.$$

By heating the product at a high temperature with a further quantity of carbon, with the addition of salt to prevent frothing and facilitate the reaction, the cyanamide is converted into cyanide as follows: $CaCN_2 + C = Ca(CN)_2.$

The crude mixture so formed has been used as a substitute for potassium cyanide under the name of "Cyankalium surrogat," and is equivalent in cyanogen contents to about 30 per cent. KCN. [See Erlwein and Frank, U. S. patent, 708,333.]

An improved method more recently introduced is to convert the calcium cyanamide into sodium cyanide by the following series of reactions:

(1) By leaching with water, a crystallizable, easily purified salt is obtained, known as dicyan-diamide:

$$2CaCN_2 + 4H_2O = (CN \cdot NH_2)_2 + 2Ca(OH)_2.$$

(2) This, when fused with sodium carbonate and carbon, is largely converted into sodium cyanide:

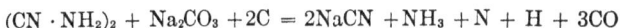

$$(CN \cdot NH_2)_2 + Na_2CO_3 + 2C = 2NaCN + NH_3 + N + H + 3CO$$

A portion of the dicyan-diamide sublimes and polymerizes; this is recovered and re-treated with Na_2CO_3 in a subsequent operation. The cyanamide and di-cyan-diamide are also utilized as sources of products valuable as manures, as they can be readily converted into ammonia, ammonium carbonate, urea, etc.

Cyanides may also be formed by the action of metallic sodium and carbon on atmospheric nitrogen (Castner); but it is preferable to use ammonia as the source of nitrogen in this reaction (see below).

(c) Production of Cyanides from Ammonia or Ammonium Compounds

When ammonia is passed over mixtures of heated alkali and carbon, only small quantities are converted into cyanide; better results are obtained by passing CO and NH_3 through a molten mixture of KOH and carbon, but even by this means much of the ammonia is dissociated into N and H, owing to the great heat which is necessary. It is supposed that potassamide is an intermediate product:

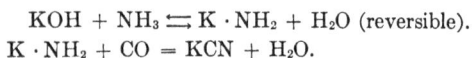

$$KOH + NH_3 \leftrightarrows K \cdot NH_2 + H_2O \text{ (reversible)}.$$
$$K \cdot NH_2 + CO = KCN + H_2O.$$

In Castner's process (Brit. patents, 12,218, 12,219, of 1894), molten sodium is allowed to flow through a mass of heated coke while ammonia gas is passed upward, the reaction being

$$2NH_3 + 2C + 2Na = 2NaCN + 3H_2.$$

The reaction takes place at a much lower temperature than in the previous process with KOH, and the losses by dissociation of NH_3 and volatilization of the cyanide are consequently smaller. It probably takes place in two stages, forming sodamide as an intermediate product:

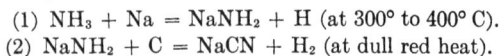

(1) $NH_3 + Na = NaNH_2 + H$ (at 300° to 400° C).
(2) $NaNH_2 + C = NaCN + H_2$ (at dull red heat).

In a modification of this method used by the Deutsche Gold Silber Scheide-Anstalt, metallic sodium is melted with carbonaceous matter in a crucible, and then ammonia is led in at 400° C. — 600° C.; this forms disodium cyanamide:

$$Na_2 + C + 2NH_3 = Na_2CN_2 + 3H_2.$$

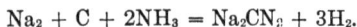

By then raising the temperature to 700–800° C., the excess of C. interacts, forming sodium cyanide:

$$Na_2CN_2 + C = 2NaCN;$$

the whole operation being conducted in the same crucible

Methods have also been proposed by J. Mactear, H. C. Woltereck, and others, in which cyanogen compounds are produced by the action of ammonia gas on gaseous carbon compounds at a high temperature.

In Mactear's method (Brit. patent, No. 5037, of 1899) the reaction

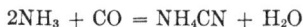

$$2NH_3 + CO = NH_4CN + H_2O$$

is supposed to take place in a closed chamber at 1800° to 2000° F., the products being condensed and absorbed in alkali hydrate and the ammonia liberated for reuse. Instead of CO, a mixture of CO with N and H (producer gas) may be used.

In Woltereck's method (Brit. patent, No. 19,804, of 1902) "perfectly dry ammonia and a volatilized or gaseous carbon compound, also perfectly dry, are passed together with hydrogen, in equal volumes, over a strongly heated catalytic agent, such as platinized pumice. One volume of NH_3 and two volumes of 'water-gas' ($CO + H_2$) make a convenient mixture. The HCN produced is absorbed in an alkaline solution."

Cyanide has also been made from the trimethylamine $(CH_3)_3N$ obtained by the distillation of beet-root molasses at a high temperature. This, at a red heat, decomposes, giving NH_3, HCN, and H.

Another source of cyanogen compounds is the crude illuminating gas from the distillation of coal. In Knublauch's method (Brit. patent, No. 15,164, of 1887) the gas is passed through a solution of an alkali or alkaline earth containing ferrous hydrate in suspension. The gas carries with it ammonium cyanide and thiocyanate, which are absorbed by the mixture and converted into ferrocyanide.

In Rowland's process (U. S. patent, No. 465,600, of 1891) the gas is passed through a solution of an iron salt, thus forming ammonium ferrocyanide. This is converted into the calcium salt by boiling with lime. The calcium ferrocyanide may then be converted into the required alkali ferrocyanide by decomposing with an alkaline carbonate.

Bueb's process (Brit. patent, No. 9075, of 1898) is a modification of the above, in which the cyanogen is separated in the form of an insoluble double compound by using a concentrated iron solution ($FeSO_4$); the reactions said to take place are:

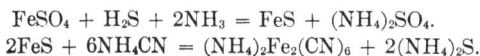

$$FeSO_4 + H_2S + 2NH_3 = FeS + (NH_4)_2SO_4.$$
$$2FeS + 6NH_4CN = (NH_4)_2Fe_2(CN)_6 + 2(NH_4)_2S.$$

The insoluble product, known as "cyanide mud," is then treated to obtain marketable cyanogen compounds.

Many other modifications have been suggested.

Cyanides may also be obtained by desulphurizing thiocyanates by means of iron, or by zinc and carbon.

Manufacture of Cyanide from Beet-sugar Residues. — C. A. Browne describes this process in Columbia School of Mines Quarterly (Abs. Mining Magazine, Sept. 1913, p. 226).

The residue, containing 12 to 15 per cent. of potassium and 4 per cent. of nitrogen, is heated in retorts and yields a number of volatile products, including ammonia and methylamine. These gases are further heated in tubes to a temperature of 1000° C. whereby the nitrogenous compounds are converted into ammonium cyanide. After cooling and purifying, the gases are passed through sulphuric acid, thus yielding ammonium sulphate and hydrocyanic acid. The latter is absorbed in water, redistilled and collected in sodium hydroxide. This solution is evaporated and crystallized to obtain sodium cyanide.

The remaining combustible gases are led back to the furnace for heating the retorts. About three fourths of the nitrogen in the residues is recovered as ammonium sulphate and sodium cyanide, the remainder escaping as nitrogen gas.

About 5000 tons of sodium cyanide are produced annually by this process by two factories in Germany, the product being exported to the Transvaal.

PART III

PREPARATORY TREATMENT OF ORE FOR CYANIDING

ONLY a brief outline can be given here of the mechanical principles and processes which form a necessary preliminary to cyanide treatment or are involved in the carrying out of the process itself. Details of the construction and mode of action of the machines used will be found in any text-book of metallurgy, and much information on these points is usually given in the catalogues supplied by the makers.

Whatever metallurgical process is to be adopted, it is evident that the ore as taken from the mine will require to undergo a certain amount of crushing before its valuable content can be extracted, so that much of what follows does not specifically concern the cyanide process. We shall endeavor, however, to point out more particularly those considerations in the preparatory treatment which have a direct bearing on subsequent cyanide work.

SECTION I

CRUSHING MACHINERY

(A) Rock-Breakers

THE reduction of an ore to particles of a size suitable for any metallurgical process is almost always carried out in at least two stages, for which different types of machines are required. These may be distinguished, somewhat arbitrarily, as *Crushing* and *Grinding*. In the first, or crushing, stage, the ore is broken by a succession of blows delivered at intervals by a hard moving body, such as the jaw of a rock-breaker or the shoe of a stamp; the rock being crushed against another (generally stationary) hard body, as the die of the stamp battery. In the grinding process the comparatively coarse lumps produced by crushing are pressed more or less continuously between two hard surfaces to reduce them to a smaller size, although in many machines (*e.g.*, Gates crusher, Ball mill, Tube mill, etc.), both actions go on to some extent simultaneously. For economical work, it may be stated as a nearly universal rule that crushing must precede grinding, and further, that separate machines should be used for different stages of crushing.

The first, or coarse, crushing is carried out in a machine known as a rock-breaker, the object of which is to break the ore as it comes from the mine, which may be in pieces the size of a man's head or larger, to lumps of $\frac{1}{2}$ in. to $\frac{1}{4}$ in. diameter, suitable for further crushing in stamps or other fine crushers.

The rock-breakers in general use are of two types: (1) those with a fixed and movable jaw, the opening between which is regulated as required, such as the Blake, Blake-Marsden, and Dodge crushers; (2) those with a gyrating vertical shaft, working against plates on a stationary circular shell surrounding the shaft, as in the Gates and Comet crushers.

In the first type (jaw crushers) the fixed and movable jaws are inclined at an angle to one another, the smaller opening being

141

at the bottom. The ore is fed in at the top, between the jaws.
When the machine is at work, the movable jaw rapidly advances
and recedes about half an inch, so that the rock gradually slips
down and is crushed between the jaws, which are faced with
removable plates of hard steel. The Dodge crusher is said to
be suitable for somewhat finer crushing than the Blake. (Fig. 1.)

In the second type (gyratory crushers), the top end of the
shaft is usually pivoted, while the bottom describes a circle. A
conical piece of chilled iron or steel (breaking head) is fitted to
the shaft, the surface of which is inclined at an angle to the crushing faces fixed to the shell or hopper surrounding the shaft. As
the shaft gyrates, the surface of the breaking head alternately
advances toward and recedes from the fixed crushing faces, thus
performing the same functions as the movable jaw in the jaw
crushers. The crushing surfaces are generally corrugated, and
are composed of chilled iron, chrome steel, manganese steel, or
some similar material combining hardness and toughness. (Fig. 2.)
In a typical case, a Blake rock-breaker will crush 300 tons of
rock in 24 hours to $1\frac{1}{2}$-in. size, running at 250 r.p.m. and requiring
14 h.p.[1]

The   accompanying   illustration,   Fig. 1,   shows   a   slightly
modified form of Blake crusher manufactured by Hadfield's
Steel Foundry Co., of Sheffield, England. The rock is broken
between a fixed jaw plate and a movable jaw plate actuated by a
powerful toggle movement communicated to it from the driving
pulley through an eccentric shaft and pitman.

Fig. 2 illustrates the Gates rock and ore breaker (Style K),
one of the gyratory type manufactured by the Allis Chalmers
Company, of Milwaukee. The main shaft (25) is suspended in
the upper part from spider arms (44) that span the top of the
opening, and gyrates without revolving, in such a way that the
ore is crushed between the concave shell (19) and a removable
block or "head" (18). This head has a circular and rolling
movement, and approaches successively every point in the
interior of the throat, owing to the movement of the lower end
of the axis of the shaft around a small circle. The rock is
broken rather than crushed. It falls on an inclined diaphragm
which protects the eccentric and gear, and thence passes to
the discharge spout (32). The main shaft is adjustable,

[1] Rose, "Metallurgy of Gold," 4th edition, p. 103.

and can be raised or lowered to regulate the size of product required.

Fig. 1. — Modified Blake Crusher.

(B) Stamps

The finer crushing of gold ores is almost always performed by means of stamps, except in cases (1) where dry crushing is desirable, or (2) where the ore is unusually soft. In modern practice, crushing with stamps is frequently followed by grinding in pans or tube mills, as described below. The stamps consist of sets (usually five in each set) of heavy iron hammers, 700 to 1250 lb. in weight. Each stamp has a vertical iron rod (stem), provided at the lower end with a cylindrical block of cast iron (head), into which the stem fits. The lower end of the head has a socket to receive a replaceable block of hard iron (shoe) which forms the crushing surface. The stamps are alternately raised and dropped by means of a horizontal cam-shaft. As the shaft

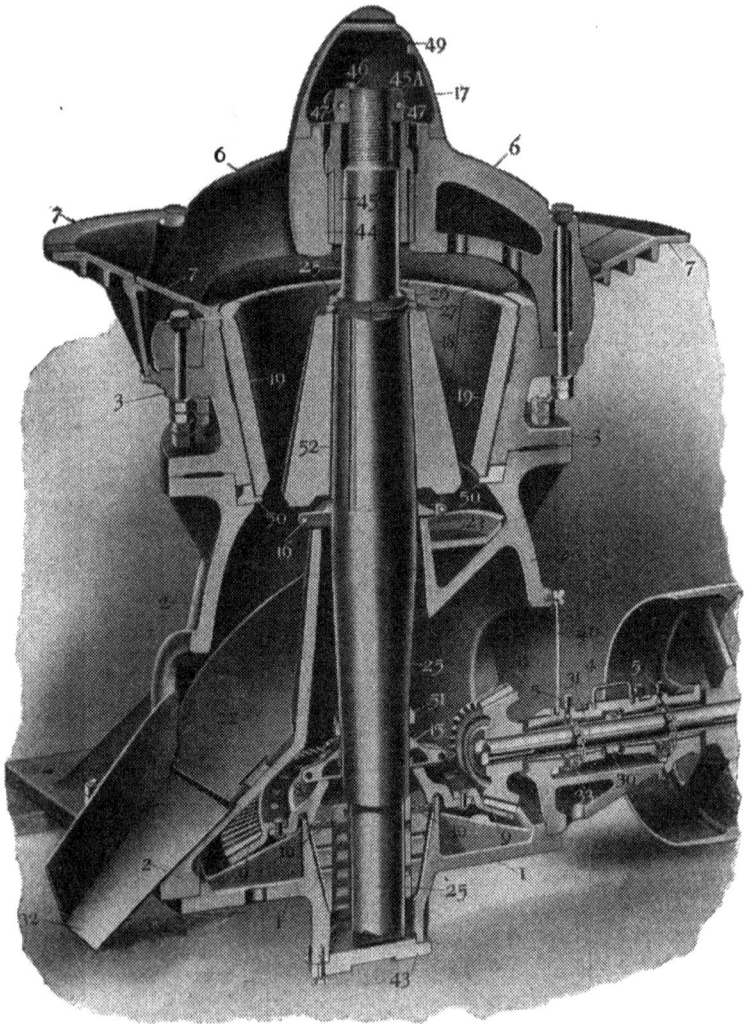

FIG. 2. — Gates Rock and Ore Breaker.

revolves, the cams (set at different angles on the shaft) strike against a block (tappet) fixed to the upper part of the stem, so that the stamp is raised a few inches as the cam revolves, and drops by its own weight when the point of the cam passes its highest position. The cams are set in such a way on the shaft that each stamp is lifted and dropped in succession, in a par-

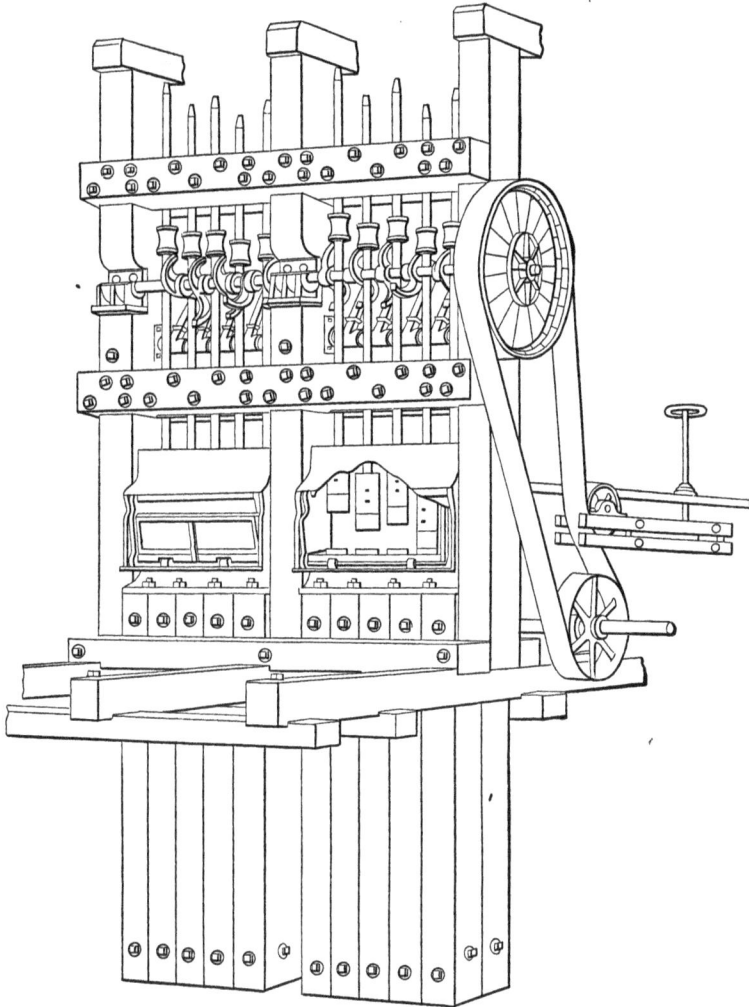

FIG. 3. — Elevation of 10-stamp Battery, showing Mortar Box, etc. [Fraser and Chalmers.]

ticular order. As the stamp falls, the lower surface of the shoe strikes the layer of ore resting on a hard iron or steel block (die), which thus acts as an anvil. The dies corresponding to each set of five stamps are set in an oblong cast-iron box (mortar-box). (See Fig. 7.) The ore is fed in on one side of the row of stamps,

FIG. 4. — Interior View of 200-stamp Mill erected by Fraser and Chalmers.

and discharged on the other side through a screen fitted to an opening in the mortar-box, the lower edge of which is a few inches above the top of the die. (Figs. 3, 4, 5, 6.)

After leaving the rock-breakers and before passing to the stamps or other mill, the ore is generally sifted, through a coarse screen consisting of parallel steel bars, known as a " grizzly " (see Fig. 16), the oversize returning to rock-breakers. When stamps are used, as is generally the case, for wet crushing, a stream of water is fed into the mortar-box along with the ore, and a further supply of water is usually added, outside the battery,

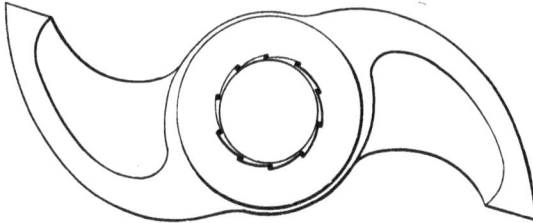

Fig. 5. — New Blanton Cam.   [Fraser and Chalmers.]

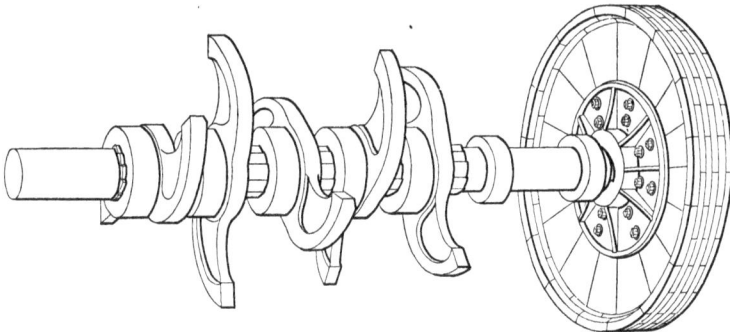

Fig. 6. — Cam Shaft, showing 5 Cams in Position. [Fraser and Chalmers.]

to wash away the ore as it passes through the screens. The fineness to which the ore is crushed, which is the principal point of interest with regard to subsequent cyanide treatment, depends chiefly on the (1) size of openings in screen used in mortar-box; (2) relative amount of ore and water fed in; (3) hight of discharge above dies; (4) hight of drop of stamps; (5) rate at which ore is fed into mortar-box.

Where the object is to treat as much of the product as possible by percolation, it is of course desirable to crush the ore only so fine as may be necessary to render a sufficient percentage of the values accessible to the solvent; this will be attained by (1) using a somewhat coarse screen; (2) using much water in proportion to ore, so as to wash out the crushed ore as soon as it is reduced to the required size; (3) having the discharge as close as possible to the top of the dies, to avoid crushed ore being repeatedly

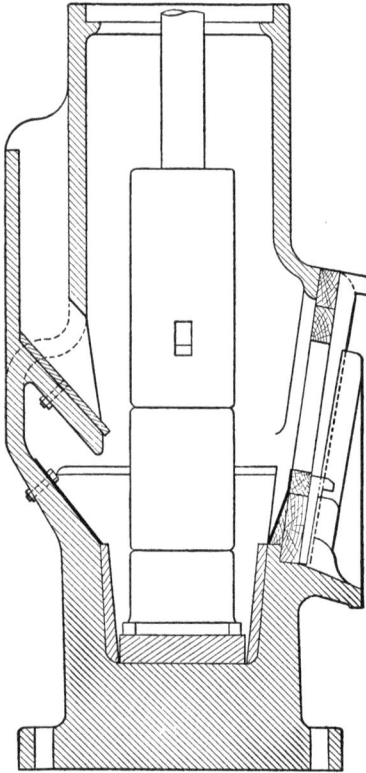

FIG. 7. — Section of Mortar-box.
[Fraser and Chalmers.]

thrown back under the stamps; (4) using heavy stamps running at a high speed, with the least hight of drop possible in practice; this results in a maximum quantity of material being crushed for a given expenditure of power; (5) keeping the rate of feed low, so as to avoid banking of the ore against the screen, which would lead to unnecessary recrushing of the same material.

In many cases, however, it is more advantageous to crush finer, even although the output of ore per stamp may be reduced, as the extraction, both by cyanide and by amalgamation, is usually greater the finer the state of division to which the ore particles are reduced. Frequently the ore contains two constituents of different degrees of hardness, quartz and pyrites, one of which carries practically the whole of the value. If, as is the case on the Rand, the quartz is barren, while the whole of the gold is associated with or imbedded in the pyrites, it would seem that the best method would be to crush coarsely at first, then separate the pyrites by some form of concentration or other mechanical means, and regrind the pyritic portion alone as finely as possible.

In general, it may be said that when the ore is to be subsequently treated by cyanide, less trouble and loss will be caused by crushing too fine than by crushing too coarse, since in the coarse particles the gold or silver is absolutely inaccessible to the solvent; whereas the difficulties in the way of the mechanical handling of slimes in order to extract their dissolved values have been greatly reduced if not entirely overcome by modern methods. Another point of great importance for cyanide work is uniformity of the mill product, both as regards fineness of the ore particles, and as to proportion of water. The whole system of treatment must be regulated in accordance with these factors, and any unforeseen variation may give rise to bad extractions and heavy losses. Where "spitzlutten" (see below) are used for separation of sands and slimes, they are arranged to give the required product by regulating the size of outlet, etc., and any variation in the ratio of ore to water will result either in slime passing into the sand tanks or in fine sand being carried away with the slime. It is obvious that the whole metallurgical treatment should be regarded as one problem, and the preliminary operations of crushing and grinding should be made subservient to the extraction processes which are to follow, and not carried out merely with a view to passing through the maximum quantity of ore in a given time.

As an example of typical stamp battery practice under modern conditions, the following is given, on the authority of T. Roskelley,[1] as having been used in the crushing of banket ore at the Robinson Deep G. M. Co.'s battery:

[1] " Proc. Chem., Met. and Min. Soc. of South Africa," III, 338.

Standard Screen Used	Hight of Drop: Inches	Speed: Drops per Minute	Water Used per Ton Crushed	Tons Crushed per Stamp per 24 Hours	Slimes Produced Per Cent.
600	8½	96	6	5.45	25.4
700	8½	96	6	5.28	29.0
900	8½	96	6½	5.07	31.8 ·
1000	7½	100	7	5.06	35.3
1200	7½	100	7½	4.95	35.0

Fall of Plates = 1 $\frac{1}{16}$ in. per foot.
Hight of discharge = 2½ in.

Other authorities (e.g., F. Alexander) recommend a greater hight of discharge, such as 6 or 8 in., which, with other conditions remaining the same, would result in a finer product and diminished output.

A full account of this subject will be found in the following papers, to which reference should be made for further details as to mill practice and its effect on cyanide treatment:

"The Stamp Milling of Gold Ores in its relation to Cyaniding," by E. H. Johnson, in "Proc. Chem., Met. and Min. Soc. of South Africa," II, 176 (1897).

"Notes on the Common Practice of Quartz Milling on the Rand," by F. Alexander. Ibid., III, 298 (1903).

Much useful information as to the effectiveness of the crushing with a view to cyanide treatment may be obtained by sifting a sample of the product through a number of screens of varying mesh, and assaying the portions passing through successive screens, similar tests being also made on the residues after cyanide treatment. When the battery pulp, before cyanide treatment, is examined in this way, it is generally found that one portion of the crushed ore is richer than another, and frequently that the finer products are the richer. With the residues after cyanide, however, it is almost invariably the case that the coarser portions are the richer, showing that finer crushing of this part of the ore would result in higher extraction.[1] Finer crushing usually results also in improved amalgamation, not only because more gold is brought into actual contact with the mercury, but because me-

[1] Tests on these lines made weekly for a number of years at the Redjang Lebong mine, Sumatra, showed scarcely an exception to this rule, both as regards gold and silver values.

chanical losses of amalgam, due to the passage of coarse material over the surface of the plates, are much lessened. Any amalgam carried off in this way is practically lost, as but little dissolves in cyanide during the time of treatment usually given. This matter will be further discussed under amalgamation. In this connection the following remarks by P. S. Tavener[1] are very much to the point:

" Would it not be better to take the pulp leaving the mortar-boxes and pass it direct to a grinding machine, or to separate, by means of classifiers, the particles requiring regrinding before amalgamation or afterwards? It would appear to be better to do so before, for if done afterward the reground portion would have to pass over the plates for a second time. Is it not likely that it is the coarser particles that help to scour the plates?"

As the stamp mill is not adapted for crushing effectively and economically to a very high degree of fineness, it is desirable to pass a part at least of the product from·the battery through some grinding machine; hence the modern practice is to use the stamp battery for comparatively coarse crushing, supplementing it by one or other of the grinding machines described in the following section.

[1] " Proc. Chem., Met. and Min. Soc. of South Africa," III, 329.

SECTION II

GRINDING MACHINERY

THE type of machine to be used for grinding will depend largely on the hardness and general character of the ore and on the treatment which it has previously undergone. In some cases the ore may pass direct from rock-breakers to the grinding machines, but this can in general only be done with comparatively soft material, or in cases where the ore is to be crushed dry. The grinding machines principally used in connection with cyaniding are rolls, grinding pans, ball mills, and tube mills.

(A) ROLLS

These consist of two parallel cylinders with their axes placed horizontally. (See Fig. 8.) The tires, or crushing faces, are of forged steel or chilled iron, usually smooth, and are replaceable when worn. The rolls are mounted so that they revolve at a distance of $\frac{3}{8}$ to $\frac{1}{30}$ in. apart, according to the fineness of the product required. The usual sizes are 12 to 16 in. across the face, and 22 to 40 in. diameter. The product obtained varies from $\frac{1}{2}$ in. size to about 70-mesh, depending on the distance between the crushing faces of the two rolls. They are used almost exclusively for dry crushing, the ore being fed between the revolving rolls by means of a hopper. After leaving the rock-breakers and before reaching the rolls, the ore is passed over screens, and the oversize, too large to be fed direct to the rolls, is returned to the rockbreakers. In some cases a succession of rolls is used to reduce the ore to varying degrees of fineness, with suitable screens between each pair. The speed may be from 60 to 150 r.p.m., the velocity at the circumference being (say) 750 to 1000 feet per minute. In some cases the two rolls are made to revolve at different velocities. A detailed account of modern practice with rolls will be found in a paper by P. Argall, "Sampling and Dry Crushing in Colorado." [1] (See Fig. 9.)

[1] Trans. I. M. M. X, 234-273.

In general, it may be stated that rolls are the most suitable machines for dry crushing where the object is to produce as little fine dust as possible; when set to a particular opening, they pro-

Fig. 8. — Smooth Crushing Rolls, manufactured by Hadfield's Steel Foundry Co., Sheffield. The frames are of toughened cast steel and the shells (crushing surfaces of the rolls) are of "Era" manganese steel or, when used for fine crushing, of best quality toughened cast steel. The rolls are driven by means of belts running on two heavy fly-wheel pulleys.

duce a minimum of material crushed to a greater degree of fineness. They are not usually suitable for very coarse or very fine crushing; the material fed to the rolls should have been previously reduced at least to 2-in. size by rock-breakers, and if very fine grinding is necessary, the product from the last set of rolls should be passed to grinding pans or tube mills. Screen analyses of the product obtained by rolls on Cripple Creek ore are given by Argall (loc. cit., pp. 269, 270), as follows:

I. Size of opening = .0212 in. (andesitic and granite ores).

Ore	(a) Per cent.	(b) Per cent.	(c) Per cent.
Remaining on 40-mesh	21	13	31
" " 40 to 60-mesh	17	20	16
" " 60 to 120-mesh	41	54	44
" " 120 to 200-mesh	15	12	6
Passing 200-mesh	6	1	3

II. Size of opening = .02 in. (base ore).

		(d) Per cent.	(e) Per cent.
Remaining on	50-mesh	31.5	32.5
" " 50 to 100-mesh		26	26
" " 100 to 200-mesh		16.5	16.5
Passing 200-mesh		26	25

It is seldom advantageous to use rolls for crushing finer than about 30-mesh. When ore requires to be roasted previous to

FIG. 9. — Argall Rolls, furnished by Fraser and Chalmers. These are belt-driven, and one roll is fixed while the other is movable. The bearings of the movable roll are connected by a heavy cast-iron yoke, insuring that the face of this roll will always remain parallel to the face of the fixed roll.

cyanide treatment, it is very generally crushed by means of rolls, as it is desirable to avoid producing an excessive quantity of fine material. In some cases the whole or part of the roasted product is ground finer before undergoing treatment.

(B) GRINDING MILLS

These are of many types. Those used in the process of pan-amalgamation consist essentially of a heavy shoe or muller, revolving about a vertical axis, the ore being ground between the under surface of this muller and the bottom of the pan.

Descriptions of these machines will be found in any work on the metallurgy of silver.[1] The bottom of the pan is lined with replaceable cast-iron " dies." The muller revolves at about 70 r.p.m., and the ore, previously crushed to a sufficient degree of fineness,

FIG. 10. — Evans-Waddell Chilian Mill, manufactured by Fraser and Chalmers. This machine consists of three crushing rollers mounted on horizontal arms, connected with a central vertical driving-shaft, the latter being driven by means of a horizontal pinion-shaft and bevel-gear attached to the lower part of the vertical shaft. The mortar or pan is lined inside with hard iron wearing-plates, and the die-ring and roller shells are of the best quality forged steel. There are six screen openings for discharging the crushed material.

is fed in with sufficient water to make a thick pulp; after the ore has been ground for some hours, mercury is added and the grinding is continued, sometimes with the addition of other chemicals.

Other forms of grinding machines which are in use, either with or without the addition of mercury, may here be mentioned.

[1] See also Rose, *loc. cit.*, pp. 169–173.

Among the best known are the Arrastra, Chilian mill, Griffin mill, Bryan mill, and Huntington mill. These machines, although of widely different construction in detail, have certain features in common. In all of them the grinding part is made to revolve about a central vertical shaft, and acts against a circular or annular surface forming the bottom or side of the machine.

The *Arrastra*, probably the oldest and most primitive machine used in ore treatment, is still employed to a large extent in some parts of America. It consists of a bed or pavement of hard stone, with a wooden shaft set in the center, bearing a horizontal revolving arm or arms, to which heavy stones are attached; the ore, previously broken in small pieces, is ground between the revolving stone and the pavement. It is capable of very fine grinding, but the output is of course very small.

The *Chilian* mill consists of large wheels or rollers, which may be as much as 8 ft. in diameter and 6 in. wide, revolving on horizontal axes in an annular trough or mortar. (Fig. 10.) The *Bryan* mill is of similar construction, consisting of three rollers mounted on horizontal axles, which revolve about the central axis of the machine. The rollers run in an annular mortar lined with segmental steel dies; the central shaft makes about 30 r.p.m., the pulp at the periphery running round at over 300 feet per minute. The capacity of a 4-foot mill is about 15 to 20 tons of quartz ore per day through a 40-mesh screen.[1] These mills are suitable for both hard and soft ores.

The *Huntington* mill is somewhat similar, but the rollers are mounted on vertical axles suspended from a circular horizontal carrier revolving about a central vertical axis. The grinding is due to the pressure of these rollers against the sides of the mill, which are lined with an annular replaceable die of steel or cast iron. Above this is an annular screen for the discharge of the crushed ore. A hopper is also placed at the side of the mill for feeding in the ore, which should be previously crushed to nut size in a rock-breaker. The rollers are suspended about one inch from the bottom of the pan, and being free to swing radially as well as to revolve on their own axes, they are driven by centrifugal force against the ring-die when the carrier revolves. The speed is 45 to 90 r.p.m., usual diameter 5 ft. This machine is particularly suitable for soft ores, which it crushes much more

[1] Rose, *loc. cit.*, p. 165.

effectively than stamps, and owing to the freer discharge it produces less slime than the latter. Enough water is added in the
feed hopper to make a thick pulp, and in cases where the mill is

FIG. 11. — Improved Huntington Mill manufactured by Fraser and Chalmers.
It is claimed that this mill can be run with less expenditure of power per
ton of ore crushed than any other similar mill. The ore is discharged
through the screen as soon as it is crushed fine enough, thus avoiding
the production of slime and giving a product very suitable for concentration.

to be used as an amalgamating machine, from 17 to 25 lbs. of
mercury are added, forming a layer at the bottom just beneath
the rollers. The cost is about two-thirds that of stamps for the
same crushing capacity; less power is required per ton crushed,[1]
and the cost of repairs and renewals is less. There is also less
floured mercury than is given by stamps with inside amalgamation.[2] (Figs. 11 and 12.)

[1] The power required is :

For mill 3½ ft. diam., speed 90 r.p.m., 4 h.p.
       5 "   "   "   70   "   6 "
       6 "   "   "   55   "   8 "

[2] *Ibid.*, p. 161.

(C) Ball Mills

These consist essentially of cylindrical revolving drums containing a number of steel balls of various sizes. The interior of the drum is lined with steel plates, arranged so that a portion of each projects inward, forming a shelf which has the effect of

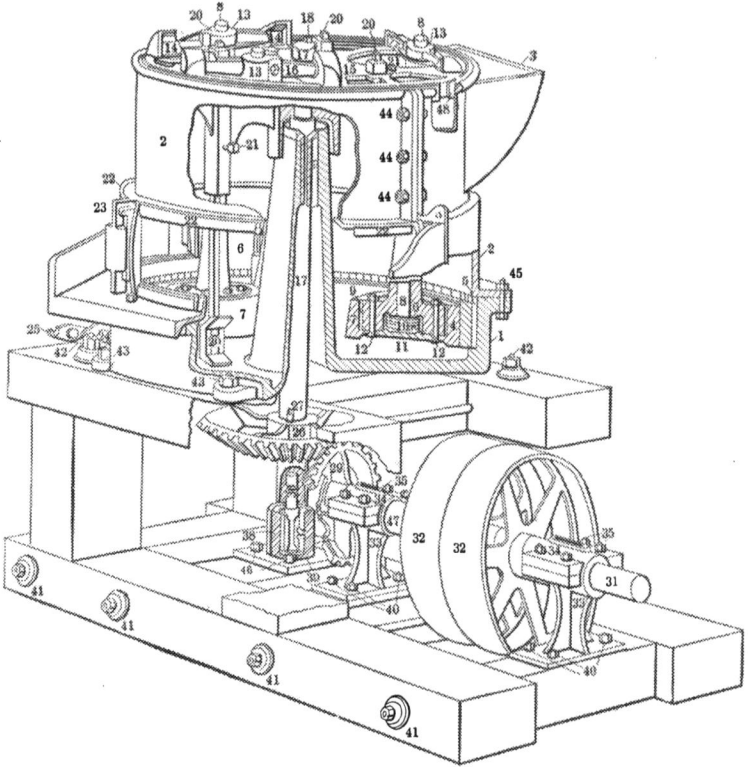

Fig. 12. — Sectional view of a Huntington Mill as made by Fraser and Chalmers.

raising the balls and dropping them during the rotation of the drum. The other portion of the lining plates is perforated. The ore is fed in through a hopper at one end of the drum, and is crushed by the impact of the balls against the grinding surfaces of the lining and against each other. The action is in fact a combination of crushing and grinding. The crushed ore passes through the perforations to a set of inner and outer sieves concentric with the circumference of the drum. The first or inner screens consist of perforated steel plates which serve to protect the outer wire screens. The oversize returns automatically to the interior of the drum through openings between the lining plates. The material passing the sieves goes through a hopper forming part of an outer casing of sheet iron, surrounding the revolving drum, which discharges the crushed ore at the bottom. (Fig. 13.)

FIG. 13. — Ball Mill, No. 8 size. [From photograph furnished by Cyanide Plant Supply Co.]

These machines may be used either for wet or dry crushing, but are more commonly used for the latter purpose. Where the object is to crush a large percentage of the product to a tolerably high degree of fineness, ball mills are more effective than stamps or rolls. Julian and Smart [1] state that ball mills of the Krupp-Grusonwerk type are generally preferable to rolls or stamps for dry crushing, and recommend crushing coarsely in the first instance, say 12 to 20 mesh, and passing the product through a second machine for the finer grinding. The following screen-

[1] " Cyaniding Gold and Silver Ores," 2d edition., p 47.

analysis given by the same authorities (*loc. cit.*, p. 199) will serve to illustrate the nature of the product obtained by crushing in ball mills. An ore from the Kalgurli mines, West Australia, fed in 2½ in. size to a No. 5 Krupp ball mill with 35-mesh screens, gave a product:

35 to 80-mesh............	30 per cent.
80 to 120-mesh............	30 "
Passing 120-mesh	40 "

Another example is given by Dr. Simon ("Trans. I. M. M." X, 290), from No. 4 Krupp ball mills, crushing dry with 30-mesh screens, on ore from Norseman Mine, West Australia:

Remaining on 30-mesh	0.6 per cent
30 to 40-mesh	16.1	"
40 to 60-mesh	17.1	"
60 to 80-mesh	25.2	"
Passing 80-mesh	41.0	"

These may be compared with the figures given for the product obtained by crushing with rolls, given by P. Argall for Cripple Creek ore (see above), where the percentage passing 120-mesh was from 9 to 21 per cent. Where ball mills are used for wet crushing, the water is led into the interior of the drum and directed in a spray against the sieves, the discharge hopper being arranged as a spitzkasten, for separating sands and slimes. The power required is 8 or 9 h.p. for an output of 200–300 lb. per h.p.-hour, crushing to 40-mesh (Argall, *loc. cit.*, p. 296).

Ball mills are also extensively used for crushing lime, cement, slag, and other materials, and sometimes for mixing the fluxes and precipitate from the zinc-boxes previous to smelting.

(D) Tube Mills

These machines, which have been very extensively introduced of late years for the fine grinding of ore before cyanide treatment, consist of sheet steel drums 13 to 22 ft. in length, having an internal diameter of 3 to 5 ft. They are lined internally with replaceable blocks of some hard material, such as hardened steel or "silex," and the remaining space is generally rather more than half filled with rounded flint pebbles. The cylinder is mounted on trunnions, with its axis slightly inclined downwards from the feed to the discharge end. The ore or battery pulp, reduced to a suitable consistency by removal of superfluous water by means of spitzkasten or otherwise, is fed in by means of a hopper through an opening in one of the ends of the drum. During the rotation of the machine the pebbles cling to the lining

FIG. 14. — Davidsen Wet Crushing Tube Mill. Feed end.

for some distance up the ascending side, and then drop to the bottom of the drum, crushing the ore by their impact. The crushed ore is discharged through an opening at the opposite end of the drum, and usually passes to spitzlutten, from which the coarser product is returned to the feed end of the mill and reground. (Figs. 14 and 15.)

To obtain the maximum efficiency from tube mills, great care is necessary in regulating the speed, the rate of feeding, consistency of pulp, and charge of pebbles. According to W. R. Dowling,[1] the tube mills of the Robinson Deep G. M. Co. are run at a speed of 29 r.p.m.; about 230 tons of sand are fed to each mill per 24 hours; the pulp should be as thick as possible (in the case cited it contains about 40 per cent. of water to 60 per cent. of solids); the load of pebbles is rather more than half the internal volume of the mill.

The nature of the liners is also a matter of considerable importance. It appears to have been generally found that " silex " linings give more satisfactory and economical results than manganese steel. The silex consists of hard flint cut in blocks about $2\frac{1}{2}$ in. thick, and secured to the inside of the shell by a setting of Portland cement. Blocks of 4 in. have been used and are considerably more economical, as they last longer. The pebbles are

[1] " Tube Mill Practice," in Journ. Chem., Met. and Min. Soc. of South Africa, VI, 308 (April, 1906).

2 to 3 in. in diameter and are mostly imported from the Danish coast, but in South Africa the practice has been introduced of using pieces of the banket ore itself as the crushing medium; lumps 3 to 6 in. in diameter are quickly reduced by the action of the mill to the form of a round pebble. The size of pebbles required seems to depend on the size of the product to be crushed; the finer the particles entering the mill the smaller the pebbles needed.

Fig. 15. — Davidsen Wet Crushing Tube Mill. Discharge end. [Furnished by the Cyanide Plant Supply Co.]

The feed is regulated by reducing or enlarging the discharge opening of the spitzkasten used for dewatering the pulp before it enters the tube mill. According to Dowling (*loc. cit.*, p. 310), "the best indication that a tube mill is being given a sufficient feed is the continual presence of a few inches of sand at the bottom of the dewatering spitzkasten." The action of the mill is thus described: "The thicker the pulp, the better the grinding and the less the wear on both liners and pebbles, since the sluggish pulp clings to both pebbles and liners and forms a layer to take the blow of the falling and the abrasion of the rolling pebbles. The clinging action of the thick pulp can be seen at the discharge end of the mill, and this is a ready means of telling whether the feed is of approximately the right consistency."

The crushing efficiency of a tube mill is generally measured by comparing the percentages of coarse and fine sand in the pulp fed to the mill and in the product delivered by it. A somewhat arbitrary rule is to take the sum of the decrease in material remaining on 60-mesh and the increase of material passing 90-mesh as representing the work done by the mill. The nature of the products entering and leaving the tube mill, in ordinary working conditions, is shown by the following table given by Dowling (*loc. cit.*, p. 313) at the Robinson Deep.

Grade	PULP LEAVING BATTERY (Entering Tube Mill)		PULP ENTERING CYANIDE WORKS (From Tube Mill)		WORK DONE	
	Per Cent. of Total Wt.	Tons	Per Cent. of Total Wt.	Tons	Per Cent.	Tons
Remaining on 60-mesh	28.66	8,599	5.28	1,584	23.38	7,015
60 to 90-mesh........	17.43	5,230	16.50	4,951	......	......
Passing 90-mesh	53.91	16,175	78.22	23,469	24.31	7,294
	100.00	30,004	100.40	30,004	47.69	14,309

A more detailed analysis of the gradings at the Waihi mine, New Zealand, is given by E. G. Banks:[1]

			Before Tube Milling Per cent.	After Tube Milling Per cent.
Remaining on	30-mesh	....	5.32	0.03
	30 to 40-mesh	....	9.77	0.12
	40 to 60-mesh	....	15.94	1.13
	60 to 100-mesh	....	13.96	7.43
	100 to 150-mesh	....	12.29	18.42
Passing	150-mesh	....	42.72	72.87

A comparative test between crushing in a tube mill with gold quartz (Rand banket ore) and Danish pebbles is given by K. L. Graham,[2] showing results slightly in favor of the quartz. In both cases the mills were lined with silex blocks, 6 × 6 × 4 in., laid on edge. The wear on the linings was certainly not greater when quartz was used. The results of a month's work (January, 1907) are given as follows:

[1] "Trans. A. I. M. E," XIII, p. 63 (Jan., 1907).
[2] "Journ. Chem., Met. and Min. Soc. of South Africa," VII, 318 (April, 1907).

Product:	No. 1 Mill Quartz fed 171.5 tons.		No. 2 Mill Pebbles 10 tons.	
	Entering: per cent.	Leaving: per cent.	Entering: per cent.	Leaving: per cent.
Remaining on 60-mesh	68.68	9.73	68.68	11.32
60 to 90-mesh	19.57	26.28	19.57	27.24
Passing 90-mesh	11.75	63.99	11.75	61.44

The lumps of ore, up to $4\frac{1}{2}$ in. diameter, are introduced by means of a feeder consisting of a hopper leading to a bent cast-iron tube, 5 in. in diameter at the narrowest part, and passing into the hollow trunnion at the feed end of the mill. The pulp from the spitzkasten is fed into the same tube near the bottom, opposite the outlet of the bend (where it enters the trunnion).

At Waihi, New Zealand, liners have been used consisting of cast-iron frames divided into square compartments about $3\frac{1}{2}$ in. deep; these are filled in with hard quartz or quartzite, embedded in Portland cement mixed with ground flint fragments. These "honeycomb liners" appear to be as efficient as silex or quartzite blocks, and cost much less. At the Waihi, 4 mills, 22 ft. in length, are run at 27.5 r.p.m. Each mill has a load of 5.5 tons of flints and requires 50 h.p. to operate it.

(E) Comparative Utility of Various Crushing and Grinding Machines

In discussing the crushing machinery suitable for small cyanide mills, Algernon del Mar (E. M. J. *96*, 769, Oct. 25, 1913) considers that for intermediate crushing, the choice lies between stamps and Chilean mills. Either of these will take the product delivered by the rock breaker (say 1 in. cubes) and reduce to a size suitable for feeding to tube-mills or other fine grinders. Rolls are excluded as unsuitable for moist or clayey ores, and ball mills have not been found as efficient in practice as the other machines. The Chilean mill has probably the widest range of usefulness, and is lower in first cost than stamps, but requires a somewhat finer feed than the stamp mill and delivers a finer product.

For fine grinding, the choice lies between pans and tube-mills. Although the Chilean mill has been used for fine grinding, it is not economical for reducing to 200-mesh. The conical type of tube mill is the most economical for grinding from 40 to 100 mesh, while the cylindrical type is more efficient between 100 and 200 mesh. Pans are much lower in first cost than tube-mills, but require a thickened pulp for efficient work, about $2\frac{1}{2}$ parts of water to one of ore being generally suitable, but the conditions both as regards dilution and speed which are most suitable for grinding are not the best for amalgamating. Where used as a grinding machine only, a suitable speed is 45 r.p.m. For amalgamation the speed should not exceed 30 r.p.m., and the pulp must be more dilute than that suitable for grinding. Costs with pans are lower than with tube-mills, when grinding not finer than 150 mesh, but it is more difficult to obtain products of 200 mesh or finer.

The Hardinge Conical Mill. — In the ordinary cylindrical tube-mill the pebbles are distributed indiscriminately throughout the whole length, so that although a large part of the feed is sufficiently crushed within the first few feet, it must travel through the remaining length of the mill before it can be discharged, causing a loss of power by forming a cushion which interferes with the action of the pebbles.

To remedy this defect, the Hardinge mill has been introduced, in which the body consists of a conical shell, the form being that of two cones of different angles, united at the periphery by a short cylindrical section. In a vessel of this form the pebbles arrange themselves automatically according to size, the larger pebbles remaining in the zone of greatest diameter, so that they diminish gradually in size from the base to the apex of the cone. Thus when coarse material is fed to the mill, it is ground by successively smaller pebbles as it is reduced in size and travels towards the outlet.

Two types of mill have been introduced which embody this principle:

(1) The Hardinge Ball Mill, which is lined with plates of chrome steel, and contains forged steel balls 2 to 5 inches in diameter. This machine is designed to take the product from rock breakers and reduce it to a 4 to 10 mesh product without screening. It can be used for wet or dry crushing.

(2) The Hardinge Conical Pebble Mill (see Fig. 15a) which is similar in construction, but lined generally with silex blocks, and filled with Danish flint pebbles. This machine is designed to take the product from the Ball Mills and reduce it to whatever

degree of fineness may be necessary, thus filling the place of the ordinary tube-mill.

H. A. Megraw (Eng. and Min. Journ., Nov. 1, 1913, 823) points out that, although the Hardinge mill is highly successful in producing a fine granular product suitable for concentration, it is not economical for producing extremely fine slime. For this purpose the Chilean mill or the straight tube-mill is much more suitable.

FIG. 15a. — The Hardinge Conical Pebble Mill.

The Lane Slow Speed Chilean Mill. — The *Chilean mill* is one of the oldest contrivances used for crushing ore preparatory to metallurgical treatment. In its primitive form it consists of a heavy stone wheel, mounted on a horizontal axis, and revolving in an annular track about a central vertical pivot. The motive power is furnished generally by mules. The ore is crushed between the slowly revolving wheel and the surface of the track, and although the capacity of such a machine is necessarily small, the grinding is very thorough, and, with addition of quicksilver, the conditions are favorable for efficient amalgamation. A common type of Chilean mill consists of three wheels revolving about a common center, on a track 5 or 6 ft. in diameter. The mills are driven by bevel gears, and the axles are hinged so as to reduce the effect of any strain caused by an obstruction on the track.

Many efforts have been made to adapt this machine to modern requirements. Hardened steel is substituted for stone as the material for the wheels, and mechanical power replaces the mule. Attempts to increase capacity by increasing the speed of rotation result in several drawbacks, whereby some of the most valuable features of the original mill are lost. To obtain a reasonably large output, such mills must be run at 35 to 40 r.p.m. The rapid motion of the pulp causes a large part of it to revolve by centrifugal force, without coming between the grinding surfaces; amalgam is scoured off and reground, thus producing "float" or floured material that finally escapes with the residue; excessive power is required to drive the mill; the wear is also very large and uneven.

To obviate these difficulties, the *Lane mill* combines the features of large size, slow speed, and heavy grinding pressure. The *diameter* of the annular track, in the two standard types of mill, is 7 and 10 ft. respectively. See Fig. 15b.

The *speed* is 8 to 10 r.p.m., which reduces wear and tear, and gives rise to a product suitable for amalgamation or cyanidation.

The *crushing weight* transmitted from the mill itself to the wheels in a 10 ft. mill is 13,000 lb., which is increased by the use of a "weight tank" containing a load of rock or scrap iron, by which the total crushing weight may be adjusted as conditions require. The total crushing weight usually obtained with the 7 ft. mill is 18,000 lb., with the 10 ft. mill, 25000.

The *power required* with the two types is

> 7 ft. mill, 8 h.p.   Capacity, 20 to 30 tons.
> 10 ft. mill, 10 h.p.   Capacity, 40 to 50 tons.

The ore to be crushed is fed either on to the track or into a central distributor which feeds into pipes discharging ahead of each wheel, the latter method being preferable as it causes more even wear.

The ordinary type of Chilean mill is only suitable for comparatively soft ores, but the use of the weight tank as above described enables very hard ores to be crushed in the Lane mill.

The ore should have a preliminary crushing in rock breakers of the jaw or gyratory type to a fineness of 1 to ½ in., but finer crushing before feeding to the mill is undesirable, as the machine is not suitable for regrinding material which has already been crushed, say, to 30 mesh or finer.

The general appearance and construction of the mill is illustrated in Fig. 15b. It consists of a cast-iron base, surrounded by a circular die of manganese steel. The tires of the wheels are of rolled or hammered steel, lathe turned, chrome steel being sometimes used. There are six wheels, revolving on steel rollers, and the friction caused by the centrifugal throw of the wheels is reduced by ball bearings.

FIG. 15b. — The Lane Slow-speed Chilean Mill.   10 ft. model, 1911.
Capacity, 40–50 tons.   Power to operate, 10 h.p.

In both types of mill the speed of the wheels is about the same, but the 7 ft. mill makes 10 r.p.m., whereas the 10 ft. mill makes 8, the greater length of track compensating for the smaller number of revolutions per minute. The speed of drive pulley for both is 250 r.p.m.

The mill may be used with or without screens, the fineness of the product being regulated by the height of the discharge. For amalgamating or concentrating ores the screen used is not finer than 10 mesh.

Arrangements are made for adjusting the wheels so as to secure even wear of the tires. Each wheel has an upward movement independent of its fellows. Should a bolt or other iron object drop into the mill pan, each wheel can pass over it without affecting the mill, owing to the compression of the spring.

The *amount of water* required is estimated at about two thirds that required for stamps. By settling and pumping back, the mill can be operated with 160 to 200 gallons of water per ton of ore milled.

In *cleaning up*, the ore-feed is shut off, and the mill run till very little remains in the pan. The amalgam and remaining pulp are then swept to the discharge opening, and drawn off into the receptacle provided to receive them, the entire clean-up occupying not more than 30 minutes.

A very interesting feature of this method of crushing is the possibility of reducing the ore in only two stages, to a pulp passing 100-mesh screen, and suitable for "all-slime" treatment.

The Musgrove Mining Co., of Salmon, Idaho, have installed a Lane mill, to crush in cyanide solution and discharge the product direct to agitators without regrinding, the product from the mill being simply classified and the oversize returned to the mill. The final product delivered to the agitators is said to be such that practically all will pass a 100-mesh screen. The feed for the mill is taken direct from the crushers, thus eliminating stamps or other intermediate crushers, and also obviating the necessity of tube-mills for fine grinding.

In trial crushings made at the Chunderloo Gold Mine, W. Australia, using a discharge 7 in. above the mill track, a pulp was obtained of which 90 per cent. passed 100 mesh; using a 9 in. discharge, 92.75 per cent. passed 100 mesh and 73.25 per cent. passed 200 mesh.

High Speed Chilean Mills. — According to H. A. Megraw (Eng. Min. Journ. *96*, 822, Nov. 1, 1913) high speed Chilean mills are more generally used in crushing ores for cyaniding than slow speed mills. Those of the "Trent" and "Akron" types make 32 to 34 rev. per min. with an expenditure of 50 to 75 h.p. They are from 4 to 6 ft. diameter across the dies. The crushing wheels are 2 to 3 ft. diameter with a face of 4 to 6 in.

In cases quoted, the Akron mill crushes about 120 tons per day from ⅜ in. size to 30 mesh.

At the Goldfield Consolidated, the Trent mill takes the product from the stamps, which passes a 4-mesh screen, and delivers a product of which nearly 50 per cent passes a 200-mesh screen, crushing 75 tons per day.

It appears that the efficiency of the high speed mill, as measured by the number of tons crushed per h.p. per day, is somewhat greater than that of the slow speed mills, but the product obtained is not so fine.

Neither high nor low speed mills are suitable for regrinding fine material, say 40 to 80 mesh ore.

SECTION III

MECHANICAL HANDLING OF MATERIAL FOR CYANIDE TREATMENT

DURING the early years of the cyanide process it was applied almost exclusively to the treatment of accumulations of already crushed ore, which had previously undergone treatment by amalgamation or other metallurgical process. Such material was usually collected in heaps or had been allowed to settle in pits or natural hollows; in either case it had been exposed to atmospheric influences frequently for long periods of time. Later, when these old accumulations had been largely worked out, the process was adapted for the immediate treatment of material as delivered from the crushing plant, generally after plate amalgamation. Further developments of the process are its application in certain cases to the direct treatment of ore from the mine, without other extraction process, and in the treatment of special classes of material, such as concentrates, and by-products of other processes. The methods adopted in handling the material naturally differ considerably in these various applications of the process. We shall consider them in the above order.

(A) HANDLING OF OLD ACCUMULATIONS

Effect of Natural Settlement. In all cases the crushed ore consists of particles differing in size and in specific gravity. When these are carried along in a stream of water of diminishing velocity, the particles tend to settle, the larger and heavier particles coming to rest first, while the smaller and lighter particles are carried farther. As a consequence of this, the material is roughly sorted into a layer of grains diminishing in size as they are deposited farther from the crushing plant. As, however, the point of inflow and the direction of the stream is varied from time to time, the final effect is the production of layers of coarse and fine material, superimposed and intermixed in a most irregular way. When

171

any attempt is made to treat such material by the ordinary percolation method, difficulties at once arise. If the material be dug out of the pits and dumped indiscriminately into the leaching vats, the charge will consist of masses of particles of varying degrees of fineness; the solution, instead of flowing uniformly through the whole charge, will form channels wherever there is least resistance, that is, in those parts consisting of coarse particles. The very fine material (slimes) forms layers which are practically impervious to liquids; when these slime layers are broken up and dumped into the vats together with the sand, the lumps of slime remain practically untreated. If, however, the slime lumps have become partly dry before charging, they may absorb solution, which remains in them when the tank is discharged; if the solution so absorbed contains gold extracted from other portions, this may lead to considerable losses.

Difficulties in Treatment of Unsized Tailings. — Moreover, the percolation of a charge consisting of particles of various sizes is always more difficult than when the particles are uniform, even if small, because in the first case the interstices of the larger particles are partially filled with the smaller particles; whereas in the latter case the interstices are chiefly filled with air or liquid. It therefore becomes desirable to separate, at least roughly, those portions which can readily be treated by percolation from those which cannot be so treated. It is generally impossible, in digging the material from the dumps or pits, to separate the sandy and slimy portions sufficiently, as the slime occurs usually in thin layers intermixed with layers of sand, but one of the first suggestions was to separate the slime-lumps by passing the whole of the tailings over a coarse screen or grizzly. (Fig. 16.) When the quantity of slime is not too large, it can in many cases be treated by partially drying the material sifted out as above described, and mixing thoroughly with a large proportion of sand, sifting again to break up any lumps. Experiments, however, show that the extraction from the sands diminishes in proportion to the amount of slime mixed with them.[1]

Attempts have sometimes been made to mix the sand and slime *in situ*, by plowing and harrowing the surface of the tailings in the dumps or pits, and allowing them to dry as much as possible by exposure before charging into the vats. The material

---

[1] Julian and Smart, *loc. cit.*, p. 60.

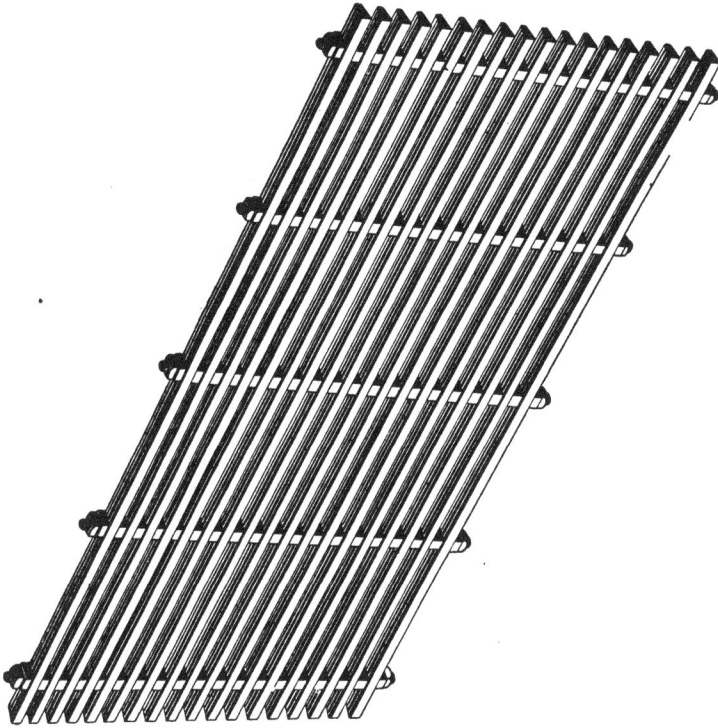

FIG. 16.

is then passed through a screen or disintegrator to obtain a uniform mixture for treatment. (Figs. 17 and 18.)

The Callow Screen (Fig. 18) consists of a perforated endless traveling belt of wire cloth, mounted on revolving rollers. The ore and water are fed on from above. The belt travels continuously at 25 to 125 feet per minute. The coarsest and largest particles strike the screen ahead of the smaller ones, thus leaving a free space for the passage of particles small enough to go through the meshes of the belt. The undersize is discharged into a hopper beneath the upper surface of the traveling belt.

In treating old accumulations much trouble is sometimes caused by the presence of masses of vegetable matter, such as grass-roots, etc., derived from the surfaces on which the tailings were originally deposited; such foreign matter, especially if partially decomposed, causes serious difficulty in the treatment,

FIG. 17. — Revolving Steel Screen, showing feed-hopper. This screen is mounted on friction rollers, so that no central shaft is needed. [From photograph furnished by Hadfield's Steel Foundry Company, Sheffield, Eng.]

not only by preventing uniform percolation, but by introducing organic compounds into the solution, which act as powerful deoxidizing agents, and may even cause local precipitation of gold in the tanks. Whenever possible, such impurities should be carefully screened out. Another source of trouble is the partial oxidation by the influence of air and moisture of iron pyrites and other sulphides contained in the tailings, giving rise

FIG. 18. — Callow Screen, supplied by Fraser & Chalmers.

to compounds, soluble in water or cyanide solution, which attack and destroy considerable quantities of cyanide. This matter will be more fully considered in a later section of this work; it may here be mentioned that this difficulty is partially overcome by preliminary water-washing and alkali treatment, applied to the tailings before running on the cyanide solution.

Methods of Transfer. In general, the tailings are transferred from dumps or pits to the treatment tanks by means of metal trucks or cars running on rails, which are adjusted from time to time to bring them into the most convenient position for loading. The cars are drawn by mules or pushed by native laborers, according to local conditions. Where circumstances allow, this work may be carried out much more economically and conveniently by mechanical haulage, using conveyor belts, bucket elevators, or similar devices. (Fig. 19.) The rails are generally carried over the top of the treatment tanks, and the loaded cars are

FIG. 19. — Bucket Conveyor, furnished by Hadfield's
Steel Foundry Co., Sheffield, Eng.

tipped so as to discharge their contents on to a wooden framework or grizzly, to assist in breaking up any lumps and to secure a uniform product for treatment. When the tank is filled, the surface is leveled off and raked over before applying the first wash.

(B) Handling of Current Mill-Product

If the tailings leaving a battery be allowed to run direct into filter tanks, it is found in most cases, after the superfluous water has been allowed to drain off or has been decanted from the settled sand, that the material so collected is quite untreatable by cyanide, owing to its mechanical condition: a partial separation of grains, according to size, takes place in the tank as in the tailings dams, and the result is a dense, almost impervious, mass. If an attempt be made to treat this, the solution will either not percolate at all, or will only pass through the coarser material deposited near the inlet.

Settling Pits. — The first suggestion for overcoming this difficulty was to construct comparatively small dams or pits which would collect sufficient clean sand for one or more tank charges, and allow the water and the bulk of the slime to run to waste. The sand so collected was then dug out and transferred to the treatment tanks, the stream of battery pulp being meanwhile diverted to another pit. The outlet of the collecting pit was provided with a "slat gate" consisting of two upright grooved posts, boards being placed in the grooves so as to raise the hight of the overflow from time to time during the filling of the pit. These pits were frequently lined with masonry and the sides inclined toward the center. The objections to this method, which is still used in many places, are that much fine sand escapes along with the slime and water at the exit end (or, on the other hand, if the pit is made very large, the tailings which settle near the exit end are too slimy to be successfully treated by percolation), and the transfer of the collected sand from the pits to the treatment tanks is an operation which would be eliminated by a successful method of direct filling.

Direct Filling by Means of Hose. — As the chief difficulty in the percolation of a charge of tailings which is allowed to run direct into the leaching tank arises from the uneven distribution of particles of different sizes, it was suggested that this trouble might be overcome by passing the stream of battery pulp through a hose, which could be moved by hand to different parts of the tank as the filling proceeded. This system was introduced on the Rand by Hennen Jennings about 1894. The tanks were provided with adjustable slat gates, over which the slime and water passed to the exit pipe or launder. This method is still in extensive use, but it requires a good deal of supervision, as, if the hose be not constantly moved from place to place, there will be local deposits of coarse sand and slime and consequent bad percolation. In some more recent applications of this system, the slat gates are replaced by canvas blinds, which are unrolled upward as the filling proceeds.

Intermediate Collecting Tanks. — A third method, introduced about the same time as the above described hose-filling system, was applied by Charles Butters, and consists in the use of special tanks for collecting the material to be treated. These tanks are filled with water before use, and when the battery pulp is turned into them, the slime and water overflow at the edges into a circular peripheral launder, whence they are carried away for storage or further treatment. When a sufficient quantity of leachable sand has collected, this is transferred, through discharge doors in the bottom of the collecting tank, either direct into leaching tanks placed beneath, or into trucks by which it is conveyed to the leaching tanks. Belt conveyors may be advantageously used for this purpose. The collecting tanks are provided with filter bottoms to assist in draining off superfluous water. In order to secure uniform mixture and distribution of the material, the battery pulp entering the collecting tank usually passes through an automatic distributor, designed by Charles Butters and Captain Mein. This consists of a conical bowl mounted on a pillar placed at the center of the tank, and provided with radial arms, consisting of tubes of different lengths bent at the exit end; the flow of liquid and the reaction of water leaving the bent pipes causes the apparatus to revolve, and so produces a uniform distribution of the pulp. (Fig. 20.)

Double Treatment. — In some cases the product collected in this way in the collecting tanks is sufficiently leachable for direct cyanide treatment, but in most cases it is found advantageous, after adding sufficient cyanide solution in the collecting tank to saturate the charge, to transfer the latter for further treatment to another tank. This method of double treatment has the advantage of exposing the material while saturated with cyanide to the action of atmospheric oxygen, under conditions favorable for solution of the gold. Occasionally charges are transferred

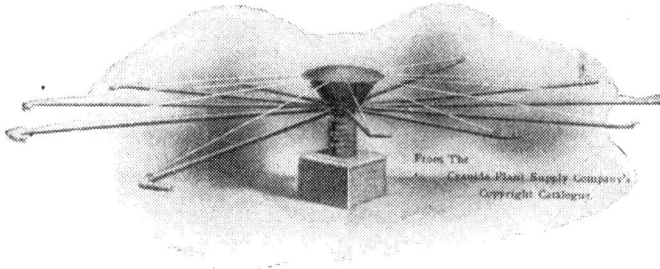

FIG. 20. — Revolving Sand Distributor for feeding tailings pulp from battery into collecting tanks. With conical hopper, revolving arms and mouthpieces, ball bearings, spindle and flanged pillar.   [From photograph furnished by the Cyanide Plant Supply Co.]

a second or third time from one tank to another during the course of treatment. (See Part IV., Section I.) (Figs. 21 and 28.)

(C) Transfer of Material to Higher Levels

As the ore and tailings undergoing treatment have to pass through a number of machines and appliances, it is obviously an advantage, from the point of view of economy, that this transfer should take place as much as possible automatically. Gravity is utilized by placing each successive apparatus at a lower level than the preceding one. It frequently happens, however, that the conformation of the site on which the works are erected does not allow of sufficient fall for every transfer to take place by gravity, and in this case means must be found to raise the material mechanically to a higher level.

Conveyors, both of the belt and bucket type, are often used to convey ore from the bins to the crushers (see Fig. 22); and also for transferring tailings from dams or settling pits to treatment tanks, for transferring from one tank to another, for transporting residues to the dump, etc.

Tailings wheels (see Fig. 23) are very commonly adopted where wet material, such as battery pulp or the coarse discharge from hydraulic separators, has to be raised to a higher level. These

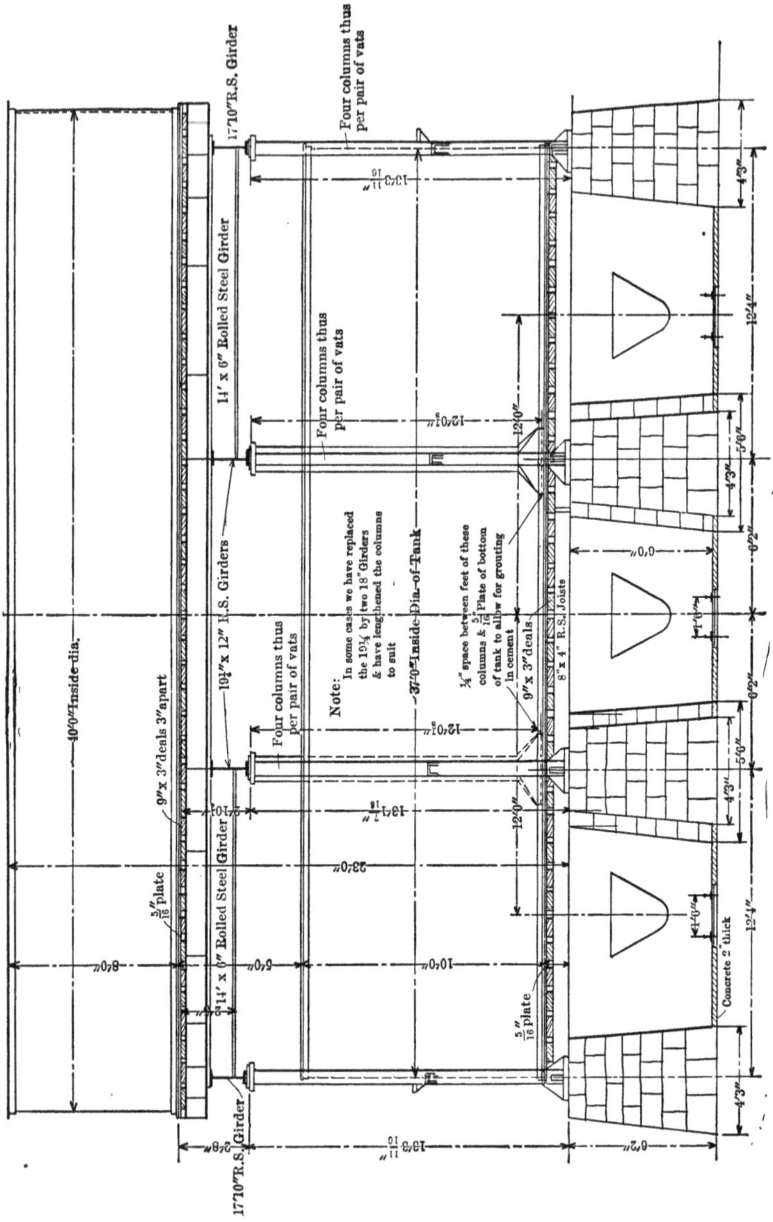

FIG. 21. — End Elevation of Iron and Steel supports on masonry walls of large Double Treatment Plant erected by Cyanide Plant Supply Co.

are steel structures, similar in general construction to a bicycle wheel, of a diameter depending on the vertical hight to which the material must be raised — say 20 to 70 ft. The rim is furnished with a frame divided into compartments or buckets, generally on the inside, the walls of which are inclined all in one

Fɪɢ. 22. — Ore Conveyor (belt system) as supplied by Fraser and Chalmers, transporting rock from ore bins to crushers.

direction, so that liquid or pulp may be fed into them at the bottom from a wooden trough or launder and discharged into another launder situated just below the top of the wheel. The speed is generally from 3 to 5 r.p.m.[1]

Sand and Tailings Pumps. — In some cases tailings are elevated by means of ordinary plunger pumps, running at a high speed, and with special contrivances to prevent the settling of sand in any part of the pump or delivery pipe. The Frenier

[1] For further details, see Julian and Smart, *loc. cit.*, p. 352.

Spiral Sand Pump [1] is a well-known apparatus for this purpose. It consists of a spiral tube which is caused to revolve so that at each turn the open end scoops up some of the tailings from the launder, in which it dips. As the spiral is only partially immersed, it takes up air during part of its revolution. On reaching the center of the spiral, the pulp is discharged into a delivery pipe, to which the spiral is connected by means of a detachable tube and stuffing-

FIG. 23. — Tailings Wheel, Sons of Gwalia Mine. [Furnished by the Cyanide Plant Supply Co.]

box; through this delivery tube air and pulp are forced by the hydrostatic pressure of the liquid in the spiral. The speed is about 20 r.p.m., and as the construction is very simple and the cost for power comparatively small, the apparatus presents considerable advantages over the centrifugal pump. The lift is from 14 to 24 ft. On the other hand, it requires considerable care, both in erection and operation, particularly in starting and stopping.

(D) PREPARATION OF ORE FOR DIRECT CYANIDE TREATMENT

In some cases it has been found advantageous to treat the ore by cyanide without previous amalgamation. When this system

[1] Julian and Smart, *loc. cit.*, p. 362.

is adopted the ore is usually crushed dry by means of rock-breakers, rolls, ball mills, etc., as previously described (Sections I and II). With exceptionally porous ores, such as that of the Mercur mine, Utah, the material may be successfully treated after merely reducing to about $\frac{1}{2}$ in. size by rock-breakers, and charging direct into the treatment tanks. A similar method was adopted at the George and May mine, near Johannesburg. Frequently it is necessary partially to dry the ore, so as to reduce the percentage of moisture below 2 per cent., before it can be successfully handled by dry-crushing machinery.

In the case of refractory ores, containing large amounts of base metal sulphides, arsenic, tellurium, etc., it is generally necessary to roast before cyanide treatment, this operation being carried out either by hand, in reverberatory furnaces, or mechanically in one or other of the numerous types of revolving furnace, or in furnaces provided with mechanical stirrers. A description of some of the principal types of furnace used for this purpose will be found in Rose's "Metallurgy of Gold" (4th ed., pp. 246–262).

Belt conveyors or similar devices are frequently used for transferring the crushed ore from the rolls or ball mills to the roasting furnaces, and from the latter to the treatment tanks. Special appliances are also used for cooling the roasted ore; thus it may be passed through a revolving drum containing tubes of cold water, or the conveyor itself may be water-jacketed.

In roasting for cyanide treatment it may be observed that a partial roast is sometimes worse than none at all, as it may convert comparatively harmless sulphides, etc., into soluble oxidized salts, such as base metal sulphates, which act as rapid and powerful cyanicides. In general it may be said that where roasting is adopted at all, a dead roast is essential. Many ores formerly treated by roasting have been found to be amenable to direct cyanide treatment in the raw condition, if finely enough ground. In some cases, however, the effect of roasting is to render the ore more porous, so that heavily mineralized ore, which could only be treated raw after very fine grinding, may be roasted after comparatively coarse crushing and treated direct. The subject of roasting for cyanide treatment will be more fully discussed in a later section. (Part VI, Section III.)

SECTION IV

HYDRAULIC CLASSIFICATION, SEPARATION, AND SETTLEMENT OF SLIMES

(A) Hydraulic Classification by Means of Pointed Boxes

It has already been pointed out in Section III that it is desirable to separate the finer from the coarser portions of the crushed ore prior to cyanide treatment, as a different procedure is necessary for each in order to get the best results. This mechanical separation can be more effectively and cheaply carried out in water than by any other means. Various devices have been used from time to time with this object, such as flat shaking screens or jigs, revolving screens or trommels, etc.; but the simplest and most effective for the present purpose is the pyramidal box with pointed bottom, introduced in its original form by Rittinger, about the middle of the last century.

These boxes are of two principal types: those which merely have an outlet for the coarse material at the bottom being known as *Spitzkasten*, while those which have, in addition to this outlet, a supply of water introduced near the bottom so as to cause an ascending current are distinguished as *Spitzlutten*. When a stream of water carrying particles of ore of different sizes and specific gravities is allowed to flow in at one side of such a box and out at the opposite side, the question whether a given particle will sink to the bottom and be discharged through the outlet, or whether it will be carried away by the overflow at the top, depends, (1) on the area exposed at right angles to the direction of flow; (2) on the specific gravity of the particle; (3) on the velocity of the stream carrying the particle through the box; (4) on the velocities of any opposing currents which it may meet with.

In very shallow water, the velocity of a falling particle varies approximately as $g\left(1 - \dfrac{\delta}{D}\right)$, where

g = acceleration due to gravity;
δ = density of water;
D = density of particle.

184

In sufficiently deep still water, the velocity is uniform and varies as $a \, (D - \delta)$, where $a^2 =$ area of section of particle at right angles to line of fall.

When the particle falls in opposition to a vertical ascending current, the pressure opposing its fall varies as $f \, v^2$, where

> $f =$ area of surface opposed at right angles to current;
> $v =$ velocity of ascending current.

It is evident, therefore, that these machines are not concentrators in the strict sense of the term, since a large particle of a light mineral and a small particle of a heavy mineral might fall at the same rate and be discharged together. The mathematical questions involved in the construction of these boxes cannot be discussed here.

A full discussion of the theory of bodies falling in water will be found in Rittinger's "Lehrbuch der Aufbereitungskunde"; an abstract of his results is given by Julian and Smart, "Cyaniding Gold and Silver Ores" (2d ed., pp., 369–375). See also Rose, "Metallurgy of Gold" (4th ed., p. 176). It should be noted that the shape of the particles influences their rate of fall, bodies presenting a pointed (wedge-shaped or conical) surface to the opposing current falling faster than bodies of the same weight and volume presenting a flat surface.

In a case where the material treated consists entirely of fairly large particles of the same specific gravity, the hydraulic separation will take place between fine and coarse particles, that is, the material will be classified according to size. In a case where the particles are approximately of the same size but of different specific gravities, the separation takes place according to density, and the apparatus becomes a true concentrator. A large spitzkasten or spitzlutte with a relatively small orifice will yield, with ordinary material consisting of particles varying both in size and specific gravity, a product consisting of approximately equal falling grains, that is, grains which fall at an equal rate in still water.

The use of hydraulic separators in cyanide practice is twofold. In the first place, it is desirable to collect for separate treatment the very coarse portion of the crushed ore together with a large proportion of the heavy mineral; this is done usually by passing the pulp through a small spitzlutte having a fairly strong up-current. In the second place, the very fine (unleachable) part of the ore, generally described as slime, has to be separated from the leachable sandy portion. In this operation the conditions governing the separation are totally different from those concerned in the separation of coarse particles. Gravity has little influence on the result, and other factors — the viscosity of the pulp, surface tension and electrostatic repulsion — come into play. For the separation of slime from fine sand a much larger spitzlutte with a relatively slow ascending current must be used.

Details of the construction of pointed boxes are given in many works on metallurgy.[1] We shall here draw attention only to a few points of special interest in cyanide work.

In order to prevent the settling of sand near the top of the apparatus, where the ascending current is least effective, it is customary to make the upper part of the box with vertical sides, the lower part being made either pyramidal, with four triangular faces meeting at the outlet, or in the form of a triangular prism, with inclined rectangular sides in the direction of flow and triangular vertical sides joining these.

In order to prevent surface currents from carrying the pulp direct to the overflow, a board is placed across the stream near the inlet end, either vertical or inclined, so as to force the stream of pulp downward toward the outlet and bring it in contact with the ascending current of clear water. According to Julian and Smart (*loc. cit.*, p. 375), "for a good separation it is necessary that the rising stream should have, as nearly as possible, the same velocity at every part of its cross-section. This desideratum is attained by passing it through a rectangular passage, one of whose dimensions is very small relatively to the other." This is attained in some forms of spitzlutte by making the whole apparatus in the form of a triangular prism, with a smaller prism of similar shape forming the partition, so as to leave a V-shaped passage, rectangular in cross-section, between the two. An adjustable form of spitzlutte has also been introduced, in which the inner V can be raised or lowered to vary the width of the passage and so regulate the nature of the product obtained. The clear water is introduced by means of a two-way pipe at the bottom of the apparatus; the continuation of the pipe forming the

---

[1] See diagrams given by Julian and Smart (*loc. cit.*, Figs. 172 to 176, pp. 375--379).

outlet is bent upward and provided with a flexible hose, so that the level of outflow can be varied as required. The inclination of the pointed bottoms should be not less than 55° with the horizontal, otherwise sand may collect at the angles, eventually falling and blocking the outlets.

The diameter of the discharge opening required is proportional to the square root of the depth; the minimum size necessary to prevent choking is about ¾ in. diameter. The pressure of the ascending clear water current in spitzlutten must be rather greater than that of the descending pulp stream in order to ensure a clear outflow. By raising the level of outflow as above described, a back pressure is set up, which enables a product free from slime to be obtained without excessive waste of clear water. When a number of pointed boxes are used in series so as to obtain a set of products of varying fineness, each successive box must be made wider than the last, the width of each box being from 1½ to 2 times greater than the preceding one. For the first box, from 4 to 6 ft. of width (according to different authorities) should be allowed for each cubic foot of pulp flowing per second through the box. The length of the successive boxes should also be increased, though this is a matter of less consequence. Where large quantities are handled, the stream of pulp is sometimes divided between a number of boxes placed in parallel, so as to avoid the necessity of constructing very deep boxes.

Cone Classifiers. — A form of hydraulic separator which offers some advantages over the ordinary spitzkasten and spitzlutten consists of a hollow inverted cone of wood or sheet iron having an outlet at the apex. The pulp is fed in either by means of a distributor with revolving arms which dip just below the surface of the water with which the cone is filled, or through a centrally placed vertical pipe which descends some distance beneath the surface. The slimes overflow into a peripheral launder at the circumference of the cone. The underflow may be made to pass to a second cone, or a number of cones arranged in series for more complete removal of slimes. The apparatus may be arranged either as spitzkasten, without fresh water supply at outlet, or as spitzlutte with ascending current introduced at the bottom (apex) of the cone. It is essential to have the apparatus quite level, so that a uniform overflow takes place in all directions. The cones have an angle of about 55°. Owing to the even and regular flow, a very perfect separation can be attained by proper adjustments. (See Fig. 24.)

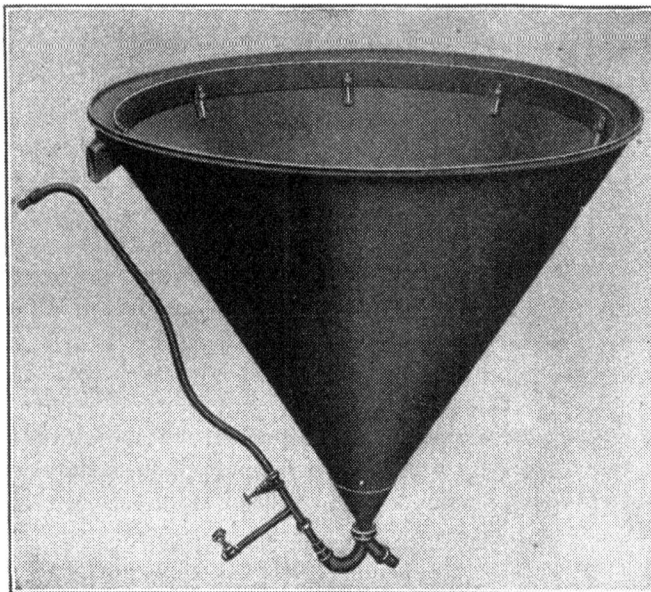

FIG. 24. — Callow Settling and Pulp-thickening Tank, furnished by the Utah Mining and Machinery Co., Salt Lake City, Utah. This is a form of cone classifier which may be adapted to serve either as spitzkasten or spitzlutte.

(B) Separation and Settlement of Slimes

When the ore is very finely divided, in the condition generally described as slime, frequently it will not settle in any ordinary time, and artificial means must be employed to remove the superfluous water. Filtration, except under pressure, is not usually successful, and may often be too costly for practical use. Since the gold values in such suspended material are generally very easily dissolved in cyanide, it might be supposed that settlement and separation of the mineral matter could be dispensed with. Usually, however, it is very difficult to recover the dissolved gold in a concentrated form suitable for smelting, unless the suspended ore particles are removed, although from time to time attempts have been made to dissolve and precipitate in one vessel, as for example in the Pelatan-Clerici process. It is generally simpler to collect the slimy material, after causing it to settle in a more or less concentrated form, by the addition of some coagulating agent. The substance universally used for this purpose is lime, one part of which is said to be capable of clarifying between 6000 and 7000 parts of water containing clay in suspension.

There are other substances which are equally or even more effective in causing settlement, but their use is precluded either by their cost or on account of some chemical property which interferes with the proper working of the process by precipitating the precious metals or by setting up other injurious reactions with the ore or solution. Among substances having a coagulating effect on suspended mineral matter may be mentioned many salts of calcium, magnesium, iron, and aluminium, notably the alums, and, in a lesser degree, neutral salts of the alkali metals, such as sodium chloride and various mineral acids. The following figures are taken from a table given by Julian and Smart (*loc. cit.*, p. 220), representing the results of their experiments on the relative coagulating effect of certain salts:

	Weights Required to Produce Equal Coagulating Effect.
Aluminium sulphate	100
Potash alum	143
Lime	654
Magnesia	748
Calcium chloride	1095
" carbonate	1215
" sulphate	2870
Magnesium sulphate	3460
Sodium chloride	45900
" sulphate	61700

It is thus seen that, weight for weight, alum has about five times the coagulating power of lime.

Application of Lime in Preparatory Treatment of Slimes. — In modern practice a certain amount of lime is usually added in the battery, sufficient to dissolve any grease from the crushing machinery which might find its way onto the plates. This has the further effect of coagulating the slime and thus bringing it into more intimate contact with the amalgamated surface. Hence it has been observed that where lime is used in the battery, the extraction by amalgamation is higher, and the assay value of the slimy portion of the tailings is lower, than when the same ore is crushed without addition of lime. As a rule, also, a further quantity of lime is added to the pulp in the launder carrying away the overflow from the collecting tanks or spitzlutten used in the separation of the sand. (See Sections III and IV, A.)

An ingenious device has been introduced by P. S. Tavener,[1] by which lime is fed automatically at a regular rate into this launder. The overflow carrying the slimy portion of the pulp is led into a large spitzkasten, in which the coagulated slime settles in practically still water. If sufficient lime has been added, the overflow from this spitzkasten is clear enough to be used at once in the mill, and is pumped back for that purpose, while the underflow, that is, the discharge from the outlet at the bottom, consists of slime carrying perhaps fifteen times its weight of water. This material is led into a collecting-tank having a conical bottom, where further settlement takes place. By using a sufficiently deep tank for this purpose, the settled slime finally obtained may contain not more than 40 per cent. of moisture. The overflow from the collecting tanks also consists of clear water which can be returned to the mill.

Factors which Influence Rate of Settlement of Slime. — Experiments made by Leo D. Bishop (Colo. Scient. Soc., Nov. 7, 1908), show that slime settles more rapidly in pure water than in either an alkaline or acid medium. Lime produces flocculation, and the settled liquor is consequently clearer than in cases where settlement takes place without flocculation, but the actual percentage of (more or less) settled liquid which can be drawn off is less, as the flocculent material does not settle so compactly. However, flocculation is essential for rapid filtration, hence lime is in general preferable to other settling and neutralizing agents. A large excess of lime diminishes the rate of settlement.

The rate of settlement increases with temperature, as shown by tests made with a range of 70° to 120° F., and the rate of filtration is also increased by increasing temperature. The rate of settlement at 120° was nearly double that at 70°, and the rate of filtration 3 times as great.

Diaphragm Cone Classifier. — This apparatus was introduced by W. A. Caldecott in 1908 for the purpose of delivering a thickened coarse product for feeding to tube-mills, and also for separating fine leachable sand from slime in material to be cyanided. The apparatus for both purposes is practically the same, differing

---

[1] For details see Julian and Smart, *loc. cit.*, p. 222.

only in size and in the position in the cone at which the diaphragm is placed. The diaphragm consists of a disk of wood or sheet steel, placed horizontally and resting by four supports against the sides of the cone, so as to leave an annular space through which the settled sand passes to the apex of the cone, whence it is discharged. When in use, the cone is always partly filled with settled solids. The pulp is fed from a launder delivering at the top, at the center, where it enters the classifier through a short baffle tube. The finer particles overflow at the periphery of the cone. The chief object of the diaphragm is to prevent the formation of a single vertical channel through which the pulp descends without classification. If the annular space becomes bridged over so as to leave only a single channel, the diaphragm must be raised higher in the cone. When the pulp contains much slime, it is in general necessary to leave a wider space around the diaphragm than when coarse, granular pulp is to be classified.

The standard cone used on the Rand for feeding a $5\frac{1}{2}$ by 22 ft. tube-mill is 6 ft. diameter by 9 ft. deep, having a flat disk diaphragm 8 in. diameter with an annular space about $2\frac{1}{2}$ in. wide, the latter varying according to the more or less slimy nature of the ore. The best position for the diaphragm is easily found by a few trials. This yields an underflow product carrying 27 per cent. moisture. A grading analysis of the products delivered by such a cone gave results as follows:

	Underflow	Overflow
+ 60 mesh.	67.1 per cent.	7.4 per cent.
60 to 90 "	20.7 " "	15.9 " "
− 90 "	12.2 " "	76.7 " "
	100.0 per cent.	100.0 per cent.

Before entering the tube-mill the pulp is diluted to about 40 per cent. moisture.

The cones used for separating leachable from unleachable material are 8 ft. in diameter by 10 ft. deep, are provided with two diaphragms at different heights, and deliver a sand underflow with about 30 per cent. moisture.

The following grading analysis shows the nature of the products obtained, the cone delivering about an equal tonnage of each.

	Underflow (Sand)	Overflow (Slime)
+ 60 mesh.	8.4	—
60 to 90 "	33.3	—
90 to 200 "	46.3	2.21
− 200 "	12.0	97.79
	100.0	100.00

Merrill Hydraulic Classifier. — The ordinary cone classifier is imperfect on account of the disturbance created by admitting the water, which prevents proper separation of the material.

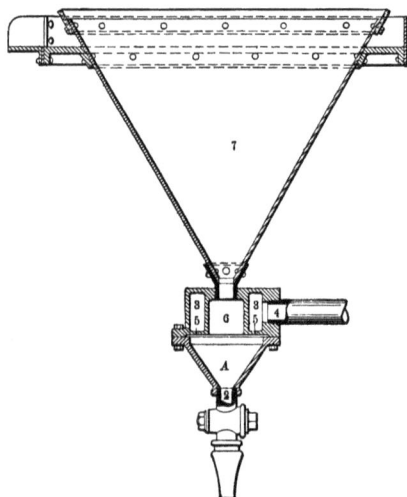

FIG. 24a. — The Merrill Hydraulic Classifier.

In the Merrill Classifier this difficulty is overcome by using a smaller secondary cone beneath the water inlet. The construction and operation of the device is as follows (see Fig. 24a):

The pulp containing minerals of different sizes and specific gravities, which are to be separated, is delivered into the upper part of the primary cone (7) and diluted with water if necessary. Water is admitted through the passage (4) into an annular chamber (3) and is thence discharged downwards through an annular channel (5) into the secondary cone (A). This water displaces the more slimy water which would otherwise be carried down with the coarse, heavy minerals. The heavy material flows down through the central opening (6), and through the lower cone (A), passing out through the discharge at (2). The effect of this structure is to admit water in a continuous surrounding sheet, so that it converges from all sides toward the center.

These cones are fitted with lead leveling-strips to insure uniform peripheral overflow.

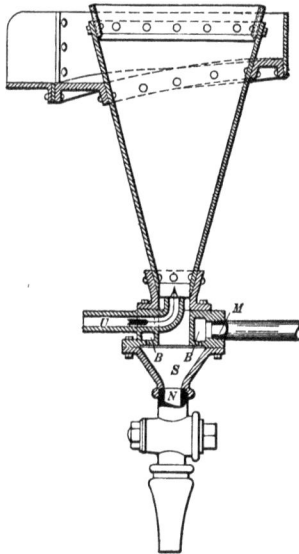

Fig. 24b. — The Merrill Concentrating Cone.

Merrill Concentrating Cone. — This device is similar in arrangement to the classifier, but the pulp is fed from the side, near the bottom of the primary cone, and the sides of the hopper are somewhat steeper than in the classifier. In this case the slime together with the coarser and lighter part of the sand, rises to the peripheral overflow, while the concentrates are discharged at the bottom free from slime and coarse silica. Water under low pressure is admitted into the annular chamber and is distributed as described in the classifier. (See Fig. 24b.)

The Dorr Classifier. — This apparatus was invented by J. V.

N. Dorr in 1904, at the Lundberg, Dorr and Wilson Mill, Terry, S.D., and first brought into general use in 1906. In principle it is very simple, consisting of an inclined trough open at the upper end, on the bed of which operate one or more rakes, by means of which the quicker settling particles of the material to be classified are pushed upward, the slime overflowing at the closed bottom end. The pulp is fed across the center of the trough, insuring a uniform flow of liquid to the overflow end. The rakes are suspended from bell cranks connected by rods to levers which terminate in rollers. The latter press against cams attached to the crank shaft. The rakes are lifted and lowered at opposite ends of the stroke by the action of the cams transmitted through the levers and rods to the bell cranks. The cranks produce the horizontal motion. A special device allows the lower end of the rakes to be raised 10 in., so that the machine may be started again when half full of sand after a sudden shut-down, without undue strain.

The sand is discharged with 20 to 30 per cent. moisture. When necessary, spray pipes are added for washing the sand from very slimy ore free from the last particles of slime.

Regulation of the point of separation is obtained by varying the speed of the rakes, the flow through the machines and by the use of baffles.

The Duplex Classifier, which is the size generally used, requires $\frac{1}{8}$ h.p. and handles 80 to 200 tons daily, depending on dilution of feed and point of separation desired   Less than one foot of head is required, and the machine when placed alongside of a tube-mill equipped with scoop feed can be operated in a closed circuit with no pumping required. (See Fig. 24c.)

The operation of the machines has shown very little repairs required. In one instance a machine was operated over 4 years with a total repair cost under $5.

The triple washing classifier shown in the cut allows the washing of sand three times to remove gold-bearing solution.

The Dorr Thickener. — The Dorr Thickener was invented by J. V. N. Dorr in 1906 at the Mogul Mill, Black Hills, S.D., and its use has spread over the world in cyanide and concentrate plants.

This apparatus consists of a flat-bottomed cylindrical tank, provided with radial arms usually inclined slightly upwards from the center. The pulp is fed in at the top through a short cylinder placed over the center of the tank. As the slime settles, the arms, which revolve slowly, constantly draw it towards the center, where it is discharged through an opening into a pipe or launder. The clear liquid resulting from the settlement, overflows at the periphery, into an annular collecting trough. Arrangements are provided for raising the arms above the level of the settled pulp in case of a shut-down, from which position they can be gradually lowered again after restarting. (See Fig. 24d, p. 196.)

Fig. 24c. — Standard Duplex Dorr Classifier, Model "C," Belmont Type.

The usual size of tanks used for this purpose is from 16 to 35 ft. diameter by 8 to 12 ft. deep.

The speed varies according to the conditions, but is very slow, say about 5 rev. per hour. The power required is $\frac{1}{8}$ to $\frac{1}{4}$ h.p.

The thickened product usually varies from 25 to 50 per cent. solids.

The settling area required is about 5 sq. ft. per ton of solids (as discharged from tube-mills) thickened in 24 hr., for silicious ore, while for clayey material the figure is 7 to 10 sq. ft. per ton.

As the operation of the machine is practically automatic, little attendance is required.

The continuous nature of the process greatly reduces the operating costs, and increases the rate of settlement, as the settled pulp is withdrawn as fast as it forms. The revolving rakes prevent any accumulation of solids which might choke the discharge, and as the sides are vertical there is no settlement on them such as occurs generally in cones.

Fig. 24d. — Dorr Continuous Thickener.   Installed in steel tank.

Automatic alarms have been devised to warn the operator if the machine is overloaded or is overflowing slimes.

Recent inventions of Mr. Dorr have resulted in means for increasing settling capacity of a thickener from 50 to 75 per cent.

SECTION V

AMALGAMATION AND CONCENTRATION

(A) Amalgamation in Relation to Cyanide Treatment

A CONSIDERATION of concentration, properly so called, might naturally follow the discussion of hydraulic classification, but as in practice some process of amalgamation almost invariably precedes concentration, these operations will be considered in the same order. In its earlier applications, the cyanide process has been looked upon merely as an adjunct to amalgamation. It has been applied simply as a means of recovering some part of the gold which is lost in the older process, or which could only be obtained by amalgamation with much trouble and expense. This view still very largely prevails, and probably governs the design of metallurgical plants at the majority of mines in most parts of the world. As a consequence, while much care is taken to secure the maximum recovery on the plates, little consideration is given to the question whether the product leaving the plates is in a suitable condition for yielding the best extraction by cyanide treatment. The two processes are looked upon as more or less antagonistic, and are usually under separate managements, between which there is often considerable rivalry. In a few localities, notably in the Rand, it has been recognized that the metallurgy of the ore should be regarded as a single problem, and that method or combination of methods adopted which is found by experience to yield the maximum profit. Thus, in some instances.it has proved more advantageous to aim at a lower extraction by amalgamation, as this may allow of an increased tonnage crushed per month, the subsequent cyanide treatment being relied on to recover values which otherwise might have been saved in the mill. Generally speaking, however, it may be said that it is cheaper to recover gold by mercury than by cyanide, especially when the gold is in fairly coarse

197

particles, and it is therefore good policy to make the amalgamation as efficient as possible, consistently with leaving the tailings in such a condition that the values not recoverable by amalgamation may be readily extracted by cyanide. Fortunately, the conditions which promote good amalgamation are in general also beneficial to cyanide treatment, as, for example: (1) sufficiently fine crushing of the ore (already referred to in describing the action of stamps: Section I, B); (2) a slight alkalinity of the water in which the pulp is suspended, obtained by the addition of lime in the battery; (3) absence of soluble and easily decomposed compounds of the base metals, which stain or "sicken" the mercury and also act as cyanicides; (4) absence of oil, grease, and similar matter.

Amalgamation is carried out in one of the two following ways: (1) by causing the crushed ore, suspended in water, to come in contact with surfaces coated with a layer of mercury (plate amalgamation); (2) by grinding ore and mercury together in an iron pan, with sufficient water to form a paste, other chemicals being added if necessary (pan amalgamation). Other modifications of less importance need not be considered here.

In the case of ores in which the value consists principally of gold, plate amalgamation is almost universally used; when the material to be treated is essentially a silver ore, pan amalgamation is more usual.

In plate amalgamation sheets of copper coated with mercury are placed inside the mortar-box, generally both on the feed and discharge sides, attached either to the mortar-box itself or to wooden blocks. Amalgamated plates are also placed on a sloping table immediately in front of the screens, so that the pulp flows over them as it leaves the mill. The practice of amalgamation inside the mortar-box is open to much criticism, on the ground of difficulties in regulating the work, losses of mercury involved, etc., and has been entirely abandoned in many places. It is, however, defended by some metallurgists (see Rose, "Metallurgy of Gold," pp. 149, 150). It is probable also that the practice of placing the outside plates close to the screens may be largely abandoned in the near future, as it seems desirable for various reasons that the amalgamation should be conducted in a room or building entirely separate from the crushing machinery. The plates used for amalgamation consist of pure copper, of copper previously coated by electroplating with a thin film of silver, or of Muntz metal, an alloy of copper and zinc. They are from $\frac{1}{16}$ to $\frac{3}{8}$ in. in thickness, thicker plates being required for inside than for outside amalgamation. For the latter, the plates are laid on an inclined wooden table, usually 4 or 5 ft. wide and 12 to 14 ft. long, having raised edges to retain the pulp.

It is found that plates previously amalgamated with gold or silver are more effective in catching the precious metals than copper plates amalgamated only with pure mercury; hence the use of silvered plates alluded to above.

Effective amalgamation depends chiefly on keeping the surface of the plate as clean as possible, and in so regulating the amount of mercury that it forms a pasty amalgam with the gold and silver which it extracts from the ore; when too little mercury is added, the amalgam becomes hard and ceases to take up fresh quantities of the precious metals; when too much is used, it runs off in drops and is carried away with the tailings, together with any gold and silver which it may have absorbed. Other important points are: (1) the due regulation of the water-supply so that the pulp passing over the plates may be neither too thick nor too thin; (2) proper inclination of the table, so that the pulp comes into effective contact with the amalgamated surface without allowing a deposit of sand or mineral to settle permanently on it. This inclination is from 1 to $2\frac{1}{2}$ in. per foot, according to the nature of the ore, a steeper grade being naturally required when much heavy material is present. At intervals during the day each battery of five stamps is stopped and the plates connected with it cleaned. A part of the amalgam is removed at the same time by means of a rubber scraper and fresh mercury applied. In many mills a certain quantity of mercury is also fed at intervals into the mortar-boxes. The total amount of mercury used varies, of course, according to the nature and richness of the ore treated, but may be put at about 1 to 2 oz. per ounce of gold amalgamated. Most of this is recovered in retorting the amalgam, but there are some losses which must now be considered, as they are of importance in connection with cyanide treatment.

The practice of feeding mercury into the mortar-boxes causes some of it to be broken up, by the action of the stamps and the agitation to which it is subjected, into very small globules known as "floured mercury." This passes through the screens and may be carried off the tables by the stream of pulp, ultimately finding its way into the cyanide vats. If ordinary base metals, such as copper, zinc, lead, or tin, be present as impurities in the mercury used for amalgamation, or occur in the ore in the form of easily decomposable compounds, these metals give rise to films which coat the surface of the mercury and entirely prevent amalgamation. When this occurs the mercury is said to be "sickened." These films consist of the oxides, sulphides, sulphates, carbonates, or other compounds of the base metal. In some cases actual alloys of mercury with another metal may be formed, and occasionally sulphates or other salts of mercury.[1] *Lead* forms an amalgam which separates as a frothy scum, carrying with it any gold which may be present; this easily becomes detached from the surface of the mercury and is then carried away by the pulp. *Arsenic* and *antimony* are particularly harmful. Arsenic, whether in the metallic state, as sulphide (As_2S_3), or as mispickel (FeAsS), produces a black coating which is a mixture of metallic arsenic and finely divided mercury, no amalgam being formed. Antimony separates from its compounds rapidly in a similar manner, but the metal forms an actual amalgam, and another part of the mercury is converted into sulphide. *Bismuth* has a similar but less rapid action. All these compounds readily break up into small particles, so that sickened mercury is very liable to be carried away by the stream of ore and water and hence to find its way into the cyanide plant, giving rise to various undesirable reactions.

All metallic sulphides except clean, coarse, undecomposed iron and copper pyrites have some action on mercury. Gold in iron pyrites largely escapes amalgamation, unless the mineral be very finely crushed and treated in pans, in which case much sickening and loss of mercury takes place. The difficulty of amalgamating gold in pyrites appears to be due to the fact that the metal occurs in thin layers or plates on the surfaces of the crystals of pyrites, or occupies fissures in it, the gold itself being possibly coated by thin films of the mineral or of sulphur, so that contact with mercury is prevented. Certain gangue minerals may also cause losses of mercury; among these may be mentioned heavy spar (barium sulphate), talc, steatite, and similar greasy

[1] Rose, *loc. cit.*, pp. 133, 143, 144.

hydrated silicates of magnesia and alumina. The latter cause frothing and form a coating on the gold which prevents amalgamation. Occasionally, the gold itself, although apparently free, is coated with a thin film, consisting of sulphur, oxide of iron, silica, arsenic, metallic sulphides, etc., which prevents contact with the mercury surface. Such gold is said to be "rusty," and is not recoverable by amalgamation unless means be taken to remove the film.

When the ore or the water used in the battery contain any base metal compound or any acid capable of acting on copper, yellow, brown or green stains are observed on the surface of the plates, which interfere with amalgamation, and if not removed, will increase until all contact between the gold and mercury is prevented. These consist usually of carbonates, oxides, and sulphides of copper, and perhaps in some cases of mercurial compounds.

From the present point of view, the matters of chief interest are those practices which affect the subsequent cyanide treatment of the tailings, and some reference must therefore be made to the chemicals used to promote amalgamation. It has already been mentioned that lime is frequently added to the ore in the mortar-boxes, or to the battery water. This neutralizes any acid substances, such as soluble salts of iron, which may be present, and which might tarnish the plates, and also helps to dissolve any oily matter introduced into the ore or feed water from the grease used in lubricating the crushing machinery. Any kind of grease has a very deleterious effect on amalgamation. The addition of lime, moreover, causes flocculation and settlement of the fine suspended ore particles, thus bringing them in contact with the amalgamated surfaces. This treatment renders the material more suitable for cyaniding, both from a mechanical and from a chemical point of view, but in certain cases it may render the amalgamation so effective that the values remaining in the tailings are reduced below the point at which cyanide treatment is profitable. (See Journ. Chem., Met. and Min. Soc. of South Africa, II, 87.) (Proc. 2, 657.)

In order to remove the stains on the plates to which reference is made above, and to produce a bright clean surface for amalgamation, certain chemicals are used, the chief being *sal ammoniac* and *cyanide*. Both of these are solvents of the copper compounds which cause the discoloration. When they are to be applied, the battery is stopped and the spots scrubbed with the required solution. The chemical is then washed off, and generally a fresh quantity of mercury is added before allowing the pulp again to flow over the plate. It is obvious that the use of cyanide for this purpose is open to very serious objections. It is applied in the form of a fairly strong solution, perhaps 0.5 per cent. KCy, and even though the mill is stopped during the dressing of the plates, some part of the solution used must almost inevitably find its way into the launder which conveys the tailings from the mill, where it will dissolve some gold. Unless the whole of the battery water is returned to the mill, a part of these dissolved values will be lost. If used at all, the plates should be carefully washed with fresh water immediately after the cyanide has been applied, and such water should be run into a separate receiver.[1] The practice of adding cyanide in the mortar-boxes is, of course, still more objectionable, and could only be defended where complete arrangements are made for crushing with cyanide solution, which involves passing all the liquid used in the mill through the precipitation boxes. Where no lime is used in the battery and the water contains acid iron salts, these would probably destroy in most cases any excess of cyanide used in cleaning the plates before any appreciable amount of gold had been dissolved from the tailings.

Cyanide is also to some extent a solvent of oily and fatty matters, but probably only by virtue of any free alkali it may contain. For this purpose a solution of caustic soda or of sodium carbonate is much more effective; and as already mentioned, lime acts beneficially in this way.

Clean floured mercury may be collected and recovered by agitation with a large mass of fresh mercury, especially if a little alkali be added to dissolve any film of grease which may be present. The sickening of mercury may often be remedied by the addition of sodium amalgam, which contains 2 to 3 per cent. of metallic sodium. The action of this substance is to reduce the metallic oxides, forming caustic soda, which at once dissolves in the water and is incidentally beneficial as a solvent of grease and some other impurities. The base metal is liberated in the metallic form and

[1] See paper by A. von Gernet, "Losses of Gold in Mill Water," in "Proc. Chem., Met. and Min. Soc. of South Africa," II, 529.

usually amalgamates with the mercury. In the case of antimony, however, the sulphide of mercury present in the film is attacked, forming sodium sulphide, which acts upon a further quantity of the antimonial mineral, forming more antimony amalgam and liberating sulphureted hydrogen, so that in this case the use of sodium amalgam probably does more harm than good.

Hydrochloric, or still better, nitric acid, will dissolve some of the impurities in sickened mercury; the latter readily dissolves some of the mercury itself, exposing a bright metallic surface, so that the particles will coalesce on agitation. Nitrate of mercury also removes the stains on the plates, forming metallic mercury and copper nitrate.

The foreign substances introduced during the crushing and amalgamating processes, and which may influence the results of subsequent cyanide treatment, may be here enumerated:

(1) Metallic mercury, due to the use of excessive quantities in dressing the plates, or to flouring or sickening, as described above.

(2) Gold and silver amalgam, scoured off the plates by coarse particles of ore.

(3) Base metals, originally present in the mercury and ultimately converted into oxides or alloys, coating the sickened mercury.

(4) Lime, in solution in the water carrying the pulp from the battery.

(5) Oil and grease from the bearings of machinery, from feed water, or from mine candles.

(6) Fragments of metallic iron from the shoes, dies, and other parts of the crushing machinery.

(7) Particles of brass and copper from plates, screens, and the percussion caps used in blasting.

(8) Chips of wood, rubber, leather, and other miscellaneous materials which may accidentally fall into the stream of battery pulp.

(B) Concentration in Relation to Cyanide Treatment

General Principles of Concentration

Concentration may be defined as any process by which a mass of material is separated into two portions, one containing a greater proportion by weight of the valuable constituents than the other. This separation is generally carried out by utilizing some difference in the physical nature of the particles which constitute the mass, with a consequent difference in behavior when subjected to the same forces. The methods of hydraulic separation by means of spitzkasten and spitzlutten have already been discussed. In these methods the forces involved are principally gravity, inertia, resistance to hydrostatic pressure, and (in the case of the finer particles) viscosity and probably electrostatic repulsion. Concentration by machines of this class must always be imperfect from a metallurgical point of view, since, as pointed out above (Section IV, A), the product obtained consists of particles differing in size and specific gravity, small particles of heavy material being collected together with large particles of light material; whereas an ideally perfect concentrator would separate the whole of whatever mineral or substance it was designed to save, and the product yielded by it would therefore consist entirely of particles having the same specific gravity. No machine actually fulfils this condition. What is aimed at in practice is the separation of the material to be dealt with into two products:

(a) *Concentrates*, consisting of a relatively small total mass, containing a relatively large part of the total valuable constituents originally present, and a relatively small part of the worthless ingredients.

(b) *Tailings*, having a relatively large total mass containing as little as possible of the original valuable contents.

The forces which are utilized in the ordinary systems of concentration, and which effect their purpose owing to the difference of their action on different kinds of particles, are: (1) gravity; (2) inertia and centrifugal force; (3) adhesion between surfaces of concentrating machine and the class of particles to be separated; (4) Adhesion between the fluid used in the apparatus and the particles to be separated.

When applied to gold and silver ores, concentration as a rule follows amalgamation. The pulp, or crushed ore suspended in water, as it leaves the plates, passes to the concentrating machines, which are of three or four principal types.

Old Types of Concentrator

The oldest and most primitive types of concentrator depend for their action solely on the fact that in a flowing stream the heavier particles carried by the water tend to settle and collect against any obstacle or irregularity in the bed of the stream. This is the principle of the "Long Tom" and other forms of riffled sluices, in which the material is washed down an inclined trough or launder, either stationary or subjected to a rocking motion, the coarser particles of gold and the heavier minerals being caught against strips of wood (riffles) nailed at intervals across the bed of the launder. Another primitive device is to allow the pulp to flow over a sloping table covered with blankets or some kind of rough cloth, the heavy particles which settle to the bottom of the stream being caught and retained by the hairs and interstices of the cloth. These cloths are removed at intervals and the heavy material, known as "blanketings," which has collected on them is brushed or washed off, by hand or mechanically, into any suitable receptacle.

In many concentrators some device is used for keeping the pulp in a state of agitation, so that the lighter particles may be thrown into suspension and carried off by the flowing stream, while the heavier particles are moved only a short distance along the bed of the machine. In the various types of "buddles" the pulp flows over a stationary inclined bed, on which it is stirred by means of revolving arms carrying brushes or similar contrivances; or the same effect is produced by revolving the inclined bed and using stationary brushes. These machines generally have a conical form; in the convex buddle the pulp is fed at the apex and the tailings discharged into a launder at the circumference; in the concave buddle the feed is at the circumference and the tailings flow away through an opening in the center. The heavy mineral collects near the feed and the lighter particles are carried off as tailings; but in all such devices a large quantity of ore is spread out over the surface between the feed and discharge, consisting of a mixture of different kinds of particles, gradually diminishing in richness from the feed to the discharge. To obtain a satisfactory concentration, this "middle product" must be collected from time to time and again fed onto the machine, necessitating repeated handling of the same material, so that the capacity of the concentrator is very small in comparison with the labor and power required to work it.

Modern Types of Concentrator

Modern concentrating machinery, as applied to gold ores, apart from the pointed boxes already described, is almost entirely of two types: percussion tables, and endless-belt tables.

(a) *Percussion Tables.* — In these machines the material to be concentrated is fed onto a sloping table having a smooth surface of wood, iron, sheet copper, linoleum, etc., together with sufficient water to make a thin pulp, some device being often used for securing a uniform distribution and regular feed of ore and water onto the table. The ore, thus spread out in a thin layer, is kept in a constant state of agitation by a series of blows delivered against one end of the table by a set of cams or some similar contrivance. These blows cause the heavier particles which have settled out of the flowing pulp to travel gradually toward one end of the table, where they are collected in a box or other receptacle; the lighter particles are thrown into suspension and carried off by the stream of water, either to the opposite end of the table or over the side into a launder, by which they are discharged. In some types of percussion table, the ore spreads itself out into a fan-shaped layer in which the different minerals are sorted according to their specific gravity; the heaviest particles form a line at the extreme edge and may be collected separately; then follow in succession bands of (say) galena, lead carbonate, pyrites, blende, black oxide of iron, coarse and fine sand, and finally slime. The middle products, if necessary, may be collected in a separate launder leading to an elevator, which returns them to the table for reconcentration.

Among the older types of machine based on the percussion principle may be mentioned the Gilt Edge concentrator; among modern improved types are the Wilfley, Woodbury, James, and many other concentrating tables of similar design. Through the courtesy of the Wilfley Mining Machinery Co., the following brief description of this well-known machine is given.

The latest form (No. 5, 1906) of the *Wilfley Concentrator* consists of a flat deck or table mounted on rockers and supported by cast-iron plates bolted rigidly to a timber frame. The movement is imparted at the upper end of the table by means of a toggle operated by an eccentric shaft, the length of stroke being adjusted by turning a hand wheel, which raises and lowers a wedge block at one end of the toggle, varying the stroke from ½ to 1 in. The shaft makes about 240 r.p.m. When in use the table is set so as to be slightly inclined from the feed to the discharge side. It is generally constructed of redwood, in narrow strips laid diagonally to avoid warping, and strengthened by steel stringers extending the full length of the table, to obviate any bending of the table surface. It is covered with linoleum, over a part of which are tacked parallel strips of wood, forming riffles. This kind of surface readily holds the ore particles, and is also durable, impervious to water, and easily replaced. (See Figs. 25 and 26.)

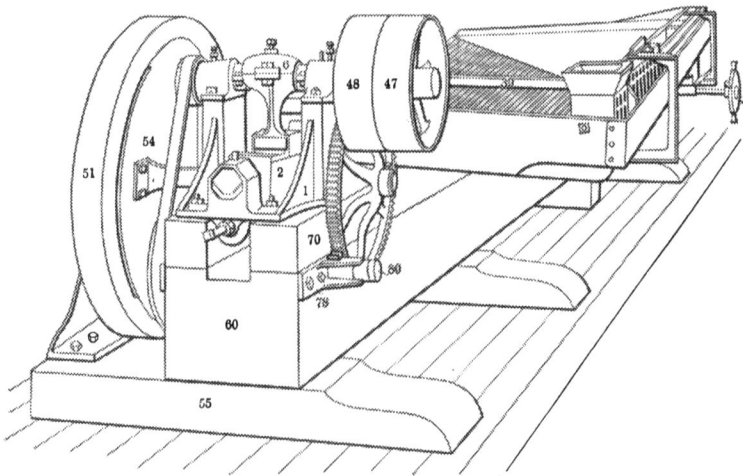

FIG. 25.—Wilfley Concentrator. (Showing driving gear. From cut furnished by the Wilfley Mining Machinery Co.)

The pulp is fed onto the table through holes in the upper side of the pulp-box, placed near the upper corner on the feed side of the table. Water is delivered by an open, perforated box extending the whole length of the feed side, below the pulp-box. The tailings are discharged over the edge of the table on the discharge side (opposite the feed) into a launder. The concentrates pass off at the lower end of the table (opposite the driving gear). Any required portion of the tailings may be returned as middlings, by a separate launder, to an elevating wheel connected with the same eccentric shaft which moves the table, and passed back to the feed box. This arrangement prevents losses of mineral which might otherwise occur through irregular feeding of pulp or wash-water. The bank of mineral formed by the return elevator tends to prevent a rush of material toward the tailings launder. The finest slimes are discharged near the head of the table; the remaining minerals arrange themselves in the order

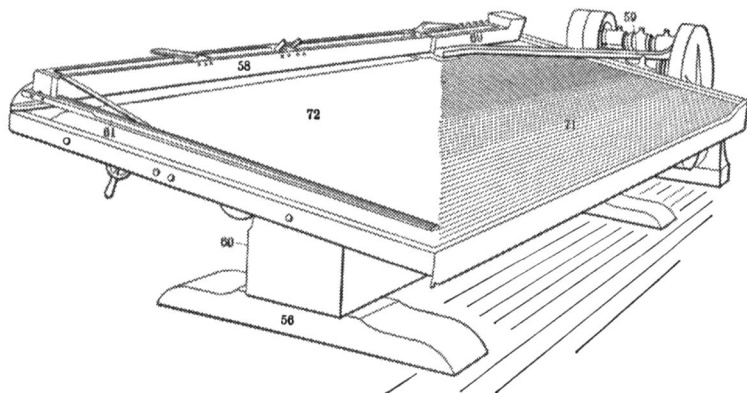

FIG. 26. — Wilfley Concentrator. General view showing distributing box and riffles. [From cut furnished by the Wilfley Mining Machinery Co.]

of their specific gravity, the heaviest passing off at the foot of the table. It is sometimes desirable to reconcentrate the middlings after further crushing, on a separate table, instead of using an elevator as above; in this case the feed to the table should be regular.

(b) *Belt Tables.* — Belt concentrators usually consist of an endless, revolving, rubber belt, flanged at the edges, which travels slowly over a couple of rollers, placed in such a position that the upper part of the belt forms a slightly inclined surface, having an area of about 12 × 4 feet. At the same time a slight but rapid shaking motion is imparted to the machine, either laterally or longitudinally. Ore and water are distributed as uniformly as possible across the width of the traveling surface, near its upper end, and the belt is caused to revolve in the opposite direction to the flow of the pulp; that is, it travels up the incline while passing above the rollers. By means of another pair of rollers placed beneath the machine, the belt on its return journey is made to pass through a tank or trough containing water, which serves to collect the concentrates. The heavy particles of mineral fall to the bottom of the pulp stream and cling to the surface of the rubber; as the belt ascends, it carries them past the pulp distributor, behind which a perforated pipe delivers a number of small jets of water. These serve to clean the concentrates still further by washing away the lighter particles, while the heavier material is carried past them and delivered to the collecting box underneath. The light portions are carried by the stream of water down the slope of the belt and discharged into a launder at the lower end. To this type of machine belong the Frue Vanner (Fig. 27), the Embrey Concentrator, the Lührig Vanner, etc. The lat-

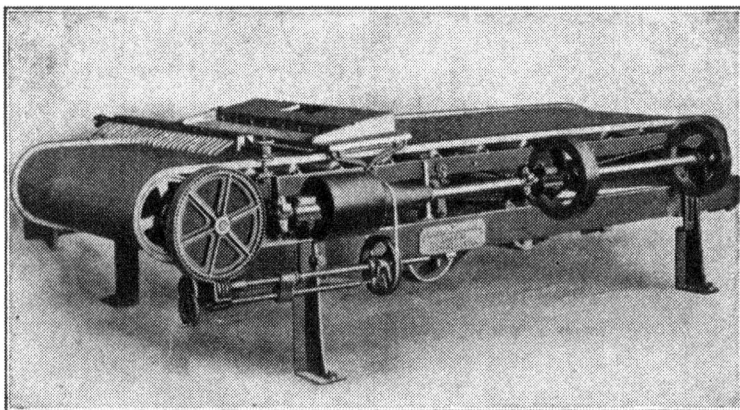

FIG. 27. — Frue Vanner, supplied by Fraser and Chalmers.

ter delivers a middle product at the side of the belt, and receives a series of blows from a cam-shaft, in these respects resembling the Wilfley table. (See above.)

Other Systems of Concentration

Other types of concentrator are used in special circumstances such as jigs, shaking screens, dry blowers, centrifugal separators, etc.; but these need not be described here. Some reference, however, may be made to the recently introduced methods of "oil concentration" and "flotation," in which gravity plays no part whatever.

In the Elmore oil concentration process, the mixture of crushed ore and water is fed into a slightly inclined revolving cylinder together with a considerable quantity of crude mineral oil. Owing to some physical action at present imperfectly understood, certain minerals, particularly those having a bright "metallic" surface, adhere to the oil.   On discharging the contents of the cylinder into tanks containing water, the oil floats to the top, carrying with it the adhering minerals, which are then separated from it by means of a rapidly revolving centrifugal extractor. By this means gold, galena, copper, and iron pyrites and other lustrous minerals may be separated from earthy carbonates, sulphates, oxides, etc., and from sand and siliceous minerals generally.

The "flotation" process utilizes the surface tension of water as a means of separating the same class of minerals, namely, those having a lustrous surface.   The pulp is distributed by means of a revolving cylinder, rifled internally, so that it falls at a certain angle on the surface of water.   Those mineral particles which under the circumstances are not "wetted," float on the surface and are carried to a receptacle where they are caused to sink by agitation.   The remainder of the pulp sinks at once in the collecting tanks and is discharged from the bottom.

Concentration Previous to Cyanide Treatment

In the early days of the cyanide process it was thought that the solvent could not be successfully applied to material containing coarse gold or refractory minerals having considerable chemical action on cyanide; hence the system was generally adopted only for tailings after concentration.   The metallurgical scheme commonly carried out in dealing with gold ores was as follows: Crushing by rock-breakers and stamps; amalgamation of coarse gold in the battery; concentration of battery tailings, generally by means of Frue Vanners (the object being to obtain the maximum gold value with the minimum of other substances); cyanide treatment of the tailings from the concentrators in cases where the values still remaining in them were sufficient to pay the cost. The concentrates were generally roasted in reverberatory or other furnaces and treated by chlorination, yielding a high percentage of extraction, but at such a heavy cost that the operation could only be conducted in large establishments treating the accumulated product from a number of mines.

This system has gradually been replaced by the method of

hydraulic classification described above, in which the tailings after amalgamation undergo only a rough concentration in spitz-lutten, the object being not to obtain a close saving of the values by concentration, but to obtain a product carrying the bulk of the coarser and heavier mineral particles, so that this product may receive special treatment, leaving the main portion of the pulp as fine sand suitable for ordinary treatment. Roughly speaking, the pulp leaving the battery is finally obtained in three products: (1) a comparatively small amount of coarse heavy sand, requiring a long time of treatment with rather strong solutions; (2) a much larger amount of fine sand, amenable to cyanide treatment with weak solutions, not over 0.25 per cent. KCN, and (say), 3 to 4 days' contact; (3) slime, capable of treatment with very weak solution by some system of agitation, followed by settlement and decantation, or by filter-pressing.

Under ordinary circumstances this more recent system is found to yield a greater profit, though probably a lower percentage of extraction than the earlier method. It also has the advantage that the whole of the gold recovery can be made in a metallurgical plant under direct control of the management at each mine, none of the products being necessarily sold to smelters or reduction works.[1]

A still more modern system, largely adopted in Western Australia and coming into general use in other fields, consists in reducing the whole of the ore to such a fine state of division that it can be treated (with or without amalgamation and hydraulic classification) by agitation with cyanide solution, the gold-bearing liquor being finally separated by filter-pressing. This system has only been rendered possible by the introduction of cheap devices for fine grinding, such as the tube mill already described. (See Section II, D.)

Cyanide Treatment of Concentrates

Where the heavy mineral consists almost entirely of clean unoxidized iron pyrites, the concentrates, even when carrying a large percentage of sulphides, can generally be treated successfully by cyanide, but a long period of treatment, sometimes amounting to three or four weeks, is necessary, and some arti-

[1] See also Rose, *loc. cit.*, pp. 221-222.

ficial means must generally be adopted for securing the amount of oxygen needed for the reaction. The concentrates must be turned over or discharged from one tank to another from time to time during the treatment, and it is sometimes advantageous to aerate the solutions. In special cases oxidizing agents or other chemicals may be added as an aid to solution. In cases where the concentrates have become partially oxidized by exposure, difficulties arise owing to the action of soluble sulphates of iron and of insoluble basic sulphates on the cyanide. These difficulties may be largely overcome by giving a preliminary treatment with a mineral acid (H_2SO_4, or HCl).[1] The reactions involved are discussed in the section of this work dealing with the chemistry of the process.

The treatment of spitzlutte concentrates presents less difficulty, both because the percentage of refractory material is smaller and because the coarse sand renders the mass readily leachable; moreover, since the material is collected by settlement from a large bulk of water, all soluble salts will have been removed before the cyanide is applied, and partial oxidation of the pyrites to basic sulphates is not likely to occur unless the treatment of the product be unduly delayed.

(C) Recent Developments of Amalgamation in Conjunction with Cyanide Treatment

The modifications in the crushing process involved in the use of tube-mills have led to various changes in the methods of amalgamation. In the first place it has been found that the finer grinding of the ore renders a further quantity of gold accessible to amalgamation, and the practice of passing the discharge from the tube-mills over amalgamated plates has been extensively adopted. W. R. Dowling (Journ. Chem. Met. and Min. Soc. of South Africa, March, 1911, p. 414) states that the recovery of gold by amalgamation in the Transvaal has been raised since 1904 to 65.7 per cent., of which some 10 to 15 per cent. is caught on the tube-mill plates.

This has suggested the advisability of doing away with the battery plates altogether, and the following reasons are advanced in favor of this practice:

(1) Since the battery is now used only for comparatively

[1] See " Proc. Chem., Met. and Min. Soc. of South Africa," I, 98–103, 133–134.

coarse crushing, there is much scouring action and consequent loss of amalgam on the battery plates.

(2) As all coarse particles are returned to the tube-mill from the classifiers and reground, the whole of the amalgamable gold is eventually brought into a condition to be readily recovered on the tube-mill plates.

(3) The amalgamating process may be kept quite separate from the crushing department, thus insuring better attention in both departments, and allowing closer supervision with reduced risk of theft.

In many plants, as for example in some of the leading American and Mexican mills, amalgamation has been eliminated altogether. Fine crushing and careful classification are relied upon to render the whole of the value accessible to cyanide. As pointed out by Dowling (*loc. cit.*, p. 417), the elimination of amalgamated plates for the final pulp can only be considered safe where good tube-mill classification obtains and an equally efficient cyanide treatment follows. Any choking of the classifiers causes an overflow of coarse sand along with the slime going to the agitation plant, and a loss of gold which might have been recovered by amalgamation.

In some cases shaking plates have been introduced, but it is doubtful if these have any advantage over stationary plates as regards recovery, and they of course involve additional cost for power.

PART IV

THE DISSOLVING PROCESS

The extraction of gold and silver from a mass of pulverized material by means of cyanide or other solvent may be brought about in one of two ways: (1) By causing the solution to pass, either by gravity or by hydraulic or atmospheric pressure, through a stationary body of the material to be treated. (2) By agitating solvent and material together by suitable appliances, so as to bring them into intimate contact, and then separating them, either by settlement and decantation or by filtration.

The term "percolation" is used simply to signify the passage of a liquid through porous material, whether anything is dissolved by the liquid or not. The term "leaching" means the extraction of soluble matter from a mass of material by the passage of a liquid through it. The two terms, however, are often used in technical literature as though they were interchangeable. The words "vat" and "tank" also are commonly used as interchangeable, although some attempts have been made to restrict the use of the word "tank" to vessels holding liquids.

SECTION I

PERCOLATION

(A) Leaching Vats

In modern practice, percolation is almost invariably carried out in circular vats of wood or iron, having a flat bottom on which rests a filter-frame, composed of thin parallel strips of wood covered with canvas or other suitable cloth to form a filter. In the early days of the process, square wooden vats were commonly used, but these presented many drawbacks, both in construction and use, and usually caused heavy losses by leakage. Cement-lined masonry tanks have also been used occasionally. The vats should rest on a solid foundation, supported above the ground on masonry piers, so as to allow a free passage underneath. Wooden vats are constructed of a number of upright staves, beveled at the edges, so that when fitted together and tightened by the hoops they form a water-tight structure. Each of the staves has a notch or "check," a few inches from the bottom, to receive the floor or bottom of the tank. The latter is constructed of planks with tongues and grooves, so that they may be tightly joined, and is cut in the form of a circle of the required diameter, allowing for the check in the staves. The hoops are made from lengths of round iron, the ends being tightly screwed together.[1] Steel vats are constructed of plates $\frac{3}{16}$ in. to $\frac{3}{8}$ in. in thickness, riveted together. Those used for the sides of the tanks are bent to the required curve by passing them through rolls. The vats are strengthened by angle-iron rings at top and bottom.[1] When carefully constructed there is little tendency to leak. Small leaks may be stopped with fine slime, paraffin wax, or vaseline.

Either type of vat has its advantages and disadvantages.

[1] For details of construction see Julian and Smart, "Cyaniding Gold and Silver Ores," 2d edition., Ch. XXXIV and XXXV.

Wooden vats are more liable to leakage, and a certain quantity of solution containing gold and silver values may be absorbed by the wood. Iron vats, on the other hand, are more expensive,

FIG. 28. — General view of New Kleinfontein Tailings Plant, showing double treatment vats. [From illustration furnished by the Cyanide Plant Supply Co.]

both in first cost and maintenance; leaks, when they do occur, are more difficult to stop, especially at the bottom of the tank, after the filter has been laid. They are also liable to distortion under internal pressure when the ore in the vat is irregularly distributed. This causes cracks through the sand charge and consequently bad percolation, as the liquid forms channels. The latter defect can be remedied by using sufficiently heavy material and bracing with angle-iron. The accompanying illustration (Fig. 28) shows a large South African Plant with double treatment tanks constructed of steel.

With regard to the absorption of solution by wooden vats, experiments by F. L. Bosqui[1] and others would seem to show that this is not a very serious matter. Somewhat greater absorption appears to take place when the immersion in solution is intermittent than when continuous. Thus a piece of redwood having an area of 180 sq. in., immersed continuously for 3 weeks in a gold cyanide solution, absorbed gold to the value of $1.07; whereas a similar piece of wood of the same area, treated for the same length of time, but immersed for 16 hours and dried in the sun for 8 hours each day, absorbed gold to the value of $2.45 per ton of wood. Tests on a screen frame of rough pine, which had been in contact with solution for 12 months, gave, however, the following result:

Bottom of frame, continuously immersed: Gold absorbed, $5.32 per ton of wood.

Top of frame, alternately exposed and immersed, about half time under solution: Gold absorbed, $2.87 per ton of wood.

S. H. Williams[2] states that a launder through which 63,000 tons of auriferous solution carrying 0.07 per cent. cyanide had passed was burned, and that the ashes were found to contain 2.766 oz. gold and 32.464 oz. silver.

Filter-frames. — These are formed by a number of parallel wooden slats 6 to 9 in. apart, above which, and at right angles to them, are nailed wooden strips of, say, 1 sq. in. section, laid 1 in. apart. Large frames are made in sections, and round holes are cut over the discharge doors.

Filter-cloths. — These are generally of cocoanut matting, sometimes covered with an upper cloth of duck or canvas or, better, hessian. The edges of the cloth are tucked tightly between the sides of the vat and the filter-frame, and the joint is sometimes made water-tight by wedging a rope into this space.

[1] "Min. Sc. Press," April 14, 1906.
[2] "Trans. I. M. M.," Bul. No. 11, Aug. 17, 1905.

Parallel strips of wood are sometimes laid above the cloth and at right angles to the strips below, to protect the cloth during the discharging of the vat. These are fastened lightly, so as to be easily removed.

Discharge Doors. — Some description of these is given below. Holes for the doors are cut in symmetrical positions in the bottom of the tank and through the filter-frame and cloth, and are so arranged that the discharge of the tank can be carried out with the minimum of labor.[1]

Solution Outlet. — This is placed in any convenient position below the filter-frame, and is furnished with a cock and a pipe leading to the precipitation department.

Dimensions of Vats. — Tanks have been constructed to hold any quantity up to 600 tons of tailings, and may be as much as 60 ft. in diameter. There is no difficulty in constructing larger ones if required. The depth is generally considerably less than the diameter. Ordinary sizes are 20 to 30 ft. in diameter, and 8 to 14 ft. inside depth. Tanks for storing solution are commonly made deeper in proportion to diameter than leaching tanks, as in the latter the efficiency of percolation has to be considered.

(B) THE LEACHING PROCESS

Conditions for Effective Percolation. — As already pointed out, it is very desirable to separate the slime as much as possible from the material which is to be treated by percolation; and better results are obtained when the grains of sand are more or less of the same size than when all sizes are mixed together. (See Part III.) To ensure good results it is necessary to lay the material uniformly throughout the vat, without undue compression, and finally to level the surface by raking it over before introducing cyanide solution. By this means the solution percolates evenly through all parts without forming channels. The tank is commonly filled to within an inch or so of the top, but the material settles considerably when moistened with solution. The valves at the outlet pipes are closed until all bubbling (due to displacement of air) has ceased; this ensures a better aeration of the charge than if the valves are left open in order that the solution sinking into the mass may drive the air out through the exit pipe.

[1] Julian and Smart, *loc. cit.*, p. 297.

In certain cases, as in the treatment of accumulated sand which has not undergone hydraulic separation, or where sufficient lime has not been added in the battery or elsewhere to neutralize the latent acidity of the material, it is necessary to add lime to the charge as the tank is filled. This is sometimes spread in a layer over the surface of the tank, and raked over before adding the first wash of solution.

Preliminary Water-wash. — This is occasionally given for the purpose of removing soluble cyanicides, that is, substances which would consume cyanide during the treatment. It is generally unnecessary when the sands have been obtained by hydraulic separation or have settled from water in collecting tanks. When used for the purpose of removing soluble salts of iron, copper, etc., it is best not to add lime or other alkali to the charge or to the water used in washing, as these salts would be wholly or partially precipitated by the alkali.

Alkali Wash. — It is a common practice to give a preliminary treatment with an alkaline solution, commonly of lime, the chief object being to secure the neutralization of insoluble cyanicides, such as basic ferric sulphate, before adding the strong cyanide solution, and thus avoid any undue consumption of cyanide in the latter. In practice this amounts to a treatment with very weak cyanide solution, as a certain amount of cyanide from previous charges finds its way into the water used for the alkali wash, for which a special storage tank is commonly reserved. The alkali wash is pumped on and allowed to leach through continuously until the effluent is distinctly alkaline. The first portions of liquid passing through are sometimes run to waste, but as they commonly contain gold, it is better to allow the whole of the effluent to return to the alkali storage tank until it shows sufficient strength in cyanide for effective precipitation; then it is diverted to the precipitation boxes, the small quantity of cyanide thus introduced into the alkali wash being no serious disadvantage. The tank is allowed to drain as much as possible, so as to avoid undue dilution of the strong solution which is afterward applied. Occasionally the strong solution is preceded by a weak cyanide wash, also with the object of diminishing the dilution which would otherwise take place in the strength of the strong solution; but the more usual practice is to add the latter as soon as the alkali wash has sufficiently drained off, it being then assumed that latent acidity in every part of the charge has been properly neutralized.

Strong Solution. — Before adding the strong solution the outlet cock is closed. The solution is then pumped on to the vat from the strong solution storage tank, in which it has previously been made up to the required strength in cyanide, usually by adding the calculated amount of liquor from the "dissolving tank," a small vat placed above the storage tanks and used for dissolving the solid cyanide from the cases. In some plants the cyanide is dissolved directly over the leaching tanks, by placing the solid lumps in a perforated box into which weak solution from the storage tanks is pumped. This method is not to be recommended, as it must lead to the treatment of some parts of the charge with unnecessarily strong solution and consequent waste of cyanide. The solution is pumped on until the charge remains well covered with liquid to a depth of two or three inches. It is then allowed to stand, frequently for twenty-four hours or longer, so as to allow the charge to soak thoroughly in strong solution. It is a better plan, however, to draw off the solution as soon as all bubbling has ceased, say after one or two hours, and allow it to run off until the effluent shows a sufficient strength in cyanide for effective dissolution of the values; the outlet valve is then closed and more strong solution is pumped on, until the charge is again covered to the required depth. The charge is now left standing under strong solution as long as may be considered necessary. The object of this procedure is to ensure the saturation of the whole charge with solution of sufficient strength, since but little diffusion takes place in the leaching tank, one solution displacing another without mingling with it to any considerable extent.

The strength of solution to be used depends on the nature of the material to be treated. Roughly speaking, the richer it is in gold, the stronger will be the cyanide solution required, but this rule is by no means universal. Where silver is to be extracted, much stronger solutions are commonly required than for gold alone, partly owing to the greater weight of metal which has to be dissolved and partly to the presence of accompanying minerals which consume much cyanide. This matter is discussed in the sections dealing with the chemistry of the process. Under normal conditions a strength of 0.25 per cent. KCN[1] (= 0.1 per

[1] 5 lb. potassium cyanide or 3.8 lb. sodium cyanide per ton of solution.

cent. CN) is a good standard to work with, though in the case of clean tailings free from coarse gold, a strength of 0.1 per cent. KCN (= 0.04 per cent. CN) is often sufficient to give a good extraction of the values. (See pp. 105–107.) When the charge has stood long enough under strong solution, the outlet cock is opened and the liquid allowed to drain off to the precipitation boxes; the effluent is tested for cyanide and also for alkali to ensure that it is in good condition for precipitation. This will generally be the case if the preliminary alkali wash has been properly carried out. For the treatment of silver ores and concentrates, strengths of 0.3 per cent. to 0.5 per cent. KCN (6 to 10 lb. per ton) are commonly used, and in some cases as high as 1 per cent.

Weak Solution. — When the strong solution has sufficiently drained off, weak solution is pumped on from the storage tanks. It is often desirable to allow the charge to remain exposed to the air for some time before pumping on weak solution, and occasionally the surface is raked over, so as to allow access of oxygen to the material, to enable the weak solution to dissolve any gold not extracted by the strong solution. The liquor used for weak solution is commonly that which passes from the pre-cipitation boxes, used again without further addition of cyanide; but it may in some cases be necessary to bring up its strength, say to 0.15 per cent. KCN (= 0.06 per cent. CN, *i.e.*, 3 lb. KCN or $2\frac{1}{4}$ lb. NaCN per ton of solution), by adding the required amount of stock solution from the dissolving tank.

The weak solution is generally added in a number of succes-sive washes, gradually diminishing in strength to say 0.05 per cent. KCN (= 0.02 per cent. CN); these are either drawn off im-mediately or left in contact only a short time, a fresh wash being added as soon as the previous one has sunk well below the sur-face of the charge. More effective extraction, however, is gen-erally obtained by allowing each wash to drain as completely as possible before adding the next, though this is not always prac-ticable, owing to considerations of time. Unless the sand treated is absolutely free from slime, the rate of percolation gradually diminishes, in consequence of the settling of slime particles in the interstices of the charge and on the filter-cloth. The main object of the weak solution treatment is to displace the strong solution and thus remove from the charge the values which have already been dissolved.

The amount of weak solution necessary varies greatly according to circumstances; the total volume used is generally considerably greater than that of the strong solution, and the time occupied is also greater. In some plants a medium solution is used after the strong solution, a special storage tank being employed for this.

Final Water-wash. — As the liquid with which the charge is saturated after the last weak wash has been drawn off still contains gold in solution, it is the custom where practicable to displace this by addition of water. It will be obvious, however, since the liquid so displaced is added to the general stock of solution in the plant, that the amount of water which can be used in this way is very limited. The amount of solution in stock is a fixed quantity, depending upon the storage accommodation. The amount of final water-wash, therefore, cannot exceed the difference between the moisture originally present in the charge and that in the tailings as discharged, except in cases where a part of the solution is run to waste or where losses occur by evaporation. In practice, the final wash is generally given with the weakest cyanide solution available.

Quantities of Solution Used in Percolation. — No definite rule can be given as to the quantities of each class of solution that should be used, nor as to the time occupied by each in contact and leaching. These considerations depend partly on the arrangement of the plant and partly on the nature of the material treated. In an ordinary case, about $1\frac{1}{2}$ tons of liquid will pass through the vat for every ton of sand (dry weight) which it contains; say $\frac{1}{3}$ ton of strong solution, 1 ton of weak solution, and $\frac{1}{6}$ ton of alkali wash and final water-wash. It is convenient to have the size of the tanks such that one tank will contain the whole amount of sand produced per day. Thus, with a plant consisting of seven leaching tanks, one might be filled and one discharged every day, allowing five days for the actual cyanide treatment — two days for strong solution and three for weak. Some such arrangement as this greatly simplifies the routine of operations in the plant.

Upward Percolation. — In the case of material containing slime, which is apt to settle on the filter-cloth and impede filtration, one or more of the solutions may be introduced from below and forced upward through the charge by hydraulic pressure. When this is done slowly and carefully, the formation of channels is avoided and the filter-cloth kept free from slime. After the charge has stood covered with solution for a sufficient time, the liquid is again drawn off from below, and subsequent washes generally added from above and allowed to percolate downward.

(C) Double Treatment

The system of intermediate filling has already been described (Part III, Section III), and it has been noted that the collecting vats are sometimes used as preliminary treatment vats. Usually, the final treatment vats are placed immediately beneath the collecting vats, so that the material may be readily transferred from one to the other by bottom discharge doors. The general method is to give the alkali wash, if any, and at least a part of the strong solution, treatment in the upper vat, so that when the transfer takes place the material saturated with strong solution is freely exposed to the air. The remainder of the strong solution treatment and the final washing with weak solution and water takes place in the lower vat. In some cases, especially in the treatment of concentrates, it is found advantageous to transfer the charge a second or third time.

The following scheme of treatment, based on a method proposed by W. R. Feldtmann for treatment of tailings at the Luipaards Vlei Estate,[1] will illustrate the method:

FIRST TREATMENT: CHARGE, 165–170 TONS TAILINGS

	Tons Solution put on	Strength: KCy	Time Leaching: Hours	Remarks
Strong ...	27	0.25 per cent		Drained off
Medium ..	27	0.20 " "	66 to 70	without
Weak	27	0.15 " "		standing

Extraction by first treatment, 67 per cent.

SECOND TREATMENT

	Tons Solution put on	Strength: KCy	Time Leaching: Hours	Remarks
Medium ..	20 to 25	0.20 per cent.		In successive
Weak	75	0.15 to 0.10 " "	179	washes
Water	20 to 30	....		

[1] F. White, "Trans. I. M. M.," VII, 124 (1899).

It is stated that in some cases a difference of over 20 per cent. in the extraction has been obtained through the introduction of double treatment. The method is extensively adopted in the treatment of pyritic ores, and in cases where refractory minerals are present, as in most silver ores. (See below.)

(D) DISCHARGING OF TREATED MATERIAL (RESIDUES)

Where facilities exist for the purpose, the treated sand may be rapidly and economically discharged by sluicing, using a strong jet of water. This method is frequently adopted for discharging treated slimes. (See below.) When the first cyanide plants were erected in South Africa, the only method of discharging the vats was by shoveling over the side into trucks, this procedure involving great waste of labor, even with the shallow square tanks then in use. In 1891 the system of "bottom discharge" was introduced by Charles Butters, whereby the cost of discharging was reduced to about a quarter of that for the previous system. Tanks have also been constructed with doors in the side for discharging, but since it is evident that the material to be discharged must be thrown a greater average distance when the door is at the side than when at the bottom, the bottom doors have been almost universally adopted.

The type of discharge door first designed by Butters and still in general use consists essentially of a circular steel or cast-iron

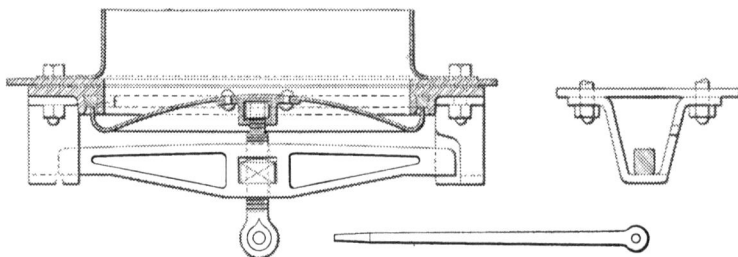

FIG. 29. — James Patent Discharge Door as furnished by the Cyanide Plant Supply Co.

disk, fitting into a flanged cast-iron tube which passes through the filter-cloth, frame, and bottom of the tank. These doors are 9 to 16 in. in diameter, and have a bolt passing through the center, carrying a butterfly-nut underneath the tank, by which the door can be fastened or unfastened as required. (See Fig. 29.)

A more recent type of door is hinged on one side and secured by a bolt and nut on the opposite side. In either case certain precautions are necessary to obtain a water-tight joint. When a tank is to be filled, the door is closed and the discharge tube (passing through the filter) is filled with sand, with a covering of clay or slime. In discharging, which is generally done by hand, the door is first unscrewed, a rod is pushed upward into the tank, and a hole is dug down to the opening. The material is then shoveled through the latter into a truck standing on rails beneath the vat. When the tanks are very deep an additional steel tube is sometimes attached to the upper edge of the discharge tube before the tank is filled, and when needed other sections can be added above this. It is then only necessary to dig down to the uppermost section before beginning to discharge. The extra tubes are removed in succession as the discharging proceeds.

Where labor is costly, various mechanical appliances have been introduced for discharging tanks. One of these, the Blaisdell vat excavator, consists of a number of revolving radial arms carrying steel disks inclined at an angle with the arms. The machine is placed above the tank to be discharged, and as it revolves the material is pushed by the disks toward the center and passes through a door to trucks or belt-conveyor beneath. A shorter pair of radial arms, also carrying disks, counteracts the tendency of the material to pile up toward the center.[1]

(E) Skimming Tailing Dumps.[2]

"After tailing has been discharged from cyanide treatment, whatever gold or silver it may contain is usually considered economically irrecoverable, but some of these valuable metals are now being reclaimed. The principle of recovery is the concentration of previously dissolved gold and silver by evaporation from surface of dumps. The remaining salt forms a white crust, so thin that it is usually recovered by lightly sweeping the surface. A product having a value of more than $100 per ton has been recovered in this way from both gold and silver tailing. The process may be repeated indefinitely by sprinkling the dump with water or running on additional tailing and waiting for the moisture to evaporate. In cases where the recovered material is to be retreated in neighboring mills, it is not necessary to take off such a thin skim of material. Recovery of about half an inch of surface material is quickly and cheaply made by shoveling. It is put back through the mill treatment. The system is used in the Oatman and Gold Road district of Arizona."

[1] For details, see Julian and Smart, *loc. cit.*, p. 192.

[2] Abstract from E. M. J. *95*, 618. March 22, 1913, "Det. of Metall. Practice."

SECTION II

AGITATION

(A) AGITATION BY MECHANICAL STIRRERS

IN the first plant erected in South Africa — the experimental plant near the Old Salisbury battery, which was in operation in the early part of 1890 — a system of agitation with paddles was used, combined with subsequent filtration of the pulp by suction. It was considered that a better contact of solution and ore, and consequently more rapid dissolution of the gold, would be obtained by the agitation system. This view was no doubt correct, but the method was very soon abandoned in favor of percolation, which required so much less expenditure of power and was far simpler in execution. Modern practice, however, shows a tendency to revert to agitation methods, and consequently percolation is a far less important factor in cyanide work than it was a few years ago. Many ores have been shown to yield a much higher percentage of their values when every portion is crushed at least fine enough to pass a 150-mesh sieve than when crushed moderately fine, say to 30 or 40 mesh, and a separation made of sands and slimes for different treatment. Ore crushed to 150-mesh cannot, strictly speaking, be regarded as slime, as it is generally possible to separate from it a considerable amount of fine sand which is perfectly leachable. It may, however, be conveniently treated by agitation, and in many cases this system would be more profitable than hydraulic separation of the sands for percolation.

The overflow from the spitzlutten or collecting tanks — in some plants the entire mill-product reduced to a sufficient fineness — generally goes first to a large spitzkasten, the overflow from which, consisting of practically clear water, is returned to the battery. The underflow from this, consisting of thickened slime, passes either direct to the agitation tanks or into special collecting tanks with conical bottoms, whence a further quantity

of water is withdrawn. The moisture in the pulp withdrawn
from the bottom of these tanks may in some cases be reduced
by this means to 50 per cent. The agitation tanks are provided
with paddles revolving about a central vertical shaft. The latter
is supported by a platform running across the centers of the vats;
at the upper end of the vertical shaft is a bevel-wheel, operated
by a horizontal countershaft. When a sufficient charge has been
collected, the agitator is set in motion, and a sample taken of the
well-mixed pulp. Lime is added, if necessary, and then sufficient
cyanide solution to give a pulp of about 4 to 5 tons liquid for every
ton of dry slime present. When the material is to be subsequently
treated by filter-pressing, a much thicker pulp than this is commonly used, say $1\frac{1}{4}$ to $1\frac{1}{2}$ tons liquid per ton of slime.

The strength of solution depends on the material to be treated,
but it is commonly much less than that required for percolation
treatment of sands. In many cases a satisfactory percentage
of the gold may be dissolved with solutions of 0.01 per cent. KCN
(= 0.004 per cent. CN, *i.e.*, 0.2 lb. KCN or 0.15 lb. NaCN per ton
of solution, or 4 to 5 times these amounts per ton of dry slime).
Slimes from pyritic or refractory ores, especially if silver or copper
minerals be present, may require much stronger solutions.

The agitation is continued until all soluble gold or silver may
be assumed to be dissolved. The time required for this also
varies greatly, from 3 to 36 or even 48 hours. In an ordinary
case, 6 hours will be sufficient. In a thick pulp the values generally dissolve more slowly than in a thin one, and the strength
of solution also affects the result. An ordinary strength for the
treatment of clean slimes is 0.05 per cent. KCN (= 0.02 per cent.
CN), but twice or three times this strength may be needed for
refractory material.

The agitator is then stopped; in some cases the stirring gear
is so arranged that the paddles can be lifted above the level to
which the settled slimes will reach. The tank is allowed to stand
without further agitation till a sufficiently complete settlement
has taken place, which may require from 9 to as much as 60
hours. As soon as a clear liquid shows above the surface of the
settling pulp, decantation is begun. This is generally carried
out by means of a pipe, jointed near the bottom of the tank and
with the upper end supported by means of a float just below the
surface of the liquid. The exit pipe passes out near the bottom

of the tank. The pipe is lowered as the settlement proceeds, so that clear solution is continuously drawn off and passes on to the precipitation plant. When filter-presses or some form of suction filter are to be used, the pulp is transferred direct from the agitation tanks, without settlement, by means of hydraulic or atmospheric pressure, or by suitable pumps, to the filtering apparatus. The apparatus known as "montejus" is frequently used for this purpose.

(B) Agitation by Circulating the Pulp and Injection of Air

A method often used in place of, or in conjunction with, mechanical stirrers is to withdraw the pulp continuously from the bottom of the tank and return it to the top by means of centrifugal or other pumps. This method has the advantage of securing a more complete mixture of the particles, especially when a little coarse material is present, than could possibly be obtained by any form of paddle agitator, and the further advantage that during the transfer the pulp is thoroughly exposed to the air and the action of the cyanide thereby accelerated.

In cases where further oxidation is considered necessary, arrangements have been made for causing the pulp as it returns to the agitation tank to fall from a hight and to be scattered in the form of spray. Another device, due to H. T. Durant, which seems to be still more effective in securing oxidation, is to place a small suction valve on the intake pipe of the centrifugal pump; this draws in air, in the form of minute bubbles intimately mixed with the pulp, though of course with some loss in the lifting power of the pump.

Sometimes agitation by injection of compressed air through a perforated pipe is also used. In this case the pipe is bent around the bottom of the vat, so as to allow air to pass through the charge at as many points as possible. In carrying out this method of agitation, the best results are obtained by using tanks that are very deep in comparison with their diameter. Tanks for this purpose erected at Pachuca, Mexico, are 15 ft. in diameter and 45 ft. high;[1] they are constructed with a conical bottom, the air being delivered through the apex of the cone under sufficient pressure to prevent settling of the pulp on the conical sur-

[1] M. R. Lamb, " Notes on Air-agitation," " Eng. and Min. Journ.," LXXXVI, 901 (Nov. 7, 1908).

face. The pressure required is from 22 to 35 lb. per sq. in.; 15 to 20 cu. ft. of air are used per charge, which it is stated gives a vigorous agitation and excellent extraction. Slime, fine sand, and concentrates are treated by this method. The apparatus used, known as the "Brown Agitator,"[1] was first introduced at Komata Reefs, New Zealand. It acts on the principle of the air lift. The vat used is a long, narrow, vertical, cylindrical vessel, with conical bottom, from 40 to 55 ft. high and 10 to 15 ft. diameter, with a central column 1 in. in diameter for every foot of vat diameter, both ends of which are open and immersed in the pulp. Air is introduced through a narrower pipe opening at the bottom of the center column, at just such a pressure as will overcome the weight of a column of slime-pulp at the point of introduction, thus establishing a circulation of the slime. The power required is $2\frac{1}{2}$ h.p. per charge of 50 tons, which would be about the slime content of a 40 × 10 ft. vat. Apparatus on this principle has been introduced at the Waihi and Waihi Grand Junction, New Zealand; also, as above mentioned, at Pachuca, Mexico.

(C) ANDREW F. CROSSE'S SLIME TREATMENT PROCESS

The slime, after collection by any suitable method, with addition of the necessary amount of lime and removal of excess of water, is washed by weak cyanide solution into a vessel consisting of two concentric steel cones, each having an angle of 45°. The inner cone has a comparatively large opening at the bottom; its sides are parallel with those of the outside cone, and its diameter about half that of the outside cone. This inside cone acts as a baffle between the pulp, which is being agitated within it, and the outside portion, which is quiescent on the surface; it is suspended on two girders resting on the top of the outside cone. An air-lift in the center of the apparatus reaches nearly to the bottom of the outside cone; the air pipe for working this passes through the bottom of the cone, at which point there is also a suitable discharge pipe and cock.

The upper end of the air-lift pump is several feet above the surface of the pulp, and the overflow runs into a cylinder having two or more horizontal discharge pipes, so arranged that the discharged pulp causes the pulp in the inside cone to revolve

[1] "Min. Sc. Press," XCV, 689.

and mix with the clear solution returning from the zinc-box. The slime-pulp, mixed with cyanide solution, is pumped or run into the inside cone, and as soon as this pulp reaches half-way up the air-lift pump, the air valve is opened, to prevent any settlement of slime. As the slime-pulp rises in the vat, the outside portion tends to settle and leaves several inches of quite clear liquid on the surface. The pulp is allowed to flow in until the cone is nearly full, then the two decanting arms are lowered, and the clear liquid drawn off and run through a zinc-box adjoining the vat at a slightly lower level. The effluent from the zinc-box is returned by a small pump to the inside cone, the whole process being thus very simple. A by-pass is provided to allow the clear decanted solution to circulate without going through the zinc-box, as the first liquid drawn off would contain no gold or silver. The time required for solution and extraction depends on the nature of the material treated. The whole process is practically automatic and one shiftsman can attend to a number of vats. After treatment, the pulp is discharged into a settling tank and the clear liquid decanted from the settled pulp. If necessary, the strong cyanide may be displaced by a weak solution by allowing the liquid leaving the zinc-box to run into a sump and running in a very weak solution at the same rate, so as to displace the stronger solution. This would be necessary in treating silver ores.

Successful experiments with this method have been made at Frankfort (near Pilgrim's Rest, Transvaal) and at Johnson & Sons' smelting works, Finsbury, London. The advantages claimed are: high percentage extraction of gold and silver; simplicity and cheapness in working; very great saving in capital outlay.

(D) Treatment of Slimes by Settlement and Decantation

This system was for several years the only one adopted on the South African gold-fields, and even to-day it is, in many parts of the world, the only method economically possible for handling very low-grade material that cannot be treated by percolation. It involves the use of large volumes of liquid, and consequently is not well adapted for countries like Western Australia, where water is very scarce.

The method consists of successive agitation with cyanide solution, settlement and decantation, the cycle of operations being repeated twice or three times, and in exceptional cases oftener. The extraction obtained by each cycle depends (1) on the proportion of the total value which goes into solution under the given conditions; (2) on the relative volumes of the clear-settled liquid and the moisture retained in the residual pulp after decantation, it being assumed that the dissolved values are distributed uniformly throughout the whole mass of liquid; (3) on the values originally present in the solution added; as the precipitation is not absolutely perfect, some gold and silver will always be present in the liquid used. Suppose, for example, the slime to carry originally 5 dwt. per ton, of which 4.5 dwt. are actually dissolved during the agitation and settlement stages of treatment, and that the cyanide solution used in the proportion of 4 tons solution to 1 of dry slime carries originally 0.6 dwt. per ton. The settled slime after decantation retains, say, 50 per cent. of moisture. The total value dissolved consists of 4.5 dwt. per ton of dry slime, distributed over 4 tons of solution, or 1.125 dwt. per ton of solution, which in addition to the 0.6 dwt. originally present gives 1.725 dwt. The total gold drawn off is therefore $3 \times 1.725 = 5.175$ dwt. per ton of dry slime treated. The value left in the residue consists of 1 ton dry slime carrying 0.5 dwt., with 1 ton solution carrying 1.725 dwt., or 2.225 dwt. per ton of dry slime. The extraction, based on original contents of dry slime, is therefore $5 - 2.225 = 2.775$ dwt., or 55.5 per cent.

Suppose now that a second treatment is given with 3 tons of solution carrying 0.4 dwt. per ton. This, with the 1 ton of solution carrying 1.725 dwt. remaining after the first decantation, gives after agitation and settlement 4 tons of solution carrying 0.731 dwt. per ton, assuming that no further dissolution of gold takes place. Then if, as before, the settled residue retains 50 per cent. of moisture, the result is 3 tons of solution drawn off, carrying $3 \times 0.731 = 2.193$ dwt., and a residue consisting of 1 ton dry slime at 0.5 dwt. and 1 ton solution at 0.731 dwt., or a total of 1.231 dwt. per ton of dry slime, bringing the total extraction by two treatments up to 75.4 per cent.

This example will serve to illustrate some of the weak points in the decantation system. It is obvious that a high extraction can only be obtained by a number of successive decantations, which involves the handling of a very large volume of liquid per ton of material treated. Ample storage room must be provided for the solutions, so that the first cost of the plant is very considerable. Good extractions, also, can only be obtained by using solutions from which the values have been very perfectly precipitated before use in the slime tanks.

The percentage of moisture in the residues may be materially reduced and the extraction correspondingly improved by the use of very large, deep tanks, generally with a pointed bottom. After one or two treatments by agitation and decantation, as already described, the pulp is transferred to the large tank. When several charges have been thus introduced, the settled material at the bottom is under very great hydrostatic pressure and may be withdrawn as a very thick pulp carrying 40 per cent. or less of moisture. There is also an additional economy in the use of these deep tanks; as the final settlement takes place in them instead of in the agitation tanks, the latter are set free for the treatment of fresh charges sooner than would otherwise be the case. This system was first suggested by Charles Butters, and has been carried out on a large scale, both on the Rand and in other districts.

By making the final settlement tank sufficiently large and introducing the pulp by means of a pipe passing some distance below the surface of the liquid contained in the tank, the process may be made continuous, the apparatus thus acting as a spitzkasten and allowing clear liquor to overflow at the periphery, while the thickened slime is constantly drawn off at the bottom. Several systems of treating slime by continuous settlement have been devised, in which this principle is utilized. One has been described by E. T. Rand.[1] After removal of superfluous water by means of spitzkasten, the thickened pulp is first agitated with cyanide solution, then pumped to a settling-vat of suitable size and form, the thickened underflow from which, mixed with fresh solution, goes to a second settling-vat. The clear solution overflow from the first settler goes direct to the precipitation box; that from the second settler goes to the launder that feeds the agitation vat, and, with addition of fresh cyanide, forms the dissolving solution. The precipitated liquor is pumped to the launder by which the pulp passes from the first to the second settler, and forms the final wash.

[1] "Proc. Chem., Met. and Min. Soc. of South Africa," II, 686 (1899).

The Clancy system [1] is very similar in general design, but the agitation tank is dispensed with and more settlers are used. None of these methods appears to have come into general use, probably owing to the difficulty of obtaining constantly a residue sufficiently low in solution contents to be economically discharged; it is not likely that the final residue would average as little as 50 per cent. moisture, a condition easily attained with the ordinary settlement and decantation process.

Continuous Agitation and Decantation. — An agitation device on much the same principle as Crosse's tank has been designed by Spaulding (Min. Sci. Press, March 1, 1913) and modified by Algernon del Mar. This consists of a cylindrical tank with conical bottom, and air-lift, as in the "Brown" or Pachuca tank, the clear liquid being decanted from an annular space at the periphery. This space is 15 in. wide and 12 ft. deep for a tank 18 ft. high and 10 ft. diameter. The air-jet and central lift are placed somewhat above the bottom. If a centrifugal pump be used for agitating, the pulp may be delivered below the surface and so that the mass has a circular motion. (Eng. and Min. Journ., *96*, 770, Oct. 25, 1913.)

Tanks on this principle may be used for a continuous system, in which the pulp and solution travel through the series of tanks in opposite directions. Thickened pulp from a Callow cone is fed to the first of the series, and diluted with the overflow from the second. The overflow from the first goes to precipitation together with overflow from the Callow cone. Barren solution is used to dilute the pulp in the last of the series and the overflow from each tank to dilute the pulp in the preceding tank.

The same method may be used with a series of Dorr thickeners, as described by J. Simmons (Eng. and Min. Journ., *95*, 627, March 22, 1913). Pulp from the mill, carrying 3 parts solution to one of ore goes to the first thickener, whence the thickened pulp of a consistency 1:1 passes to an agitator, and the overflow goes to the precipitation boxes. From the agitator the pulp passes to a second thickener, thence to a third and fourth, and finally to the filter from which it is discharged after water-washing. The overflow from the second thickener is returned to the mill, that from the third is used to dilute the pulp in the second, that from the fourth goes to the third, while the pulp in the fourth thickener is diluted by the effluent from the filter together with the barren solution

[1] " Proc. Chem., Met. and Min. Soc. of South Africa," II, 741.

leaving the precipitation boxes and as much water as may be necessary to give the required consistency. The water used as a wash in the filter serves to keep the volume of liquid in the system constant.

In Spaulding's system, introduced at the Veta Colorada Mine, Parral, Mexico, agitation is produced in flat-bottomed tanks by injecting barren solution into 4 vertical transfer pipes with elbow-joints at the top arranged so as to discharge the pulp tangentially. In the center of each vat is a cylinder of half the diameter of the vat, open at both ends, and containing baffle plates. The clear solution which settles in this cylinder is decanted to a clarifying press and thence to the zinc boxes. Thus a constant circulation of solution is maintained in each vat, while the slime pulp passes successively through a number of vats.

It is claimed that the cost of pumping in this system is much less than that of agitating with compressed air as in the Pachuca or Parral tanks. The power required for a vat 20 ft. diameter by 20 ft. deep, with four 10 in. transfer pipes is estimated at $2\frac{1}{2}$ to 5 h.p. as compared with 10 to 15 h.p. with compressed air.

Objections to Continuous Agitation. — One of the objections usually raised against continuous agitation is that all parts of the material treated do not receive the same length of treatment. Leon P. Hills (Min. Sci. Press, Feb. 8, 1913, quoted from Colorado Sch. of Mines Mag.) remarks: "There is no guarantee that the particles will stay in the apparatus long enough, or that they will be discharged when they have parted with their gold. In fact the heavier and coarser particles pass through in the shortest time, the part of the ore requiring the longest treatment getting the shortest, and vice versa."

Also the moisture in the discharged pulp is necessarily higher than that of settled pulp as discharged in the intermittent system.

Combined Agitator and Settler Suitable for Intermittent System. — An agitator recently introduced at the United Mines Plant, Tuolumne, California, is a modified Pachuca, the central fixed column being surrounded by a vertically adjustable column of a few inches greater diameter. The height of this column above the conical bottom can be regulated as desired, always keeping the upper extremity below the surface of the pulp. During settlement, the adjustable column is lowered, so that the pulp is prevented from settling at the bottom of the cone over the air inlet, and when agitation is restarted the column is raised, allowing pulp to enter into circulation without the need of extra air-pressure. A baffle is provided so that the settled liquid may be withdrawn from the circumference (as in Crosse's and Del Mar's tanks) at the same time that pulp is being run in and agitation going on in the central part of the tank. During agitation, barren solution may be run in to replace the gold-bearing solution and also to vitalize the solution in the tank. Washing in this way is repeated as often as the richness of the ore requires.

The use of the vertically adjustable protective column is patented by Leon P. Hills, P. H. McHugh and J. F. Thomasson.

(E) The Parral Tank

This is a modification of the standard " Pachuca " tank introduced by Bernard Macdonald at the Veta Colorada plant, Parral, Chihuahua, Mexico, and also used in cyanide plants at Zacatecas and Guanajuato. Mr. Macdonald (Mex. Min. Journ., March, 1912, p. 27) enumerates the following defects in the ordinary Pachuca tank:

(1) Owing to its tall, narrow shape the capacity of the tank is small in comparison with the amount of steel used in construction.

(2) The tall tank also necessitates expense in elevating the charge unless the site is very steep.

(3) The air pipe, with valve as commonly constructed, is liable to choke with pulp when air is shut off.

(4) Owing to large diameter of central lift-pipe, a high pressure and volume of air are required, hence much power is consumed.

(5) The constituents of the pulp are imperfectly mixed, the coarser grains circulating near the air-lift while the finer particles are thrown towards the outer walls of the tank.

(6) It is difficult to start agitation when compressed air has been shut off, owing to the settlement of solids in the cone.

To remedy these defects, the Parral tank has the ordinary cylindrical flat-bottomed shape, and is provided with four or more air-lifts in different parts of the tank. The standard Parral tank is 25 ft. in diameter by 15 ft. high and has a capacity of about 7670 cu. ft. of pulp, having four lift-pipes 8 in. in diameter as against one lift pipe 16 in. diameter in the standard Pachuca. Instead of admitting air from the compressor through a perforated pipe provided with a tightly fitting rubber sleeve (as is the arrangement in the standard Pachuca) the air-nozzle is provided with a ball-valve, which closes the orifice during a shut-down and is easily restarted without excessive pressure. Less air is needed, for the agitation of a given amount of pulp. It is stated that 250 tons of dry pulp were agitated in a Parral tank having four 12 in. lift-pipes with the same amount of air as was needed in a Pachuca tank to agitate 83 tons, using a 16-in. central lift-pipe.

In order to lift the pulp effectively, the air must form a disk across the whole diameter of the lift-pipe; such disks are, of course, more easily formed in a narrower pipe, and it is claimed that the construction of the ball valve, with horizontal slots for exit of air, facilitates their formation.

Another feature of the Parral tank is an elbow at the top of the lift-pipes by which the stream of pulp is directed horizontally. This sets up a rotary motion which, it is stated, extends throughout the whole charge and prevents the accumulation of pulp at the bottom, which is a constant source of trouble with most forms of Pachuca tank.

These tanks may be adapted for the continuous system, pipes being arranged so that the pulp flows by gravity from a point near the top of one tank to a point at a somewhat lower level where it enters the next tank in the series.

Owing to the rotary movement, the solids in the pulp travel a much greater distance than in the Pachuca tank with vertical circulation, under otherwise similar conditions, hence a less violent agitation is needed. It is claimed that the extraction is at least $2\frac{1}{2}$ per cent. higher than in a Pachuca tank treating the same pulp under comparable conditions.

(F) Paterson Hydraulic Agitator

This apparatus combines the principles of circulation with centrifugal pump and decantation of more or less clear solution obtained by the use of a baffle (as in Crosse's Tank. See text, p. 229).

In general construction it resembles the "Brown" or "Pachuca" tank, having a conical bottom and central circulating tube. Instead of an air-lift, however, the solution delivered by the centrifugal pump is injected at the bottom of the circulating tube and directed upward, forming a hydraulic elevator. The upper part of the tank is provided with an annular baffle extending several feet below the surface, the upper part of the baffle being flared inwards to prevent splashing. This produces an outer calm zone, from which nearly clear solution passes to the intake of the circulating pump, to be returned as described, in a jet at the base of the lift-tube. At the top of the lift-tube is a circular distributor which discharges the pulp in a radial sheet, thus securing aëration. (See Fig. 29a, p. 238.)

Among the advantages over air-lift tanks claimed for the Paterson system are lower costs for construction and maintenance owing to substitution of centrifugal pumps for air-compressors, and lower power consumption, lower cyanide consumption owing to avoidance of excessive aëration, absence of foaming, simplicity of construction and ease of operating.

These tanks are supplied in various sizes, from 5 ft. diameter by 10 ft. high, with a capacity of 1.5 tons dry ore to 15 ft. diameter by 45 ft. high, with a capacity of 50 tons. The power required is from 1 to 4 b.h.p.

The apparatus is adapted for the continuous process, a number of tanks being connected in series by a system of discharge pipes and valves, the pulp flowing successively through each tank and being finally discharged near the top of the last. The pulp in each tank is maintained at a slightly lower level than in the preceding one, so as to cause a flow equal to that from the mill.

The tank can also be used for treatment by decantation and settlement, using barren solution or water as required, this being delivered by suitable connections through the agitation pump. After settlement the clear solution is decanted. In order to stir up settled pulp, the top of the center lift is closed by a damper; the action of the pump then forces the solution out through the bottom of the lift and stirs up any settled material.

(G) THE TRENT AGITATOR

This appliance, invented by L. C. Trent of Los Angeles, consists of a flat-bottomed vat into which solution is introduced under pressure into a distributor placed at the bottom of the vat. This distributor is mounted on ball-bearings and carries radial arms with bent nozzles, so that the issuing pulp causes the mechanism to revolve as in the Butters and Mein distributor. The solution is delivered by a centrifugal pump with sufficient force

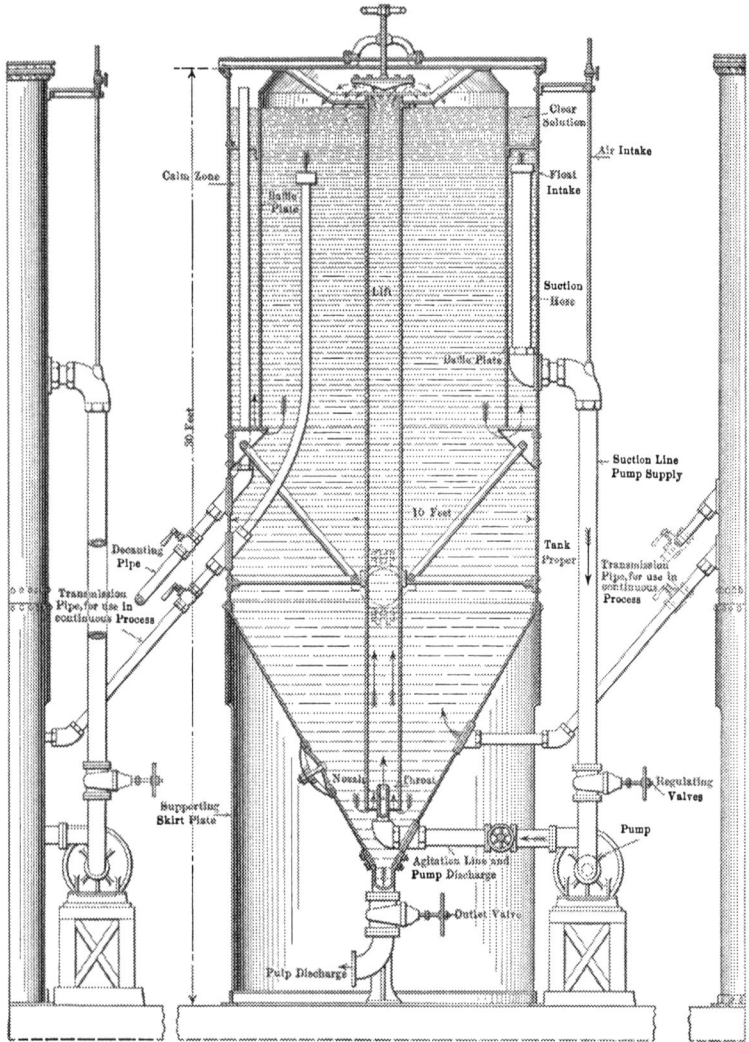

Fig. 29a. — Paterson Hydraulic Agitator. (Section.)

to cause the arms to turn. The same pump continually draws nearly clear solution from the top of the agitator, and it is necessary to operate so that the pulp remains thick at the bottom and thin at the top of the tank.

To avoid this disadvantage, the pumps at the West End Mill, Tonopah, Nevada, where these agitators are in use, are provided with chilled iron or manganese steel liners and runners, and are driven at slow speed, so that the wear is nominal and the pulp is kept at nearly the same specific gravity throughout, giving thorough agitation and avoiding settlement of heavy pulp when stopping. (Jay A. Carpenter, Min. Sci. Press, *106*, 646, May 3, 1913.)

These agitators have been used for a system of continuous agitation, the pumps drawing from one vat and discharging into the next of the series. For successful treatment it is necessary that the pulp entering and leaving a given vat should remain of the same specific gravity and fineness, and that there must be no opportunity for pulp entering a vat to leave it immediately without sufficient agitation. The latter may sometimes occur when Pachuca tanks are used in series. With the Trent agitator there is little danger of this, as the pulp is introduced at the bottom and there are no rapid upward currents.

As compared with the Pachuca tank the first cost of the Trent agitator is somewhat less, but operating costs are greater.

These tanks are in use for the intermittent system of treatment at the Tonopah Extension and Macnamara plant in the same district.

(H) The Dorr Agitator

The Dorr Agitator was invented by J. V. N. Dorr in 1910 and put on the market in 1912.

This machine combines the principle of the air-lift with that of the mechanical agitator. It consists of a flat-bottomed tank, having two radial arms provided with scrapers, which serve to bring the pulp which settles on the bottom towards the center, as in the thickeners. The central column of the tank is hollow, and is provided with a jet of air introduced from below. The pulp, as it is raised, is distributed at the top of the tank by revolving launders, perforated at intervals, so that it is evenly distributed, and circulates uniformly. (See Fig. 29*b*, p. 240.)

The pulp is fed in at one side and discharged on the opposite side of the tank. By adjusting the position of the intake to the discharge the composition of the pulp leaving the tank may be made different from the average mixture agitated, thus securing a longer period of agitation for the coarser particles.

FIG. 29b. — Dorr Agitator.   Installed in steel tank.

The sweeps at the bottom of the tank are hinged, so that they may be drawn up so as to stand close to the central cylinder when the machine is stopped. On restarting they are gradually lowered to their normal position, thus avoiding any difficulty in restarting after a shut-down.

The extraction is said to be somewhat more rapid than in Pachuca tanks.

The mechanical scraper avoids the possibility of the tank gradually filling with solids, as occurs in most forms of cone agitators, thus rendering the apparatus very suitable for continuous systems of treatment.

Any sand which can be kept in suspension by the mechanism may be treated.

The following are the data given for an installation of this agitator:

Dimensions ....16 ft. diameter by 16 ft. deep
Speed ....1 rev. per minute.
Power ....Less than ⅓ h.p.
Air for lift ....8 cu. ft. about 20 lb. pressure.

At another plant:

Dimensions ....30 ft. diam. by 12 ft. deep.
Speed ....3.6 rev. per min.
Power ....1.2 h.p.
Air for lift ....30 cu. ft. at low pressure.

SECTION III

FILTRATION BY PRESSURE AND SUCTION

(A) Advantages of the System

As already pointed out, the methods of treating slime or finely divided ore by settlement and decantation involve the use of tanks of large capacity for the settlement of the pulp and for the storage of solutions. It has been shown also that a high extraction could be obtained only by repeating the cycle of operations several times. By using some method of filtration by pressure or suction, the greater part of the dissolved values may be extracted in one operation, in far less time and with the aid of a much smaller quantity of solution than is required for the decantation system. In the latter system, much more time is consumed in separating the solution carrying the values from the insoluble residue than in actually dissolving these values; whereas by the system now to be described the separation of solution from pulp may be made with little delay as soon as a sufficient percentage of the values has been dissolved.

On the other hand, the expense of maintenance and repairs and the cost of labor and supervision are probably very much higher for any system of pressure or suction filtering than for the decantation process. Consequently the latter is almost invariably adopted where finely divided low-grade material has to be dealt with. In the filter processes the quantities of material dealt with in one operation are necessarily very much smaller than in the decantation process.

(B) Filter–Presses

General Principles. — Filter-presses consist essentially of a number of cells or elements into which the material to be filtered is forced by hydraulic or atmospheric pressure. These chambers are lined internally with filter-cloths, through which the filtered liquid is forced as a result of this pressure, passing out through channels in the walls of the cells, either to separate outlets or to a common solution-outlet channel.

Details of Filter-press Construction. — The following account is condensed from a description given by Clement Dixon.[1] "A filter-press consists of a number of hollow frames (from 30 to 50 in a 6-ton press) placed alternately between solid flanged plates, with filter-cloths of strong duck material between each. The hollow frames in a 6-ton press would be 3 ft. 6 in. or 4 ft. square inside, and 2, 3, or 4 in. in depth or thickness, as required for the particular ore treated. It will thus be seen that we have in reality a number of little vats standing on end, the hollow frames forming the sides, and the solid plates and filter-cloths the bottoms. Each of the hollow frames (which receive the slimes) is connected by a slot with the slimes inlet passage, and every alternate solid plate is connected by a slot with the wash-solution inlet passage; these solid flanged plates act as filter-bottoms for admitting wash solutions to the slime cakes. The remaining solid plates are connected with the wash-solution outlet passage and serve to carry away the wash solution after it has passed through the slime cakes. On each of the solid flanged plates is a cock, used for draining the press when filling same with slimes."

This description applies to the Dehne and other similar types of press, which, however, show various modifications in the less important details of construction. Illustrations are here given of the plates and frames used in the "Cyanippus" press, furnished by the Cyanide Plant Supply Co.; in this type the use of drainage cocks for the separate chambers is dispensed with. In these figures *a* is the channel through which the slimes pulp is introduced and which is connected only with the frames (Fig. 32); *b* is the channel for introducing wash solution, and is connected only with each alternate plate (those known as high-pressure plates (Fig. 31); *c* is the channel which receives the solution that has been forced out through the filter-cloths in the case of the high-pressure plates; *d* is a separate channel serving as an outlet for the filtered solution on the low-pressure plates only (Fig. 30). A general view of filter-presses in use for treatment of slimes at Lake View Consols, West Australia, is also given (Fig. 33).

[1] "Proc. Chem., Met. and Min. Soc. of South Africa," III, 13.

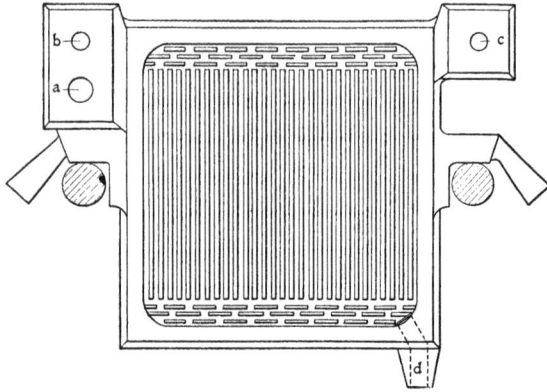

FIG. 30. — Cyanippus Filter Press, Low-pressure Plate.

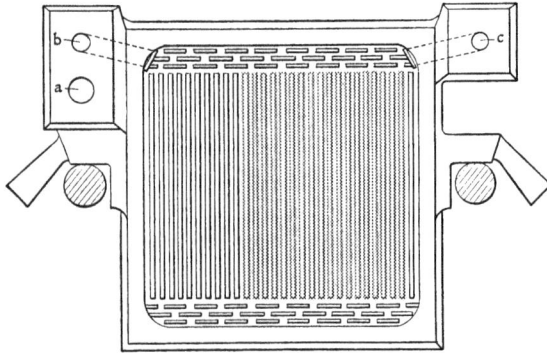

FIG. 31. — Cyanippus Filter Press, High-pressure Plate.

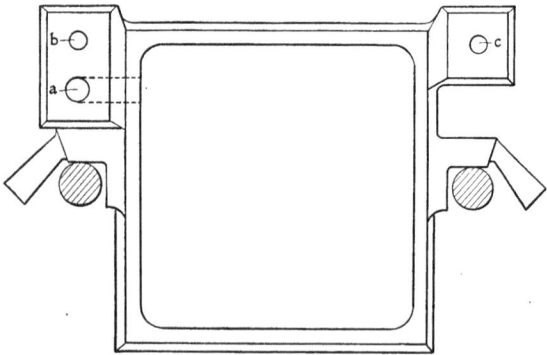

FIG. 32. — Cyanippus Filter Press.   Frame.

To facilitate the passage of solution through the cloths, the surfaces of the plates are grooved or corrugated. Some mechanical arrangement is also required for tightening the plates and frames while the press is in use. This is generally done by means of screws and levers at one end of the press.

FIG. 33. — Filter Press Slimes Plant at Lake View, Consols, W. Australia. Showing type of press furnished by the Cyanide Plant Supply Co.

Methods of Filling Filter-presses. — Although some solution of gold and silver may take place during the process of pressing, it is generally better practice to make sure that a sufficient percentage has been dissolved before transferring to the presses. Hence some system of agitation is commonly employed, after which the pulp, thickened to the proportion of about one part dry slime to one part of solution, is forced by compressed air or otherwise into the filter-presses. For this purpose a montejus, an airtight cylindrical vessel, is often used. This is provided with inlets at the top for the pulp and compressed air, and an outlet pipe, reaching nearly to the bottom, for conveying the pulp to the slime-pulp feed channels of the presses. The pressure required may be from 30 to 75 or even 100 lb. per square inch.

In other cases the presses are filled by means of a pump; the pumping is continued until the cells of the press are filled with slime under a sufficient pressure to give an efficient filtration — say 75 to 80 lb. per square inch. It is important to increase the pressure steadily and uniformly; hence some form of air chamber is commonly necessary.

Treatment in the Press. — After the solution originally present in the pulp has been forced out as much as possible, the pulp inlet channel is closed and fresh solution is forced in under pressure by means of a pump. This passes by the solution inlet channel (*b*, Fig. 31) into the chambers of the presses and through the slime cakes, passing out through the outlet channels *c* and *d* (Figs. 30 and 31). Wash-water may subsequently be introduced in the same way, and continued until the effluent is sufficiently low in values. In some presses a special inlet channel for wash-water is provided. Arrangements are also made by which compressed air may be forced into the press, in order to remove as much liquid as possible at the end of the operation. It is stated that by this means the residual moisture may be reduced in some cases from 25 to about 14 per cent.

Discharging Filter-presses. — When the operation is complete, the closing-screw of the press is loosened and the plates and frames drawn apart. The slime-pulp remains usually in the frames in the form of a square cake, which is discharged by tilting or tapping the frame, dropping the pulp through a discharge hopper into trucks running beneath the presses. After each set of operations, the plates, frames, and filter-cloths are washed and cleaned, and the press again closed ready for a fresh charge.

Capacity of Filter-presses. — The size varies greatly according to the work required. There may be 20 to 50 frames, with a total capacity equal to 1 to 5 tons of dry slime per press charge. In West Australia an average of 6 charges are treated per day in each press, one cycle of operations occupying 4 hours.[1]

(C) AUTOMATIC SLUICING PRESSURE FILTER FOR ORE SLIMES
(MERRILL SYSTEM)

Where conditions allow, the discharge of the treated slimes may be made by sluicing. In *The Merrill Press*,[2] designed by C. W. Merrill and introduced at various plants in the Black Hills, South Dakota, arrangements are made for introducing

---

[1] Julian and Smart, *loc. cit.*, p. 258.
[2] " Eng. and Min. Journ.," LXXXI, 76 (1906).

water under pressure, whereby the treated slime may be sluiced out of the frames without opening the press. The following particulars relating to this type of press are summarized from an account given by F. L. Bosqui.[1] The press is of a common flush plate and distance-frame pattern, but consists of larger units, the dimensions being:

Number of frames	92
Size of frame	4 × 6 ft.
Length of press	45 ft.
Capacity	26 tons
Weight of press	65 tons
Thickness of cake	4 in.

In addition to the ordinary channels for introducing slime-pulp and solution, there is provided at the bottom of the frames a continuous channel, within which lies a sluicing pipe with nozzles projecting into each compartment. This pipe can be revolved through an arc of any magnitude, so as to play a stream into any part of the cake, washing it down into the outlet channel. When the press is being filled, and during the cyanide treatment, the discharge ends of the pipe are sealed.

The mode of operating is as follows: The slime-pulp, 3 parts water to 1 of solids, is charged by gravity to the presses under about 30 lb. pressure. The cyanide treatment is carried out in the press itself, the effluent solution going to four precipitating vats, where the gold is recovered by zinc-dust. There is no power cost for agitating or elevating slime-pulp, but only for elevating solution to the press. The quantity of solution required is 0.6 ton per ton of dry slime, of which only 0.3 ton is precipitated. The power required is $\frac{1}{10}$ h.p. per ton of dry slime treated. Four tons of water per ton of slime are required for sluicing. All filtering is done by gravity, at a cost of 2 cents per ton. In experimental tests on this system a recovery of 91 per cent. of the original value of the slime was obtained.

The Merrill Metallurgical Company indicates the following as the essential requirements of an efficient slime filter:

(1) The operations of cake-forming, leaching, washing and discharging must be as nearly automatic as possible and must be conducted with a minimum expenditure of power, labor and supplies.

(2) The operation of washing the cake in the filter must be rapid, to avoid the effect of osmosis or mixing of the soluble matter

[1] " Min. Sc. Press," Dec. 15, 1906.

in the cake with the liquid used as a wash. Hence it is desirable to operate at as high a pressure as the strength of the filtering medium will allow.

(3) The recovery of dissolved values in the slime cake must be practically complete.

(4) The filtrate must be practically free from solids in suspension.

(5) The first cost of the filter per unit of daily capacity must be as low as consistent with good workmanship and satisfactory operation.

Construction of Filter. — In its general details, the Merrill press resembles the Flush plate and Distance-frame type, such as the Dehne press already described (p. 243).

The essential difference is in the method of discharging. In the ordinary types it is necessary to draw apart the plates and frames each time the chambers are to be discharged, while in the Merrill press a rotatable sluicing pipe extends throughout the length of the press and is fitted with small nozzles, one projecting into each frame. When the press is ready to be emptied, discharge cocks at the bottom are opened and water under pressure is admitted to the sluicing pipe. As this pipe slowly revolves back and forth through an arc of approximately 180°, the streams of water break up the slime cakes and sluice them out of the press. (See Fig. 32a.) This method of discharging greatly reduces the operating cost.

In some types, the press is opened and closed by hydraulic pressure, when necessary to replace a broken cloth.

The chambers are filled from a continuous channel extending along the median line at the top of the frames and communicating with each filter compartment.

Each plate is provided with a separate solution outlet fitted with a gauge glass and stopcock, so that should any filter cloth be defective, the muddy solution may be detected and shut off until the leak has been repaired.

Treatment or wash solutions are admitted either through a channel at the lower corner of the frames or through the main pulp feed channel.

Through the median line at the bottom of the press, and passing through the plates and frames, is a continuous channel, within which, with ample room, is the sluicing pipe for discharging, above referred to.

Two Types of Filter. — According to the nature of the material to be filtered, one of two different types of filter is used:

(1) Where the slime is comparatively granular and admits of the formation of cakes 3 to 5 in. in thickness in a reasonable time and at a comparatively low pressure, the *Solid Filling* press is used. The containers are 4 to 6 in. in cross section; the wash solution enters through a main solution channel and is admitted back of the filter cloth on each alternate plate. The wash solution thus passes through the full thickness of the cake, forcing before it the moisture originally contained in the slime. Diffusion is negligible, and no cracks or channels can form in the cake. The wash water required for a practically perfect displacement is usually from $\frac{1}{2}$ to $\frac{3}{4}$ of a ton per ton of dry slime washed.

(2) Where the slime is of a talcose or hydrated nature it is practically impossible to form cakes more than 1 to $1\frac{1}{2}$ inches in thickness. Under

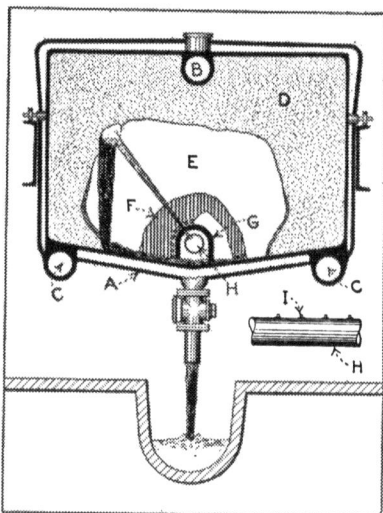

FIG. 32a. — Arrangement for Sluicing out Slime Cake in Merrill Press.

A. Standard container or frame.
B. Feed channel through which the slime pulp enters each frame.
C. Channels from which water or solution is drawn off during process of filtration.
D. Partially sluiced slime cake.
E. Filter cloth, with portion removed showing corrugations of filter plate.
F. Filter plate.
G. Horseshoe clamp for holding filter cloth against filter plate.
H. Sluicing pipe, containing water under 60 to 90 lb. pressure admitted at either or both ends.
I. Sluicing nozzles located at center of each container or frame.

these conditions the *Center-Washing* or *Partial-Filling* type of press is used. (See Fig. 32b.) The containers are 3 in. in cross-section, and the pulp and wash solution are introduced from a channel communicating with the center of each frame. When the press is being filled, the flow of slime pulp is shut off before the chambers are completely filled. The cake always forms first on the filter cloths and extends inwards towards the center; it thus consists of two equal portions with a vertical space between them extending from the top to the bottom of the frames, and varying

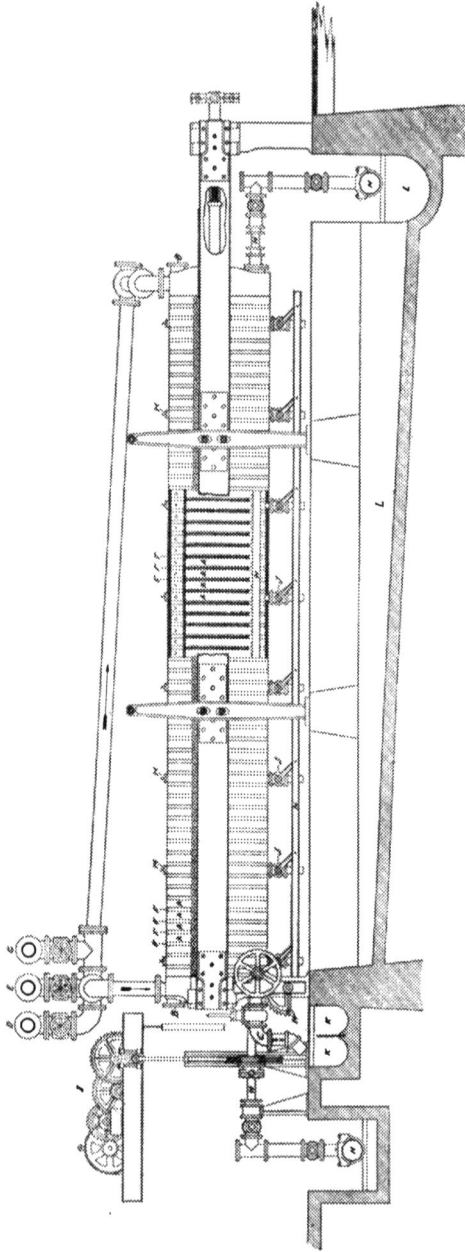

FIG. 32b. — Merrill Slime Filter (Automatic Sluicing). "Partial Filling," or "Center Washing" Type.

A. Standard container or frame.
B. Feed channel through which the Slime Pulp enters each frame.
C. Pipe from which water or solution is drawn off during process of filtration.
D. Slime or Pulp line.
E. Barren solution wash line.
F. Filter Plate.
G. Water-wash line.

H. Sluicing Pipe, containing water under 60 to 90 lb. pressure, admitted at either or both ends.
I. Sluicing motor and gearing.
J. Residue discharge cocks.
K. Effluent solution Launders.
L. Discharged residue Launder.
M. Peep Gate.
N. Lever System for residue discharge cocks.

in width according to the conditions. As the flow of slime pulp is shut off, the wash solution is turned into the same channel without allowing the pressure to drop materially, the wash solution being thus forced through the cakes from the center outwards.

In this type of press the displacement is less perfect than with the "solid filling"; usually 2 tons of solution per ton of dry slime are required for a perfect wash. The cycle of operations, however, is very short and a large capacity per unit of filter area is obtained.

In discharging, air is sometimes introduced behind the filter cloths to loosen the slime cakes.

Advantages of Merrill Filter. — (1) Much of the auxiliary equipment needed with ordinary filters is avoided. Slime from collectors may be fed direct to presses without passing through agitating tanks. Also there is no necessity for tanks and pumps for handling excess slime pulp.

(2) The pressure for filling the press is obtained wholly or partially by gravity. The leaching and wash solutions, and in some cases the sluicing water also, are handled the same way.

(3) The discharging of the press is done with an economy of power, the sluicing mechanism requiring about one-half horse power per press, with the further advantage that the labor of opening and closing the press is eliminated.

(4) Operating costs are low on account of the large capacity per unit of area and the avoidance of handling and pumping. One man can operate from two to four units.

(5) Values are dissolved and extracted in the press by preliminary washing of the cake with barren solution, which would not be obtainable by longer treatment in agitators. This is probably due to the intimate contact obtained by forcing solution under high pressure through the cake.

(6) The recovery of dissolved values is high, owing to the facts, (*a*) that the cake is never exposed to air after formation and hence does not develop cracks; (*b*) that there is no excess pulp or wash solution to be returned and circulated, so that enrichment of returned solution cannot take place, as may happen with other types of filters.

Disadvantages. — (1) High first cost. This is offset to some extent by the elimination of subsidiary appliances referred to above.

(2) Considerable water required in sluicing, amounting to

four parts water to one of slime. A portion of this, however, can be recovered for re-use, the actual consumption of water approximating one ton per ton of slime treated.

Cycle of Operations (time required). — The following figures are published by various plants in which this type of filter is in use:

Homestake Slime Plant (S. Dakota), (no preliminary contact with cyanide solution; all extraction made by leaching the cake in the press).

	Pressure lb.	Time hours	min.
Filling with slime	30	1	10
Aërating	15	2	45
Leaching strong solution	25	1	20
Leaching weak solution	25	2	5
Washing water	˙25	0	45
Discharging water	70	1	15
		9	20

Experimental Trial at Santa Gertrudis, Pachuca.

	Pressure lb.	Time hours	min.
Filling (caking)	30 to 35	0	30
Solution washing	40	0	40
Water washing	40	0	5
Sluicing	90	0	15
		1	30

Pittsburg — Silver Peak, Blair, Nevada.

	Pressure lb.	Time hours	min.
Filling press	30	0	30
Leaching with solution	25	2	30
Water washing................	30	0	20
Discharging water	75	0	45
		4	5

Esperanza, El Oro, Mexico.

	Time hours	min.
Filling	0	30
Washing	0	35
Discharging	0	45
	1	50

Costs per ton dry weight	Homestake (complete treatment) Slime	Pittsburg — Silver Peak Slime	Ore
Labor...............	0.0653	0.079	0.031
Power	0.0196	0.036	0.014
Supplies............	0.1103	0.148	0.059

(D) The Burt Slime Filters

These machines were designed and introduced by Edwin Burt, at the El Oro Mining and Railway Co., El Oro, Mexico.

There are two types, in both of which the material to be filtered is introduced under pressure into a closed cylinder, containing a filtering apparatus in its interior. The cylinder is closed at one end by a powerful hydraulic press operating against a door by means of specially arranged toggles.

In the *Rapid Cyanide Filter*, designed for treating colloidal, slow-settling slime, the cylinder is stationary, and inclined at an angle of about 45°. The filtering medium consists of a number of mats of cocoa fiber, framed by a perforated bent pipe, each mat being inclosed by two thicknesses of canvas sewn together. The pipes communicate through the shell of the cylinder with a common discharge pipe, by which the filtered solution is withdrawn. Compressed air may be admitted both to the solution discharge pipe and to the interior of the cylinder.

Mode of Operation. — The slime is forced in under 25 to 40 lb. pressure until a cake of sufficient thickness has formed on the mats. Then the slime valve is closed and the air and discharge valves are opened, forcing out all the excess slime, which returns to the supply tank. The discharge valve is then closed, wash solution admitted and the air valve closed. The excess solution is then removed in the same manner as the excess slime; a water-wash given in the same way, and finally after excess liquid has been removed, air pressure is raised until finally the moisture in the cakes is reduced to about 30 per cent. The air and solution valves are then closed, the door opened by means of the hydraulic cylinder and air admitted through the solution discharge pipe and thence to the interior of the mats, causing the cakes to drop off and slide out of the cylinder. The entire cycle of operations occupies from 20 to 90 minutes.

As in other types of filter, the cloths are cleansed from time to time by soaking in acid solution and scrubbing.

The filter may also be used as a *dewatering* machine in plants where the values are too low to justify washing the cake, in which case the installation is simplified and capacity increased. With slight modifications it may also be adapted for *clarifying* solution, in which case it is placed nearly horizontal, with a fall of about 6 in. There is a pipe connection at the lower end for pumping in the solution, and another for a blow-out pipe. The solution filtered through the mats passes by a launder to the precipitation room. To clean the cakes off the mats an air-nozzle is held for a second or two into each connection to the filter mats, which causes the slime to drop from the cloths. A small stream of water under high pressure is also introduced through holes in the shell to assist in cleaning the cloths. A filter of this type 54 in. diameter by 20 ft. long, with 20 h.p. motor and a centrifugal pump is used at El Oro, Mexico, filtering 4000 tons of solution per day. The pressure used is about 25 lb. per sq. in. and costs $2.58 (gold) per day. The filter takes the place of three sand filters each 40 ft. diameter by 10 ft. 6 in. deep.

Fig. 32c. — The Burt Revolving Filter, showing discharge of filtered solution.

In the *Revolving Filter* the cylinder is supported horizontally by a trunnion bearing at the feed end and a roller bearing near the discharge end. In general appearance the machine resembles a tube-mill, but is provided with filter mats fastened to the inner surface of the shell by means of specially shaped angles and bolts. During formation of cake the filtered solution is discharged on to a cement floor, from which it is conveyed to a receiving sump. (See Fig. 32c.)

The discharge end is closed by a cast-iron door operated by hydraulic pressure, being opened and closed by means of a piston working in a cylinder as shown in Fig. 32d. This door slides on six rods, which connect the end of the filter-shell with a cone-shaped casting. On these rods are springs which aid in closing the discharge door.   At six intermediate points on the door, between the rods, are fastened one end of each toggle joint; the other end is fastened to the cone-shaped cross head, and is provided with an arm extending to the center piston rod, which carries a large iron collar.   By this device, when the door is closed, the toggles are straightened out and great pressure exerted: the amount is easily

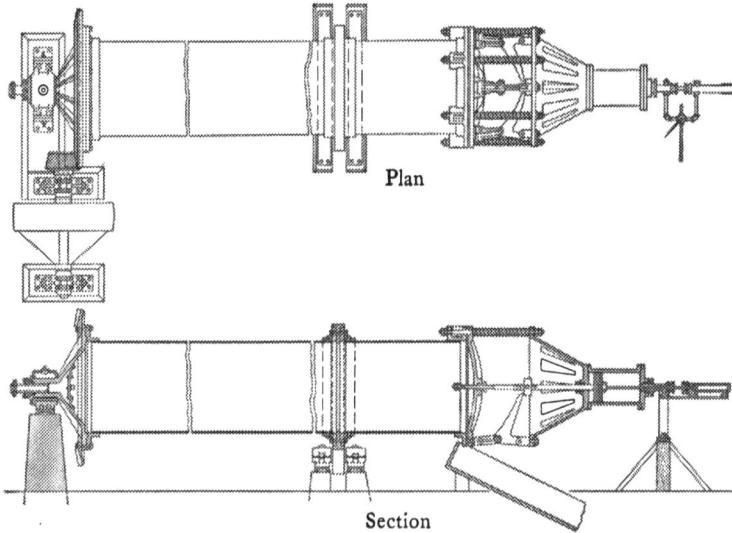

Plan

Section

FIG. 32d. — Burt Revolving Filter.   Plan and section.

regulated by moving the collar forward or backward on the thread provided for the purpose.   The joint between the door and the flange on the cylinder is made by means of a square rubber gasket. A recent improvement is a locking device on the door mechanism to lock the toggles when the door is shut, so that it is unnecessary to maintain a pressure in the cylinder to keep the door closed.

The *filter cloths* are 28 in. wide by 5 ft. long, one cloth being used for every 5 foot length of the filter cylinder.   The cloths are interchangeable and are held in place by angle irons which are slipped over studs fastened in the shell at the point where two cloths meet.   Each cloth has its individual outlet at the center for the filtered solution.

Method of Working. — The filter is revolved continuously at about 15 r.p.m.   The discharge door being closed, the slime feed valve is opened to admit a sufficient charge, then closed and the air valve opened.   Pressure is maintained at 25 to 45 lb., until all the slime has gone to form cake, which is signalized by the filter blowing. The air valve is now shut and wash solution introduced, opening the escape valve if necessary to reduce the pressure so as to allow the wash solution to enter. When the filter starts blowing, wash water is admitted in the same manner, and when finished the pressure is dropped to zero, and the door opened.

Advantages. — This type of filter is especially adapted for granular slime which admits of the formation of a thick cake, 3.5 to 4.5 in. in thickness.

In this type, there is no excess slime, solution or water wash to be returned to storage tanks; everything introduced passes through the filter except the slime residue. The cake when finally formed contains 20 to 22 per cent. moisture. Very little water is required for discharging, only sufficient to make the canvas slippery so that the loosened cake moves towards the discharge door. This water is added in a small amount at one time, just as the cake falls, and about 100 lb. are required per ton of dry slime.

The power consumption is stated to be from $\frac{1}{4}$ to $\frac{1}{2}$ that required to operate a vacuum filter.

The washing of the cake is said to be more perfect, and is accomplished with less wash solution than in any other type of filter. The cake is perfectly uniform throughout; owing to the fact that the cylinder is constantly revolving and the pulp kept in motion until the cake is formed. In an ordinary vacuum filter the cake formed on the lower portion of the leaf is thicker and contains the coarser particles, while the light, flocculent particles form a more impervious layer in the upper portion, making it almost impossible to give an even wash to the entire cake.

As the cloth in the revolving filter is exposed only for a very short time to the action of the air, there is very little chance for the formation of lime deposit. The wear on cloths is very little compared with tonnage handled. Acid wash is given without removing the cloths, and much less labor is required for repairing and renewing than is the case with vacuum filters.

Any cracks which may form in the cake are filled by the scouring action of the wash solution or water, which remains practically stationary in the cylinder and washes sufficient solids from the cake to fill the cracks and give an even wash.

The following *data* were obtained with the Burt Revolving Filter at *El Oro, Mexico.*

No. of charges put through..13

Average thickness of cake....................................4.3 inches

Average dry metric tons per charge............................5.24 tons

Average pressure per sq. in. during caking ....38 lb.

"   "   " " "   "   washing ....48 lb.

Moisture in cake as formed ....21.1 per cent.

"   " discharged sludge ....25.7 "   "

Screen analysis of pulp fed to filters:

Mesh	Percentage
+ 100	2.6
100 to 150	23.2
150 to 200	8.4
− 200 (sand)	27.4
− 200 (colloid)	38.4
	100.0

	Minutes
Average time, charging pulp	3.9
" " forming cake....................	25.5
" " adding wash	7.4
" " washing	41.7
" " discharging....................	2.9
" " total cycle	81.4

	Gold $	Silver oz.	Total Value $
Heads to press, unwashed	$3.14	3.30 oz.	$4.79
" " washed	0.88	1.90	1.83
Tails discharged, unwashed	0.69	0.74	1.05
" " washed.............	0.66	0.73	1.03
Unwashed value discharged.........	0.03	0.01	0.04
Extraction in press [1]	0.22	1.17	0.80

(E) The Kelly Filter

This apparatus is a pressure filter in which the filtering frames are inclosed in a cylindrical shell or pressure tank, the frames being mounted on a movable carriage which telescopes into the shell. (See Figs. 32e and 32f.)   The filter consists of the following parts:

(1) A supporting frame, on which the carriage moves on a slight incline.

(2) The pressure tank or shell, the ends of which are concave, the front end or head being movable.

(3) The filter carriage, which supports the filter frames; the front end forms the head of the filter tank; the rear end is mounted on wheels which travel on rails inside the press shell.

[1] Difference between washed heads and tails.

(4) The filter frames, attached to the head of the filter carriage. These are arranged longitudinally inside the pressure tank, and hang vertically at regular intervals. The frames are rectangular and all of the same length, but of different widths.

FIG. 32e. — Kelly Filter Press (closed for filtering).

(5) The head mechanism, which forms the front end of the filter carriage and locks to the front end of the shell, while the press is being charged, by means of an arrangement of radial

FIG. 32f. — Kelly Filter Press (open for discharging cake).

locking arms. Around the inner face of the head is an annular projection which fits in a groove of the ring forming the front end of the pressure cylinder. In this groove is a gum gasket which forms an absolutely tight joint.

The shell is further provided with inlets for slime pulp and compressed air.

Each filter leaf has a separate solution outlet at the lower front corner. These all communicate with a 4-in. manifold which empties into a distributing box.

The upper front corner of each filter leaf is connected by a separate cock with another manifold, which communicates with a water-tank placed above the press, the pressure derived from this head of water being used for discharging the cakes.

The mode of operation is as follows: The pressure chamber is first closed and the head locked. The valve is then opened to admit the slime pulp to be filtered. When sufficient cake has formed, the surplus pulp is forced out by compressed air into a montejus, whence it may be drawn for a later cycle. Solution, or wash water, is then introduced into the press shell and forced through the cake under pressure. The surplus water is then withdrawn and the cakes dried by compressed air.

The head is then unlocked, and the carrier runs down the inclined track, bringing the loaded frames out of the shell, where they are then discharged by water-pressure, as above mentioned.

When the cakes are discharged the carrier and filter frames are returned to the press shell by means of counterweights operating at the rear end of the shell, and generally hung overhead.

The cake is formed on the outside of the filter-leaves as in a vacuum filter, the leaves being constructed of wire screen, inclosed in twill bags. The upper side of each frame consists of an iron pipe, perforated on the under side. All liquors leave the press by nozzles connected with the lower forward corners of the frames.

The press is made in various sizes with a capacity of 75 to 1150 sq. ft. of filtering area, treating 6 to 100 tons per day. It is stated that the cakes are discharged in some cases with only 9 to 12 per cent. moisture

(F) Filtration by Suction

Early Applications. — As already mentioned in Section II, A, the principle of suction as an aid to filtration was employed at an early stage in the development of the cyanide process. In the first plant where the treatment of slimes was attempted on a working scale, namely at the Robinson slime plant, erected by Charles Butters,[1] the slime pulp, after being agitated sufficiently

[1] " Proc. Chem., Met. and Min. Soc. of South Africa," II, 241 (Feb. 1898).

to dissolve the gold, was run on to a filter-vat, provided with stirring gear that could be raised or lowered; the material was allowed to settle, the clear liquid was decanted from above, while at the same time suction was applied below the filter-cloth. By this means about 10 tons of slimes could be settled, decanted and sucked dry by a vacuum pump to about 28 per cent. of moisture, and would remain on the filter-cloth as a tough leathery layer about 6 to 8 in. thick. Fresh solution was applied and the paddle gradually lowered while revolving at the rate of about 16 r.p.m., so that in half an hour the slimes were washed into a pulp again. After further dilution, the process of simultaneous settlement, decantation, and suction by vacuum pump was repeated. The method was abandoned owing to the wear and tear on the filter-cloths and the cost of keeping the suction pumps in order, together with the power and attention required for running them.

Filtration aided by suction is used also in some cases in the treatment of fine sands by percolation and in the direct cyaniding of dry-crushed ore. For this purpose the material is generally treated in comparatively shallow vats and suction applied by the use of a steel cylinder connected with the bottom of the tank, below the filter-cloth, by a pipe, furnished with a cock. This cylinder is also connected with a vacuum pump by a separate pipe. The communication with the tank is first shut off and the steel chamber exhausted by the vacuum pump. The valve connecting with the latter is then closed and the cock below the filter-tank is opened, the solution from the latter being sucked into the exhausted cylinder. The advantage of this arrangement is that the solution from the tank does not pass direct through the vacuum pump, so that the valves of the pump are not liable to injury from grit passing through the filter.

The Moore Filter. — Among the more modern appliances for filtering by means of suction may be mentioned the Moore filter the salient feature of which is that the filtering apparatus is movable and is transferred from one tank to another as required. The following description is quoted from an article by R. Gilman Brown, on "Cyanide Practice with the Moore Filter."[1] The filter consists of a frame or "basket" carrying a number of "plates." Each plate is composed of a double thickness of

[1] "Min. Sc. Press, Sept. 1, 1906.

canvas, of medium quality, 5 × 16 ft., sewed around three edges. The fourth (long) edge, forming the top of the filter, is bolted between strips of wood $1\frac{1}{2}$ × 6 in. section. In the bottom edge of each filter-plate is a $\frac{3}{4}$-in. channel-iron which serves as a launder for collecting the filtered solution. The filters are stitched vertically through both sides at 4 in.-distances, and in the compartments so formed are placed $\frac{1}{4}$ × 1 in. strips to allow circulation. Each plate is provided with a 1-in. vertical suction pipe, dipping into the launder at the bottom, the end of the pipe being flattened for this purpose. The upper end of this pipe is connected by means of suction hose with a 3-in. manifold and vacuum pump. There are 49 of these plates in each basket.

The mode of operation is as follows. The basket is lowered into a vat full of pulp until the upper edge of the plates is submerged. The suction pump is then started. If the solution drawn through is at first muddy, it is returned to the vat. In order readily to locate any leaks in the filter-plates, a glass nipple is placed on the pipe connecting each plate with the manifold; any plate showing leakage may thus be at once cut off. Suction is maintained with intermittent agitation till a sufficient coat of slime, averaging $\frac{3}{4}$ in., is obtained, the time required being about one hour. The basket, with the suction pump still running, is then raised by a traveling crane, transferred to a wash tank and lowered therein. Water is sucked through the slime layer adhering to the plates until the effluent is sufficiently low in cyanide (approximately 0.0075 per cent.), about 0.7 tons of wash-water being needed per ton of dry slime. The cake then contains 40 per cent. of moisture. The basket is now raised again and transferred to the discharge hopper; the suction is continued until excess of moisture is removed; the action is then reversed and an air pressure of 35 lb. per square inch given in successive blasts for a few seconds, causing the cakes to drop off. The residual moisture is about 32 per cent. When it is necessary to clean the filters, water pressure instead of air is applied until the pores of the cloth are freed from slime. This is done about every alternate day.

In some cases a second cyanide treatment is given before water-washing. The apparatus is charged with thick slime-pulp in the first tank, transferred to a second containing weak solution, where the suction is continued for a time, say 20 to 40 minutes, then to the water tank, and finally to the discharge hopper. In general, it is found that water pressure is preferable to air for discharging, as it brings the slime cake off in one mass instead of in patches. One of the difficulties encountered in using this system was the imperfect adhesion of the slime during transfer, owing to jarring and partial drying. This caused uneven suction and consequent poor extraction.

Among the plants where the Moore system was first adopted may be mentioned the Lundberg and Dorr Mill, Terry, South Dakota; the Liberty Bell, Colorado, and the Standard, Bodie, California. It was originally introduced by G. Moore at the Consolidated Mercur mines, Utah, in 1902 or 1903.

The Butters Filter. — This apparatus is in many respects similar to the Moore filter. The two systems have the following points in common: (1) The filtering apparatus is immersed in the pulp that is to be filtered; (2) the solution is withdrawn by suction, by the use of a vacuum pump; (3) the cakes of slime formed adhere to the outside of the filter-plates; (4) a large filtering area is obtained in a cheap and compact form.

The essential difference is that in the Butters system all the operations are conducted in one vat. The filter itself is stationary, but so arranged that the separate plates or "leaves" may be lifted out if necessary. The pulp is first brought to the proper consistency by settlement and decantation; the thickened pulp is then run into a special rectangular filter-box, in which the apparatus is suspended. Suction is applied until a sufficient coating of slime is obtained; the surplus pulp is then withdrawn from the filter-box by suitable pumps and returned to the settling tank or slime reservoir. The box is now filled with weak cyanide solution, which is sucked through as long as may be necessary for extraction of the values. The surplus solution is then pumped back to the storage tank, water substituted, and suction continued till soluble values are extracted. Finally, the suction is cut off, water pressure is applied to force off the adhering cakes, which fall into the filter-box and thence through hoppers with sides inclined at a sharp angle, whence they are discharged into a launder by means of a gate-valve at the apex of each hopper. On closing the valves, the filter-box may be again charged with slime-pulp.

The construction of the "leaves" in this system differs somewhat from that of the "plates" in the Moore system. A detailed description by E. M. Hamilton is given by Julian and Smart;[1] it need only be said here that each leaf consists essentially of an oblong frame suspended from a wooden bar forming the upper side of the leaf. Its remaining three sides are formed by a pipe bent at right angles and perforated along its upper edge in the part forming the bottom of the frame. The inner space is filled with cocoanut matting. The whole is enclosed in canvas, stitched vertically as in the Moore system, and forming the filtering surface to which the slime adheres. The perforated pipe is closed at one end and communicates at the open end with the vacuum apparatus, and also with the water tank used when water pressure is required for discharging.

The Butters filter, which is the result of experiments carried out by Charles Butters and staff, has been installed at the Butters plant, Virginia City, Nevada; at the Butters Copala Syndicate Mill, Mexico; at the Combination Mill, Goldfield, Nevada, and elsewhere. At the first-mentioned plant it is stated that 150 tons of slime are treated per day at a cost of $11\frac{3}{4}$ cents per ton.

The Cassel filter is of very similar design to the Butters filter. The chief distinction is that the filter-cloth is separately supported on a movable frame. At the end of the operation the slime-cake is detached by oscillating this frame.

The following particulars of the costs and power required for working the Butters system are given by F. L. Bosqui:[2]

At the 200-ton plant, Virginia City, Nevada, 10 cents per ton;
At the 40-ton plant, Combination Mill, Nevada, 45 cents per ton.

The latter could be reduced to 31 cents per ton by working the plant at its full capacity of 56 tons. Former cost of filter-pressing was $1 per ton.

The power required at the Combination Mill is:

For filtering .... 9 h.p.
For agitating slime-pulp.............. 3 "
For pumping and other purposes ...... 9 "

The pumps required are: one 4-in. Butters centrifugal pump; one 12 × 10-in. Gould's vacuum pump; one 2-in. centrifugal pump for raising filtered solution to clarifying filter-press; and

[1] *Loc. cit.*, p. 246.
[2] "Min. Sc. Press," Dec. 15, 1906.

one 2-in. centrifugal pump for returning slime overflow from leaching vats to slime settlers.[1]

Traveling Belt Filters. — As early as 1893 it was suggested by W. Brunton to use an endless traveling belt of filter-cloth, passing over vacuum chambers, for filtering slime-pulp, the slime being spread over the cloth at the head of the machine. Several other filters on the same or somewhat similar lines have been suggested.[2]

(G) THE CRUSH FILTERS

The courtesy of the Cyanide Plant Supply Co. permits the following description of these appliances, of which that company controls the patents. They are of two types, distinguished as the "Traversing Filter" and the "Fixed Immersed Filter."

The Crush Traversing Filter

This system was elaborated by Barry and Banks at the Waihi mine, New Zealand, and in its general features somewhat resembles the Moore filter, with the important exception that no reverse pressure of air or other fluid is necessary for discharging the cakes, and that the material to be filtered is agitated during filtration.

Each basket carries a number (generally ten) of filter-plates or leaves hung between two joists, and is suspended from a traversing crane, by which it may be raised or lowered and transferred from the pulp tank to the wash tank or to the discharge hopper, as required. Each filter-plate consists of a sheet of thin galvanized iron, carried by a wooden bar and enclosed in a frame of iron piping. The whole is enveloped in cotton twill or duck, between which and the iron may be placed cocoanut matting or other filtering medium if required. The plates are kept apart by the supporting wooden bars or by wires arranged at an angle. Each frame communicates by flexible rubber tubes with a main, this being further connected by flexible hose with a vacuum reservoir or pump. The frames are also connected to another main having a valve opening to the atmosphere. (See Figs. 33*a* and 33 *b*.)

The mode of operation is similar to that of the Moore filter.

---

[1] As the filter-cloths in time become choked with lime and other salts, it is necessary at intervals to immerse the leaves in dilute hydrochloric acid, which dissolves the deposits and thus restores the efficiency of the filter.

[2] See Julian and Smart, *loc. cit.*, p. 251.

(See above.)   When the basket is lowered into the pulp tank, the valve on the atmospheric main is left open to allow the displaced air to escape.   As soon as the filter is immersed, the valve on the atmospheric main is shut off and that on the vacuum main turned

Fig. 33a. — Crush Traversing Filter immersed in Pulp and Solution Tanks.

on.   The pulp is agitated with compressed air or otherwise to prevent settlement during the filtering operation.

The time required for formation of the cake is 30 to 45 minutes, and for extraction in the wash tank about 30 minutes.   The whole cycle of operations, including charging, discharging, and transferring, takes about 2 hours.   The cake varies in thickness from $\frac{3}{4}$ to $1\frac{1}{4}$ in.   In discharging, the vacuum is maintained until a crack begins to form at the junction of the supporting bar with the corrugated plates; the vacuum is then cut off and air admitted. The cake immediately discharges itself with a rolling motion, and after a little cleaning the basket is ready for a fresh charge.

FIG. 33b. — Crush Traversing Filter, raised for transferring or discharging.

During the drying operation before discharging, the inrush of air has the effect of lowering the vacuum and so interfering with the suction on the other baskets; it is therefore advisable either to have a separate vacuum pump for each basket or a separate connection with high- and low-pressure pumps, the latter, giving a vacuum of 5 to 10 in., to be applied as soon as the filter is raised out of the wash tank for discharging.

The frames are made in two sizes — 16 × 4 ft. and 10 × 5 ft., with 10 frames in each basket. Each basket of the first type treats 40 to 60 tons of slime per day; the filter-cloth requires renewal about once in 6 months, 1 sq. yard of cloth treating about 4 tons of slime.

The power required for the crane is 10 h.p.; the vacuum pump requires 6 h.p. per basket.   When air agitation is used, additional power is needed.   (See Figs. 33a and 33b.)

The Crush Fixed Immersed Filter

In this type the filter is stationary and the crane is dispensed with.   In principle it resembles the Butters and Cassel filters. The pulp is agitated in the filter-box by means of centrifugal pumps, and after a sufficient thickness of cake has accumulated on the plates, the surplus pulp is returned to a pulp storage-tank (a, Figs. 34a, and 34b.)   This type of apparatus is claimed to have

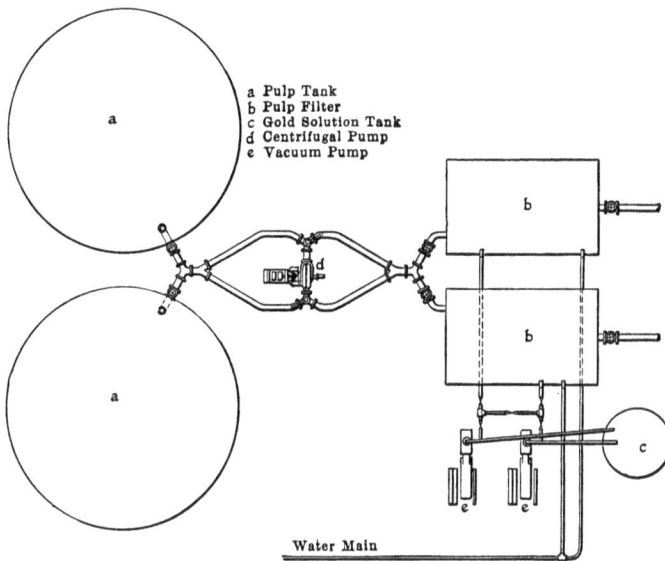

a Pulp Tank
b Pulp Filter
c Gold Solution Tank
d Centrifugal Pump
e Vacuum Pump

Water Main

FIG. 34a.—Plan of Crush Filter Plant. (Fixed immersed type.)

the drawback that there is a danger of untreated material being discharged, as the filtration and discharge of the slime both take place in the same vessel.   This can only be obviated by carefully attending to the washing out of the pulp-box (b, Fig. 34a), after removing the surplus pulp when the cake has formed. These washings should of course be returned to the general stock in the supply tank, which, however, may thus become unduly diluted.

The construction of the filter-plates is identical with that of the traversing type of filter, but no hessian or cocoanut matting is used in the leaves, and it is claimed that the latter are not liable to be choked with lime or fine particles. It is advisable in this case also to have high- and low-pressure vacuum pumps in connection with the frames, the former to be used during the actual filtering operations and the latter while filling or emptying the pulp-box. Too high a vacuum while the plates are exposed to air would cause cracks to form in the slime cakes, rendering the subsequent washing inefficient.

Another drawback pointed out by the Cyanide Plant Supply Co. is the liability of the cakes to drop off during the rather long period occupied in emptying the tank of surplus pulp and refilling with solution or wash-water. This renders the method un-

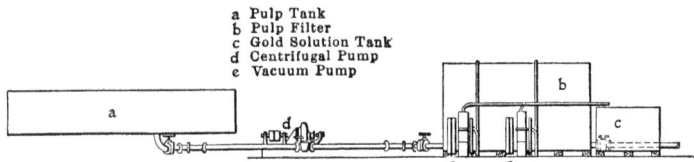

a Pulp Tank
b Pulp Filter
c Gold Solution Tank
d Centrifugal Pump
e Vacuum Pump

FIG. 34b.—Elevation of Crush Filter Plant. (Fixed immersed type.)

suitable for sandy or granular matter. The difficulty may be remedied by the use of very large pumps for transferring pulp and solution, but this adds greatly to the cost of the installation.

The capacity of a standard plant on this system is about six to eight charges per day, say 100 tons of dry slime, or about 1 ton per day for every 30 sq. ft. of filtering area. The power required is about 9 h.p. when the slime-pulp is fed by gravity to the filter-tank, and about 30 h.p. when pump circulation is used.

(H) THE RIDGWAY FILTER

A further extension of the idea of a traveling suction-filter, first introduced in the case of the Moore apparatus, already described, is seen in the Ridgway filter, of which illustrations in plan and section are here given (Figs. 35 and 36). The carriers bearing the filter-plates revolve about a central hollow axis and travel along a circular track in such a way that during part of the circuit the filter-plates are immersed in slime-pulp. The latter, containing the necessary amount of cyanide for dissolution of the values, is fed continuously into a section of the annular space between the tracks, and a portion of the pulp adheres to each plate as it passes through, in consequence of the suction.   Every plate is connected by a separate pipe with the central hollow shaft about which the apparatus revolves; this communicates with

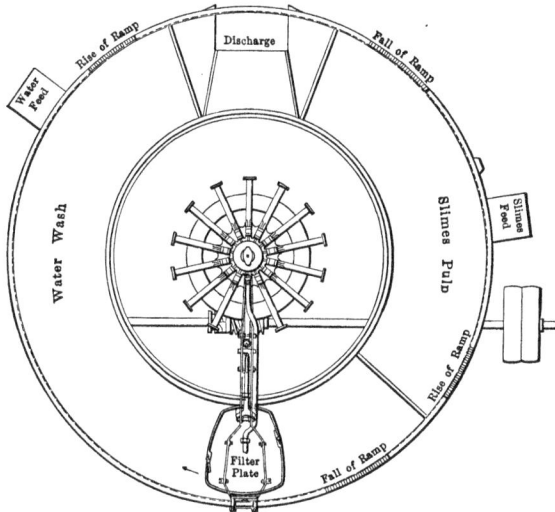

FIG. 35. — Plan of Ridgway Filter. [From illustration furnished by the Cyanide Plant Supply Co.]

the receiver of a vacuum pump, by which the filtered solution is continuously withdrawn.   At a certain point in the circuit, the outer edge of the track ascends so that the traveler rises out of

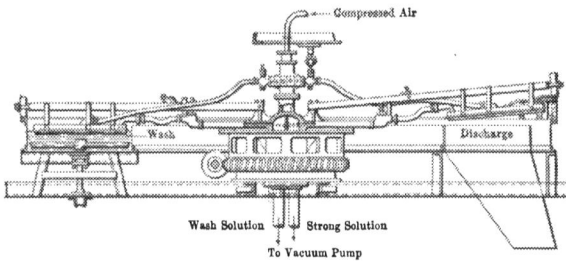

FIG. 36. — Sectional Elevation of Ridgway Filter.

the slime-pulp; beyond this point it descends again into a separate section of the annular trough, which is kept filled with water. As the filter-plates travel through this water, the latter is sucked continuously through the adhering slime cake, displacing the solution. Finally, the outer track again rises so that the carrier emerges from the liquid and at the same time the suction is automatically cut off, and compressed air applied by separate pipes connected with each filter-plate. As a result of this pressure, the slime cakes drop off and are received in a discharge hopper at this point. The carrier then descends again into the section of the apparatus containing the slime-pulp and the cycle of operations is repeated. This apparatus is also furnished by the Cyanide Plant Supply Co.

The following additional particulars with regard to the Ridgway filter may be of interest.[1] The first trials with the machine were made at the Great Boulder mine, Kalgoorlie, West Australia, January, 1906. There are 12 to 14 radial arms or levers, each bearing a flat cast-iron filtering frame, corrugated on the under surface; to these corrugated surfaces screens are attached over which ordinary filter-cloth is fixed, giving 4 sq. ft. of filtering area per frame. The central column, in its lower part, has two internal compartments, each connected separately with a vacuum pump and serving as channels for withdrawing the effluent gold solution and wash solution respectively. The upper portion of the column is either connected with an air compressor or is constructed to act as an air compressor, driven by the machine itself. Each filtering frame is connected by 3 radial pipes and rubber hose with the two compartments of the central column and with the compressed-air section.

The outer ends of the levers are provided with rollers running along a circular track at the edge of the annular trough which holds the slime-pulp and wash-water. The frames are suspended horizontally from the levers in such a way that the lower surface only is immersed; at the points between the sections of the trough, where the track is raised, the filtering frames emerge from the liquid, and at the same time one of the three valves connecting with the central column is closed; on descending again, another valve is opened. These valves are automatically operated while the machine is in motion by small rollers passing over elevations on one or other of three concentric tracks. The series of operations is as follows:

After the frame has passed the discharge chute, it descends

---

[1] "Monthly Journal Ch. of Mines," West Australia, Nov., 1906. See also "Min. Sc. Press," Feb. 9, 1907.

into the slime-pulp; at the same time the gold-solution valve is opened by a roller running up an incline on its track, opening direct connection between the filter-frame and the vacuum which is maintained in the center column.   When the frame ascends from the pulp, the gold solution valve is closed; as it descends into the wash solution, the wash valve connected with the other section of the center column is opened, causing the vacuum to draw wash solution through the adhering pulp.   The frame continues in the wash trough for half a circle; when it again rises, the wash valve is not closed till the discharge chute is reached, so as to allow the slime cake to be extracted by suction as thoroughly as possible.   When the discharge chute is reached, the wash valve closes and the air valve opens, admitting a puff of compressed air, which enters the filter-frame and causes the residue to drop off into the chute and thence into a truck.   The machine makes about one revolution per minute.   Air-tight connections between the rotating and stationary parts are secured by means of glands and stuffing-boxes.

The supply of pulp, wash solution, and water to the annular troughs is regulated by float valves in the receiving compartments, which are placed outside the peripheral track.   The pulp is kept from settling by the revolution of small agitators driven independently of the other parts of the machine.

From 0.8 to 1 ton of wash solution is used per ton of dry slime: the residues carry 28 to 33 per cent. of moisture, showing only a trace of dissolved gold and cyanide.   An extraction of 89 per cent. is claimed in the treatment of old battery slime.   The vacuum pump requires $3\frac{1}{2}$ h.p., the agitators 1 h.p., and the filter machine $\frac{1}{2}$ h.p., making a total of 5 h.p.   The capacity is 25 to 66 tons per day, according to the nature of material treated.

(I) THE BERTRAM HUNT FILTER

The following account of this apparatus is summarized from description given by the inventor.[1]   The filter-bed is composed of gravel and sand, resting on wooden slats and supported on an annular concrete structure, raised at the edges 2 in. above the filter-bed.   The slats are $\frac{1}{4}$ to $\frac{3}{8}$ in. apart.   The top layer of the filter is composed of sand of 8 to 12-mesh size, and is one

[1] "Min. Sc. Press," XCVII, 430 (Sept. 26, 1908).

inch in thickness. A vacuum filter is applied beneath, but no filter-cloth is used. A carriage bearing scrapers revolves about the central hollow pillar, and travels over the annular filter-bed, being supported on the raised edges of the concrete bottom. By means of this carriage, the material to be treated is continuously fed onto and removed from the filter-bed. The apparatus carries a scraper in front, which removes the upper layer of slime and transfers it to a screw conveyor, by which it is discharged or conveyed to the central hollow pillar, whence it is discharged by sluicing. Behind the scraper is a distributor for the sandy portion of the pulp, followed by another for the slimy portion. Revolving pipes to spray solution and wash-water follow the carriage at suitable distances.

The outside diameter of the machine is 15 ft., the annular filter-bed being 3 ft. wide, thus giving a filtering surface of 113 sq. ft. The speed is 1 r.p.m. Taking the quantity of material discharged during each revolution as represented by a layer $\frac{1}{4}$-in. thick, say 2.26 cu. ft., carrying 50 per cent. moisture, or 123 lb. of dry residue, the capacity of the apparatus would be 80 tons per day. The power required to drive the machine is about 1 h.p., and for the vacuum filter 4 h.p., or a total of 5 h.p. It is stated that a clear filtrate is always obtained.

Whenever it is desirable to clean the permanent filter-bed, the adjustable scraper is lowered and the sand above the slats removed. Clean sand is then spread to the required depth. The apparatus can be arranged with a separate vacuum for strong and weak solutions.

(J) MODIFICATIONS OF THE BUTTERS FILTER

A. *Externally Washed Leaf.* — One of the chief troubles in connection with vacuum filtration arises from the difficulty of discharging the filtered cakes rapidly and without injury to the filtering medium. This operation is usually carried out by reverse pressure of air or liquid applied from the interior of the leaf, and gives satisfactory results only when the cake is of uniform thickness and consistency, otherwise portions are apt to remain adhering to the filter surface.

In the form of leaf introduced by Chas. Butters (U. S. Pat. No. 1,078,994, Nov. 18, 1913), which has been in use since May, 1913, at the plants of the Belmont Milling Co., at Tonopah and at Millers, Nevada, the necessity of discharging the cake by internal pressure is avoided, and consequently, also, the necessity of having a uniform thickness of cake.

The framework consists as usual of a rectangle of pipes, of which the top and bottom members are perforated on the inner side, and supported on each side at the top by wooden or iron "headers" resting on a ledge at the side of the tank.   The filtering surface consists of canvas or other suitable material, the internal space being filled, preferably, with cocoa matting.   The filter cloths are stitched only at the edges, and extend as a loose flap 6 or 8 in. below the bottom of the frame.   As no reverse pressure is used, there is no need of stitching to prevent bulging of the leaf when discharging.   Two vertical strips of wood serve to give rigidity to the frame.

The perforated pipes are connected with vacuum pump in the usual way.

For discharging the cake a separate pipe is provided, along the top of the leaf, and perforated along its lower side with (say) $\frac{1}{4}$ in. holes 3 inches apart. This pipe, which is not in communication with the pipes forming the filter-frame, serves for the supply of water under any required pressure, which is thus delivered on top of the headers and discharges the cake by washing down the outside.   (See Fig. 36a.)

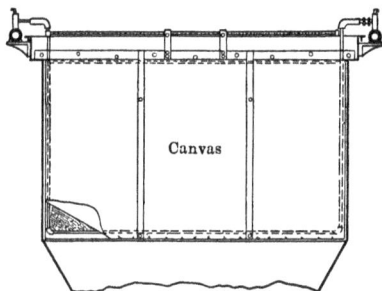

FIG. 36a. — Modified Butters Filter Leaf.   Externally washed type.

(See also Eng. Min. Journ. *96*, 1074, Dec. 6, 1913, and Min. Sci. Press, *107*, 935, Dec. 13, 1913.)

B.   *Horizontally Charged Leaf.* — This new type of filter presents several advantages over that in which the leaves are immersed in the pulp to be filtered.   In the ordinary system, the pulp tank in which the filter is immersed is kept filled with pulp until the required thickness of cake has been formed, and the surplus pulp then pumped back to the storage tank or transferred to another vessel, so that the same material is rehandled, perhaps several times, before finally discharging.   Moreover, the pulp tank must be filled before filtration can begin, and must be emptied of surplus pulp and refilled with solution before washing of the cake can begin. The present system avoids these delays and unnecessary handlings of pulp.

The filter consists essentially of a number of superimposed trays, held together in a box or frame, and mounted on a central pivot so that the whole can be inclined simultaneously at any angle. Each tray contains a filter leaf of the usual construction, having a frame of perforated pipe and a filling of wooden strips.

The pulp to be filtered is elevated by a centrifugal pump to a manifold, from which separate pipes distribute it uniformly to the upper surface of each leaf. During this operation the filter is kept in a horizontal position. The pulp overflows at the end of the frame opposite the feed, and is constantly returned by the centrifugal pump until a sufficient thickness of cake has formed. This is then drained by means of a vacuum pump connected with another manifold, which communicates by separate pipes with each of the filter leaves. Wash solution is then applied through the pulp manifold; this solution is preferably made somewhat turbid by addition of slime, which helps to fill up any cracks formed in the cake. The fact that this washing is done while the cake is exposed to the atmosphere is an advantage, as it gives the opportunity for additional aëration and a further extraction which would not occur if the leaf were entirely immersed in solution during the entire operation.

A final wash of water may then be given in the same manner.

For discharging, the filter is tilted into a vertical position, when the cakes slide off, with the aid of a little water for sluicing.

The whole system requires very little power for pumping, as compared with the old method, in which the large amount of surplus pulp has to be removed to the storage tank after each charging of the filter.

(K) THE OLIVER CONTINUOUS FILTER

This type of filter was originally elaborated at the mill of the North Star Mines Co., Grass Valley, Nevada County, California, by Edwin Letts Oliver. It has recently been extensively adopted in the western states in conjunction with continuous systems of agitation; over 100 of these filters having been installed in various parts of North America.

The machine consists of a hollow drum mounted so as to

revolve on a horizontal axis. (See Fig. 36b.) The lower part of
the apparatus is constantly immersed in a tank containing the
slime pulp to be filtered. The periphery of the cylinder is formed

FIG. 36b. — Oliver Continuous Filter.

by wooden staves and is supported by a framework of steel
spiders. The outer surface is divided into sections by divisions
parallel to the axis, and consists of a filtering medium formed by
a framework of wooden strips covered by screening, and this again
by burlap and canvas.

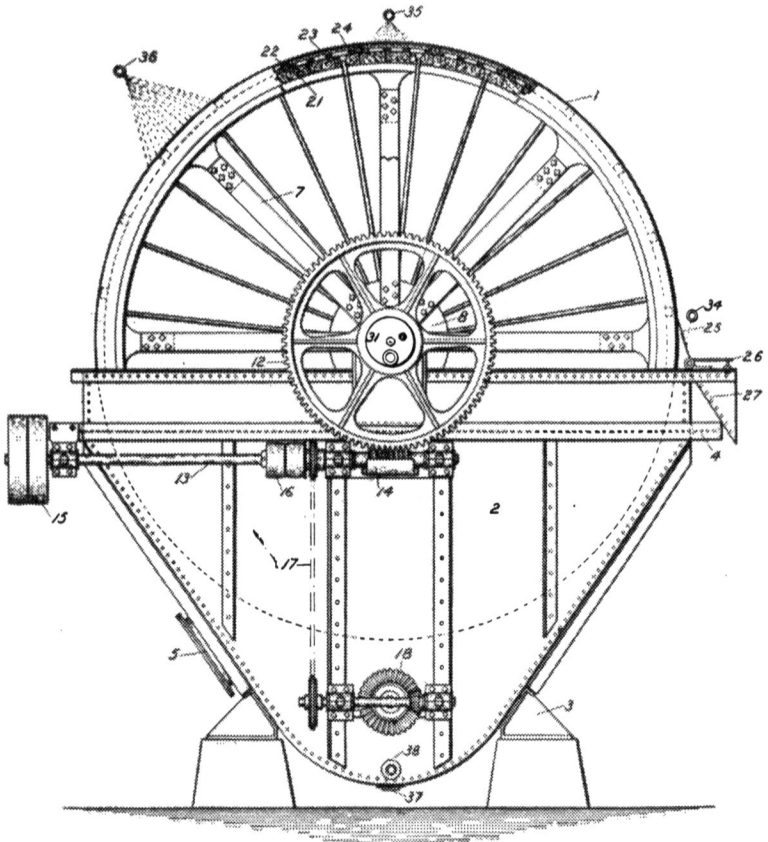

Fig. 36c. — Oliver Continuous Filter. Transverse section.

LIST OF PARTS

Nos.		Nos.		Nos.	
1.	Filter Drum.	16.	High Speed Wiring Pulleys.	29.	Compressed Air Pipes.
2.	Steel Filter Tank.			30.	Regrinding Valve Seat.
3.	Cast Iron Pedestals.	17.	Agitator Drive Chain.	31.	Automatic Valve.
4.	I Beam Frame.	18.	Bevel Gears Agitator Drive.	32.	Vacuum Connection.
5.	Manhole.			33.	Compressed Air Connection.
6.	Steel Spider Rim.	19.	Agitator Shaft.		
7.	Channel Steel Arms.	20.	Agitator Shaft Bearings.	34.	Discharge Spray Pipe.
8.	Hollow Trunnion.	21.	Wood Staves for Drum.	35.	Water Wash Atomizer Pipe.
9.	Steel Shaft.	22.	Section Division Strips.		
10.	Main Bearings.	23.	Filter Medium.	36.	Solution Wash Atomizer Pipe.
11.	Stuffing Boxes.	24.	Wire Winding.		
12.	Worm Drive Gear.	25.	Steel Scraper.	37.	Drain Flange.
13.	Worm Shaft.	26.	Scraper Adjuster.	38.	Air Agitator Pipes.
14.	Oil Well for Worm.	27.	Tailing Apron.		
15.	Filter Drive Pulleys.	28.	Vacuum Pipes.		

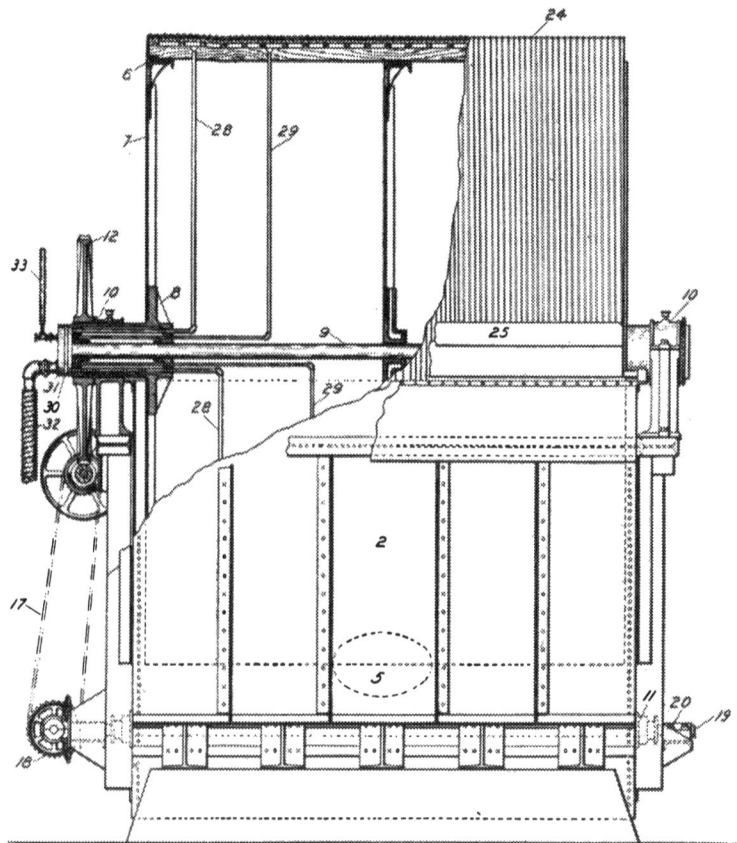

Fig. 36*d*. — Oliver Continuous Filter.   Longitudinal section.

Each filter cell thus formed is provided with two sets of pipes connecting with the hollow trunnion at one end of the drum. This latter connects with automatic valves so arranged that vacuum is applied through one set of these pipes during the major part, $(\frac{2\,3}{2\,4})$, of the revolution of the drum and air-pressure through the other pipe during another part (to release the cake only). The filter cycle is controlled by the arrangement of the automatic valves, which consist of a valve plate into which are screwed the vacuum and blow pipes leading to the various filter sections.   The valve chamber seats against this plate and is held firmly in place by vacuum, a simple valve stem being used to prevent it from turning. The application of vacuum and compressed air may be readily changed at any time to suit the necessities of treatment.

The pulp in the filter tank is kept in agitation by a mechanical stirrer, or by circulation with air-lifts, or both.

NOTE. — The old system of air injection has been discontinued.

During the passage of the cylinder through the pulp, the vacuum causes a portion of the latter to adhere to the filtering surface, and as the suction continues during the first half of the revolution, the solution is extracted, until when the pulp reaches a point (variable depending on slime) of the circumference at which only about 30 per cent. of moisture remains, the adhering slime cake meets an atomized spray of water delivered from a series of atomizers running parallel to the axis of the drum, so that a water-wash is applied, which is sucked through by the action of the vacuum as the revolution of the cylinder carries the slime cake beyond this point. As it descends, the vacuum is automatically cut off at a certain point and pressure applied, to loosen the cake, which now contains about 30 per cent. of moisture. Just before the filter-surface again reaches the pulp tank, a scraper removes the loosened cake and discharges it into a launder or on to a traveling belt, the discharge being assisted by a small water spray in the former case only. The filter surface is protected and held in place by wire winding, which also serves as a guide to the scraper. The cylinder makes one revolution in from 5 to 10 minutes, the speed depending on the nature of the material filtered. The thickness of cake varies from $\frac{1}{4}$ to $\frac{1}{2}$ in.

The power required for driving the drum and agitator is under $\frac{1}{2}$ h.p. The vacuum pumps and compressor consume an average of 1 h.p. for 10 tons of slime filtered per 24 hours. The size of the drum varies according to the service required, up to 15 ft. diameter, by 20 ft. broad. The capacity of the filter is from 25 to 150 tons dry slime per 24 hours.

Among the advantages claimed for this system are that it is absolutely continuous and automatic in operation, and that the costs for power and maintenance are low in comparison with those for other types in which the pulp must be transferred by pumps to and from storage tanks; also that it is possible to discharge the residue with a minimum of moisture, so that it may be removed by belt conveyors.

SECTION IV

HANDLING OF SOLUTIONS

(A) Sumps and Storage Tanks

In addition to the vats or other vessels used for holding the material under treatment, storage room must be provided for the solutions, before and after use. The volume occupied by the solution may be taken as about 31 to 32 cu. ft. per ton. The volume occupied by ordinary tailings varies from 20 to about 26 cu. ft. per ton, according to the density with which it is packed into the leaching vats. It is impossible to give any fixed rule as to the capacity required for storing solutions, but it may be taken as about equal to the vat capacity required for the material actually undergoing treatment at any one time. As already mentioned, there are in general three classes of solution, for which separate tanks must be provided, namely, strong cyanide, weak cyanide, and alkali.

The storage tanks are generally similar in construction to the leaching tanks, but without filter-frames, and may be of wood or iron. It is usual to make them deeper in proportion to diameter than the leaching tanks, as this form is more economical in construction, and there is in this case no limitation of depth on account of difficulties in percolation. They are generally placed either above the level of the leaching vats, so that the solution can pass from them to the latter by gravity, or below the outlet of the precipitation boxes, so that the precipitated solution flows into them by gravity. In the latter arrangement the storage tanks are commonly described as "sumps." In the first case, arrangements must be provided for continuously pumping the precipitated solution as it leaves the boxes up to the storage tanks. The solution flowing from the boxes runs into a small cement-lined depression or sump in the floor of the extractor house, from which it is pumped at the same rate. In the second case, with the storage tanks below the precipitation boxes, every solution

279

run on to the leaching tanks must be pumped up, as required, but the pumping will not be continuous.

It is obvious that the same amount of liquid must be raised in both cases; but in the first case the average hight of lift will be greater, as all the solution must be raised from the level of the boxes to the top of the storage tanks, whereas in the second case the average lift will be from half the depth of the storage tanks to the top of the leaching vats.

Dissolving Tank. — A special small tank is generally provided for dissolving the solid cyanide and preparing a concentrated solution, from which the necessary amount is drawn off as required for making up the solutions in the storage tanks.

(B) Settling and Clarifying Tanks

It frequently happens, for various reasons, that the solutions leaving the leaching vats, or decanted from slime-settling tanks, and also those drawn off from filter-presses or suction filters, are not perfectly clear. Consequently it is the usual practice to run such solutions into a separate tank, instead of allowing them to flow direct to the head of the precipitation box. When the liquid has settled clear, it is drawn off by syphon or otherwise, and the introduction of sand or other suspended matter into the boxes is avoided. The arrangement has the additional advantage that a perfectly regular flow of liquid through the boxes can be maintained, as the rate of flow need not depend on the rate of leaching, decanting, etc. In some cases it is advisable to allow the solutions to percolate through a sufficient thickness of clean sand, in what is described as a clarifying tank. Other materials, such as cotton waste, coir matting, etc., are likewise used for clarifying. Sometimes the solution is introduced from below and caused to percolate upwards through the clarifying medium.

Another arrangement sometimes adopted is to pass the liquid to be clarified through a special small filter-press adapted for this class of work. An account of this method is given by Truscott and Yates.[1] Previous to the introduction of the clarifying press, at Redjang Lebong, Sumatra, slime had been carried into the zinc-boxes, notwithstanding the use of several successive settlement tanks and the addition of lime in the launder from the filter-

[1] "Journ. Chem., Met and Min. Soc. of South Africa," VII, 3 (July, 1906).

presses, and caused much trouble and expense in the smelting of the precipitate. It was found that the solution could be effectively clarified by passing it through two small Johnson presses; these were opened twice a week and the cloths well scrubbed to free the pores from fine slime. The beneficial effect of the filter-press clarification upon the solution was at once felt in the smelting; the precipitate became clean and was easily reduced, the smelting costs being reduced from 3.07 pence to 0.87 pence per ounce of fine metal recovered. (See also Section F.)

(C) Pumps

The pumps commonly used for transferring solutions or for circulating slime-pulp are centrifugal pumps, the size varying with the work required. A 4-in. pump is a convenient size for many purposes. When the pumping required is for the continuous transfer of small quantities of solution, several small pumps, one at least for each class of solution, will be needed; this will be the case when the storage tanks are above the leaching tanks. Where pumping is only necessary at intervals and the object is to transfer large quantities of solution rapidly, one or more large pumps are necessary. This is the case when the storage tanks are below the leaching vats.

Pumps with brass or copper fittings should be avoided; all parts liable to come in contact with solution should be of iron or steel. For special purposes other types of pump are used. For transferring sandy or gritty material the "air lift" is commonly employed. Some account of the Frenier sand pump has already been given. (See Part III.) The power required for transferring slime-pulp is considerably greater than that needed for solution.

(D) Piping for Conveying Solutions

Iron pipes are used for transferring solution (1) from the storage tanks to the leaching tanks, agitation tanks, or filter-presses; (2) from the treatment tanks to the precipitation boxes or clarifying tanks; (3) from the precipitation boxes to the storage tanks. When, as is usually the case, several different kinds of solution are used, arrangements must be made to convey each class of solution as required to each treatment tank. This may be done either by having a separate pipe for each class of solution,

with a cock over each treatment tank, or by having a common
"solution main" for all classes of solution, with separate pipes
and cocks between it and the storage tanks, and also separate
branch pipes and cocks for each treatment tank. The latter
system is more economical as regards piping and the number of
cocks required, but is perhaps somewhat more complicated in
actual working, as there is more likelihood of mistaking the solution required; and it also involves the necessity of opening two
cocks for each transfer.

When a pump has to be used for filling a tank or transferring
a solution to a treatment tank, it is necessary to open the cock
before starting the pump; otherwise the pipe may burst or the
driving belt of the pump may be thrown off.

The pipes used for running solution to the tanks should be
fairly large, so that this operation may be carried out as quickly
as possible. For conveying solution from the treatment tanks
to the settling or clarifying tanks, or direct to the precipitation
boxes, a separate pipe from each treatment tank is desirable,
so that the nature of the effluent and the rate of flow from each
tank may be readily observed. Some arrangement is also necessary to divert the effluent as required into any precipitation
box. This may easily be done by means of a length of rubber
hose at the exit end of the pipe conveying solution to the boxes.
In some plants separate receivers or launders are provided for
each class of solution, with which the pipes from the tanks may
be connected as required; these receivers are connected with one
or more of the boxes set apart for each class of solution.

For conveying solution from the boxes to the storage tanks,
it is often necessary to arrange for putting any box in communication with any of the storage tanks; the system will therefore be
similar to that for connecting the storage tanks with the treatment tanks. In cases where separate boxes are reserved exclusively for one class of solution, however, it is only necessary to
have a single pipe connecting each box with its appropriate
storage tank.

In general, it may be stated that it is desirable to keep the
different classes of solution as distinct as possible throughout
the whole series of operations; where large volumes of slime-
pulp have to be rapidly transferred, open wooden launders may
often be substituted for pipes with great advantage, both owing

to reduced friction and owing to the opportunity for aeration which they afford.

The inside diameters of solution pipes vary from 1 to 4 in., according to the rate of flow required in the liquid passing through them.

(E) Arrangements for Heating Solutions

In very cold climates it is necessary to enclose the whole plant in a suitable shed or building, warmed sufficiently to prevent the solutions from freezing. In some cases it has been found that higher extractions are obtained by the use of hot solutions, at a temperature of, say, 100° to 130° F., and provision is sometimes made for heating solution or slime-pulp by the injection of steam. According to Julian and Smart's experiments,[1] the maximum solubility of gold in cyanide solutions of 0.25 per cent. KCy is observed at a temperature of 85° C. (185° F.). But as the temperature of the solution also affects the action of cyanide on other constituents of the ore, heating the solution may in some cases be a disadvantage. In most cases it is doubtful whether the improved extraction outweighs the increased cost. Experiments made by the writer in conjunction with Mr. J. J. Johnston at Minas Prietas, Mexico, showed no distinct advantage; other factors in the treatment seemed to be of greater importance, as the higher extractions were obtained sometimes with cold and sometimes with hot solutions.

(F) The Merrill Sluicing Clarifying Filter

The clarifying of slightly turbid solutions by means of sand filters has the drawback that the filter-bed quickly becomes coated on the top with a layer of slime, which greatly retards the flow. Consequently very large, shallow tanks must be used, or the filter-bed must be renewed frequently.

To meet this objection, filter-presses have been employed for the purpose, but in the ordinary types, the capacity is soon reduced, owing to the formation of a layer of extremely fine slime on the cloths, rendering the opening and closing of the press for the purpose of cleaning the cloths frequently necessary.

The use of vacuum filters with stitched leaves for this purpose

---

[1] *Loc. cit.*, p. 91.

is unsatisfactory on account of fine slime being drawn through the stitching, so that a clear filtrate is not obtained.

The Merrill clarifying filter is designed to overcome these drawbacks. In construction it does not differ materially from the ore slime filter already described, and may likewise be operated either by gravity or by pressure supplied by a pump.

The rotating sluicing pipe is operated by hand, with a pressure of water or solution amounting to 40 lb.

PART V

THE PRECIPITATION AND SMELTING PROCESSES

THE processes of precipitation and smelting have for their object the recovery of the gold and silver extracted by the cyanide solution in a marketable form.

In the precipitation process, the precious metals are thrown out of solution with as little admixture of foreign impurities as possible and obtained in the solid state, but usually in a very finely divided condition. In the smelting process, the finely divided precipitate is brought by the action of heat into a coherent metallic mass, and impurities are removed as far as practicable by the use of suitable fluxes and reagents. Many different processes of precipitation have been proposed, some of which will be described in a subsequent part of this book; the present discussion will be confined to the usual method, namely, the deposition of the gold and silver on zinc shavings or filaments.

The finely-divided precipitate obtained in this process is frequently referred to as "slimes"; here, however, this term will be restricted exclusively to its more usual sense, namely, the finely divided and generally unleachable material obtained in the crushing or grinding of ores.

SECTION I

ZINC–BOX CONSTRUCTION AND PRACTICE

(A) Zinc-Box Construction

The precipitation by zinc shavings is carried out in rectangular boxes or troughs, divided by transverse partitions into a number of compartments. The partitions are so arranged that the solution flows alternately downward through a narrow compartment, and upward through a wide one, the latter alone containing the zinc shavings. The hight of every alternate partition is one or two inches lower than that of the succeeding one, so that the liquid, after ascending through a wide compartment, overflows the lower partition into the narrow space between it and the next (higher) partition; the lower edge of the latter is raised several inches above the floor of the box, so that the solution flows under it into the next wide compartment, and so on. (See Fig. 37.)

• The dimensions of the boxes vary, according to the work required of them; the following may be taken as representing ordinary conditions: Length, 12 to 24 ft.; width, $1\frac{1}{2}$ to 3 ft.; hight, $2\frac{1}{2}$ to 3 ft.

The sides, ends, and bottoms are usually constructed of $1\frac{1}{2}$-in. boards, strongly held together by bolts passing horizontally from side to side of the box, generally through the narrow compartments. Vertical bolts are also used for securing the floor of the box to the sides. The partitions, which may be of thinner boards, fit into vertical grooves.

The box is preferably raised somewhat above the floor of the extractor house, and should have a slight fall from the feed to the discharge end. Sometimes the bottom is inclined to one side, and plugs are provided on this side just above the bottom, so that the precipitate may be easily collected during the clean-up. The advantage of this latter arrangement is somewhat doubtful, as it may lead to losses through leakage, etc.

In each of the wide compartments is a tray, resting loosely

287

Sectional End
Elevation

Elevation (Part Section)

Plan (Part Section)

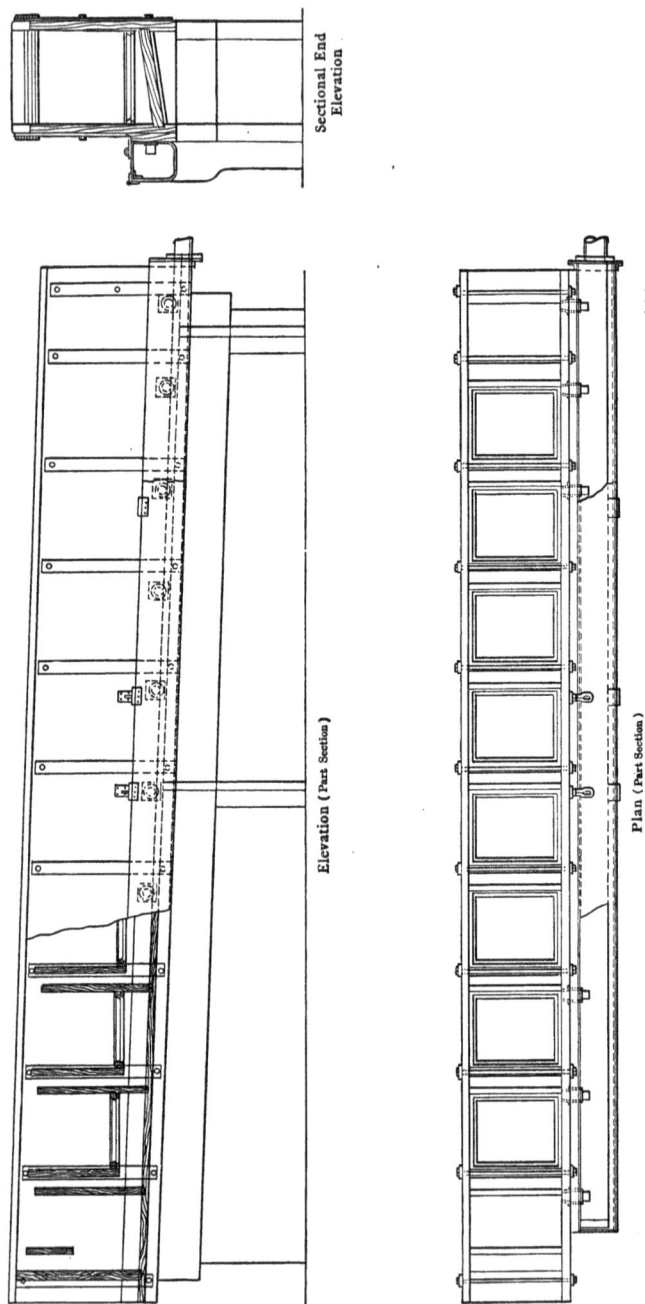

Fig. 37. — Extractor Box. These drawings are to scale, but dimensions vary with the capacity of the box. [From plans furnished by the Cyanide Plant Supply Co.]

on supports fastened to the sides of the box. The bottom of this tray is formed of wire screening, generally with four holes to the linear inch, and serves to hold the zinc shavings. The sides are of wood, with iron handles by which the tray may be lifted out of the box when necessary. The last compartment, at the exit end of the box, has an outlet near the top, from which a pipe leads either to the storage tanks or to a small sump whence the precipitated solution is pumped to the storage tanks.

Zinc-boxes of similar design are also constructed of iron or steel, painted or enameled so as to resist the action of cyanide or alkali solutions. For further details of zinc-box construction, see Julian and Smart.[1]

(B) Circular Vats for Zinc Precipitation

W. A. Caldecott[2] has advocated the use of circular vats in place of zinc-boxes for precipitation. The vat is filled with zinc shavings resting on a circular perforated tray. The solution is introduced from below and ascends evenly through the mass of shavings. The exit pipe is placed near the top of the vat, and extends to the center, being bent upward at the orifice. The advantages claimed are that the tendency to uneven flow observed in rectangular boxes is avoided, that there is less liability to leakage, and that there is economy of space as compared with ordinary boxes. In the rectangular box there is a likelihood of channels of easy flow occurring at the sides and especially at the corners of the compartments, leading to imperfect precipitation, unless great care is exercised in packing the shavings into the box.

A further development of this idea is due to P. S. Tavener,[3] who uses a circular vat with a conical bottom. A stirrer mounted on a central vertical axis has paddles which revolve in the space beneath the perforated bottom. At the clean-up, the precipitate which passes through the screen may thus be swept toward the central discharge hole. The solution to be precipitated descends through a central column and ascends through the perforated bottom, passes through the mass of zinc shavings, and overflows into a circular launder at the top of the vat.

[1] " Cyaniding Gold and Silver Ores," 2d edition, pp. 306–307.
[2] " Proc. Chem., Met. and Min. Soc. of South Africa," II, 762 (Aug., 1899).
[3] Julian and Smart, loc. cit., p. 308.

(C) Cutting and Preparing the Zinc Shavings

The shavings are prepared from sheet zinc, which is cut into disks punched in the middle, so that a large number can be set side by side on a lathe. While rapidly revolving, a chisel is held against the edges of the disks. The width and thickness of the shavings can be varied within certain limits by the manner of holding the cutting tool, and by altering the speed of revolution.

There is considerable difference of opinion as to the best thickness for the zinc shavings. As effective precipitation depends on exposing a large surface to the liquid, it would seem that the shavings should be cut as fine as possible; on the other hand very fine shavings rapidly dissolve or break to pieces, and eventually form lumps which choke the zinc-boxes, giving a precipitate full of small pieces of zinc which is very troublesome to smelt. Ordinary commercial zinc is preferable for this purpose to chemically pure zinc, owing to the electrical actions set up by the contained impurities, but very impure zinc should not be used, especially if containing much arsenic and antimony, as these metals may give rise to much trouble in the subsequent treatment of the precipitate. For effective work, the shavings should be freshly prepared, and as free as possible from oxide.

(D) Lead-Zinc Couple

In order to increase the efficiency of the zinc for precipitating, it is a very usual practice to dip the shavings, before charging them into boxes, in a 1 to 5 per cent. solution of lead acetate, which causes an immediate deposit of finely-divided lead on the surface of the zinc. This forms a powerful galvanic couple, which is capable of depositing gold from solutions much weaker in free cyanide than can be precipitated by the ordinary zinc shavings. In cases where much copper is present in the solution, the use of zinc-lead couple is often advantageous. The upper compartments, in this case, are filled with ordinary shavings and serve to precipitate gold and silver; in the lower compartments are placed the shavings which have been dipped in lead acetate; these precipitate copper much more energetically than the plain zinc shavings, and so prevent this metal from accumulating in the solution to an injurious extent. This important improvement in the zinc precipitation process was introduced by J. S. MacArthur at the Lisbon Berlyn mine, Lydenburg, Transvaal, about 1894,[1] with successful results, but does not appear to have been adopted elsewhere until it was brought forward at Johannesburg, in 1898, by W. K. Betty and T. L. Carter.

(E) CHARGING THE ZINC INTO THE BOXES

The best results are obtained when the zinc forms a homogeneous, spongy mass. The compartments should be well filled, but the shavings should not be pressed down too tightly. Especial care should be taken in filling the corners, and any parts where it appears likely that channels may form. The boxes should be inspected from time to time while the solution is passing through and any irregularities in the flow rectified by adding fresh shavings, if necessary. A plan sometimes adopted is to arrange the shavings in layers crossing each other alternately at right angles. The zinc is cut into very long threads, as thin as possible without becoming too brittle; these are laid in separate bundles on the trays in each compartment; another layer of bundles is placed over these at right angles. This arrangement is said to be very effective in preventing irregular flow.

The quantity of zinc to be used per cubic foot of space varies considerably, according to the manner in which it has been cut and other circumstances. The action of the solution on the zinc is naturally most active near the head of the box; as the zinc dissolves in the top compartments, the shavings from the lower compartments are taken out to replace them, and fresh zinc is placed in the lower compartments.

The first and last compartments of the box are frequently left empty, the first as a safeguard against suspended matter which might be carried through by the solution and deposited on the zinc, and the last as a trap to retain any fine precipitate which might be carried over by the flow of solution, especially during the transfer of shavings from one compartment to another or during any disturbance of the box. The last compartment has occasionally been filled with coke or sawdust, with the object of filtering out any finely divided precipitate which might be contained in the effluent. This compartment would, in the latter case, only be cleaned up occasionally, perhaps once in three months, when the filtering material would be burned and the

[1] " Proc. Chem., Met. and Min. Soc. of South Africa," II, 448 (Oct. 1898).

values recovered by smelting the ashes. Whenever the zinc in the boxes is to be moved or disturbed in any way, the flow of solution should be cut off or diverted to another box, to avoid the danger of suspended gold precipitate being carried out of the boxes in the effluent solution.

The boxes should be "dressed" at least twice a day, and fresh zinc added as required. Careful attention to the filling of the zinc-boxes is a very important factor in the successful operation of the process.

(F) Conditions which Influence Precipitation

The principal conditions on which satisfactory precipitation depends are: (1) the freedom of the entering solution from solid matter in suspension; (2) the presence of a sufficient amount of free cyanide; (3) the presence of sufficient free alkali; (4) the absence of excessive amounts of foreign salts, particularly of copper and iron compounds; (5) a regular and uniform flow of liquid through the box; (6) a sufficient volume of zinc per ton of solution to be precipitated per day; (7) a clean, unoxidized surface, on which the gold and silver are deposited in a loose flocculent condition, easily detached.

Influence of Suspended Matter. — The precautions usually taken to prevent the entrance of suspended matter — sand, slime, organic substances, etc. — into the boxes have already been described (Part IV, Section IV, B). When such substances are present they are not only liable to coat the zinc shavings with deposits of different kinds, thus interfering mechanically with the precipitation of the precious metals, but the foreign matter thus introduced adds greatly to the expense and trouble of the smelting operation, extra fluxes being required to remove it. Where efficient settling tanks or clarifiers are not provided, it is customary to leave the first wide compartment of the box empty with the object of collecting any sediment. In some cases this compartment is filled with coarse sand, coir fiber, or other filtering medium.

Influence of Free Cyanide. — When ordinary zinc shavings or filaments are used for precipitation, it is found that the precipitation of gold and silver is very much more effective in solutions carrying a fair amount of free cyanide than in solutions weak in cyanide. The strength of the entering solution should never be allowed to fall much below 0.2 per cent. KCN, and in some cases it is advantageous to allow a strong solution to drip slowly into the top compartment of the box, to ensure a sufficient strength for good precipitation. When the zinc-lead couple is used the matter is of less consequence and very weak solutions can in general be precipitated by it. T. L. Carter[1] found that solutions of 0.005 to 0.008 per cent. free cyanide could be satisfactorily precipitated, with an extraction of 90 to 95 per cent. of the gold, using the zinc-lead couple and a drip of strong cyanide at the head of the box, so that at the beginning the strength of the entering solution was raised to 0.025 per cent.; after 12 or 14 hours the strength was allowed to fall to 0.008 per cent. Other workers using the lead-zinc couple have found the strong cyanide drip quite unnecessary. Occasionally a drip of lead acetate solution is used.

It would appear that the presence of some other salts besides cyanide may in certain cases induce good precipitation of gold on zinc, for it has been observed that solutions which have become very "foul," that is, highly charged, with accumulated salts, chiefly complex cyanogen compounds of iron, zinc, copper, etc., sometimes precipitate their gold contents on clean zinc better than solutions otherwise pure, but weak in free cyanide,[2] and it is even possible to precipitate in the complete absence of free cyanide. Another advantage of a sufficient strength in free cyanide is that the formation of injurious deposits in the boxes is largely prevented. The so-called "white precipitate" consisting of hydrate, cyanide, and ferrocyanide of zinc, chiefly occurs in boxes where weak solutions are precipitated. Strong cyanide solutions also are less liable to precipitate copper in preference to gold.

Influence of Alkali. — A certain minimum of free alkali is found to be essential for good precipitation. On the other hand, a large excess should be avoided. Whether alkalies actually decompose the double cyanides of zinc with liberation of free cyanide appears somewhat doubtful, but their presence undoubtedly increases the solvent efficiency of these compounds, and also promotes precipitation. They are also solvents for some at least of the constituents in the white precipitate referred to above.

[1] " Proc. Chem., Met. and Min. Soc. of South Africa," II, 440.
[2] Alfred James, " Trans. A. I. M. E.," XXVII, 278.

Excessive amounts of alkali cause a violent action on the zinc, with great waste of the metal, accompanied by "foaming," due to the evolution of large quantities of hydrogen, and a tendency for the gold precipitate to be detached and carried away by the flow of solution. Too large a quantity of lime in solution may lead to the deposition of insoluble calcium compounds, carbonate or sulphate, on the surface of the zinc, owing to reactions between the lime and the soluble carbonates and sulphates present as impurities in the cyanide or extracted from the ore. During the passage of solution through the boxes an increase of alkalinity is invariably observed. This is explained in the section dealing with the chemistry of the process.

It is occasionally found necessary to add alkali, either in the zinc-boxes themselves or in the storage tanks. In some cases addition of sodium carbonate to the solution in the sumps has a beneficial effect, as it precipitates salts which might otherwise cause a deposit of insoluble carbonates in the boxes.

Rate of Flow. — Under a given set of conditions the slower the rate of flow the more complete will be the precipitation. It is found, however, that an increased rate of flow may be compensated for by increasing the surface of zinc with which the solution must come in contact; and as a rapid flow has other advantages, it is generally good practice to pass the solution through the boxes as rapidly as is consistent with efficient precipitation.

In the early days of the process it was considered necessary to use about ¾ cu. ft. of zinc shavings to precipitate the solutions used in treating each ton of ore per day;[1] according to modern practice, as much as 7 or even 12 tons of solution can be precipitated per cubic foot of zinc per day.[2] The actual rate of flow in the latter case may be from 7 to 13 cu. ft. per minute across a section of 14 sq. ft. of zinc, or 0.5 to 0.92 ft. per minute for each square foot of section. This method, however, can only give good results when much care is taken in charging the compartments.

According to Caldecott and Johnson,[3] a pound of ordinary zinc shavings exposes about 40 sq. ft. of surface; a cubic foot of

[1] W. H. Gaze, "Handbook of Practical Cyanide Operations" (1898).

[2] Julian and Smart, *loc. cit.*, p. 142.

[3] "Proc. Chem., Met. Min. Soc. of South Africa," IV, 263. See also Gaze, *loc. cit.*

the shavings, as usually packed in the compartments, would weigh from 3.5 to 4 lb., and hence exposes a precipitating surface of 140 to 160 sq. ft. Taking the volume of a ton of solution as 32 cu. ft. (Part IV, Section IV), a zinc-box of 8 compartments, each containing 4 cu. ft. actually occupied by zinc shavings, would contain at any moment 1 ton of solution undergoing precipitation and exposed to a precipitating surface of about 4800 sq. ft. If 1 ton of solution be precipitated per cu. ft. per day, 32 tons of solution must pass through the box per day; in other words, the rate of flow will be 1 ton in 45 minutes.

Nature of the Precipitating Surface. — It has already been pointed out that the best results are usually attained by cutting the zinc in very long, thin threads. These usually have a thickness of $\frac{1}{250}$ to $\frac{1}{1000}$ in., and a width of $\frac{1}{16}$ to $\frac{1}{4}$ in., but practice varies very widely in these particulars. The smaller surface exposed by thick shavings may be largely compensated for by using the lead-zinc couple and thus obtaining greater precipitating efficiency. Mixtures of lead and zinc shavings, and shavings prepared from an alloy of zinc containing 3 per cent. of lead, have also been used, and are more effective, especially with solutions weak in cyanide, than ordinary zinc shavings. These combinations, however, are less generally adopted and probably less energetic than the couple formed by immersing the zinc in a solution of a lead salt.

Zinc-copper and zinc-mercury couples have also been tried, and are as efficient as the zinc-lead couple in promoting precipitation of gold and silver; but they have certain serious drawbacks. Mercury causes the shavings to fall to pieces and rapidly reduces them to a pulpy mass, from which the mercury may be separated by squeezing through a cloth.[1] When copper is used for this purpose, a very base bullion is obtained in the smelting of the precipitate.

Great care must be taken to expose the shavings to the atmosphere as little as possible. When charged into the boxes, they should be freshly turned, clean and unoxidized; the compartments should be filled so that about an inch of solution is left over the top of the zinc, and the moist shavings especially should never be exposed to the air any longer than absolutely necessary during transfer from one compartment to another.

[1] W. H. Virgoe, "Proc. Chem., Met. and Min. Soc. of South Africa," IV, 615.

It may generally be assumed that satisfactory precipitation is taking place when the deposit of gold and silver is in a loose condition and of a brownish-black color. When a smooth, coherent, metallic deposit is observed, the precipitation soon ceases altogether, and it is probable that some alteration is necessary in the conditions.

A vigorous, but not too violent, evolution of hydrogen also accompanies good precipitation; the reduction of the gold-alkali cyanide and consequent precipitation of the gold is commonly supposed to be due to nascent hydrogen, that is, to an action taking place between the gold compound and the hydrogen at the moment when the latter is set free by the decomposition of water, as explained in Part II. On the other hand, the accumulation of bubbles of hydrogen on the surface of the zinc hinders precipitation. The gas in this form has no action whatever on the gold compound, and prevents contact between the solution and the zinc surface. Hence the occasional removal of the bubbles by shaking or turning over the zinc shavings tends to promote good precipitation.

(G) Difficulties in Zinc Precipitation

These difficulties have already been alluded to in discussing the conditions which influence precipitation; it may, however, be of advantage to recapitulate them. They are: (1) Deposition of suspended matter on the surface of the shavings. (2) Formation of insoluble compounds coating the shavings, by chemical action of cyanide, alkali, or other substances on the zinc. (3) Formation of insoluble compounds by subsidiary actions going on between various substances in the solution passing through the boxes. (4) Formation of thin metallic coatings (of gold, silver, copper, etc.), which protect the zinc from the solution and so check further precipitation. (5) "Polarization" of the zinc, by accumulation of hydrogen bubbles on its surface. (6) Imperfect deposition of gold and silver from solutions not sufficiently strong in cyanide. (7) Excessive consumption of zinc under certain conditions. (8) Irregularities of flow due to the formation of channels, thus leading to inefficient precipitation.

Deposits Formed by Action of Solution on Zinc. — These are generally due to insufficiency of free cyanide. Researches on the nature of these deposits have shown that they generally consist principally of hydrated zinc oxide,[1] presumably formed by the oxygen liberated in the electrolysis of water by the various metallic couples present in the box, in which the zinc acts as anode. Relatively small quantities of zinc cyanide may also be present. As these compounds are soluble in a sufficiently strong cyanide solution, the obvious remedy is to increase the strength of the solution entering the box by addition of cyanide, but this is not always practicable. When large quantities of iron are dissolved from the ore as ferrocyanides, there may be a precipitation in the boxes of ferrocyanide of zinc, or of some insoluble double ferrocyanide of zinc and an alkali metal. These deposits are still more troublesome as they do not readily redissolve, either in cyanide or alkali. In many cases the only remedy is, after the deposit has accumulated to a certain extent, to clean up the box and start afresh with clean zinc. Ammonium salts and sodium chloride have sometimes been added as solvents for deposits of insoluble zinc compounds.

Caldecott and Johnson (loc. cit.) appear to consider that a mixture of lead and zinc shavings is preferable to the lead-zinc couple in cases where deposition of zinc hydrate takes place; they state that "a mixture of lead and zinc shavings has a lengthy efficiency, possibly due to the fact that the zinc hydrate does not coat the whole couple to the same extent as with the usual lead-coated zinc shavings."

Other Deposits Due to Subsidiary Reactions. — The deposition of calcium carbonate and sulphate in the boxes has already been mentioned. Under certain conditions alumina, and possibly silica from soluble silicates, may separate out. Solutions containing manganese very readily throw down a brown hydrated oxide of the metal.

Sulphur, either as finely divided pyrites carried through mechanically and settling on the shavings, or as sulphide of zinc due to the reaction between zinc hydrate and alkaline thiocyanates in the solution, is also frequently present, and when iron or other base metal is contained in the precipitate may give rise to the formation of a matte during the smelting.

At Redjang Lebong, Sumatra, selenium is precipitated in the zinc-boxes in considerable quantity, but no evidence is at hand as to the form in which it occurs. It is particularly trouble-

[1] "Journ. Chem., Met. and Min. Soc. of South Africa" V, 62, 75.

some, as it is extremely difficult to eliminate it completely in the subsequent operations, and its presence in the bullion even in small quantities is very detrimental, as it renders it brittle.

Arsenic, antimony, lead, and mercury, if present in solution, will also be deposited on the zinc. Tellurium probably acts similarly to sulphur and selenium.

"*Metallic*" *Coatings.* — As already pointed out, gold is sometimes deposited, not in a loose flocculent condition, but as a bright metallic coating having the natural color of gold. This occurs chiefly when the solutions are either very rich in gold, or low in cyanide. There are considerable grounds, as pointed out by Argall, for supposing that the ordinary loose precipitate is a compound of zinc, gold, and cyanogen, while the metallic coating referred to above is probably gold in the elementary state. Similar metallic coatings may be formed with silver, copper, and mercury, and when they occur, further precipitation in those parts of the box in which they are formed is almost entirely arrested. It has been found that the activity of the coated shavings can be restored by immersing them for a few moments in dilute sulphuric acid, washing in water, and at once replacing in the boxes.

Consumption of Zinc. — The normal consumption of zinc on a cyanide plant is from 0.25 to 0.5 lb. per ton of ore treated, but in special cases it may be considerably more or less than these amounts. Only a portion of this is actually dissolved by the solution, and of that which is dissolved, only a small part corresponds with the gold or silver precipitated. A large part goes in the form of small fragments of metallic zinc into the precipitate collected during the clean-up, and a further quantity is dissolved by subsidiary reactions with cyanide, alkali, or oxygen from decomposition of water, as already described. We may thus distinguish two sources of consumption, mechanical and chemical, the former including all zinc passing in a solid form into the precipitate, and the latter all that goes into solution. When strong cyanide solutions are used the chemical loss is greater, but the mechanical loss is less than with weak cyanide solution.

The following particulars, given by W. H. Virgoe [1] with regard to two Mexican plants, treating sand and slime respectively, are of interest in this connection.

---

[1] " Proc. Chem., Met. and Min. Soc. of South Africa," IV, 615.

	Sand Plant	Slime Plant
Number of zinc-boxes	4	4
Number of compartments per box..............	8	5
Total zinc capacity in cu. ft.	40.64	60
Weight of zinc per cu. ft. (lb.)	13.5	13.5
Total zinc capacity (lb.).	584.64	810
Zinc-space per ton of solution per hour (cu. ft.) ..	25.4	20
Total consumption of zinc per ton (lb.)	1.144	1.166
Mechanical loss of zinc (per cent.)	13.9	63.2
Chemical loss of zinc (per cent.)	86.1	36.8
Tonnage treated in 4 months	2646.4	1481
Consumption of cyanide per ton (lb.)	1.5	0.6
Strength of entering solution (per cent. KCy)	0.32	0.04

The curious feature of these results is that the total consumption of zinc per ton of ore treated was practically the same in both plants, but the nature of the loss was entirely different, being chiefly chemical in the case of the sand plant, where strong solution was used, and chiefly mechanical in the slime plant, where the solution entering was very weak in cyanide. In the latter case the gold was no doubt largely deposited as a thin film which could not be detached from the shavings by slight rubbing, so that a large quantity of metallic zinc had to be smelted in order to recover the values.

Estimating the consumption of zinc merely by the amount used up per ton of ore treated may give a false idea of the work actually done by the boxes; a more exact gauge of efficiency is often obtained by considering the amount of zinc consumed per ton of solution passed through, or per ounce of gold or silver precipitated. On the latter basis, the efficiency of zinc for precipitating silver is generally much greater than for gold, chiefly because the entering solutions carry a much larger percentage by weight of precious metal. Heavy mechanical losses of zinc are generally due to errors either in the preparation of the shavings, or in the handling of them during the passage of solution or during the clean-up. Badly cut shavings are apt to give large quantities of small fragments, which pass into the precipitate, and many causes, such as uneven flow resulting in channels, exposure to atmosphere, and unnecessary handling, tend to make the zinc brittle. The deposition of gold, etc., in a thin coherent film is also, as pointed out above, a cause of mechanical loss.

The causes of chemical loss have already been fully discussed. It has been considered that losses may occur through contact of zinc with iron surfaces, as for example in the screens used to support the shavings in the compartments. This, however, is practically unavoidable, and it does not appear probable that the loss from this cause is serious, since the iron rapidly becomes coated with a film of oxide or carbonate that acts as an insulator.

Precautions against Theft. — It is customary to cover the tops of the zinc-boxes with wire screening, or with locked lids, which are only opened when it is necessary to transfer shavings or during the clean-up and refilling of the boxes. In many plants the precipitation and smelting departments are strictly isolated from the rest of the establishment, and only those directly concerned are admitted.

(H) Non-Accumulation of Zinc in the Solutions

It is a remarkable fact that although zinc is constantly dissolving during the passage of the solution through the boxes, the average amount of it in the solution remains about stationary after it has once reached a certain point. It is evident, therefore, that somehow zinc is being removed from the solution. The amount of solution thrown away in the residues and replaced by fresh water does not afford a sufficient explanation. Some portion is no doubt redeposited in the boxes as insoluble zinc hydrate, etc., from solutions weak in cyanide; some may be precipitated in the ore or in the boxes as zinc sulphide, by soluble sulphides. Soluble sulphides of the alkali metals may be present in commercial cyanide, formed by the action of alkali on sulphides in the ore, or formed by reduction in the zinc-boxes through the action of nascent hydrogen on the thiocyanates, which are almost invariably present in the solutions when the ore under treatment contains pyrites. W. H. Virgoe [1] has suggested that in presence of lime and carbonic acid there may be a precipitation of double carbonate of zinc and calcium, as follows:

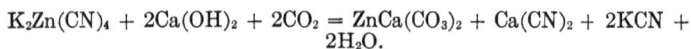

$$K_2Zn(CN)_4 + 2Ca(OH)_2 + 2CO_2 = ZnCa(CO_3)_2 + Ca(CN)_2 + 2KCN + 2H_2O.$$

(I) Tests for Regulating the Working of the Boxes

The methods used in making these tests will be fully described in a later section of this book. Samples of the solutions entering and leaving the boxes should be taken at frequent intervals. This is generally done by means of glass tubes provided with

[1] " Proc. Chem., Met. and Min. Soc. of South Africa," IV, 636.

cocks and passing through the top and bottom ends of the boxes. The cock is turned so as to allow a small uniform drip of solution, which is caught in a bottle beneath. These samples are tested at regular intervals, say every 12 or 24 hours, for cyanide, alkali, and gold (for silver also where this metal is of importance). These represent the average results for equal periods of time. Special samples for cyanide and alkali are generally required at more frequent intervals for regulating the leaching operations, the supply of cyanide for the storage tanks, etc., and a test is commonly made of each solution run on and drawn off from each charge under treatment, unless the material is so uniform and the conditions so regular as to render this unnecessary.

The strength of cyanide to be maintained depends on such a variety of circumstances that no general rule can be given. Some information on this head will be found in Part IV, Section I.

The alkalinity needed also depends to a certain extent on the nature of the material under treatment. A slight but constant alkalinity should be maintained in the entering solution, this being secured by the addition of suitable amounts of lime to the ore or pulp to be treated, or, occasionally, by the addition of caustic soda to the solutions themselves. For slimes plant solutions, W. A. Caldecott recommends an alkali strength equivalent to 0.006 to 0.01 per cent. NaOH,[1] but frequently as much as 0.05 per cent. or more is admissible.

[1] " Proc. Chem., Met. and Min. Soc. of South Africa," II, 101.

SECTION II

CLEAN-UP OF ZINC-BOXES

(A) Preliminary Operations

THE frequency of the clean-up will depend on the nature and amount of the precipitate which is formed. In large plants it may take place once a week, in small ones once a month, the common practice being twice or three times a month. When much white precipitate has formed, it may be necessary to clean up oftener than would be the case with a clean gold deposit containing the same values. Before cleaning up it is usual to displace the solution in the box with water, to which alum may be added to promote settlement of the precipitate during the process. In some cases the strength of the entering solution is raised considerably for an hour or so before it is displaced with water. This is for the purpose of loosening the deposit which adheres to the zinc. After all cyanide has been displaced, a little acid may be added, which also assists settlement.

(B) Separation of Precipitate from Coarse Zinc

The trays are first lifted in the compartments, and shaken up and down to allow as much of the precipitate as possible to pass through the screen. The operator, wearing rubber gloves, generally also shakes the shavings under water in the compartment, and then holds them over an adjoining (lower) compartment where they are sprayed with a jet of water. In some cases the tray and contents are lifted bodily from the box and placed over the clean-up tank, into which the fine adhering precipitate is washed. The washed shavings are then either set aside in any suitable water-tight vessels, such as enameled-iron pans, buckets or trays, or are transferred to the lower compartments of another zinc-box. Each compartment in succession is treated in this manner, beginning at the head and proceeding to the foot of the box.

302

(C) Removal of Precipitate from Box

After the whole of the coarse zinc has been lifted out, the trays are turned over, washed and scrubbed to detach any adhering precipitate.

The liquid and precipitate remaining in the box are then transferred to a special settling tank, usually about 4 ft. diameter by 3 ft. deep, known as the clean-up tank. This is generally of wood, sometimes lined with lead. As this tank is often used for the subsequent acid treatment of the precipitate, it is sometimes provided with a wooden paddle mounted on a central vertical shaft, or with other means of stirring. When there are no outlets to the compartments, the transfer is made by means of enameled-iron buckets. In some plants, the contents of the box are allowed to stand and settle for some time, and the clear liquid syphoned off. The residue is then scooped out and carried to the clean-up tank. When plugs are provided at the bottom of the compartments, these are pulled out and the liquid allowed to run into a launder beneath, which leads to the clean-up tank. The sides of the compartments are then washed down with a hose until the whole of the precipitate has been cleaned out and run into the tank. When the box is perfectly clean, the coarse zinc is replaced on the trays in the same order, unless it has been already transferred to another box. The box is at once filled up with solution, so as to avoid oxidation of the zinc, and is ready for use after any fresh zinc that may be necessary has been added.

In some modern plants, the whole of the liquid drawn off during the clean-up is passed through a small filter-press. The precipitate, after passing through a fine screen (60- to 80-mesh), is also filter-pressed, and the material retained on the screen treated separately. Formerly the fine precipitate was squeezed by hand through filter-bags.

(D) Settlement and Sifting of Precipitate

Where filter-presses are not used, the liquid is allowed to stand for 24 to 48 hours in the clean-up tank, sometimes with addition of a little alum, until the precipitate has completely settled. The clear liquid is then carefully syphoned off and run to waste. The precipitate during transfer to the settling tank is frequently run through a fine sieve, 40- to 80-mesh, and any material retained on the sieve (generally known as "shorts" or "metallics") is set aside for separate treatment. When sufficiently coarse, it may be returned to the top compartment of the box and placed on a layer of clean shavings; if there is too much of this product, however, or if it is too fine, it would be liable to choke the screens of the boxes, and it is usually treated with acid to dissolve as much as possible of the zinc, as described below.

If acid treatment is not to be used, the settled precipitate, if free from shorts, may be at once pumped to a filter-press. Lead-lined tanks should not be used when the precipitate, as is often the case, contains mercury, as this metal amalgamates with the lead and rapidly corrodes the lining.

In some plants vacuum filters are used. As the precipitate is very finely divided, the filtering medium must be very closely woven, such as flannel or very thick canvas. The liquid is drawn off by a suction pump and a wash of fresh water added, and this is again sucked through to extract as much as possible of any soluble matter which may be present. Any wash-waters drawn off in the above processes should be assayed before finally running to waste, as they occasionally carry sufficient values to be worth recovering. In that case they should be returned to the boxes, after addition of alkali, if necessary.

SECTION III

Zinc Dust Precipitation (Merrill System)

This method was introduced in 1897 at Marysville, Montana, in the plant of the Montana Mining Co., and subsequently at the Homestake Mine in South Dakota, where 130,000 tons of ore are treated per month and 4000 tons of solution precipitated every 24 hours.

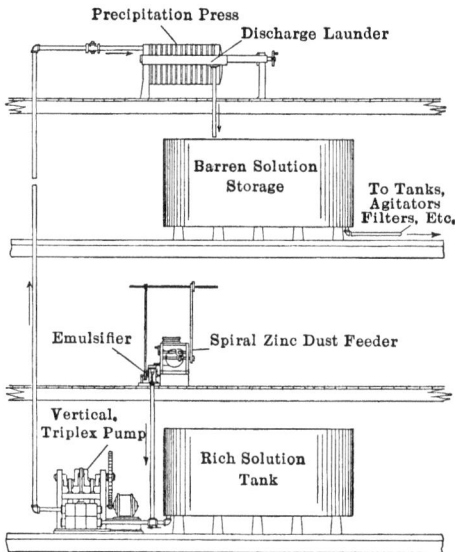

Fig. 37a. — Arrangement of Merrill Precipitation Equipment. Ordinary system.

With very few exceptions this process has been installed in all of the large mills built during the last two years in the United States, Mexico and Canada, and has recently been introduced in South Africa, where 5 installations are now (1913) in operation.

The general arrangement of the plant is shown in Figs. 37a and 37b.

The rich solution to be precipitated is collected in one or more

sumps. A pump, preferably of the plunger type, is placed with
its suction connected to this rich solution sump and its discharge
leading to a filter press, usually at a higher level.

The zinc dust is mixed with the solution at a point in the
suction line about midway between the pump and the tank.

The apparatus for controlling the supply of zinc consists of a
hopper suitably mounted and provided with a worm discharge.
The dry zinc powder from this feeder is mixed and ground with
solution in a small tube-mill or mixer and then flows directly to
the suction of the pump.   Both feeder and tube-mill are operated

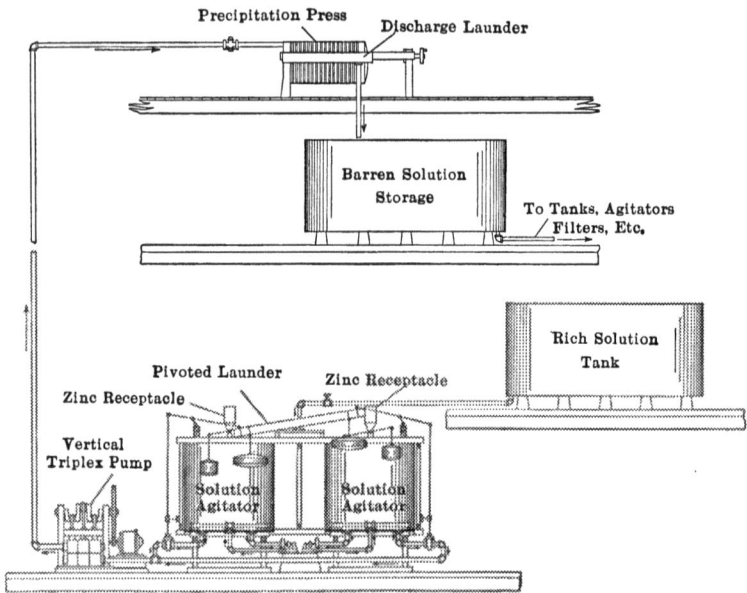

Fig. 37b. — Arrangement of Merrill Precipitation Equipment for
long contact.

from a line shaft or small motor and, by means of an adjustable
friction drive on the worm, the discharge of zinc may be regulated
to suit the flow of solution which is drawn from the sump and
forced through the filter-press.

The frames of the Merrill press are triangular in section (see
Fig. 37c), and the feed pipes are so arranged that the solution and
zinc enter from the bottom or apex of each compartment.   In
consequence of this form the precipitate as it accumulates in the

frames is kept constantly agitated and the full precipitating efficiency of the zinc is obtained.

The clear barren solution leaving the filter press flows to a storage tank and may be used as required in the mill.

In cases where a long contact between zinc and solution is required, as in solutions weak in cyanide and low in gold, the arrangement shown in Fig. 37b is used.

The rich solution is collected in a main sump or storage tank and from this flows alternately to one of two smaller tanks. Each of these small tanks is connected to the suction of a pump and in turn to a filter-press, as before described. By means of an arrangement of floats and levers, the filling and emptying of these small tanks from the main sump is entirely automatic, one being pumped through the filter-press, while the other is filling. By means of two additional floats a zinc-dust feeder is made to discharge into each tank, every time it is filled, a given amount of zinc.

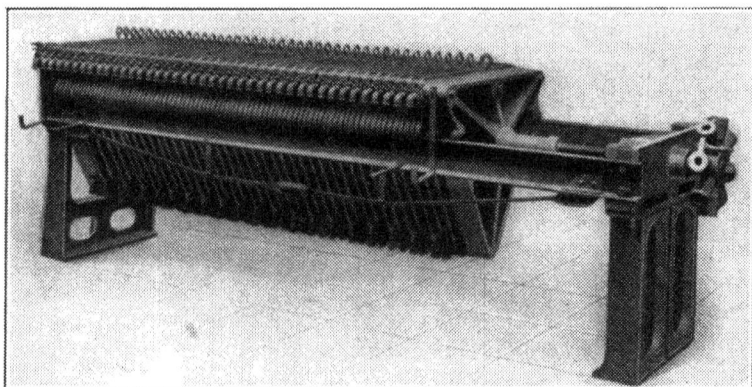

FIG. 37c. — Merrill Precipitation Press for use in zinc dust precipitation.

Each of the small tanks contains an agitator by means of which the mixture of zinc or precipitate and solution is taken from a point at the bottom of the tank and circulated by means of a centrifugal pump through revolving arms provided with a series of small nozzles. In this way the contents are thoroughly mixed, and by regulating the time at which the floats operate, any desired length of contact between zinc and solution may be obtained.

This system of using two tanks alternately provides an accurate measure of the tonnage of solution precipitated and thus furnishes a check on the actual bullion recovery, which should correspond with the difference between head and tail assays of the solution.

Advantages of the System. — The principal advantages claimed for the zinc-dust precipitation process are:

(1) More uniform and efficient precipitation, owing to the extremely fine state of division and consequent large precipitating surface exposed. Since the filter-cloths become coated with a layer of fine zinc, the whole of the solution comes in intimate contact with the precipitant. Hence a very short time of contact is necessary for effective precipitation.

(2) Lower consumption of zinc. Owing to the short time of contact, but little zinc is dissolved beyond that required for actual precipitation.

(3) Cost of zinc-dust is 20 to 25 per cent. less than that of shavings.

(4) Less labor and handling of precipitate. The clean-up with zinc dust is much simpler and quicker than with shavings. The presses are dried with air to about 30 per cent. moisture, then opened and the cakes of precipitate dumped into cars, which are sampled and weighed to determine the actual dry weight of precipitate.

(5) Purer product for smelting. In cyaniding silver ores with zinc-dust the precipitate obtained may contain from 75 to 85 per cent. of pure silver, which may be fluxed and melted without preliminary treatment.

(6) Complete clean-up possible. With shavings precipitation a large part, say 25 to 40 per cent., of the total precipitated values are returned to the boxes after a clean-up, along with the shavings not consumed. With zinc-dust all the values are converted into bullion.

(7) Less danger of theft, since the precipitate is locked in the heavy cast-iron filter-press until this is opened for the clean-up. Moreover, the uniform ratio of zinc-dust to bullion serves as an additional check on theft.

(8) Special methods may be used to increase extractive power of solutions. When sodium sulphide, for example, is added for this purpose, it causes a boiling over in ordinary zinc shaving boxes, owing to violent action of the alkali on the zinc. With the system of agitation and filter-pressing this trouble does not occur.

The chief objection to the zinc-dust method of precipitation is that it requires more skill in operating than the shavings method. Formerly it was supposed that the consumption of zinc would be greater and the grade of bullion lower with dust than with shavings, but under proper conditions the contrary is found to be the case.

Comparative Results with Dust and Shavings. — The following figures were obtained at the plant of the Homestake Mining Co., South Dakota (A. J. Clark, Journ. Chem. Met. and Min. Soc. of South Africa, Jan. 1909), the comparison being between 12 months' use of zinc shavings in 1906 and 9 months' use of zinc-dust in 1908. It is to be noted that solutions below .03 per cent. KCN could not be satisfactorily precipitated by shavings, whereas solutions as low as .015 per cent. were successfully handled with zinc-dust.

	Shavings	Dust
Tons precipitated	5909	5585
Average cyanide strength	.030 per cent.	.024 per cent.
Average barren assay	$0.0214	$0.0178
Per cent. precipitated	91.7 per cent.	94.4 per cent.
Pounds of precipitant used per ton of solution	0.1724	0.1285
Pounds of precipitant used per dollar recovered	0.6735	0.4192
Value of precipitate per lb.	$2.35	$4.03
Total cost of precipitant	129.75	51.15
All costs, per ton solution	0.0412	0.0138
" " per dollar recovered	0.1608	0.0449
Total cost of labor and supplies	$243.28	$76.93
Total gold discharged	126.45	99.41
Value of cyanide wasted	709.00	540.60
Total cost	$1078.73	$716.94

Comparative results at El Oro, Mexico.

	Shavings	Dust
Tons milled per 24 hr.	475	700
Tons solution precipitated per 24 hr.	3000	2100
Men employed	13	6
Value of head solution	$1.85	$4.38
Per cent. gold in raw product	8 per cent.	10 per cent.
Per cent. silver in raw product	49 per cent.	55 per cent.
Ratio zinc to bullion	4.75:1	2.01:1
Fineness of bullion { Gold	110	160
Silver	710	730

Comparative results at El Oro, Mexico.

	Shavings	Dust
Zinc used per ton of ore...................	1.45 lb.	0.5 lb.
Cost of zinc per ton of ore	$0.105	$0.033
" " labor per ton of ore	0.030	0.016
Fuel and miscellaneous	0.066	0.030
Total cost per ton of ore	$0.201	$0.079

The following was the consumption of zinc-dust at various plants using the process described above:

	Zinc consumed per ton of ore treated lb.
Homestake, 1908 (9 months)	0.1285
Homestake Slime Plant, 1910 (4 months)........	0.098
Goldfield Consolidated, Nevada................	0.500
Pittsburg — Silver Peak, Nevada	0.257
El Oro, Mexico	0.500

	Lb. zinc per lb. of bullion recovered
Santa Gertrudis, Pachuca, Mexico..............	0.90
La Blanca y anexas, Pachuca, Mexico	0.90

SECTION IV

PRECIPITATION WITH ALUMINUM DUST

ONE of the most successful recent developments in connection with precipitation has been the use of aluminum dust in place of zinc dust. The use of aluminum as a precipitant was patented by C. Moldenhauer in 1893, and has occasionally been tried in the form of plates or shavings, as, for example, by H. F. Julian at the Rand Central Plant in 1893, but was abandoned, owing to difficulties arising from the accumulation of coatings of alumina, which impeded precipitation and caused trouble in smelting.

The advantages claimed are:

(1) That the reaction involves no formation of a cyanogen compound, precipitation being entirely due to the action of the nascent hydrogen evolved.

(2) The double cyanides of gold and silver are decomposed with recovery of the cyanide in an active form (as $NaCN$ or KCN).

The reactions involved are:

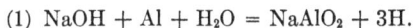

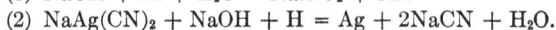

(1) $NaOH + Al + H_2O = NaAlO_2 + 3H.$
(2) $NaAg(CN)_2 + NaOH + H = Ag + 2NaCN + H_2O.$

From this it will be seen that the aluminum goes into solution as sodium aluminate, and that free caustic soda must always be present in excess.

The substitution of aluminum dust for plates or shavings overcame the difficulties due to deposits of alumina, and within the last few years the process has been introduced extensively in the treatment of Canadian silver ores, both at Deloro and Cobalt, Ontario.

Experiments were made in 1906 in the treatment of Timiskaming ores, by S. F. Kirkpatrick, who in 1910 patented a special tank for mixing the dust with solution and introduced the process on a working scale in the plant of the Deloro Mining and Reduction Co.

The following details are summarized from descriptions pub-

311

lished by S. F. Kirkpatrick (E. M. J. *95*, 1277, June 28, 1913) and E. M. Hamilton (E. M. J. *95*, 935, May 10, 1913).

It was found that, when using zinc-dust as a precipitant, the Nipissing solutions rapidly lost their efficiency; this effect was traced to the accumulation in them of compounds of zinc and arsenic, and it was finally decided to eliminate the use of zinc altogether.

One of the difficulties encountered in the use of aluminum dust was its great tendency to float, so that it is not easily wetted. Even when wet, the hydrogen bubbles produced again raise it to the surface. Kirkpatrick overcame this trouble by using a precipitation tank fitted with a vertical revolving shaft carrying a set of propellers, thus creating a vortex which draws down the light powder at the center and throws it out towards the periphery at the bottom of the vessel. In the Nipissing plant agitation tanks are used, 6 ft. deep by 5 ft. diameter, having a central revolving shaft to which are bolted vertical planks extending from the bottom to the surface of the liquid. This agitator, revolving at 60 r.p.m., keeps the dust immersed and renders the propellers unnecessary.

More time of contact is required with aluminum than with zinc-dust, hence the equipment for the Merrill zinc-dust process was not effective and was modified in various ways. Aluminum dust could not be fed as an emulsion into the stand-pipe connected with the pump suction, on account of the tendency to float already mentioned.

At the Deloro plant, the solution to be precipitated was drawn from the storage tank in charges of 20 to 30 tons, and agitated in a special tank with propellers as described, the required amount of aluminum dust being added. It was then transferred by a 3-throw plunger pump to one of two filter-presses, the filtered solution going to the barren storage tank. The time of agitation required is 10 to 15 minutes. A single precipitation may yield as much as 7000 oz. of silver. The precipitate in the press is water-washed, air-dried and melted with a flux consisting of borax 16 per cent., sodium nitrate 8 per cent., fluor spar 4 per cent. and sand 5 per cent., yielding bullion 999 fine.

As will be seen from reaction (2) above, there is a regeneration of cyanide in the process, and in fact the amount of silver in the solution may be approximately determined by titrating the free cyanide before and after agitation with aluminum dust.

The costs of precipitating 1000 oz. of silver are given by Kirkpatrick as follows:

35.6 lb. soda at 2¢ per lb.	$0.712
9.1 lb. aluminum at 32¢ per lb.	2.91
	$3.622

At the Nipissing plant the pregnant solution passes first through a clarifying press, thence through a launder, into which the aluminum dust is fed by an automatic device with precautions to avoid loss. This launder conveys it to the agitation tanks already described, of which there are two in series with a special mechanism for regulating the flow of solution and maintaining a constant level. After a total agitation of 5 to 10 minutes the emulsion is pumped to a precipitate press, the filtrate passing to the barren solution tank.

Hamilton (*loc. cit.*) points out that lime must be avoided in this process, as it forms an insoluble calcium aluminate, which collects in the precipitate press giving a low-grade product almost impossible to melt by ordinary methods. As lime is sometimes preferable to caustic soda in the extraction process, this factor might in some cases interfere with the successful application of the method.

The efficiency of precipitation at Nipissing is given by Hamilton as 97 to 98 per cent. and the consumption of aluminum as .02 lb. per ounce of silver precipitated.

The regeneration of cyanide amounts to 0.64 to 0.72 lb. KCN per ton of solution, and in addition to this the loss by direct action of cyanide on the metal, which occurs in the case of zinc, is avoided by the use of aluminum.

The precipitate produced at Nipissing carries generally 85 to 92 per cent. of silver, but, when lime is present in the working solutions, may be very low grade, sometimes not more than 10 per cent. silver. The product is briquetted with a little soda and borax and melted direct in an oil-fired reverberatory producing bullion 997 fine (see E. M. J., Dec. 7, 1912).

The following comparison of costs is given by E. M. Hamilton (E. M. J., May 10, 1913).

Zinc Dust

Price laid down at Nipissing..7¢ per lb.

Consumption ....0.10 lb. per oz. silver

Amount used per ton ....2 lb. at 7¢ = $0.14

Aluminum Dust

Price, laid down at Nipissing ....38¢ per lb.

Consumption ....0.02 lb. per oz. silver

Used per ton:

$$0.4 \text{ lb. of Al. at } 38¢ \quad = \$0.152$$
$$1.5 \text{ lb. caustic soda at } 2\tfrac{1}{4}¢ = 0.033$$

$$\$0.185$$

Cyanide recovery — 1.6 lb. KCN at 15¢ per lb. = $0.24 per ton.

Thus although the actual cost of precipitation with aluminum is greater than with zinc, there is a difference, due to cyanide regeneration, in favor of aluminum, amounting to 19.5 cents per ton.

In addition to this there is a further saving, due to absence of action between the precipitant and cyanide, and also a saving in the cost of melting and refining.

Chas. Butters proposes the use of aluminum granules as a substitute for the dust; these can be produced at 24 cents per lb., and it is proposed to place them in a revolving tube-mill, through which the solution is to be passed, the attrition keeping the metal surfaces bright so that precipitation remains effective.

ADDITIONAL DATA ON ALUMINUM PRECIPITATION

Through the courtesy of Mr. J. J. Denny and of the Nipissing Mining Company, I am enabled to give the following additional particulars with regard to aluminum precipitation.

᾿ ⸗When the Nipissing working solutions are agitated with aluminum dust, soluble sulphides are always produced, the formation of which is an indication of complete precipitation of the silver. They appear to arise from the reducing action of nascent hydrogen on the thiosulphates present in the solution. If the silver precipitation is incomplete, the addition of an alkali sulphide to the solution causes a darkening; this test should always be made in doubtful cases. (See p. 519.)

In order to oxidize these soluble sulphides, the barren solution is agitated with air immediately after precipitation, otherwise thiocyanates might be formed, with consequent loss of cyanide. Lead salts cannot be used with these solutions for removal of sulphides, as they appear to have a prejudicial effect on the extraction, the reason of which has not been definitely ascertained.

Temperature has an influence on precipitation; thus it was found that at 40° F. the results were very poor; at 60° F. precipitation is effective, but proceeds more rapidly at higher temperatures.

The *proportion of Caustic soda* to be added is important. In higher grade solutions more is needed for good precipitation owing to the larger amount both of silver and base metal carried. It has been found that aluminum precipitates copper more readily than zinc does.

In titrating the solutions for alkali, it is necessary to distinguish between lime and caustic soda, as only the latter is effective in aluminum precipitation. Sodium carbonate is effective only in hot solutions.

The *proportion of Aluminum dust* must also be regulated with some care. When too much is added at once, the presses may get so hot that the cloths are burned, as the action becomes very violent. This may perhaps be partly due to the formation of aluminum amalgam by the reduction of mercury compounds in the solution. This amalgam decomposes water readily at the ordinary temperature of the air, forming alumina and generating hydrogen. Owing to the copious evolution of hydrogen, no naked lights can be used in the neighborhood of the precipitation tanks and presses.

The physical condition of the dust must also be taken into consideration. Various brands differ widely as to the bulk occupied by a given weight, which of course affects the rate at which the dust must be fed into the mixing tank to produce the required effect. Thus a Canadian brand requires to be fed at about double the rate of a particular English brand. Some kinds are mixed with stearine, which gives the powder a polished appearance and causes it to float on the solution, so that violent agitation is necessary for mixing. This form is the most suitable for continuous feeding, since the unpolished dust has a tendency to form lumps.

The impurities commonly present in aluminum dust are iron, silica, carbon, copper and zinc.

It has been supposed by some writers that an accumulation of aluminum would occur in the solutions through continued use. This has not so far been observed at the Nipissing. Three determinations of aluminum in the solutions, made at considerable intervals, showed the following successive results: Per cent. Al in Barren Solution: (1) .053 (2) .093 (3) .083.

The non-accumulation of aluminum is explained by the formation of an insoluble calcium aluminate already referred to. A certain amount of lime is fed with the ore, and reacts on the sodium aluminate formed in the precipitation process as follows:

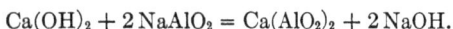

$$Ca(OH)_2 + 2\,NaAlO_2 = Ca(AlO_2)_2 + 2\,NaOH.$$

The Battery solution, since the desulphurizing process has been in use, has shown a maximum of .03 per cent. Al, and there is no tendency to increase.

The following is an analysis of a solution from the Nipissing High-Grade Mill, used in treatment of Table Concentrates, which assayed originally 12,256 oz. silver. The material, crushed to − 200 mesh, had a sp. gr. 3.333. The solution contained free cyanide equivalent to .3 per cent. KCN and alkali equivalent to .25 per cent. NaOH. In order to obtain total precipitation of the silver, the alkali had to be raised to .60 per cent. NaOH.

		Per cent.
Nickel	Ni	.1574
Copper	Cu	.0385
Iron	Fe	.0266
Cobalt	Co	.0242
Arsenic	As	.2960
Sulphocyanide	KCNS	.0290

EFFICIENCY OF ALUMINUM DUST PRECIPITATION

The latest available figures are as follows:

Low-Grade Mill, Nipissing Mining Co., 1913.

Aluminum dust consumed	43,921 lb.
Silver precipitated	1,989,743 oz.

or 1 oz. Troy of dust for 3.106 oz. silver.

Average assay of pregnant solution 8 to 9 oz. January to March, 1914:

1 oz. Troy of dust for 3.24 oz. silver.

Average assay of pregnant solution 12 oz.

SECTION V

ACID TREATMENT AND ROASTING OF PRECIPITATE

AFTER the precipitate has been separated as far as possible from the coarse zinc, there is considerable divergence in the subsequent procedure.

The methods adopted may be briefly classified as follows: (1) Direct fusion. (2) Roasting and fusion. (3) Mixing with niter, roasting and fusion. (4) Acid treatment and fusion. (5) Acid treatment, roasting, and fusion. (6) Roasting, acid treatment, and fusion. (7) Smelting with litharge and cupellation. Since acid treatment generally precedes roasting, the operations will be considered in that order.

(A) ACID TREATMENT

The precipitate, having been transferred to the acid tank with or without filtering, is allowed to settle somewhat, and as much as possible of the clear liquid drawn off. Sulphuric acid is then added, in quantity more than sufficient to dissolve the zinc supposed to be present, and diluted with enough water to form a 10 per cent. solution. The heat produced by the mixture of acid and water, with that generated by the chemical action of the acid on the zinc, is generally sufficient to maintain the reaction with sufficient vigor till the greater part of the zinc is dissolved. In some cases it may be necessary to apply artificial heat, by injection of steam or otherwise. The mixture is stirred from time to time, mechanically or by hand.

At first there is a very violent disengagement of gases, chiefly hydrogen, but mixed also with hydrocyanic acid and, where arsenic is present, with arseniureted hydrogen. As the latter are extremely poisonous, it is necessary to provide a hood with a good draft, for carrying off the fumes. The tank must be sufficiently large to avoid the danger of frothing over. When all action has ceased, the tank is filled up with water, preferably hot, as the zinc sulphate formed is more soluble in hot water. The mixture is stirred, allowed to settle, and the clear liquid drawn off, this operation being repeated several times. The acid tank may be of wood, with or without a lead lining, or of enameled iron.

Sodium bisulphate (a by-product in the manufacture of nitric acid) is sometimes used as a substitute for sulphuric acid. An 8 per cent. solution of the salt is an effective strength. The temperature should not be above 33° C., as the solubility of the salt decreases with rise of temperature.[1] This material is cheaper and more easily transported than sulphuric acid.

In cases where much arsenic is present, it is advisable to add nitric acid in the proportion of about 1 part HNO_3 to 2 parts H_2SO_4; this has the effect of oxidizing the arsenic to arsenic or arsenious acid, and at the same time converts hydrocyanic into cyanic acid, thus avoiding the evolution of large quantities of poisonous fumes. It must be mentioned, however, that under certain circumstances nitric acid may lead to a loss of gold. Experiments have shown that this loss is greater with nitric than with hydrochloric acid, and greater with hydrochloric than with sulphuric acid. Nitric acid has sometimes been added for the purpose of dissolving the lead introduced by the use of the lead zinc couple in precipitating, but it seems preferable to remove this metal by suitable fluxes in the subsequent fusion.

At the Metallic Extraction Co.'s mill near Florence, Colorado, a treatment with hydrochloric acid was given after roasting, the roasted precipitate being treated in a lead-lined sheet-steel pan with a mixture of one part hydrochloric acid to two of water, the residue being afterward filtered, washed, and dried.

After acid treatment, the precipitate is frequently pumped to a filter-press, through which hot water is forced to extract as much as possible of the zinc sulphate still retained, and finally dried by forcing air through the cakes. The product may also be extracted and dried by vacuum pans.

The following is a description of the method employed at the Myalls United Mines, Australia, given by C. J. Morris.[2] The zinc is scrubbed, and the precipitate is placed, together with the

---

[1] J. E. Thomas and G. Williams, "Journ. Chem., Met. and Min. Soc. of South Africa," V, 334.

[2] "Trans. I. M. M.," XV, 543 (1906).

shorts, in a wooden acid tank 5 ft. deep by 4 ft. diameter, provided with a 4-armed revolving paddle, plugged holes for draining
off the wash, and a locked cover. Twenty pounds of acid are
added to the charge at one time, and a further quantity when the
violent action has ceased, until the required amount has been
added. The charge remains under treatment all night, with
occasional stirring, and exhaust steam from the solution pump
is led in through the cover by means of an iron pipe. In the
morning all action has ceased, and twelve times the bulk of water
is added, with agitation. After settling, the wash-water is drawn
off. This contains only 1 to 2 dwt. of gold per ton, which might be
recovered by long settling. The washed precipitate is then dried
with addition of 5 per cent. of niter, roasted and fluxed. It is
claimed that acid treatment, as described, gives far better results
than addition of niter, roasting, and fusion without acid treatment.

In many cases it is found advisable to keep the shorts and fine
precipitate separate, as they may be advantageously treated by
different fluxes. At Redjang Lebong, a considerable difference
was noticed in the composition of the bullion obtained from these
two products. As the shorts contain very much more metallic
zinc than the fines, they will naturally require a larger amount
of sulphuric acid. In some cases it is found that the solution of
the zinc by sulphuric acid is imperfect. This may be due to the
metal being present as an alloy insoluble in that acid, or to the
presence of protective coatings of insoluble sulphates ($PbSO_4$ or
$CaSO_4$).

Composition of Acid-treated Precipitate. — The zinc-gold precipitate as generally obtained is a very complex mixture, sometimes containing, in addition to gold, silver and zinc, nearly all
the common metals, with silica, carbonates, sulphates, sulphides,
and various cyanogen compounds; of these ingredients the proportions vary enormously in different samples, so that no fixed
rules can be given as to the best method for its reduction. An
analysis of acid treated and roasted precipitate by A. Whitby[1]
gave the following percentages:

Gold 34.5	Ferric oxide 3.65	
Silver 4.75	Zinc oxide 7.0	
Lead 12.5	Sulphuric acid 6.95	(SO_3)
Copper 2.55	Nickel oxide 1.0	
Silica 21.0		

[1] " Proc. Chem., and Met. and Min. Soc. of South Africa," III, 46 (1902).

(B) Roasting of Zinc-Gold Precipitate

The precipitate, whether acid-treated or not, is best treated by filter-pressing or vacuum filtration to remove as much of the moisture as possible. By the use of hot air, the drying may be practically completed in the filter-press, but more usually the cakes of precipitate are broken up and placed on sheet-iron trays about 1½ in. deep, on which the material is first carefully dried over a slow fire and finally roasted, either over an open grate or in a muffle or reverberatory furnace. It is a common practice to mix about 5 per cent. of niter with the dried and powdered precipitate before roasting, especially in cases where acid treatment is not used. The mixture, when perfectly dry, may be ignited and burns of itself, all carbonaceous matter being thus removed and a complete oxidation of the zinc obtained. Caldecott recommends mixing 30 parts of silica and 30 parts of niter with every 100 parts of dried precipitate; this yields a product which does not undergo losses by dusting, and the sand serves as a flux for the zinc oxide in the smelting. Other operators saturate the precipitate, before drying, with a strong solution of niter.

The roasting should be done slowly, at a dull red heat, with careful regulation of the draft, since heavy losses of gold may occur when copious fumes of zinc are given off. The exact amount of this loss has never been ascertained, but it is the general experience that the best recovery of the contained values is obtained when the zinc is eliminated as completely as possible before roasting. Stirring may be occasionally necessary during the roasting, but should be avoided as much as possible, as it may lead to mechanical losses.

In some works the precipitate is placed in shallow iron pans, which fit inside the muffle, and allowed to remain over night at a moderate heat, with little or no stirring. When cool, the trays are withdrawn and the lumps ground and mixed with the necessary fluxes for smelting. When the acid treatment is very carefully and thoroughly performed, the roasting operation is of less importance, and may in some cases be advantageously omitted. As a precaution against "dusting" during the early stages of the roasting, the precipitate is generally charged into the furnace in a slightly moist condition, and is often compressed or briquetted.

Probably more loss of gold and silver occurs in the roasting than in any other stage of the process. When metallic zinc is volatilized, it carries with it considerable quantities of the precious metals, probably mechanically. In contact with the air, the zinc vapor burns with a greenish flame, producing dense fumes of white oxide of zinc. These are deposited in the flues and cooler parts of the furnace, but large quantities escape into the air, carrying with them gold and silver in a state of minute subdivision. Zinc oxide itself is non-volatile, and if the zinc can be oxidized by chemical means before roasting, this loss does not occur. This is the chief object of adding niter in the manner described above.

SECTION VI

FLUXING, SMELTING, AND REFINING OF THE PRECIPITATE

(A) FLUXES FOR ZINC-GOLD PRECIPITATE

THE fluxes to be used will depend to a large extent on the nature of the materials contained in the precipitate, and on the preliminary treatment which it has undergone. The following are given as examples, but in any case the most suitable flux should be ascertained by trial. They are tabulated according to the preliminary treatment of the precipitate.

1. For direct fusion without acid treatment or roasting:

| | Parts by weight | |
	(a)	(b)
Precipitate	100	100
Borax	35	60
Bicarbonate of soda	50	7
Sand	15	11.5
Niter ....2 to 4		19

(a) Julian and Smart, "Cyaniding Gold and Silver Ores," 2d ed., p. 168.

(b) T. L. Carter, "Trans. Chem., Met. and Min. Soc. of South Africa," II, 444.

The method generally yields a very base bullion containing much zinc, and of irregular composition; (b) was tried for treatment of precipitate from zinc-lead couple.

2. Roasting with niter, without acid treatment:

	Parts by weight (a)
Precipitate 100	
Borax 40	
Bicarbonate of soda 40	
Sand 15	

(a) Julian and Smart, *loc. cit.*, p. 168.

When Caldecott's system of roasting with sand and niter is used, the silica is omitted from the flux and the following is used:

	Parts by weight (b)
Precipitate	100
Borax	40
Carbonate of soda	10

(b) "Journ. Soc. Chem. Ind.," XVII, 3 (1897).

3. Acid treatment without roasting:

	Parts by weight		
	(a)	(b)	(c)
Precipitate	100	100	100
Borax	25–30	66	10
Carbonate of soda	50	9	40
Fluor-spar	–	9	2
Sand	–	–	15

(a) T. H. Leggett, "Trans. I. M. M.," V, 147.

(b) E. H. Johnson, "Eng. and Min. Journ.," p. 220 (1899).

(c) J. Gross, "Trans, A. I. M. E.," XXXV, 616.

4. After acid treatment and roasting: Clay liners used when much lead is present:

	Parts by weight			
	(a)	(b)	(c)	(d)
Precipitate	100	100	100	100
Fused borax	20–35	60	70	50
Carbonate of soda	–	25	–	50
Manganese dioxide	20–40	–	15–20	–
Fluor-spar	5	10	0–10	10
Sand	15–40	5	10	44

(a) E. H. Johnson and W. A. Caldecott, "Trans. Chem., Met. and Min. Soc. of South Africa," III, 51 (1902).

(b) B. W. Begeer, "Metallurgy of Gold on the Rand," 1898.

(c) Used at Abosso G. M. Co., Gold Coast, West Africa, 1907–08, yielding bullion about 940 fine. Previously roasted with 5 per cent. niter.

(d) The present writer has found this flux satisfactory for a base precipitate containing much copper and lead, together with sulphur and a little arsenic.

(B) Mixing of Precipitate and Fluxes

The various materials used for the flux should be thoroughly dried, especially the borax and soda, which are liable to contain considerable amounts of water. In small plants, the necessary quantities for each charge are weighed out, mixed on an iron tray or other suitable receptacle with a scoop, and fed into the previously heated crucible. Where large quantities are dealt with at a time the mixture is put through a small ball mill specially reserved for this purpose, so that the different ingredients are thoroughly incorporated and all lumps broken up. The flux must be well mixed with the precipitate before transferring to the crucible. The mixture is generally fed in by means of a long iron scoop, the lower end of which is formed into a tube to avoid loss by scattering of the charge while feeding into the pot. A wide-necked iron hopper or funnel is also used for this purpose.

(C) The Fusion Process

The fusion is almost invariably made in graphite crucibles (actually made from a mixture of graphite and fire-clay), the kind known as "salamander" being frequently used. The size of crucible will depend on the amount of fluxed precipitate which can be conveniently smelted at one time in the furnace. Size No. 60 is largely used. From 10 to 20 lb. of dried precipitate can be smelted in one crucible charge. In cases where lead is present, or when for any reason the graphite of the crucible might act injuriously in the fusion, a "liner" is used, consisting of a fire-clay crucible that fits inside a large graphite pot. Both pots and liners should be carefully dried and annealed by gradual heating outside the furnace, and raised to a dull redness in the furnace before charging in the mixture to be smelted.

The furnaces used are square, and generally similar to the wind fusion-furnaces used in assaying, but somewhat larger and built so that the mouth of the furnace is flush with the floor. It is a good plan to have the building so arranged that fuel can be supplied to the furnace and the ashes removed by openings outside the smelting room proper. The floor of the latter may thus be kept clean throughout the operation, and no unskilled workmen need be admitted while the smelting is going on, thus diminishing the chances of loss or theft. The ash-pit should, however, be accessible from the smelting room in case an accident should occur, such as the breaking or upsetting of a pot in the fire.

The fuel generally used is coke; where coal or charcoal is employed, the furnace must be somewhat larger to allow of a sufficient quantity of fuel being packed round the crucibles. Furnaces have also been built for gasoline or other liquid fuel. Where large quantities of precipitate have to be smelted, a reverberatory furnace, as introduced by Charles Butters, is sometimes used. Another appliance very convenient for this purpose is a tilting furnace, similar in design to that proposed by H. L.

Sulman,[1] consisting of a rectangular furnace cased with iron and lined internally with brick. It is mounted on trunnions and contains an inner (replaceable) retort of graphite or fire-clay, which serves the place of the crucible and holds the charge to be smelted. Arrangements are provided, so that when the fusion is complete the entire furnace may be tilted and the contents of the crucible poured into a mold. A furnace of this type was used by the Redjang Lebong Mining Co., for running down acid-treated and roasted-zinc precipitate containing a large percentage of silver and some selenium. The latter was partially expelled as oxide by introducing into the furnace a blast of air from an air-compressor. There is considerable saving of labor and probably less risk of loss in using some such arrangement, but the cost of installation and repairs is greater than with the ordinary pot furnace.

The crucible, in the usual square furnace, should be supported firmly by a fire-brick resting on the furnace bars. It should be well packed all around with coke, and covered with a graphite or fire-clay lid. At the beginning it should not be more than three-quarters filled with the mixture of flux and precipitate, for the charge swells considerably through the disengagement of carbonic acid, etc., as it becomes heated. After it is thoroughly melted it subsides, and a further quantity of the mixture may now be added. Toward the end of the operation, when the fusion is nearly complete, the contents of the pot are stirred with an iron rod and any infusible dross which remains on the surface removed by skimming. When the zinc has not been perfectly removed by acid treatment, dense fumes will be given off during the fusion, and some of the zinc will be volatilized and burn at the mouth of the crucible with a greenish flame. Under these circumstances some loss of gold and silver by volatilization undoubtedly takes place, but opinions differ as to the extent of this loss. Deposits of zinc dust and zinc oxide carrying high values are sometimes recovered from the flues, and dust chambers are occasionally used for collecting this material.

When the fusion is complete, the pot is lifted from the furnace by means of basket tongs; when large pots are used, a block and tackle arrangement should be provided for lifting the pots. The contents are then poured into a conical mold which should be previously oiled and heated. As soon as the bullion has set, the

[1] "Journ. Soc. Chem. Ind.," December, 1897.

mold is inverted onto an iron tray and the slag detached from the button of metal by one or two blows of a hammer. The slag should be clear and uniform; it is generally of a light greenish-gray color, and is set aside to be crushed and panned, as it usually contains sufficient value in the form of small shots of metal to be worth treating.

When the whole quantity of fluxed precipitate which is to be smelted has been run down in this way, the buttons from the conical molds are remelted in a clean crucible with a little borax and cast into an oblong ingot-mold.[1] It is usual to take a sample of this bullion just before casting, by means of a dipper formed from a piece of an old graphite crucible. The molten metal should be well stirred and the dipper heated before sampling. (See Part VII.)

(D) Use of Special Fluxes

Niter is used to oxidize any metallic zinc which may still be present in the precipitate after the preliminary treatment. It is generally unnecessary when acid treatment has been given, as in that case any zinc remaining is probably present as sulphate. It is ineffective for oxidizing lead, as it decomposes and gives off its oxygen below the temperature required for the formation of litharge.

Manganese dioxide is used as a substitute for niter when lead is present; it is not advisable to use it when much silver is present, as it has a tendency to carry the silver into the slag.

Fluor-spar is used chiefly to give fluidity to the charge. It is also of some use in fluxing calcium sulphate.

Sand is added for the purpose of forming a fusible silicate with the zinc and other bases present. It is omitted when practically all the zinc has been removed by preliminary operations.

Carbonate of soda is used only in small quantity and is some-times omitted. When silica is present in the precipitate or has been added to the charge, it helps by forming a fusible double silicate of zinc and sodium.

Borax is the principal flux for all bases present. It also

[1] The following flux answers very well for remelting the buttons into bars.

Weight of metal remelted	9 to 10 kilos.	
Fused Borax	1.7	"
Fluor-spar	0.2	"
Niter (added at moment of fusion)	0.15 to 0.2	"

increases the fluidity of the charge, apart from any chemical action.

(E) Treatment of Matte

It has already been noted that sulphur, from various sources, is liable to be present in the deposit collected from the zinc-boxes. Unless the precipitate after acid treatment is thoroughly roasted, this sulphur, in the subsequent fusion, will form a " matte," generally carrying gold, silver, iron, zinc, and lead, also copper if the latter metal is present in the material treated. This matte forms a brittle but strongly adherent layer above the button of metal, between it and the slag, and as it frequently carries high values, some means must be found to reduce it to a marketable bullion. Selenium, tellurium, arsenic, and antimony form similar products when they occur in the material smelted.

A method of recovering fine bullion from matte is described by A. E. Drucker [1] as follows. The fluxes used for the reduction are borax and cyanide. The matte, borax, and cyanide are put separately through a rock-breaker and crushed fine. Alternate layers of borax, matte, and cyanide are charged into a No. 60 graphite crucible until the pot is nearly full, when a layer of borax is added as a cover. The crucible is now put into the furnace and maintained at a white heat for two or three hours, until the charge subsides and bubbling ceases. Large volumes of sulphur are given off at the end of the first hour, and burn at the mouth of the pot. When the action is complete the slag becomes quite thick and must be removed by a skimmer. The remaining contents are then poured into a conical mold. If excess of cyanide has been used it forms a crust just above the gold button and can be broken off with a blow from a hammer. The matte is completely decomposed, and 85 to 94 per cent. of the total values contained in it are recovered as fine metal. Only a light porous slag remains, which may be remelted with a subsequent charge.

The usual method of treating this matte is to remelt with additional flux and scrap-iron, niter being sometimes added to assist in oxidizing the sulphur. This method, as stated by Drucker, is slow and incomplete, and when graphite crucibles are used they are rapidly corroded by the niter.

[1] " Min. Sc. Press," May 18, 1907.

(F) Smelting of Zinc-Gold Precipitate with Litharge

In the early days of the cyanide process, the suggestion of melting the zinc-precipitate with lead, and afterward cupeling, seems to have been carried out.   G. H. Clevenger [1] states that at the Balbach Smelting and Refining Co.'s plant, Newark, New Jersey, the precipitates, tied up in paper sacks in parcels of 1 to 5 lb., were charged from time to time on a bath of molten lead in a cupeling furnace.   The gold and silver were quickly absorbed by the lead, and until the mass was well melted, the precaution was taken of keeping all drafts closed.   The slag was then removed by skimming and the bullion cupeled and refined.

For this method it would seem that there is considerable probability of metallic zinc being volatilized with consequent loss of gold and silver.   This is avoided in P. S. Tavener's process, first introduced at the Bonanza mine, Johannesburg, in August, 1899, in which the precipitate, after filter-pressing (but without acid-treatment or roasting), is mixed with litharge, slag from previous operations, slag from assay fusions, sand and sawdust, and smelted in a reverberatory furnace.   The zinc is fluxed off without being volatilized to any appreciable extent, and lead bullion is obtained which is afterward refined on a bone-ash "test," in an ordinary cupellation furnace.   The process may be likened to a scorification assay on a large scale.

The following account of the process is summarized from a detailed description given by Tavener.[2]   The fine precipitate from the filter-press and the zinc shorts are dried separately on trays in an oven for 15 minutes.   The dry precipitate is then rubbed through a sieve of four holes to the linear inch, roughly weighed, and mixed with the fluxes.   These are as follows:

	Parts by weight.
Precipitate	100
Litharge	60
Assay slag	10 to 15
Slag previously used	10 to 15
Sand	10 to 20

After adding the precipitate, the mixture is again sifted to ensure

[1] "Trans. A. I. M. E.," XXXIV, 891 (October, 1903).

[2] "Proc. Chem., Met. and Min. Soc. of South Africa" Vol. III., p. 112 (October, 1902).

thorough mixing, and shoveled into the furnace, which is as yet unlighted. The fine zinc is then mixed as follows:

	Parts by weight.
Fine zinc (shorts)	100
Litharge	150
Slag	20

This is put into the furnace on the top of the precipitate to prevent loss by dusting and also to ensure the greater part of the litharge being on the top of the charge. A slow fire is now lighted and the charge allowed to dry for a couple of hours. The temperature is then raised, and in four or five hours the charge is reduced; any sweepings or by-products which have to be worked up are now added, and are quickly absorbed. When the slag has again become fluid, it is well stirred with a rabble, and sawdust thrown in to reduce the excess of litharge. The slag is now tapped off through the slag-door above the level of the lead, and the latter is skimmed; a shovelful of lime is thrown in and a final skimming is given. The clean lead-bath is then stirred and sampled, the tap-hole is opened, and the lead bullion run into molds.

The lead bullion is now cupeled on a test, consisting of an oval cast-iron frame filled with bone-ash ground to pass a 20-mesh screen, and mixed with 3 per cent. caustic potash and 10 to 11 per cent. water. The mixture is again sifted to break up lumps. The test-frame is placed on a cast-iron plate and filled with the bone-ash, which is then tamped in. The center is then hollowed out, leaving a rim round the sides. About 300 lb. of bone-ash are used for one test, which may be used about four times, cupeling altogether over $7\frac{1}{2}$ tons of lead bullion. The test should be dried slowly for some weeks before use.

When a new test is put in the furnace, a slow fire is kept up for 3 or 4 hours. An iron blast-pipe, 3 in. diameter, flattened and turned down at one end to allow the blast to strike on the molten lead, is now fitted to the back end of the cupel. Temperature is raised, and the lead bars fed in one at a time through the working-door of the furnace onto the test. A channel is cut in the rim of the latter $\frac{1}{4}$ to $\frac{1}{2}$ in. deep and $1\frac{1}{2}$ in. wide, and communicating with a hole by which the molten litharge runs down to a suitable pot placed beneath. When the test is filled with molten lead nearly to the level of this channel, the temperature is increased to the melting-point of litharge, and when the lead is covered with molten litharge the blast is turned on, and the litharge allowed to flow away along the channel, which is deepened as the operation proceeds. Fresh lead bars are fed in until all are melted. The temperature is raised as the proportion of gold in the alloy remaining on the cupel increases. At the finish it is necessary to add a little assay-slag, which is melted and run off. When the operation is complete, the gold freezes or solidifies. It is then broken in pieces, remelted in crucibles, and cast into bars.

This method possesses many advantages over the ordinary systems of reducing zinc-gold precipitate. The cost of material is much lower; a large part of the litharge is recovered for re-use, and the slag used as flux in this process is a waste product from which the contained values could otherwise be obtained only at a high cost. It also affords a means of easily disposing of a number of troublesome by-products of the mill and cyanide works which cannot be economically treated by other methods. The cost of installation, however, is so high that the method could only be economically carried out on a large scale; moreover, skilled labor is required, especially in the cupellation process. The method would seem in general to be more suitable for customs works than for individual mines. In spite of the fact that comparative tests made with great care showed a much higher recovery by the litharge-smelting process as compared with the ordinary acid-treatment, roasting, and smelting the system has not been generally adopted. In some plants, where the ordinary method of smelting is used, it is customary to remelt all the slags from the first fusion of precipitate, with litharge and a little additional flux, adding a reducing agent if necessary or adding a small quantity of metallic lead at the finish. The resulting lead bullion is then cupeled.

(G) Nature and Properties of Cyanide Bullion

The bullion bars, after the final smelting, are frequently turned out of the ingot molds into a vessel of water as soon as they are set. The adhering slag is detached by hammering, and in some cases the bar is " pickled ", by immersing in dilute nitric acid and cleaned by scrubbing with a hard brush. If dip samples have not been taken, it is commonly sampled by machine drill as described below, in Part VII, and then immediately packed for shipment.

The nature and composition of the bullion varies according to the proportion of gold and silver, and to the method by which it has been obtained. When much zinc is present, it is of a pale yellowish-green color, and shows much variation in composition of different parts of the bar, the general tendency being for the gold value to concentrate toward the center of the bar.[1] This refers chiefly to bars of 650 to 800 fineness; with richer bars the liquation is less marked. It is quite possible to obtain cyanide gold which is little, if at all, inferior to battery gold in fineness. This is mainly a question of care in the clean-up and smelting, more particularly in the thorough extraction of metallic zinc by acid treatment and in the selection of the proper fluxes in the fusion process.

The following analyses, given by Dr. T. Kirke Rose,[2] will illustrate the character of bullion produced when direct smelting is used without preliminary treatment to remove the zinc:

	(1) Per cent.	(2) Per cent.	(3) Per cent.
Gold	60.3	61.7	72.6
Silver	7.3	8.1	9.2
Zinc	15.0	9.5	7.1
Lead	7.0	16.4	4.9
Copper	6.5	4.0	4.8
Iron	2.2	0.3	1.4
Nickel	2.0	-	-

By acid treatment and roasting the amount of impurity may be reduced to 5 or 6 per cent., particularly if manganese dioxide be used for fluxing off the lead.

(H) Refining of Bullion

On account of the deductions made by bullion buyers, considerable efforts have been made to devise a means of economically purifying the low-grade bullion produced by the cyanide process. It may be remarked, however, that it is generally easier and more satisfactory to remove the impurities before conversion into bullion than after.

The method of remelting with borax is very ancient, as is also the practice of sprinkling borax on the metal at the moment of fusion.

[1] See paper by F. Stockhausen, " Proc. Chem. Met. and Min. Soc. of South Africa," II, 46.

[2] " Metallurgy of Gold," 4th edition, p. 331.

Refining by leading a current of chlorine gas through the molten bullion was practised for many years in dealing with battery gold in Australia, where it was introduced by Miller in 1867 and used at the Sydney mint. The process was in use at the Pretoria mint at one time for refining cyanide bullion, but has the drawback that it involves a separate operation for the recovery of the silver, as the scoria formed contains this metal as chloride.

Dr. T. Kirke Rose [1] has experimented on a method of refining by injecting oxygen gas into the molten metal. Air may also be used, and is equally effective in removing base metals, but the action is slower. The method is also applicable to zinc-gold precipitate without previous smelting; sand, borax, and charcoal are added, the mixture fused in a graphite pot with clay liner, and air or oxygen injected into the molten mixture through a $\frac{1}{8}$-in. clay pipe, by means of a Root's blower.

(I) Electrolysis of Low-grade Bullion

The following method was used at the Lone Star Mine, Nicaragua, for refining bullion carrying a large amount of copper, having been introduced by T. W. Bouchelle (Eng. and Min. Journ., *95*, 238, Jan. 25, 1913), after other methods had been tried and abandoned on account of high freight rates on chemicals, and duties on imported materials and exported bullion.

Current was furnished by a 3 kw. direct-current generator capable of delivering 500 amp. at a pressure of 2 to 6 volts at a speed of 1800 r.p.m., power being derived from a Pelton water-wheel.

The electrolyte was contained in a wooden cell 36 × 24 × 15 in., painted with P & B paint, and supported on insulated legs. An air-lift and perforated distributing launder were used for circulating the electrolyte.

The anodes consisted of the bullion to be refined, carrying gold 14.6 per cent., silver 10 per cent., zinc 2 per cent. and copper 73.4 per cent.

The cathodes were of sheet lead $\frac{1}{32}$ in. thick, but thicker material would be preferable.

---

[1] " Trans. I. M. M.," XIV, 378–441 (April, 1905).

The electrolyte consisted of copper sulphate 17.6 per cent., sulphuric acid 5 per cent. in aqueous solution.

The anode plates were inclosed in tight-fitting canvas sacks, to retain the slime formed, and suspended from the busbars by copper hangers. The cathodes were folded over the bars and secured by clamps.

It was found essential to maintain the electrolyte at a concentration of not less than 10 per cent. copper sulphate, and to keep the temperature at 150° F.

The anode slime assayed 920 to 970 fine in gold, and was fused in clay crucibles, using manganese dioxide in the flux, yielding bullion 970 fine.

The copper and silver precipitated on the cathodes, the amount of copper recovered paying the cost of the entire operation of refining.

Silver could be recovered by using the cathodes from the above process as anodes, and employing a lower current density, in which case the silver would remain as anode slime. Any small percentage of silver passing into the electrolyte could be precipitated by adding sodium chloride.

The current employed was 50 amp. at 2 volts.

(J) Smelting of Zinc Precipitate

Many attempts have been made to improve the methods of dealing with the product of the zinc boxes, so as to simplify the process and eliminate expensive and tedious operations such as acid treatment and roasting.

The Tavener process, described in the text, has found only a limited application, probably on account of the skilled labor required, particularly in connection with cupellation. (See p. 328.)

In places where oil fuel is available, good results have been obtained by the use of tilting furnaces, specially constructed with a view to this class of work.

Arthur Yates (Journ. Chem. Met. and Min. Soc. of South Africa, June, 1909) describes the installation at the Redjang Lebong Mine, Sumatra. The retorts used were 30 in. long, with a diameter varying from $6\frac{1}{2}$ in. at the mouth and $9\frac{1}{2}$ in. at the bottom, to a maximum of $13\frac{3}{4}$ in., and are placed at an angle of 30° with the horizontal. Special arrangements were introduced to allow of turning the retort, to avoid cutting through by the action of the metal on one spot, and to allow of easy renewal of retorts. The top of the furnace was closed by three removable arches of firebrick, and a hole in the furnace bottom, plugged with ore slimes, served as an outlet for metal and slag in case a retort should crack. The furnaces were filled in with firebricks, leaving a space of only 2 in. around the sides and bottom of the retort.

The furnace was heated by means of kerosene oil delivered from an oil-storage tank through "Billow" atomizers, of which two were used for each furnace, just below the retort, being placed in pieces of 3 in. pipe built into the brickwork at the center of each side of the furnace, the brass shells of the atomizers being protected by a cover of 2 in. piping. Oil was delivered through a flexible steel tube, and air under 40 lb. pressure, previously heated by passing through the furnace flues, was also delivered to the atomizers. The two jets point slightly upward so as to meet just below the retort and distribute the heat where most required.

Each such furnace would melt about 600 lb. of fairly clean precipitate in 10 hours, consuming 1 gal. of oil for 15 lb. of roasted precipitate.

The time required is less than half that with mixed coke and charcoal on similar material. As the furnace is charged from the outside, and the pouring is simply performed by turning a handwheel which tilts it to the required position, the whole operation is cleaner, safer and less troublesome than with reverberatory and pot furnaces.

In places where electric power can be cheaply supplied, the electric furnace may be successfully applied for this purpose. H. R. Conklin (Eng. and Min. Journ. *93*, 1189, June 15, 1912) gives an account of the installation at Lluvia de Oro, Chihuahua, Mexico. The cost of oil and charcoal being prohibitive, it was decided to utilize power from a hydro-electric plant owned by the company. The furnace used is rectangular in section, 16 in. square inside and 4 ft. deep. During use the section increases to 24 or 26 in. and becomes nearly circular. The furnace is lined with firebrick, all joints next the charge being filled with fireclay. This lining lasts, in use, about 200 hours. Tap holes are formed by a piece of $\frac{1}{4}$ in. iron plate in which a $\frac{7}{8}$ in. hole is drilled, but special castings provided with a spout were to be tried for the purpose.

An alternating current of 60 cycles is used at 110 volts. A direct current would answer equally well.

Graphite electrodes are used, but iron can be used as a substitute, though it has the disadvantage that the ends burn off, forming shot which collect at the bottom of the furnace and run out with the bullion. One side of each furnace circuit is grounded through the bottom electrode to avoid shocks when tapping the furnace with iron tools, and an independent transformer is used for each furnace. The current used for melting precipitate is between 250 and 350 amp. and about 400 kilos are melted in 24 hours.

Fluxes vary, but the following is generally used: precipitate 100 parts, lime 5, sand 15, borax 10, soda 10. The slag is returned to the mill.

All manner of impure by-products are easily reduced, and concentrates are also treated in this furnace, so that only pure bullion, 800 silver, 100 gold and 100 base is marketed. Zinc is largely volatilized during the smelting.

The method of operating is as follows: In starting the furnace all resistance is cut into circuit by opening the regulating switches; the upper electrode is lowered to the bottom of the furnace in one corner, 5 or 6 in. from the bottom electrode. A handful of powdered graphite is dropped on the bottom to conduct the current. Borax and powdered slag are added until a layer of melted slag covers the bottom of the furnace. The upper electrode is then raised and the switches closed one after the other. When a layer of slag about 1 ft. deep has accumulated, the precipitate, mixed with flux, is fed in, and the bullion tapped, one or two bars at a time, as it melts, avoiding too large an accumulation of bullion in the furnace, as this would reduce the resistance and chill the metal.

Short zinc is previously roasted in a small reverberatory furnace, and then smelted in the electric furnace, the charge being: roasted zinc 100, sand 30, borax 10, soda 10, the metal formed being remelted with precipitate.

Concentrates are reconcentrated to a value of about $3000 per ton, then roasted in the reverberatory furnace and fluxed as follows: concentrate 100 parts, lime 15, sand 30, slag 15. The concentrate consists of pyrite mixed with much metallic iron. By smelting the above mixture in the electric furnace, bullion, matte and a heavy iron slag are obtained. The tap hole used in $1\frac{1}{8}$ in. diameter. The current consumed is 400 to 600 amp., to smelt 800 kilos of concentrate per 24 hours. The bullion is re-melted with precipitate.

It is suggested to treat the matte by melting with precipitate, using a jet of compressed air to burn off the excess of sulphur.

PART VI

SPECIAL MODIFICATIONS OF THE CYANIDE PROCESS

THIS part of the book will be devoted to the discussion of certain departures from the normal course of cyanide treatment adopted in particular cases and under special circumstances. We shall refer briefly to certain new developments of the process which have not yet stood the test of experience long enough to demonstrate their practical value. Some account will likewise be given of obsolete or nearly obsolete methods which possess a historical interest for cyanide workers. As the admission of such descriptions may be criticized, it may be pointed out that all improvements in the cyanide process are merely applications of well-known principles employed in other branches of industry; that much may be learned by studying the causes of past failures; and that the experience of the past often points out the directest road to advance in the future.

These special processes will be considered under the following heads: (1) Direct treatment after dry or wet crushing. (2) Crushing with cyanide solution. (3) Roasting before cyanide treatment. (4) Use of auxiliary dissolving agents. (5) Electrolytic precipitation processes. (6) Other precipitation processes (7) Special treatment of cupriferous ores.

SECTION I

DIRECT TREATMENT AFTER DRY OR WET CRUSHING

DIRECT treatment may be defined as any process in which the ore is crushed dry or with water, and the cyanide solution applied at once to the crushed product, the latter undergoing no amalgamation, hydraulic separation, concentration, roasting, or other intermediate process previous to cyanide treatment. When this system is adopted, dry crushing is nearly always employed in preference to wet, chiefly because the wet-crushed product is seldom in a condition suitable for cyanide treatment as a whole, being generally unleachable without some form of hydraulic separation; also because the cost of plate amalgamation is generally so trifling in comparison with that of crushing, that there is no sufficient advantage in omitting it in the case of wet-crushed ore.

For dry crushing, Chilian, Griffin or Ball mills and rolls are preferable to stamps. Some particulars with regard to these machines are given in Part III. The system is particularly suitable for cases in which the ore is exceptionally friable or porous; in such cases good extractions can sometimes be obtained even with very coarse crushing, as at the Mercur mine, Utah, where the oxidized ore was treated direct after dry crushing to ½-in. size. At the George and May mine, Johannesburg, a porous oxidized ore was crushed coarsely in a Gates crusher and treated direct by cyanide, giving an extraction of 70 per cent. of the gold.[1] At the Lisbon Berlyn mine, Lydenburg, Transvaal, a 10 dwt. ore was put through a Blake rock-breaker, Marsden fine crusher, and Gates rolls; the product gave a 68 per cent. extraction by direct cyanide treatment.

It is generally necessary to dry the ore sufficiently to reduce the percentage of moisture below 2 per cent. previous to dry crushing, care being taken in this operation to avoid partial roast-

[1] Julian and Smart, *loc. cit.*, p. 202.

339

ing, which might lead to the formation of cyanicides. To avoid the formation of large amounts of fine dust, the crushing is best done in several stages, using screens and returning the oversize for recrushing. As a rule, shallow tanks must be used for the leaching of the dry-crushed product, and in many cases filtration is aided by suction. Since no moisture has to be displaced by solution, it is possible to give a more thorough final water-wash than is the case when wet crushing is used, without increasing the stock of solution in the plant.

SECTION II

CRUSHING WITH CYANIDE SOLUTION

GENERAL CONSIDERATIONS

THE crushing of ore with cyanide solution instead of water was attempted at least as early as 1892, when experiments in this direction were made at the May Consolidated Battery near Johannesburg.[1] The system was introduced at the Crown mine, Karangahake, New Zealand, in 1897, and at Central City, South Dakota, in 1899. Five mills on this principle were established in South Dakota in 1904, since when the method has been extensively adopted in the United States and Mexico.

The early attempts at crushing with cyanide seem to have been abandoned on account of difficulties in handling slimes. The proportion of ore to liquid in the material crushed is necessarily much greater than when water is used, and it is of course impossible to run the slimes to waste without losing the greater part of the dissolved values. Owing to the thicker pulp, the ore is also crushed finer than when water is used under ordinary conditions, and the product is therefore less adapted for direct leaching. In modern practice, however, these difficulties are overcome by a system of hydraulic separation, in which the clear overflow is returned to the mill and the sand and slime collected for separate treatment. To avoid the necessity of handling enormous volumes of solution, the amount of liquid used in the battery is generally from $1\frac{1}{2}$ to 1 ton of solution per ton of ore crushed, but as much as 7 tons is sometimes used.

The advantages of the system are briefly as follows: (1) The solution of gold begins from the moment the ore enters the battery. (2) There is no necessity for special appliances for agitation and aeration, these being sufficiently secured by the transfer of the pulp from the battery to the classifiers and treatment tanks.

[1] " E. and M. J.," Oct. 8, 1892.

341

(3) Less water is required than when the crushing is done in water. (4) There is no necessity for running any solution to waste, as must be generally done when final water-washes are given in the ordinary system of treatment.

The objections are: (1) There is danger of loss, owing to the fact that gold in solution is transferred for a considerable distance and through a number of different appliances. (2) Owing to the thickness of the battery pulp, the capacity of the mill is reduced, and, as already pointed out, a larger percentage of fines is produced. (3) The action of the cyanide hardens the amalgam on the plates, and corrodes the plates themselves, owing to the solvent action of cyanide on copper. (4) It is impossible to give preliminary water or alkali washes. This is a serious objection in the case of acid ores, or of ores containing soluble cyanicides, which might be removed by preliminary treatment in the ordinary process.

The first objection is not of much consequence in a well-constructed plant, where proper means are taken to prevent leakage; the second is of little importance in the case of ores which must be crushed fine in any case in order to obtain a good extraction. When very dilute solutions (.03 to .07 per cent. KCN) are used, as is generally the case, the action on the plates is not serious, and it may possibly be avoided by using Muntz metal or some other alloy instead of copper.

Illustrations from Practical Working

On the Rand, the system of crushing with cyanide solution forms part of the scheme of treatment introduced by G. A. and H. S. Denny.[1] In this system the ore is crushed with a solution containing .03 per cent. cyanide (as KCy) and .004 per cent. alkali (calculated as NaOH), in the proportion of 6.6 tons of solution to 1 ton of ore. After passing over amalgamated plates, the pulp, mixed with the returned coarse product from the spitzkasten following the tube mills, goes to hydraulic classifiers, where coarse sand and concentrates are separated from fine sand and slime. The coarse sand and concentrates are passed through tube mills, and thence over shaking amalgamated plates to spitzkasten, the overflow from which goes to the conical slime settlers, while the underflow (coarse product) is elevated and

---

[1] "E. and M. J.," LXXXII, 1217 (Dec. 29, 1906).

returned to the pulp leaving the battery plates. The fine sand and slimes go through another set of spitzkasten, from which the sandy product goes to percolation tanks, while the slimy portion goes to a large conical tank, whence the clear solution overflows and is returned to a solution tank supplying the battery. The thickened slime-pulp, together with the slime overflow from the tube-mill product, goes to conical slime settlers and is then pumped to filter-presses. The precipitated solutions are also returned to the mill-supply tank.

This system could only be successfully applied in cases where the gold in the finely crushed ore is very rapidly dissolved; on the Rand it is claimed that 98 per cent. of the gold recovered from the slime is dissolved before the slime is settled, and that 70 per cent. of the gold recovered from sands is in solution before percolation begins. In the pulp leaving the mortar-boxes 12.65 per cent. of the total gold is already in solution.

The results of a month's treatment at the Meyer and Charlton G. M. Company's plant, working on this system, are given as follows:

```
Ore treated  ........................10,740 tons
Average assay before treatment ........$10.90 per ton
Average assay of residues ..............$ 0.51  "   "
Extracted on battery plates   ..........  43.85 per cent.
    "      on shaking plates ...........   3.01  "    "
    "      in sand treatment  ..........   9.76  "
    "      from ore in transit ..........  38.67  "    "
           Total extraction ............  95.29
```

Of the total ore milled, 52.3 per cent. is reground in tube mills; approximately 70 per cent. is treated by percolation as fine sand, and 30 per cent. as slime. It will be noticed that comparatively little of the gold is recovered from the sand in the percolation process, most of it having been already dissolved and carried off with the solution in the slime overflow. This suggests a further modification of the process, in which percolation is eliminated altogether, and the whole of the ore ground fine enough to be treated by settlement, decantation of clear solution, and filter-pressing of the settled pulp. It is doubtful, however, whether the reground sand would in all cases yield its gold values to the solution as rapidly as is the case with true slime. In ores where the values are less rapidly dissolved, some form of mechanical agitation and aeration will generally be necessary.

In the United States, similar methods have been adopted at the Liberty Bell mine, Colorado, and at various plants in the Black Hills, South Dakota, and in Nevada. In cases where the finely ground pulp is treated by suction filters on the Moore or Butters principle, it is found advisable to pass the overflow from the first set of hydraulic cone classifiers through a second set, returning the coarse products from each set for regrinding, or delivering them to the leaching tanks, and allowing only the fine product from the second classifiers to go to the slime filter tank. It is found that these filters are best adapted for treating slime which is as free as possible from fine sand.[1] The presence of a little slime in the sand treated by percolation is of less consequence, as the material is very thoroughly mixed by the use of the Blaisdell sand distributor and by double treatment.

In the Black Hills,[2] on the other hand, where the Merrill filter-press is in use, the practice is to make as clean a sand as possible. The stream of pulp is passed through a succession of cones, the final ones being arranged as spitzlutten — that is, they have an upward stream of solution introduced at the bottom. The fine product treated by filter-pressing contains 15 to 20 per cent. of sand passing a 150-mesh screen. The battery pulp is delivered to the cones by means of sand pumps, and solution is added in the launder which carries the sands to the distributor, so as to dilute the pulp in the ratio of 5 to 1. The battery solution has a strength of .06 to .065 per cent. KCN and .04 per cent. NaOH; the solution after precipitation carries .075 to .08 per cent. KCN and .05 to .06 per cent. NaOH. (These figures refer to the practice at the Maitland properties, South Dakota, in 1904.)

At the Desert mill (Tonopah Mining Company), Millers, Nevada,[3] the ore is crushed with a .15 per cent. solution, using 7 tons solution per ton of ore. The pulp is concentrated on Wilfley tables, the tailings from which pass to cone hydraulic classifiers, and clean slime is separated from them as already described. The *sand* goes to collecting vats, in which it is drained. It is then excavated by the Blaisdell apparatus and carried by conveyors

---

[1] This is contrary to the experience in other localities, where a small proportion of fine sand in the slime is found to be essential to the successful working of suction filters.

[2] J. Gross, " Trans. A. I. M. E.," XXXV, 616.

[3] A. R. Parsons, " Min. Sci. Press," XCV, 494.

to leaching vats; lime is added in the collecting vats and lead acetate during the transfer. The solutions used in leaching contain .25 to .15 per cent. KCN. After five days' treatment, including transfer, the sand is passed to a second set of vats and similarly treated for another five days, then to a third set for three to five days, the total time of treatment being twelve to fifteen days. The residue carries 0.6 dwt. gold and 3.1 oz. silver. The *slime* is collected in vats of 36 ft. diameter by 20 ft. deep, with rim overflow, the clear solution being returned to the battery storage tank. The thickened pulp goes to agitation vats previously filled with precipitated solution, to which the requisite quantities of lime and cyanide have been added. Agitation is continued for thirty hours, assisted by injection of compressed air and circulation with centrifugal pump, the dilution of pulp being about 4 to 1. After settling for six hours, the clear solution is drawn off for precipitation and the slime is agitated with fresh solution for a further twenty-four hours, after which it goes to the Butters filter plant.

In Mexico, the process is in use at various plants treating gold and silver ores. At Butters' Copala mines, Sinaloa, the ore treated carries gold $1.96 and silver 15.8 oz. It is crushed by stamps with 12-mesh screens, with a solution containing .07 per cent. NaCN and .135 per cent. alkali calculated as NaOH, using 16 tons of solution per ton of ore. The battery product goes to cone classifiers, from which the coarse product goes to tube mills and the fine product to a second classifier. The coarse product from the latter also goes to the tube mills, the effluent from which returns to the second classifier. The fine product from the second classifier goes to a third, which makes the final separation of 42 per cent. sand 58 per cent. slime. The sand is leached with 0.3 per cent. NaCN, and the slime is treated in agitation vats and Butters filters.

SECTION III

ROASTING BEFORE CYANIDE TREATMENT

THE conditions under which roasting is necessary before cyanide treatment have been briefly referred to in Part III of this book. In general, it may be stated that roasting is only necessary or advantageous in cases where the gold is combined, or closely associated, with some element which prevents or retards its solution in cyanide. The chief applications of roasting in practice are: (1) to telluride ores, such as those of Cripple Creek and Western Australia, where some part of the gold is probably in chemical combination with tellurium, the compound being unattacked by ordinary cyanide solution, even in presence of excess of dissolved oxygen; (2) to arsenical sulphide ores, such as those of the Mercur district, Utah.

Two types of furnace are in general use, viz.: (1) furnaces with single or superimposed hearths and revolving rabbles, such as the Edwards and Merton furnaces; (2) furnaces consisting of an inclined revolving cylinder, such as the Argall roaster. Roasting by hand in reverberatory furnaces was formerly employed as a preliminary operation in the cyaniding of concentrates on the Rand and elsewhere, but is now superseded in most parts of the world by mechanical contrivances.

FURNACES WITH REVOLVING RABBLES

The *Edwards furnace* consists of a single hearth enclosed in an iron framework, and supported above the ground on central pivots, which enable the hearth to be tilted to any required angle to aid the passage of the ore along the bed of the furnace. The furnace is kept at the desired inclination by means of screw-jacks. The rabbles are supported on vertical shafts passing through the roof of the furnace, and are driven by means of a horizontal shaft and gear-wheels above the furnace. To these vertical shafts are attached a number of removable plows or rakes, so arranged that each revolves in the opposite direction to the succeeding one. The ore is thus made to travel across the furnace in a zigzag direction. In some cases the five rabbles nearest the fire-box are water-cooled; the shafts are made hollow and a stream of water is allowed to descend through the central pipe, pass through a cavity in the rabble itself, and escape into a circular channel surrounding the vertical shaft above the furnace arch.

The ore is fed in at the receiving end of the furnace by means of fluted rolls, consisting of cast-iron cylinders having eight V-shaped channels on the circumference; these rolls are driven by toothed gearing. The rabbles then gradually work the ore toward the discharge end (nearest the fire-box), where the heat is greatest. The roasted ore is finally discharged through an opening on to a "push conveyor." This consists of a semicircular trough provided with transverse blades free to move only in one direction. The conveyor moves horizontally to and fro on rollers, and at each forward movement the ore is pushed by the blades a distance of 20 in.; on the return movement the blades are free to swing and pass back over the heaps of ore formed by the forward movement. The ore is thus gradually pushed to the end of the trough, and at the same time turned over by the blades, thus exposing fresh surfaces and allowing an opportunity for improving the final roast. The roasted ore is then ready for cyanide treatment, and may be discharged directly from the conveyor into a stream of solution.[1]

The *Merton furnace* has three superimposed horizontal hearths connected by vertical discharge holes. The rabbles, as in the Edwards furnace, are carried by vertical shafts, of which there are four or five, each shaft passing through the whole series of hearths, and being mounted at the lower end on a footstep below the bottom hearth. Attached to each shaft are three rabbles, one on each hearth, arranged so that the radii of the circles they describe in revolving are a little less than the distance between the shafts. By this means the ore is passed successively from one rabble to the next, and eventually falls through the discharge door on to the hearth below. A sliding bar at the discharge doors regulates the rate of discharge. There is also a finishing hearth next the fire-box, with rabbles which are sometimes water-

[1] E. W. Simpson in "Trans. I. M. M.," XIII, 27–33.

jacketed. The vertical shafts are driven by bevel or worm-gearing at 1 to 2 r.p.m., and are provided with special arrangements for counteracting the effect of expansion.

The Merton furnace is more economical, both in first cost and in working, than the Edwards, but has the disadvantage that the inclination of the hearths cannot be varied. There is also more difficulty in withdrawing the shafts of the rabbles for repair. Doors are provided at the ends and sides for admitting air and for removing and renewing the rabbles. The capacity of this furnace varies from 5 to 25 tons per day, for ores containing from 25 to 6 per cent. of sulphur. The roasting is very perfect, and the percentage of sulphur may be reduced to .05 per cent. or less.[1] In Australia the fuel used is generally eucalyptus wood, but oil or gas may be used with advantage, and give a more regular temperature. The accompanying drawing (Fig. 38) shows a recent type of this furnace in plan and sectional elevation.

Revolving Cylindrical Furnaces

Several furnaces of this type have been in use for many years for roasting ores for the chlorination process, as, for example, the Brückner cylinder. Most of these, however, are costly, both in initial expense and in repairs. We shall here describe only the *Argall roaster*, which has been successfully applied in preparing for cyanide treatment the ores of the Cripple Creek district, Colorado. A similar apparatus is used as a drier for ore previous to crushing in rolls. The furnace consists of four or more parallel steel tubes lined with brick or tile, fire-brick being used in the parts exposed to the greatest heat. These tubes are mounted together, so as to revolve as a single cylinder supported on two tires. The latter revolves on friction wheels, operated by differential gear. (See Figs. 39 and 40.) The apparatus is inclined from the feed to the discharge end at a slight angle. The ore to be roasted is fed in mechanically by a shoot, and travels in a thin layer through each of the revolving tubes; the discharge end is provided with a hood with suitable openings, through which the ore drops continuously into a hopper beneath. This hood opens into a stationary cylinder of steel plate which conveys the heated gases directly from the fire-box to the hood and thence to the tubes. The fire-box is mounted on wheels, and can thus be easily withdrawn from

[1] Julian and Smart, *loc. cit.*, p. 439.

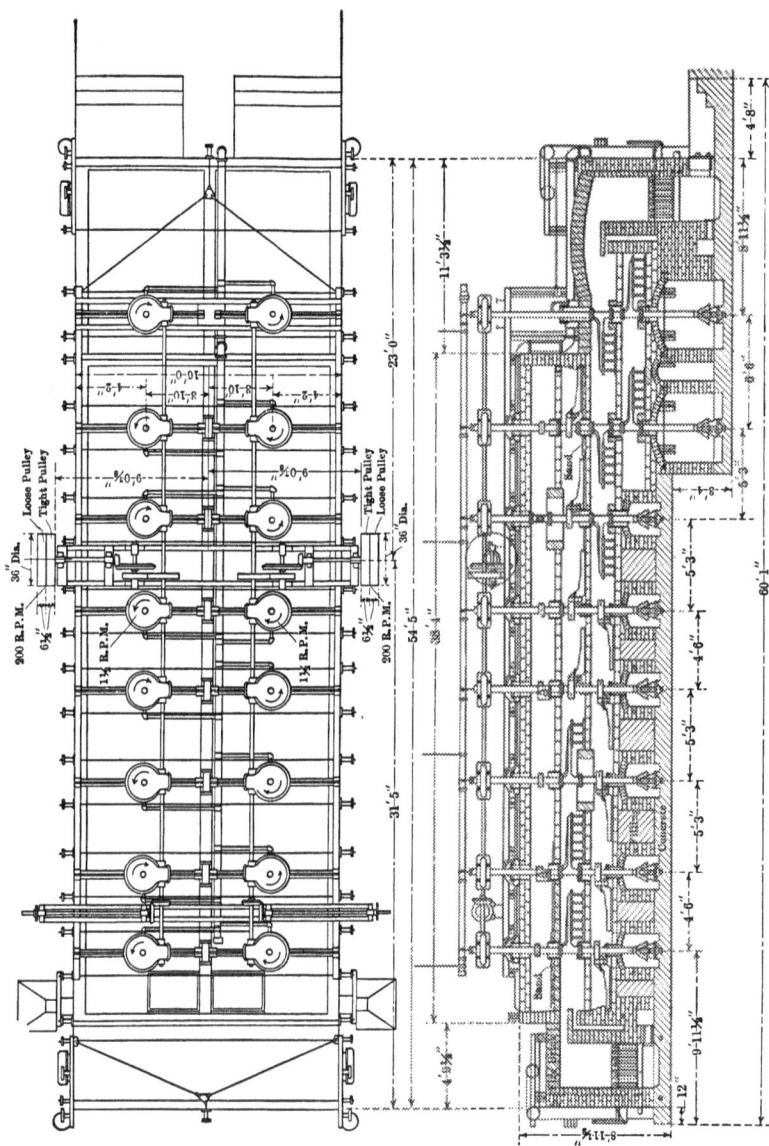

FIG. 38. — Merton Furnace. Plan and sectional elevation.
[From drawing furnished by Fraser & Chalmers.]

FIG. 39. — The Argall Roasters.   [From a photograph.]

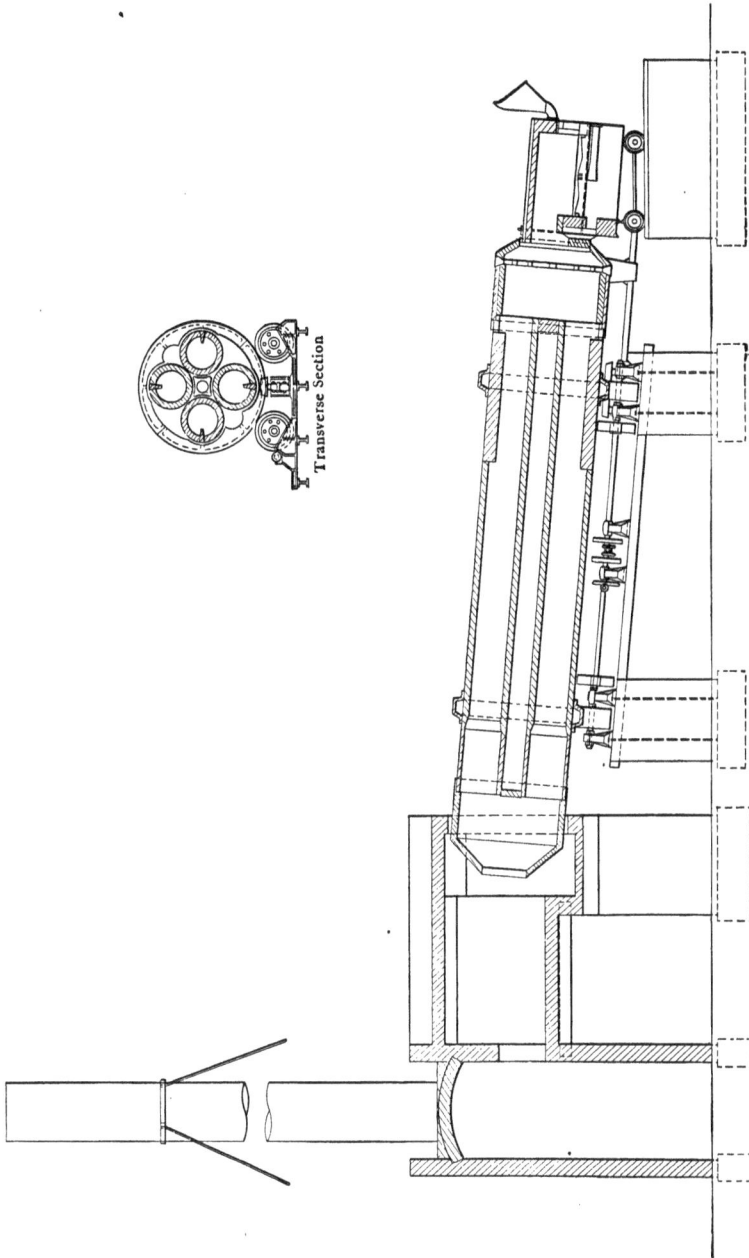

Transverse Section

Fig. 40.—Longitudinal Section of Argall Roaster.

the furnace if necessary. The layer of ore in the tubes gradually diminishes from the feed to the discharge end, so that it is thinnest where exposed to the greatest heat. It is claimed that this furnace gives a practically perfect roast, with less dust than other furnaces and small expense for repairs. The accompanying illustrations are furnished by the Cyanide Plant Supply Company.

In all processes of treating roasted ore with cyanide, it is advisable to cool the ore before charging into the treatment tanks. This is done sometimes by spreading on special cooling floors. In other cases the ore is sufficiently cool as it is discharged from the conveyors to be fed direct into a vessel or launder containing the cyanide solution. If the ore is sufficiently hot to raise the solution to boiling-point, however, a very large consumption of cyanide takes place, and the extraction is ineffective, because the oxygen is thereby expelled from the solution.

SECTION IV

USE OF AUXILIARY DISSOLVING AGENTS

Oxidizers in Conjunction with Cyanide

As soon as the function of oxygen in the reaction taking place between gold and the alkaline cyanides was recognized, the suggestion of adding a reagent to supply the necessary oxygen in an active and concentrated form naturally occurred to many investigators. A very large number of substances have been tried experimentally, and many processes involving their use have been suggested and patented; but the only oxidizers which appear to have had even a limited application in practice are: (1) Atmospheric air, injected under pressure into the pulp or solution, or beneath the filter-cloth; (2) sodium peroxide; (3) potassium permanganate; (4) manganese dioxide. Other substances, such as potassium ferricyanide, have been used, which, although not, strictly speaking, oxidizers, may aid the reaction by liberating cyanogen in an active form.

Oxidizers act in two ways: by supplying oxygen in a nascent or active condition, so accelerating the solution of gold; by oxidizing deleterious impurities that may be present in the ore or solution. The general conclusion arrived at by many experimenters in this direction is that, with a given ore crushed to a given degree of fineness, the ultimate maximum extraction of gold which can be attained is the same, whether an oxidizer be employed or not. The only advantage secured by the addition of the oxidizer is greater rapidity of extraction. In practice, this advantage is frequently nullified by the fact that much time is necessarily consumed in washing out the dissolved gold, and as this washing is done, as a rule, with cyanide solution, the dissolving process is carried to its limit in either case.

Sodium peroxide has been used to some extent in America in connection with the so-called Kendall process. The reagent is generally mixed with dry-crushed ore previous to cyanide treatment. It gives up its oxygen very rapidly in contact with water; hence, if mixed with the moist ore or tailings previous to or during the process of charging into the vat, the oxygen will be liberated and the greater part of its efficiency lost before the cyanide solution has come in contact with the ore.

Potassium permanganate has been occasionally used in South Africa in the form of a preliminary wash, especially in the treatment of concentrates. It should, however, be removed by water-washing previous to cyanide treatment, as any excess remaining reacts with and destroys the cyanide. In some instances a mixture of permanganate and sulphuric acid has been used for the preliminary oxidation and removal of cyanicides.

Manganese dioxide may be added with advantage in some cases. It gives up oxygen gradually, and thus the beneficial effect is continued throughout the treatment.[1] In this respect manganese dioxide is superior to sodium peroxide and similar reagents, but if used in excessive amounts it causes large loss of cyanide, and may also give rise to objectionable deposits in the zinc-boxes.

The use of *potassium ferricyanide* in conjunction with cyanide was suggested by C. Moldenhauer in 1892. It is not in reality an oxidizing agent, but reacts by liberating a part of cyanogen from the cyanide and forming potassium ferrocyanide

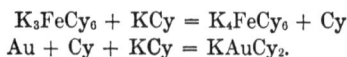

$$K_3FeCy_6 + KCy = K_4FeCy_6 + Cy$$
$$Au + Cy + KCy = KAuCy_2.$$

It thus, theoretically at least, enables the cyanide to dissolve gold in the absence of oxygen. The reaction is very rapid and effective, but other secondary reactions occur which give rise to an accumulation of salts in the solution and deposits in the zinc-boxes, both of which are detrimental.

Quantities of Oxidizers to be Added.—No definite rule can be given as to the amounts required, as this will depend on the gold or silver value of the material to be treated and on the nature and amount of the oxidizable materials present. In all cases small-scale preliminary experiments with varying quantities should be made, and the effect on precipitation tested as well as on extraction.

---

[1] As is well known, compounds of manganese exhibit a tendency to act as "carriers" of oxygen, by alternately combining with this element and giving it up to any substance capable of absorbing it.

THE BROMOCYANIDE PROCESS

An auxiliary agent which for several years was in extensive and successful use was introduced in 1894 by Sulman and Teed. This consists of the bromide of cyanogen, BrCy.   Other haloid compounds of cyanogen have been proposed, and their use was claimed in the Sulman-Teed patents, but they have never been successfully applied in practice.   The reagent is a volatile crystalline solid, soluble in water, and when mixed with an alkaline cyanide forms a very powerful and rapid solvent for gold.   Its action is discussed in Part II, but it may be repeated here that the reaction of bromide of cyanogen on alkaline cyanides is accompanied by the evolution of cyanogen or hydrocyanic acid, and that in presence of an excess of alkali the bromide of cyanogen is immediately destroyed with formation of bromide, cyanate, and (probably) bromate of the alkali metal.   It must therefore be used in solution containing little or no free alkali, and is most effective if added in small quantities at a time.   The bromide of cyanogen is not *per se* a solvent of gold.   The simplest theory of the reaction is represented thus:

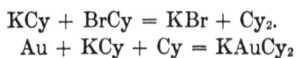

$$KCy + BrCy = KBr + Cy_2.$$
$$Au + KCy + Cy = KAuCy_2$$

BROMOCYANIDE PRACTICE IN WESTERN AUSTRALIA

The chief centers in which the bromocyanide process has been applied are the Deloro mine, Ontario, Canada, where a mispickel ore was treated without roasting, and at various mines in the Kalgurli district, Western Australia, where it was introduced in conjunction with tube milling, or fine grinding in pans, under the name of the Diehl process.

Details of the process will be found in the following papers: "The Sulman-Teed Gold Extraction Process," Sulman and Teed, in "Journ. Soc. Chem. Ind.," XVI, 961 (1897); "The Diehl Cyanide Process," by H. Knutsen, in "Trans. I. M. M.," XII, 2 (1902); "Metallurgy of the Kalgoorlie Goldfield," by Gerard W. Williams, in "Eng. and Min. Journ.," LXXXV, 345 (Feb. 15, 1908). The following particulars are summarized from these publications.

The ores to which the process is applied in Western Australia are sulpho-tellurides, the gold in which cannot be extracted by treating raw with ordinary cyanide solutions.   In the Diehl process the operations are: (1) Crushing and grinding to a high degree of fineness. (2) Treating the finely ground ore by agitation with a cyanide solution, to which bromide of cyanogen is added at intervals. (3) Separating solution from sludge by means of filter-presses. (4) Precipitation on zinc shavings. In some plants amalgamation and concentration are used, the concentrates being generally roasted and treated by ordinary cyanide.

The grinding is done in Krupp tube mills, 18 ft. long by 4 ft. diameter, the fine sands being reground until the whole product passes a 200-mesh sieve, less than 3 per cent. remaining on 220. The slimes, after removal of excess of water by spitzkasten, go to agitators, where they are treated first with ordinary cyanide. The dilution of pulp is about 2 parts ore to 3 of water. Cyanide is added in quantity sufficient to form a solution of 0.10 per cent. KCy, equivalent to 0.15 parts KCy per 100 parts of dry slime. After from one and a half to two hours' agitation with cyanide alone, the bromocyanide is added in the proportion of 0.04 parts per 100 parts of dry slime, equivalent to a strength of about 0.027 per cent. BrCy in the solution. This mixture is agitated for twenty-four hours, lime being added two hours before the finish in the proportion of 3 to 4 lb. per ton of dry ore. The sludge is then run into a receiver, from which it is forced by compressed air into the filter-presses.

The above details refer to the treatment at Hannan's Star mill, 1902. The system varies slightly at different plants, but in general the amount of bromocyanide used is from $\frac{1}{4}$ to $\frac{1}{3}$ that of the cyanide, corresponding approximately with the proportions indicated by the equations given above.

According to G. W. Williams (loc. cit.), the treatment at the mines of the Ivanhoe Gold Corporation is as follows: After concentration on Wilfley tables, the sands are ground in pans, reconcentrated, the fines separated by spitzlutten and conveyed to settlers. The slimes after settlement are pumped to agitators, where they are diluted with weak solution to form a pulp with the proportions 1:1, dry ore to solution. After two hours' agitation, 0.6 lb. bromocyanide is added per ton of dry slimes and the agitation continued for twelve hours. Lime is then added in the proportion of 1 lb. per ton of dry slimes. The pulp is then filter-pressed. The agitators are 20 ft. in diameter by 8 ft. deep, closed in and fitted with mechanical stirring gear. The method involves a separate treatment of concentrates, amounting to 18 per cent. of total tonnage. These are roasted in Edwards furnaces, mixed with spent cyanide solution, ground fine in pans, agitated with 0.1 per cent. cyanide and filter-pressed. The entire scheme of treatment comprises

Recovery: per cent. of total gold.

Amalgamation 28
Treatment of concentrates by roasting and ordinary cyanide ...13
Treatment of sands: 5 days percolation 17.5
Treatment of slimes by bromocyanide 28

Total recovery 86.5

The total treatment cost is given as \$2.20 per ton. The recovery by similar systems of treatment in other plants varies from 85 to 95 per cent. The Golden Horseshoe, Oroya-Brownhill, and Lake View Consols, treat part of their product by bromocyanide.

Owing to the instability of the reagent and the unpleasantness of handling it in quantity, mixtures of salts are shipped to the mines, which, on addition to cyanide solution, produce bromocyanide in the required proportion. Bromocyanide is usually made as required, by mixing sulphuric acid, or a bisulphate, and cyanide, bromide and bromate of an alkali. Dr. H. Foersterling, of the Roessler and Hasslacher Chemical Co., prepares a double salt by fusing together an alkali cyanide and haloid (e.g., cyanide and bromide of sodium) in molecular proportions. The bromocyanide is produced by adding to this double salt an acid and an oxidizing agent,

$$NaBr + NaCN + Na_2O_2 + 2H_2SO_4 = BrCN + 2Na_2SO_4 + 2H_2O.$$

Thus it is easy to add the exact amount of bromocyanide necessary in the treatment. (Met. and Chem. Eng., Jan. 1913.)

USE OF MERCURIC CHLORIDE

Keith and Hood have proposed addition of mercuric chloride or double cyanide of mercury to the solution. It is found that the action of cyanide on gold, and still more on silver, is generally accelerated. This reagent is occasionally employed as an auxiliary to cyanide in treatment of ores containing silver sulphide. A small amount of mercury in solution is beneficial in assisting precipitation of precious metals in the zinc-boxes. The solvent effect on sulphide ores is apparently due to the great affinity of mercury for sulphur, thus:

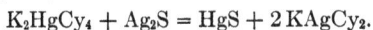

$$K_2HgCy_4 + Ag_2S = HgS + 2KAgCy_2.$$

SECTION V

ELECTROLYTIC PRECIPITATION PROCESSES

THE idea of using an electric current for precipitating the precious metals from their cyanide solutions dates at least as far back as 1840, when Elkington's patent for electroplating and electro-gilding was taken out (E. P. No. 8447; Sept. 25, 1840). In fact, the supposition that an electric current was necessary, at least for the precipitation if not for the solution of metals in cyanide, appears to have been general among investigators up to the time of the discovery of zinc-thread precipitation.

THE SIEMENS-HALSKE PROCESS

About the year 1887 the firm of Siemens and Halske, of Berlin, introduced an electrolytic process for depositing copper, zinc, gold, silver, etc., from cyanide solutions. (See E. P. No. 3533; Feb. 27, 1889.) This process was subsequently applied to the treatment of solutions resulting from the cyanide extraction of gold ores and tailings, and for several years found an extensive application in South Africa, where it was introduced by A. von Gernet in 1893.

In its original form, this process consisted in precipitating the metals from cyanide solutions by means of an electric current passed between anodes of iron and cathodes of sheet lead, the precious metals being deposited on the latter, while the iron was gradually converted into oxide, and ultimately into soluble and insoluble cyanogen compounds, such as ferrocyanides. Some account of the introduction and working of the process at the plant of the Worcester mine, Johannesburg, where it was first applied in South Africa on a working scale, is given by von Gernet.[1] The method of applying this process usually adopted in South Africa was to place the anode and cathode plates in the compartments of a box similar to the ordinary zinc-box, but of larger dimensions, so as to allow from 15 to 25 cu. ft. per ton of solution per 24 hours.

[1] "Proc. Chem., Met. and Min. Soc. of S. A." I, 28 (Aug. 18, 1894).

The plates were commonly arranged parallel to the sides of the box; in each compartment the anodes and cathodes were placed alternately side by side, the cathodes of one compartment being connected electrically with the anodes of the next lower compartment. The anodes of the top compartment were connected with the positive pole of a battery, dynamo, or other source of current, the cathodes of the bottom compartment being connected with the negative pole of the same.

The *anodes* were rectangular plates of sheet iron, $\frac{1}{8}$ to $\frac{1}{4}$ in. thick, with an effective area of, say, 5 sq. ft. They were usually held in grooves or by means of slats at the sides of the box, and were enclosed in canvas bags to insulate them from the cathodes and to collect the products of the reaction (ferric oxide, etc.) referred to above.

The *cathodes* were of thin sheet lead suspended on horizontal wires between each pair of anodes, at about 3-in. distances. As it was found advantageous to increase the cathode area as much as possible, in later practice the lead sheets were cut into long strips and the threads separated as much as possible so as to give the maximum surface for precipitation. A current of .04 ampere per sq. ft. of anode surface was usually sufficient, and was found to deposit the gold on the cathodes in a coherent form. When sufficiently coated with precious metal, the cathodes were removed and replaced by fresh ones, without interrupting the regular working of the box; they were then melted in a reverberatory furnace and cast into lead-bullion bars. These were then cupeled in an ordinary cupeling furnace. A part of the lead was recovered for re-use by reduction of the litharge resulting from the latter operation. A further quantity of gold was obtained by treating the sludge which formed inside the sacks containing the anodes.

The chief advantages claimed for the Siemens-Halske process were: (1) The possibility of precipitating gold from solutions very weak in cyanide, which did not admit of treatment by the ordinary zinc process. It was thus practicable to employ only the minimum quantity of cyanide needed for dissolving the gold. This was of particular importance in the treatment of slimes by the decantation process, where the volume of liquid used per ton of material treated was necessarily large. (2) The purity of the bullion obtained by cupellation of the lead bars as compared with that produced by the ordinary smelting of zinc precipitate.

(3) There was no necessity for interrupting the regular work of the plant during the clean-up. (4) Copper and other obnoxious base metals were removed from the solution very effectively. (5) There was no accumulation of double cyanide of zinc, etc., in the solution, which remained always in good working condition and did not become charged with foreign salts.

On the other hand there were certain evident disadvantages: (1) The necessity, when dealing with solutions low in gold value, of a large surface for deposition, involving the use of boxes of great size, and hence considerable initial outlay. (2) More skill and attention was needed to secure satisfactory working than with zinc precipitation.

With the introduction of the zinc-lead couple in 1898, the first and principal advantage of the Siemens-Halske process disappeared, as it was then found possible to precipitate effectively by means of zinc from solutions as weak in cyanide as those dealt with by the electrolytic process. By the use of the Tavener process (see Part V), or by various methods of refining zinc precipitate, as for example by fluxing with dioxide of manganese in clay-lined pots, it is also possible to obtain bullion from the zinc process having the same fineness as that furnished by cupellation in the Siemens-Halske process.

Later Developments of Electrolytic Process

As the action of the solution on the anodes gave rise to a troublesome by-product, many attempts were made to find a material for the anodes which would be insoluble in the cyanide solution. Thick sheet lead was used for some time by Charles Butters, but this had a tendency to become hard and brittle, and under certain conditions became rapidly coated with cyanide of lead, so that after use for a short time the plates crumbled to pieces. E. Andreoli and others used anodes of lead coated with a layer of peroxide of lead; this was obtained by immersing the sheets in a bath of potassium permanganate or of plumbate of soda. They were then immersed in strong cyanide solution and an electric current passed through, which caused the coating of peroxide to become firm and hard. These plates were practically insoluble in the solution, and hence no by-products were formed.

Many attempts were also made to find a cathode from which the deposit could be removed as required without destruction of the cathode itself. Sheet-iron carefully freed from oxide has been used; the gold deposit is removed by immersing the cathode in a bath of molten lead. A more satisfactory method, however, is to use a cathode of tinned iron. This was applied successfully at Charles Butters and Company's works at Minas Prietas, Mexico, and at Virginia City, Nevada. The gold and silver are deposited in a loose form, so that they may be rubbed off from time to time, to collect as a sludge at the bottom of the box. This is cleaned up at intervals and smelted direct, yielding high-grade bullion, while the cathode remains intact. When much base metal, *e.g.*, copper, is present in the solution, however, there is a dense hard deposit on the plates, which is scraped off with difficulty; also a considerable amount of gold adheres to the plate before any of the deposit begins to fall off.

ELECTROLYTIC PROCESS FOR GOLD-COPPER ORES

The following system of combined electrolytic and zinc precipitation was used by C. P. Richmond, at the San Sebastian mine, Salvador, Central America,[1] for the treatment of a complex ore containing copper, arsenic, antimony, and tellurium. The ore is crushed in ball mills and roasted. The sand and slime are then separated in cyanide solution, the sand being treated by percolation and the slime by agitation and filter-pressing. The greater part of the gold, together with some copper, is precipitated by electrolysis, after which the solutions are passed through a zinc-box to recover the residual gold. The accumulation of copper in the solution is thus prevented.

The cyanide solution, averaging 16 dwt. per ton, flows through the electric box at the rate of 150 tons per 24 hours; the box is 30 ft. long by 10 ft. wide; and 4 ft. 8 in. deep, and is inclined 1 in. per foot. There are twelve compartments, but the last two are used for settlement only, so that the effective capacity is 1166 cu. ft., or about 8 cu. ft. per ton per 24 hours. Each compartment has 25 anodes and 24 cathodes. The anodes are rolled lead plates $22 \times 48 \times \frac{1}{8}$ in. The cathodes are lead plates of similar size, but only $\frac{1}{16}$ in. thick. The anodes are peroxidized by immersing them in a solution of potassium permanganate, with or without sulphuric acid. One method is to dip them in a solution of 1 per cent. $KMnO_4$ and 2 per cent. H_2SO_4, keeping them

[1] Chas. P. Richmond, " E. and M. J.," LXXXIII, 512.

for six hours under a current of 2.5 amperes per square foot. By this means a coating of lead peroxide is obtained which lasts 8 to 12 months.

In the boxes, the anodes and cathodes of successive compartments are connected in series, with a current strength of 1 ampere per square foot. The gold and copper form a hard dense coating on the cathodes, with no tendency to fall off as sludge, but there is a gradual accumulation of low-grade precipitate on the anodes and at the bottom of the box. After passing the electric box, the solution flows through two zinc-boxes 22 ft. long, with fourteen compartments, each $2 \times 2 \times 1$ ft., which reduces the gold contents to about 0.06 to 0.10 dwt. per ton. Hardly any copper is deposited in the zinc-box. The electrical box precipitates 80 to 90 per cent. of the gold, together with much copper, which is then separated as follows: After remaining in the cyanide electric box for 20 to 30 days, the cathodes are transferred to another box, where they are used as anodes. This box contains sulphuric acid of 2 to 3 per cent. H_2SO_4, the cathodes used being lead plates $\frac{1}{16}$ in. thick. There are four compartments, each containing 5 anodes and 6 cathodes, 4 in. apart and connected in series. The current used is 5 amperes per square foot of anode surface. The anodes are hung in a wooden frame with closed bottom and open sides, over which is stretched a sack of cotton cloth. The copper dissolves and precipitates on the cathodes, where it forms a slime, which falls to the bottom. The gold remains in a loose form in the anode sacks. The anode plates are then washed and returned, as cathodes, to the cyanide electric box. Special arrangements are employed to secure circulation of the acid liquid and to prevent short-circuiting.

To recover the gold, the acid liquid is syphoned off, and the anode frames are raised and allowed to drain over the box. They are then suspended over a filter-tank, the sacking at the sides is cut away, and the gold precipitate washed into the tank. The frames and old sacks are washed in the same tank. When thoroughly drained, the precipitate is removed from the tank, dried, and smelted in graphite pots. The copper slime from the cathodes is flushed through holes near the bottom of the box, filtered, and dried. The bullion recovered is high-grade, and owing to the constant precipitation of copper there is a regeneration of cyanide amounting to 30,000 lb. in a year's run. Sufficient copper is extracted to cover the cost of refining by the acid box and smelting of gold precipitate.

OTHER ELECTROLYTIC PROCESSES

Numerous processes have been suggested and tried experimentally, but very few have achieved even a temporary commercial success. In the so-called "electro-zinc" process, zinc plates were used as anodes, and amalgamated zinc or copper plates as cathodes. The anode being soluble, a smaller current could be used than with insoluble or slightly soluble anodes, but the method was unsatisfactory owing to the rapid accumulation of zinc ferrocyanide. The effect of the current is also to produce a hard non-adherent amalgam. Many attempts have been made to use carbon as anode, but it is difficult to obtain it in a form which will not disintegrate under the current. Aluminium has been used as cathode by S. Cowper Coles, but there appears to be much difficulty in removing the precipitated gold. Mercury is not suitable as a cathode, for the reason pointed out by von Gernet, that it is not practicable to obtain the enormous area required for effective precipitation of solutions low in gold. Amalgamated copper plates have, however, been frequently tried, and an interesting series of experiments with them has been made by T. K. Rose.[1] These experiments show that with solutions rich in gold, a current of 0.03 amperes per square foot precipitates the gold as a black powder which does not amalgamate. Other objections arise from the cost of the process, the fact that the plates absorb a portion of the gold, and that they are corroded by the action of the current.

Douglas Lay[2] enumerates several drawbacks inherent to electrocyanide processes in general. A very weak current must necessarily be used, to avoid excessive action on the anodes; this in turn involves, as already pointed out, a very large cathode area when solutions low in gold are to be treated. He also states that deposition of carbonate of lime on the cathodes takes place, involving a decomposition of cyanide and so coating the cathodes as to prevent precipitation of the precious metals.

[1] "Trans. I. M. and M.," VIII, 369.
[2] "E. and M. J.," April 11, 1908, p. 765.

SOLUTION AND ELECTROLYTIC PRECIPITATION IN THE
SAME VESSEL

The idea of dissolving and precipitating the gold simultaneously in one and the same tank or receiver is an old and favorite one with inventors; it figures in J. H. Rae's apparatus, patented in 1867, and reappears in various forms in the Pelatan-Clerici, Riecken, Gilmour-Young and other processes, in which an electric current also is generally introduced, as an aid both to solution and precipitation. There appears, however, to be very little evidence that the electric current aids the solution of the precious metals in any way, and with regard to precipitation it is for various reasons more effective in solutions from which the suspended solid matter has been removed.

In the *Pelatan-Clerici* process, used at Delamar, Idaho, in 1897, the pulp was agitated with cyanide solution in vats, the bottoms of which were composed of sheet-copper, covered with mercury, and constituting the cathodes. The anodes were in the form of sheet-iron agitators, so that the gold could be dissolved and precipitated in the same vessel. The gold was recovered in the form of amalgam.

The *Riecken* process was introduced at South Kalgurli; Western Australia, in 1900. The ore is roasted, mixed to a thick pulp with cyanide solution, and fed to agitation vats, of about 13 tons capacity, having vertical ends and sloping sides. The latter are provided with amalgamated copper plates, over which a thin stream of mercury constantly flows. The coarse gold amalgamates directly. An electric current is passed through the vat by means of iron rods, which constitute the anodes, the amalgamated plates forming the cathodes. The current used is 150 amperes at 1.5 volts, with a density of 0.4 ampere per square foot. A charge of 13 tons can be treated in 18 hours with an extraction of 93 per cent. from ore assaying over an ounce per ton. The cost, exclusive of crushing and roasting, is given as 9 s. 4.8 d. per ton.[1]

In the *Molloy* process the cathode consisted of mercury, and the anode of peroxidized lead immersed in sodium carbonate, contained in a compartment which separated it from the cyanide

[1] T. K. Rose, "Metallurgy of Gold," 4th edition, p. 346.

solution. The result of the electrolysis was to form sodium amalgam which permeated the mercury in the outer vessel. The gold was precipitated and amalgamated, with regeneration of cyanide, thus:

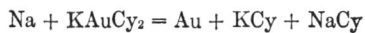

$$Na + KAuCy_2 = Au + KCy + NaCy$$

SECTION VI

OTHER PRECIPITATION PROCESSES

IN this section we propose to give a brief description of various processes not involving the use of an electric current, which have been proposed as alternatives to the ordinary method of zinc precipitation. These may be classified as follows: *(1) Methods in which the values are dissolved and amalgamated in the same vessel. (2) Precipitation with finely divided zinc (dust or fume). (3) Precipitation with charcoal. (4) Precipitation with cuprous salt in acid solution.

The *Gilmour-Young* process, employed at the Santa Francisca mine, Nicaragua, consists in agitating the gold-bearing material, which contains much clay, as a thick pulp, in pans, with addition of mercury and cyanide. After two hours of this treatment, copper and zinc amalgam are added and the pan run for four hours longer. Thus the same vessel is used for solution and precipitation. The gold and silver are obtained as amalgam, which is separated and retorted in the usual way. Extraction of 80 to 90 per cent. of the gold is said to have been obtained. For details see " Min. Ind.," VII, 334.[1]

Zinc-dust Precipitation. — This method was originally applied by H. L. Sulman [2] and H. K. Picard, at the Deloro mine, Ontario, Canada. It has been extensively used in the United States, notably at Mercur, Utah, and in the Black Hills, South Dakota. The zinc-dust consists of the finely divided metal obtained by the distillation and condensation of zinc, and generally contains about 90 per cent. of metallic zinc. A small amount of lead is usually present, which is advantageous for precipitation. The zinc is very finely divided, over 95 per cent. of the powder passing a 200-mesh sieve. In this condition it is very easily oxidized, and Sulman has proposed the use of ammonia and ammonium salts for dissolving the oxide and cleansing the surface previous to use.

[1] Also, "Trans., I. M. and M." Nov. 16, 1898.
[2] Eng. Patents Nos. 18003 and 18146, 1894.

The following description of the use of this reagent is given by W. J. Sharwood.[1] The solution to be precipitated is run into a collecting tank and agitated for a few minutes. A suitable quantity of the dust is then introduced, which may be sprayed into the tank as an emulsion. The mixture is then forced by a pump through a filter-press, from which the barren liquid flows to the storage tanks, the precipitated metals and any excess of zinc remaining in the press. These form a porous mass which offers slight resistance to the passage of solution. The extra power required, as compared with zinc-thread precipitation, is slight, but some trouble is experienced, owing to the clogging of the filter-cloths when the solutions are turbid with suspended siliceous matter. From $\frac{1}{6}$ to $\frac{1}{3}$ lb. of zinc-dust is used per ton of solution treated. About 6 tons of liquid per hour can be passed through every 100 sq. ft. of filter-cloth in the press. When a clean-up is to be made, the contents of the press are washed by pumping a little water, then partially dried by forcing air through it. When the press is opened and the frames shaken, the cakes fall out readily. At this stage, the product usually carries about one-third of its weight of water; its composition does not differ materially from that of the precipitate obtained with zinc shavings, and it may be treated in the same way. It is, however, more uniform, owing to the absence of threads of "short zinc." The blowing operation oxidizes some of the zinc and sometimes causes the presses to become warm. The evolution of hydrogen which accompanies precipitation continues slowly while the moist precipitate remains in the press. Care must therefore be taken to bring no naked lights near while the clean-up is proceeding.

One filter-press, with 16 frames 2 ft. square, working 15 to 16 hours per day, requires 2 h.p. (neglecting the additional friction in pumps) in excess of the power required for pumping when zinc shavings are used. In other cases, however, difficulties have been experienced in treating the precipitate after removal from the presses. It was found to pack so hard that it was difficult to disintegrate it for the acid treatment, so that it was practically impossible to dissolve the zinc in dilute sulphuric acid. The particles are said to become coated with a thin layer of metallic

[1] "E. and M. J.," LXXIX, 752 (from 13th annual convention California Miners' Assoc., December, 1904).

gold which protects them from the acid, but a mixture of nitric and sulphuric acids has been found to be effective.[1]

The precipitation by zinc-dust or fume appears to be rapid and complete, five minutes' agitation being sometimes sufficient for the reaction; but the difficulties in collecting and refining the precipitate have led to the abandonment of the process in several cases where it had been originally adopted, e.g., at various plants in the Black Hills.   Much ingenuity has been exercised in devising apparatus for mixing the zinc effectively with the solution, and for settling the precipitate after mixture.   Picard used a conical vessel, in which the fume was fed in through a central funnel and met an ascending current of solution which was passed through a perforated conical distributor near the bottom of the cone. The settler contained a number of transverse plates of glass or smooth wood, set at an angle of 45°, on which any particles not collected in the cone are deposited.[2]

Charcoal Precipitation. — The fact that gold is precipitated by charcoal from cyanide solutions has long been known, though no satisfactory explanation of the reaction has yet been given. It has been alleged to be due to occluded hydrogen or hydrocarbons contained in the pores of the charcoal.   The method has been used in several small plants in Victoria, Australia, on a working scale.[3]   The charcoal, crushed to a suitable size, is placed in tubs about 2 ft. in diameter, having a central vertical cylinder also filled with charcoal.   These are used much in the same way as Caldecott's vats for zinc precipitation; the solution passes downward through the central column and upward in the outer portion, the overflow passing in this way through a succession of tubs each at a slightly lower level than the last.   As about 24 lb. of charcoal are required for the precipitation of an ounce of gold, the quantity needed in a plant of any size would be very large.   To recover the gold, the charcoal is burnt to ashes in a special furnace so arranged that the products of combustion pass through water, to avoid mechanical loss.   The ash is then smelted with a mixture of sand and borax.   According to experiments made by Prof. S. B. Christy, the efficiency of the charcoal rapidly falls off with continued passage of the solution.

Precipitation with Cuprous Salts. — This method seems to have

[1] " Min. Sci. Press.," XCIII, 607 (Nov. 17, 1906).
[2] " Trans. Fed. Inst. Min. Eng.," XV, 417.
[3] J. T. Lowles, " Trans. I. M. and M.," VII, 192.

been suggested independently, about 1895, by Prof. P. de Wilde, of Brussels, and Prof. S. B. Christy, of Berkeley, California. It depends on the fact that gold is precipitated completely as aurous cyanide on acidifying the cyanide solution and adding a sufficient excess of a cuprous salt. Several modifications have been proposed, in which an attempt is made to regenerate the cyanide decomposed in the reaction, or to precipitate the cyanogen in some form from which a soluble cyanide may subsequently be recovered before precipitating the gold. It is obvious, however, that a method involving the acidification, or at least the neutralization, of 100 tons or more of alkaline liquor per day is hardly likely to find a practical application. The precipitants suggested are: (1) a mixture of copper sulphate and sulphurous acid (De Wilde); (2) freshly precipitated cuprous sulphide, to be agitated with the solution (Christy); (3) cuprous chloride dissolved in common salt. After calcining the precipitate; the oxide of copper is to be removed by dissolving in sulphuric acid, leaving nearly pure gold. The method forms the basis of the analytical methods for determining gold in cyanide solutions proposed by Christy and A. Whitby. (See Part VIII.)

TREATMENT OF CUPRIFEROUS ORES

THE causes which render the treatment of cupriferous ores by the cyanide process particularly difficult have already been discussed in the earlier parts of this book; they may be summarized thus: (1) From certain minerals, notably the carbonates and oxides, copper is readily dissolved, forming a double cyanide with the alkaline cyanide. (2) The metal coats the zinc shavings in the boxes with a thin adherent metallic film, which prevents the deposition of gold and silver. (3) The presence of much copper in the precipitate gives rise to a low-grade bullion, as the metal is only imperfectly removed in the ordinary smelting and refining operations:

The remedies which have been applied or proposed are, for the most part, as follows: (1) The copper is removed from the ore by a preliminary operation, previous to cyanide treatment. (2) The double cyanide is decomposed in such a manner that the copper is separated and the free alkaline cyanide regenerated. (3) A modification of the precipitation process is used, whereby the gold, silver, and copper may be precipitated simultaneously or successively. (4) The precipitate or the bullion undergoes special refining processes for separation of the copper and other base metals.

PRELIMINARY TREATMENT FOR REMOVAL OF COPPER

Sulphuric Acid. — In cases where the copper exists as carbonate, satisfactory extractions have sometimes been obtained by giving a preliminary wash with dilute sulphuric acid. This system was at one time applied at the Butters Plant, Virginia City, and has been used at Cobar, New South Wales, by Messrs. Nicholas and Nicols.[1] The solution used at the latter plant averaged 0.65 per cent. H_2SO_4. The liquor drawn off was passed

[1] W. S. Brown, "Trans. I. M. and M.," XV, 445.

through boxes containing scrap-iron for precipitation and recovery of the copper. The charges in the vats were then water-washed, mixed with a sufficient quantity of lime to neutralize any remaining acidity, and transferred to other vats, where they were treated with cyanide, up to 0.3 per cent. KCy, for extraction of the gold. The first cyanide solution used ran off with little or no cyanide and generally with low gold values; this was precipitated in a special zinc-box and again used for the final wash before discharging. Some interesting features in connection with this plant were the use of charcoal after the zinc-box for the recovery of gold not precipitated by zinc from weak solutions, and the use of nitric acid in the clean-up. The acid vats had a capacity of 25 tons and the cyanide vats of 75 tons, so that three charges of the acid vats were treated together in each cyanide vat. The acid treatment consisted of 10 to 12 tons of dilute sulphuric acid (0.65 per cent.), followed by two water-washes. The acid was first partially drained off, then allowed to digest in contact with the ore for an hour or so, and finally passed slowly through the scrap-iron boxes. In the cyanide treatment, 15 tons of weak solution were used to displace moisture; this solution carried little gold, and was used again as a final wash before discharging; after passing slowly through a zinc-box it was allowed to flow through about 20 cu. ft. of packed charcoal. The precipitate from the zinc-boxes, in which much lead acetate was used, was extremely base, and was first treated with sulphuric acid for removal of the zinc, then washed with distilled water, and finally treated with nitric acid to dissolve copper and lead. In order to prevent gold going into solution, any chlorides present were precipitated with silver. Bullion was produced over 900 fine.

Sulphurous acid has been tried as a preliminary solvent of copper by A. von Gernet and others, and is claimed to have some action not only on carbonates, but also on sulphides of copper.

Ammonia. — The use of ammonia as a preliminary solvent of copper has been suggested by H. Hirsching.[1] The ammoniacal solution is distilled, and the ammonia recovered for re-use, while the copper is precipitated as oxide.

Double Cyanides of Copper.—It has been shown by Scrymgeour[2] that cupriferous cyanide solutions are capable of extracting a

[1] H. Hirsching, " The Ammonia Process."
[2] " E. and M. J.," Dec. 20, 1902, p. 816.

further quantity of copper from certain ores. The copper is finally recovered from this solution by electrolysis, with regeneration of a part of the cyanides. After this preliminary wash, the ore is cyanided as usual and the solutions precipitated electrically. This process appears to depend on the instability of the cupricyanides of the alkalis, which are readily transformed into cuprocyanides by taking up a further quantity of copper.

PRECIPITATION OF COPPER FROM CYANIDE SOLUTIONS

Several methods have been suggested for precipitating copper from cyanide solutions. We have already described the electrolytic method practised at Butters' Salvador mines (see Section V, above), and the effect of copper in the zinc-boxes, with and without the lead-zinc couple, has been discussed. It is well known that certain acids precipitate copper from cyanide solutions as white insoluble cuprous cyanide. It has been proposed by H. A. Barker [1] to use sulphuric acid for precipitating copper and gold from the solutions, which are then to be filtered and made alkaline with caustic soda; it is stated that cyanide is regenerated in the process, but the portion of cyanogen precipitated with the copper is obviously lost. The reactions are probably as follows:

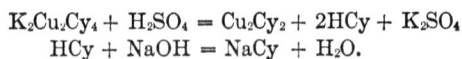

$$K_2Cu_2Cy_4 + H_2SO_4 = Cu_2Cy_2 + 2HCy + K_2SO_4$$
$$HCy + NaOH = NaCy + H_2O.$$

It is also suggested that this method of precipitation might be used as an adjunct to Scrymgeour's process, described above, the weak solution used as a preliminary wash in that process being precipitated with sulphuric acid when sufficiently charged with copper.

USE OF AMMONIA AS AUXILIARY SOLVENT

It has been found that a mixture of ammonia or ammonium salt with cyanide forms a much more effective solvent for gold in cupriferous ores than cyanide alone. When an ammonium salt is mixed with an equivalent of an alkaline cyanide in solution, decomposition takes place, with formation of ammonium cyanide, thus:

$$2KCN + (NH_4)_2SO_4 = K_2SO_4 + 2NH_4CN.$$

[1] "Trans. Inst. Min. and Met." XII, p. 399 (May 21, 1903).

A series of experiments made by Jarman and Brereton [1] appears to show that a mixture of ammonia and potassium cyanide dissolves less copper than does potassium cyanide alone. A process based on the use of this solvent has been applied by Bertram Hunt in treating tailings from the Comstock Lode, and cupriferous ore at Dale, California. The observations on which this method is based are summarized as follows by J. S. MacArthur, in a discussion on Jarman and Brereton's results: [2] (1) A weak solution of ammonia generally does more work as a solvent of copper than a corresponding quantity of a stronger solution. (2) A solution of ammonia along with a cyanide dissolves less copper than the sum of the amounts dissolved by the solvents separately. (3) For ordinary strengths of cyanide and ordinary grades of ore, the addition of ammonia equal to 0.11 per cent. of NH_3 produces a solution having a maximum solvent effect on gold and a minimum solvent effect on copper.

[1] "Trans. Inst. Min. and Met." XIV, p. 289 (Feb. 16, 1905).
[2] Loc. cit., p. 331.

SECTION VIII

TREATMENT OF ANTIMONIAL AND ARSENICAL ORES

Action of Antimony. — Antimony in ores generally occurs as stibnite, Sb_2S_3. In some cases it is a constituent of complex minerals such as pyrargyrite, Ag_3SbS_3.

The suggested methods for treating antimonial ores depend either on the use of a solvent which will remove the antimony by a preliminary operation or of one which will dissolve gold and silver without affecting the antimonial mineral. Such methods are:

(*a*) Preliminary treatment with a strong alkali, such as caustic soda, which is a solvent for antimony sulphide.

(*b*) Preliminary treatment with hydrochloric acid, which also readily dissolves most antimonial minerals.

(*c*) Preliminary treatment with a hot concentrated brine solution containing free hydrochloric acid and the chlorides of copper and iron (this latter solvent also extracts silver).

(*d*) Treatment with dilute cyanide solution containing a minimum of alkali, so as to leave the antimony as much as possible undissolved.

(*e*) Treatment with hydrocyanic acid, which extracts gold without attacking the antimony mineral.

When antimony is allowed to dissolve in the cyanide solution it causes trouble by precipitating in the zinc boxes, forming a coating on the zinc and giving rise to difficulties in smelting.

The method (*b*) is described by J. Jones and H. S. Bohm (Met. and Chem. Eng., April, 1911, p. 218). The ore is treated with HCl, which attacks the sulphides, generating H_2S.

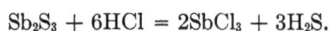

$$Sb_2S_3 + 6HCl = 2SbCl_3 + 3H_2S.$$

The solution, containing chloride of antimony, is electrolyzed with carbon anodes and antimony cathodes, and the chlorine thus liberated used to decompose the H_2S and regenerate HCl. The

374

charge in the vats is heated by exhaust steam or otherwise, and the two gases led to a regenerating chamber, where the reaction:

$$H_2S + Cl_2 = 2HCl + S$$

takes place.

It would be necessary to neutralize the residual acidity in the charge before treating with cyanide. The authors quoted suggest chlorination as more suitable for recovering the contained gold.

The method (c) was tried experimentally by W. Bettel on a complex argentiferous copper ore at the Willows Mine, Transvaal, containing sulphantimonides of copper and iron. The antimony was not dissolved, but carried over from the agitation tanks as a fine slime, which was allowed to settle; the silver was extracted from the solution by precipitating on copper. The method has probably never been used in conjunction with the cyanide process, but might be applicable as a preliminary treatment in certain cases.

The method (d), used at the Hillgrove Mines, New South Wales, is described by W. A. Longbottom (Min. and Eng. Rev. May 6, 1912, see abstract in E. M. J., Sept. 21, 1912, p. 554). The antimony is present as stibnite and cannot be entirely removed by concentration. Part of the gold appears to be coated with a thin layer of stibnite which prevents amalgamation. In cyaniding, a neutral or very slightly alkaline solution is used. Lime is added in calculated quantity during the filling of the treatment tank. The solution is occasionally purified by the addition of some powerful oxidizing agent, such as potassium permanganate. Lead acetate is sometimes used to accelerate precipitation and also to throw down any sulphide there may be in solution. Red lead is also added during the smelting, so that a base bullion is obtained which is subsequently reduced with metallic iron and cupelled.

The method (e), lately introduced in Australia as the Gitsham process, is only applicable in cases where calcium carbonate and similar minerals are absent.

Action of Arsenic. — The sulphides of arsenic behave much in the same way as stibnite; i.e., they are soluble in alkalies and in alkaline sulphides. It is not, however, possible to extract them by treatment with hydrochloric acid, as they are practically insoluble in that reagent.

As already noted, ores containing mispickel (FeAsS) have in some cases been successfully treated by the bromocyanide process.

Oxidized arsenical compounds resulting from the gradual decomposition of mispickel give trouble owing to the high consumption of cyanide. Andrew F. Crosse (Journ. Chem. Met. and Min. Soc. of South Africa, April, 1912) notes the following reaction as occurring in the case of scorodite (ferric arsenate):

$$FeAsO_4 + 2NaCN + 3H_2O = Fe(OH)_3 + Na_2HAsO_4 + 2HCN.$$

This mineral is soluble in dilute HCl and in aqueous sulphurous acid. It is also slowly dissolved by weak caustic soda with formation of sodium arsenate:

$$FeAsO_4 + 2NaOH + H_2O = Fe(OH)_3 + Na_2HAsO_4.$$

On addition of lime to the solution, an insoluble calcium arsenate is precipitated:

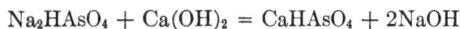

$$Na_2HAsO_4 + Ca(OH)_2 = CaHAsO_4 + 2NaOH$$

caustic soda being regenerated. The arsenic may therefore be eliminated from the solution by using a mixture of lime and caustic soda. In one sample containing scorodite it was found by Crosse that 13 lb. NaOH were required per ton of ore.

Arsenic causes trouble, not only by consumption of cyanide, as described above, but by precipitation in the zinc boxes. On treating the precipitate with sulphuric acid the very poisonous gas, hydric arsenide, is liberated, which has been the cause of several fatal accidents. This may be prevented by the use of nitric acid, as described, p. 129.

SECTION IX

NEW APPLICATIONS OF ALUMINUM IN ORE TREATMENT

Dissolving in Cyanide and Precipitating on Aluminum in the Same Vessel. — James E. Porter proposes to treat ores by agitation, after fine grinding in a ball-mill or otherwise, first with a caustic soda solution to neutralize cyanicides, then with cyanide until the values (gold, silver and copper) are sufficiently dissolved. Air is introduced through a porous bed at the bottom of the tank both during the alkali and cyanide treatment, in the form of minute bubbles, which keep the pulp in motion without violent agitation and serve to oxidize sulphides or other deleterious compounds. When this action is sufficiently complete, "plates of aluminum or an aluminum alloy are introduced into the suspended pulp and moved there through, whereby the values are precipitated and the combined cyanide regenerated." The deposit is brushed or wiped off and the plates returned to the pulp as long as values are deposited. The following reactions may be assumed:

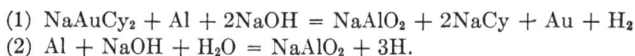

(1) $NaAuCy_2 + Al + 2NaOH = NaAlO_2 + 2NaCy + Au + H_2$
(2) $Al + NaOH + H_2O = NaAlO_2 + 3H.$

It will be noted that a regeneration of cyanide occurs; as similar reactions take place with the double cyanides of silver and copper, the use of aluminum as a precipitant in this way might effect quite an appreciable saving of cyanide.

The only novelty in this process is, of course, the combination of the operations in such a way that solution and precipitation are effected in the same vessel, and it is doubtful if this could be satisfactorily secured in practice. It is claimed that the abrading action is avoided by the fineness of the pulp and the uniform gentle agitation produced by the minute air-bubbles. It is said that the best results are obtained at a temperature of 180° to 190° F.

For further details see "Mining Science," abs. Mexican Min. Journ., March, 1912, p. 25, article by G. J. Rollandet.

Desulphurizing with Aluminum as a Preliminary to Cyanide

Treatment. — This process has lately been successfully introduced by the inventor, James J. Denny, aided by the staff of Chas. Butters and Co., at the Nipissing Low-grade Mill, Cobalt, Ontario. It depends on the reducing action on sulphides and various complex minerals of silver, arsenic, antimony, etc., of the nascent hydrogen obtained by the action of caustic alkalis on metallic aluminum. This reaction, which takes place at the ordinary temperature, may be expressed thus:

$$(1) \quad Al + NaOH + H_2O = NaAlO_2 + 3H$$

the aluminum dissolving to form an aluminate of the alkali metal.

Silver sulphide is decomposed by the nascent hydrogen as follows:

$$(2) \quad 6H + 3Ag_2S + 6NaOH = 3Na_2S + 6H_2O + 6Ag.$$

The native sulpharsenides and sulphantimonides of silver are similarly decomposed, thus *proustite*

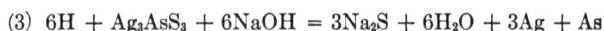

$$(3) \quad 6H + Ag_3AsS_3 + 6NaOH = 3Na_2S + 6H_2O + 3Ag + As$$

and *pyrargyrite*.

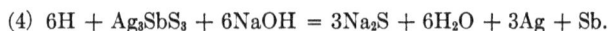

$$(4) \quad 6H + Ag_3SbS_3 + 6NaOH = 3Na_2S + 6H_2O + 3Ag + Sb.$$

The ore treated is very complex, containing, in addition to the minerals mentioned above, cobalt, nickel, copper and bismuth, in various combinations with sulphur, arsenic and antimony.

After the ore has been crushed fine by stamps and tube-mills, it is passed through a special tube-mill containing aluminum ingots, where the desulphurizing process takes place, followed by agitation in a tank containing aluminum plates. The pulp is merely filtered before passing to cyanide treatment, it being unnecessary to wash out the surplus alkali, which in fact is found to be beneficial. The small amounts of arsenical and antimonial compounds remaining in the solution have no detrimental effect. The cake carries 26 per cent. alkali solution, and is treated by agitation for 48 hr. in a .25 per cent. cyanide solution, with a dilution of 2.5 : 1. The silver-bearing solutions are precipitated with aluminum dust, and the precipitate melted in a reverberatory furnace.

The following details are given by Denny (Min. Sci. Press, Sept. 27, 1913, p. 488) as to the costs of the process: Treating 7268 tons per month, the costs per ton for collecting, desulphurizing and transferring of pulp were

	Per ton
Labor	$0.050
Supplies:	
Aluminum 0.81 lb. ⎫	
Caustic soda 1.46 " ⎬	0.347
Lime 5 " ⎭	
Power	0.027
Workshop	0.008
Total	$0.432

Alkali Solution, Filtering and Transferring.

	Per ton
Labor	$0.069
Supplies	0.006
Power	0.028
Workshop	0.002
Total	$0.105

The process is stated to save from one to four ounces per ton at a total cost of 54 cents per ton.

It is suggested that the process may be useful in the treatment of gold telluride ores, some of which, for example sylvanite, have been found to be readily reduced to their elements by this method.

SECTION X

Electrolytic Regeneration of Cyanide
(Clancy Process)

A BRIEF account of this interesting and ingenious process must be given, though from a commercial point of view it can hardly be said to have justified the expectations raised by its introduction, a few years ago, with a great flourish of trumpets, in the Cripple Creek district. A great number of claims are put forward in the patent specification of the inventor, J. Collins Clancy, most of which need not concern us here. The essential feature of the process is the *oxidation of complex cyanogen compounds, especially of thiocyanates, by means of oxygen produced by electrolysis* of the solution. The most important reaction may be expressed thus:

$$KCNS + 2KOH + 3O = KCN + K_2SO_4 + H_2O.$$

Other suggestions, such as the addition of alkaline halides, notably potassium iodide, and the use of calcium cyanamide as a source of cyanogen, are of great scientific interest, but hardly commend themselves to practical workers under present conditions.

The addition of *potassium iodide* was advocated for the purpose of generating cyanogen iodide, which, in conjunction with an alkali cyanide, forms, as is well known, a very active solvent both for gold and for the native tellurides of gold. The supposed reactions are:

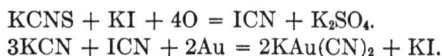

$$KCNS + KI + 4O = ICN + K_2SO_4.$$
$$3KCN + ICN + 2Au = 2KAu(CN)_2 + KI.$$

The action of alkalis transforms a portion of the iodide of cyanogen into cyanate, thus:

$$ICN + 2KOH = KCNO + KI + H_2O,$$

but this only takes place slowly, and a part is converted into iodate, hypoiodite and cyanide by some such series of reactions as the following:

$$5CNI + 6KOH = 5KCN + KIO_3 + 2I_2 + 3H_2O.$$
$$I_2 + 2KOH = KIO + KI + H_2O.$$

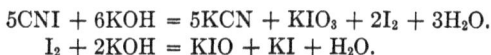

The hypoiodite is a good solvent of tellurium, thus:

$$Te + 2KIO + 2KOH = K_2TeO_3 + 2KI + H_2O.$$

Calcium cyanamide was suggested, both as a cheap source of cyanogen compounds, which are obtained by electrolyzing its solution, and as a means of reducing cyanates and recovering their cyanogen in a form available for gold extraction. The reactions are somewhat obscure, but according to Clancy, a mixture of calcium cyanamide and caustic potash yields on electrolysis as an intermediate product, potassium amidodicyanate $(CN)_2NH_2OK$, which upon further electrolysis forms a powerful gold solvent, though it does not yield any appreciable amount of cyanide which can be detected by silver nitrate. The action of calcium cyanamide on (for example) sodium cyanate may possibly be:

$$CaCN_2 + NaCNO + 2H_2O = Ca(OH)_2 + (CN)_2NH_2ONa.$$

It is also stated that a mixture of calcium cyanamide, lime and an alkali ferrocyanide forms an effective gold solvent, and it is proposed to utilize this reaction in the re-treatment of old residues. The reaction goes on without the aid of electrolysis, but if the mixture is electrolyzed, the solution of gold takes place more rapidly.

In all attempts to put this process to practical use the following difficulties were encountered:

(1) In order to get effective electrolysis of the solution so as to decompose complex cyanides and regenerate active gold solvents, it was necessary to use either a high current density or a very large electrode surface. It was practically impossible to find any substance for use as anode which would resist both the intense chemical action of nascent oxygen and the mechanical action of the suspended ore particles.

(2) The conversion of thiocyanates into cyanides by electrolysis is never complete. As long as a considerable concentration of thiocyanate exists in the solution, the reaction:

$$KCNS + 2KOH + 3O = KCN + K_2SO_4 + H_2O$$

proceeds, and the amount of cyanide formed is proportional to the amount of thiocyanate decomposed. When, however, the amount of cyanide is large in comparison with the amount of thiocyanate present, the nascent oxygen begins to attack the cyanide, probably converting it into cyanate, thus:

$$KCN + O = KCNO.$$

In all electrolytic tests with this process it is found that the production of cyanide rises to a maximum, and then ceases. If electrolysis is still continued, the cyanide gradually diminishes and finally disappears. As shown by the first equation, there is a consumption of alkali in the process, and the alkalinity must be maintained by fresh additions of lime or other caustic alkali whenever the tests show this to be necessary.

Of the many substances tried as anodes, including iron, peroxidized lead, aluminum and various forms of carbon, the harder and denser forms of graphite gave the best results, but even these disintegrated with continued use under practical conditions. Good results are *said* to have been obtained by the use of anodes of fused magnetite, formed by melting oxide of iron in the electric furnace. It is claimed that these anodes do not disintegrate under high current densities, and are not worn down by attrition of the ore particles; also they have the advantage that no carbonic acid is formed along with the nascent oxygen, as would be the case with a carbon anode.

PART VII

ASSAYING

SECTION I

SAMPLING

(A) Sampling of Ores and Similar Material in Bulk

Samples of Hard, Coarse Material. — The sampling of ore " in place" does not generally come within the limits of the work carried out by the staff of a cyanide plant. It may be necessary, however, in certain cases, to deal with large masses of coarsely broken rock, such as ore dumps, ore as delivered from the mine to the mill, slag heaps, and other similar accumulations of hard material, so that a few words on the methods of sampling them may not be out of place.

Definition of Sample. — The essential point in all sampling is to select from a relatively large mass of material, containing various ingredients unevenly distributed, a relatively small quantity which shall contain each of these ingredients in the same proportions as those which occur in the entire mass.

Sampling Large Ore Heaps. — A large pile of ore is generally sampled in the first instance by making cuts into it at various points round the base; when circumstances allow, a channel is dug right across the heap through the center, or, better, two or more intersecting channels. A plan sometimes adopted is to shovel alternately to right and left, and to throw every third shovelful into a barrow, the portion going into the barrow constituting the sample. The assumptions involved in this process are: (1) That a cut made through the heap from side to side, passing through the center, will consist of material of the same average composition as the entire heap; and (2) that every third shovelful is, on the average, of the same composition as the remaining two. The first of these assumptions appears less justifiable than the second; hence it is safer to make more than one channel.

Coning and Quartering. — The further division of the sample must generally be preceded by breaking up the larger lumps, which may be done either mechanically or by hand. For this

385

FIG. 41. — Vezin Sampler, as furnished by Hadfield's Steel Foundry Co., Sheffield.

purpose it may be spread evenly over a smooth surface, such as a cement floor, and any pieces broken up that are obviously larger than the rest. The entire sample is then mixed by turning over with a shovel; to ensure thorough mixing the process of "coning" is often resorted to. A space is swept clean, and the ore thrown into this, a shovelful at a time, each shovelful being thrown on the top of the conical heap so formed, so that the material rolls down as evenly as possible on all sides. If the heap is still very large, it may be again sampled by trenching; if not too large to be conveniently handled, it is spread out in a circular flat-topped heap and "quartered" by means of boards held edgewise across the center so as to divide it into four segments as nearly equal as possible. Another method is to form the material into a ring on the floor, and make a heap by shoveling toward the center.

The operations of coning and quartering are repeated successively a number of times; each time two opposite quarters are rejected and removed from the floor, and the remainder broken smaller before proceeding to quarter again. Care must be taken to sweep away all dust belonging to the rejected portions before mixing the other quarters.

Sizes to which Samples must be Crushed. — The degree of crushing necessary at each stage will depend on the nature of the material, and must be varied in special cases. The following figures are given by A. Harvey [1] as representing the sizes found to be safely allowable in actual work with a Vezin automatic sampling machine, cutting out $\frac{1}{8}$ of the entire mass put through at each operation. (This machine is so constructed that it deflects the entire stream falling through a chute at regular intervals of time, — say every two or three seconds.) (See Fig. 41.)

TABLE I. — ALLOWABLE SIZES OF ORE PIECES IN SAMPLING

Diameter of Largest Pieces in Sample: Inches	Minimum Weight of Sample: Pounds	Diameter of Largest Pieces in Sample: Inches	Minimum Weight of Sample: Pounds
$5\frac{1}{2}$	79,300	$\frac{2}{3}$	256
4	69,109	$\frac{1}{3}$	32
$3\frac{1}{2}$	44,958	$\frac{1}{8}$	4
$2\frac{1}{2}$	16,384	$\frac{1}{12}$	$\frac{1}{2}$
$1\frac{1}{4}$	2,048	$\frac{1}{25}$	$\frac{1}{16}$

[1] "Min. and Sci. Press," LXXXVIII, p. 78 (Jan. 30, 1904).

W. Glenn [1] gives the following rules, applicable to such material as 10 per cent. copper ore, originally in a 10-ton pile consisting of large lumps; 1 ton, obtained as a sample by "trenching" as above described, is quartered to 500 lb., breaking the larger lumps with a hammer. This is treated as follows:

Diameter of largest pieces: Inches	Weight o. sample: Pounds
1	.500
½	.250
"fine gravel"	.125
"coarse sand"	60

Proceed to mix and quarter on a sheet of paper.

"At this point the sample would weigh about 15 lb.; its larger grains would be in size like coarse sand. It would be safe now, without further breaking, to mix and quarter it twice, or until its weight did not exceed 4 lb. Run this through the mortar and then mix and quarter it twice, or down to 1 lb. weight. Grind this to something approaching powder, and for the last time mix and quarter it."

Relation of Size of Sample to Grade of Ore. — D. W. Brunton [2] discusses the whole subject of mixing and quartering from a mathematical standpoint, and shows that the size to which ore must be crushed, so that the error in sampling may be within allowable limits, depends: (1) On the weight or bulk of sample (the smaller the sample the finer it must be crushed); (2) on the relative proportion between the value of the richest minerals and the average value of the ore (with high-grade ores we may crush more coarsely than with low-grade, for the same percentage of error); (3) on the specific gravity of the richest mineral present. The higher the specific gravity of the richest mineral, the greater the value of a particle of given size and grade, and hence the greater the influence of such particle on the sample. Finer crushing is therefore required when the rich material is also of high specific gravity.

Formula for Dividing Samples. — S. A. Reed [3] gives the following formula for determining the maximum diameter allowable for the particles in any stage of the sampling of a given ore:

---

[1] *Trans.* A. I. M. E., XX, p. 155 (June, 1891).
[2] *Trans.* A. I. M. E., XXV, p. 826 (October, 1895).
[3] "Sch. Mines Quart.," VI, p. 351 (1885).

$D =$ diameter of largest pieces in inches;

$p =$ quantity of the lot in Troy ounces;

$f =$ number of parts into which p is to be divided (one part to be chosen as sample);

$k =$ percentage of metal sought (gold or silver) in the richest specimens of the lot:

$s =$ sp. gr. of richest minerals;

$m =$ average grade of ore (ounces per ton);

$a =$ number of pieces of D size and k value that can be in excess or deficit in the portion chosen as sample;

$l =$ largest allowable percentage of error.

Before cutting to $\frac{1}{f}$, we must crush or grind the lot so that

$$D = .053 \sqrt{\frac{m \, p \, l}{s \, k \, (f - 1) \, a}}$$

The following table is given for ores of different grades, assuming in all cases that $s = 7$, $l = 1$, and that with samples:

Class A, $m = 50$ $k = 1$ ⎫   Medium,

"   B, $m = 75$ $k = 10$ ⎬   high-grade,

"   C, $m = 500$ $k = 30$ ⎭   very rich.

TABLE II

Sample Reduced from	VALUE OF D, IN INCHES		
	Class A	Class B	Class C
100 to 10 tons.........	5.28	2.96	2.58
10 to 1 ton	2.46	1.38	1.2
2000 to 200 lbs..........	1.14	0.6	0.56
200 to 5 "	0.3	0.18	0.16
5 lbs. to 10 assay tons....	0.034	0.02	0.018

These figures correspond roughly with those given by Harvey. (See above.)

Sampling Ore in Mine. — As the assayer may occasionally be called upon to take samples of ore from the mine itself, the following description of the method employed on the Rand is quoted from a paper by G. A. Denny — "Observations on Sampling, Computation of Assay Averages," etc.[1]

"The sampling of the mine is entrusted to men of experience who take as their equipment a water squirt, with which to wash

[1] *Trans.* A. I. M. E., XIX, p. 294 (June 15, 1900).

down the rock at the point to be sampled, hammer, chisels (both diamond and chisel-pointed), a receptacle in which to catch the samplings, and sacks about 15 in. deep and 9 in. wide wherein to place the completed sample. . . . The sampler first proceeds to measure equal distances along the level, rise, or other working which he is to sample. The distances taken are either 5-ft. or 10-ft. intervals; in the case of thin rich leaders, always 5 ft. Having chalked or otherwise marked the intervals, he proceeds to the work of sampling. In cases where the reef is thin, say from 1 in. to 3 in. in width, the sample is taken over a width of 6 in., that width being carefully channeled out to a certain breadth and depth, so that as nearly as possible a true average may be obtained. Samples should be taken of approximately a minimum weight of 5 to 6 lb.; in general the bigger the sample the nearer it is to the true average."

When the reef is "solid" (i.e., contains no waste), and less than 18 in. wide, it is sampled as one section; if of more than 18 in. width, it is sampled in so many sections of 18 in., each section being separately sacked and recorded. This is done to avoid getting undue quantities of ore from one section, as may happen when the hanging and foot-wall sides are of different character. "In cases where the reef is divided into bands of ore and quartzite, only the ore is sampled, the bands of waste being measured and calculated on the total sampled width in the following manner":

TABLE III

Band	Width in Inches	Nature	Assay Value: Dwts.	Assay – Inches
A	6	Reef	15	90
B	12	Waste	—	—
C	10	Reef	20	200
D	30	Waste	—	—
E	2	Reef	60	120
	60		6.83	410

Where samples have been taken at *regular* intervals, the average assay is found by dividing the total assay-inches by the sum of the widths in inches of all the samples taken.

Where samples have been taken at *irregular* intervals, the length of reef represented by each sample also enters into the calculation.

It may be added that all mine samples should be taken as nearly as possible in a plane at right angles both to the dip and strike of the vein or deposit to be sampled; otherwise an erroneous impression will be given of the quantity of ore represented.

Machinery for Crushing Samples. — The work of breaking up the larger lumps, before mixing and dividing ore samples, is carried out either by hand, with a large hammer and anvil, or by machines, of which the two principal types are represented by the Blake and Gates crushers respectively. The finer crushing is performed by smaller machines of the same types, or by the pestle and mortar, or by machines imitating the action of the latter. Crushers of the Ball mill type are unsuitable for sampling work on account of the labor and difficulty of cleaning them effectively. The finest crushing is best done on a flat steel or cast-iron plate with raised edges on three sides (known as a "buckboard"), the ore being ground on this by means of a heavy iron block ("muller"), worked backwards and forwards by means of a wooden handle.

Precautions in Preparing Samples. — Attention may here be drawn to certain points of the utmost importance, neglect of which is probably the chief cause of inaccuracy in the preparation of ore samples for assay:

(*a*) When any portion has been selected as a sample and is to be passed through a sieve of particular mesh, the entire portion so selected must be made to pass the sieve. If the harder portions, which naturally resist crushing the longest, be rejected, the sample is worthless.

(*b*) After the operations of crushing, sifting, etc., have been completed on one sample, every appliance used in the process should be carefully cleaned before proceeding with another sample.

(*c*) After crushing and sifting, and before dividing, every sample must be thoroughly mixed. The operation of sifting in itself causes a certain amount of separation in the different ingredients, so that the sifted sample is never homogeneous.

When, as is frequently the case in mine or cyanide works' assay offices, the preparation of the sample is entrusted to natives not under close supervision, one or all of these precautions will probably be shirked.

Cleaning Sampling Implements. — Great care is necessary in order to prevent the mixing of one sample with material from others, or with foreign impurities. All vessels and instruments used should be so constructed that they may be easily cleaned. The absence of this condition is a fatal drawback to various otherwise effective devices which have been designed for mechanically crushing, mixing, and dividing samples.

Sieves for Samples. — The sieves can generally be cleaned effectively with a hard brush; but sometimes it may be necessary to pass through the crushers, sieves, etc., a quantity of barren rock, sand, charcoal, or other material for the purpose of removing adhering particles of a previous sample. The sieves used for this purpose are of brass-wire screening, set in a metal frame; as an additional precaution they may be soldered all round on both sides at the joint of the screen with the frame. All sheets of iron, cloth, and rubber, and all sampling floors and tables, should be carefully swept and dusted after each sample is finished; they should also be washed occasionally.

Necessity for Separate Sampling Room. — Whenever possible, the room in which the samples are crushed and sifted should be entirely distinct from the assay room. These preliminary operations inevitably cause a certain amount of dust which should be carefully kept away from the supplies of flux, and from the crucibles and other utensils used in the assay itself.

Importance of Sampling. — The importance of correct sampling can hardly be overestimated; it is obvious that the subsequent assay is valueless, however carefully conducted, unless the sample assayed really represents the average composition of the material to be examined.

(B) SAMPLING OF SAND AND SIMILAR MATERIAL

In cyanide plants we have to deal, as a rule, with material which has already been more or less finely crushed, and in most cases with ore from which the coarser metallic particles have already been removed by amalgamation.

Difficulties of Sand Sampling. — To obtain a correct average sample, representing, say, the material delivered from the mill to the cyanide tanks during a run of 24 hours, or the charge run into some particular tank, is by no means so simple a matter as might appear at first sight. The crushed ore is usually far from uniform in composition, and the methods adopted for filling the tanks tend to produce layers or rings of material differing in composition and value from the remainder.

Points where Samples may be Taken. — Samples for regulating the work in a sands plant may be taken: (1) Before entering the tanks; (2) in place, after the tanks are filled; (3) while the tanks are being discharged. Where the sand is collected in one tank and treated in another, the sampling is often done during the transfer from the collecting to the treatment tank.

Samples Taken before Charging Tanks. — It is customary in many reduction works to take two samples daily in the battery, one representing the material passing through the screens of the mortar-boxes immediately before it comes in contact with the amalgamated plates, and the other after it leaves the plates and enters the launder leading to the hydraulic separators. These samples are taken by placing a metal scoop or other suitable vessel across the whole width of the discharge, and intercepting a portion of it at regular intervals (say every hour). The portions so collected are transferred to a bucket, and at the end of the day's run the combined sample is drained, mixed, quartered down, and assayed.

Automatic Samplers. — An automatic sampler is sometimes used on the launder leading to the separators. These samplers are of various designs, the general principle being, however, that the entire stream of pulp is diverted into a special vessel for a few seconds at regular intervals of time. D. Simpson[1] states that at the New Goch G. M. Company, the method of catching the outflow from the delivery hose at regular intervals, and also that of the automatic sampler, were discarded at an early date as unreliable. In general, it may be remarked that when intercepting a stream by means of a scoop it is extremely difficult to avoid loss by splashing, so that the sample taken in this way is liable to contain an undue proportion of the coarser and heavier particles.

Truck Samples. — In cases where comparatively dry sands, such as accumulated tailings, are to be treated, and these are delivered into the tanks in trucks (cars), it is usually found that a satisfactory sample of the charge can be obtained by taking small equal portions from each truckload, provided the number of loads required to fill the tank is fairly large.

[1] "Sand Sampling in Cyanide Works," I. M. M., Bull. No. 25, Oct. 11, 1906, p. 6.

Sand Samples taken in Place by Sampling-Rod. — After the tanks have been filled it is customary to take a sample of the contents by means of a sampling-rod. This consists of a metal tube, long enough to penetrate the whole vertical depth of the material to be sampled — say from 4 to 10 ft. length and 1½ to 2 in. diameter. This tube is slit from the bottom to within a few inches of the upper end and is provided with a crosspiece at right angles, serving as a handle. The edges of the groove and the lower end of the instrument are sharpened to allow it to cut into the mass of sand.

When the rod is pushed into the charge with a slight twisting motion, the hollow fills with sand. The rod is then withdrawn and the surface scraped smooth. The core of sand is then cut out and set aside to form part of the sample, using a suitably shaped piece of hard wood or metal for cleaning the groove. A number of cores are taken in this way from various parts of the tank, and the whole mixed and quartered. The same instrument is commonly used for sampling tailings in dumps or pits, or for any similar accumulations of moist sandy material.

Objections to Rod-Sampling. — It is a frequent complaint that the sampling-rod does not really take out equal quantities of sand at different depths in the tank. In passing through the upper layers, the lower part of the tube becomes choked, especially if the material be somewhat slimy, so that the lower layers are not represented at all, although the rod has been pushed to the bottom of the tank. This difficulty may be partially obviated by taking samples to half the depth of the tank, cleaning the rod, and then pushing it down to the bottom of the tank through the hole already made, the second portion taken thus representing only the bottom layers. As it frequently happens that the bottom layer (6 in. or 12 in.) is of very different composition from the remainder, it cannot be considered that this method may be safely relied on to give accurate results, unless special precautions are taken.

Location of Samples in Tank. — The number of rod-samples taken per charge, and the points at which they are taken, are also matters of considerable importance. If the charge were absolutely uniform, a single sample taken at any point would correctly represent its average composition. This, however, is seldom, if ever, the case. When the charge is not uniform, a better result will be obtained by taking samples in such places that each one represents an equal mass of sand, rather than by haphazard sampling in any part of the charge. The tanks used at the present day are practically always cylindrical in form, with a flat bottom at right angles to the sides. Hence, by describing on the upper surface a number of circles of varying diameters, the charge may be divided into any required number of sections, each representing an equal mass of sand. Thus, a circle 40 feet in diameter may be divided into 4 equal areas by describing from its center circles having radii of 10 ft., 14.14 ft., and 17.32 ft. The general formula is as follows:

Let n = number of equal areas required;

$r_1, r_2, r_3,$ = radii of circles described from center of upper surface;

r_n = radius of the tank itself;

Then a = area of each section.

Then, $a = \pi r_1^2 = \pi(r_2^2 - r_1^2) = \pi(r_3^2 - r_2^2) = \pi(r^2_{n-1} - r^2_n)$.

Hence, $r_1 = \sqrt{\dfrac{a}{\pi}}, \quad r_2 = \sqrt{2}r_1, \quad r_3 = \sqrt{3}r_1, \quad r_n = \sqrt{n} \cdot r_1, \quad a = \pi\dfrac{r^2_n}{n}$.

These areas may again be subdivided by diameters of the 40-ft. circle inclined at equal angles. Thus, by taking 2 diameters at right angles we obtain 16 equal areas, and by taking a rod-sample as nearly as possible in the center of each we should obtain a tolerably good average of the whole charge.

Causes of Irregularity in Distribution of Sand. — From the above it will be evident that the method sometimes adopted of taking samples at equal distances along a diameter gives an undue proportion of material from the central parts of the tank, where the coarser and heavier particles generally tend to collect. The lumps of slime generally accumulate at the bottom and sides of the tank. When material is transferred from an upper to a lower tank, the lumps of slime and any lighter or finer material tend to roll down the sides of the cones formed below each dis-charge door, and form layers and walls on the filter-mat at the junctions of the cones with one another and the sides of the tank.

Section Sampling. — Where the conditions permit, according to Simpson,[1] an accurate sample may be obtained by discharging sand from a tank, leaving a vertical face in the plane of a diameter of the tank. This is divided into a number of horizontal zones of equal width, and samples taken by scraping the face of each

[1] *Ibid.,* p. 3.

zone into a separate bucket. These samples are then assayed separately, or a combined sample is made by mixing equal quantities of each.

Samples taken while Discharging Tanks. — When the tanks are discharged by means of trucks, a good sample of the residues may generally be obtained by taking with a scoop or, better, with a sampling-rod an approximately equal portion from each truckload, the sample thus obtained being afterward mixed and divided. In cases where the material is entirely composed of comparatively small grains and is fairly uniform, only a small quantity need be taken from each truck. Where the material contains large lumps, as may be the case with slimy tailings, it is necessary to take larger amounts. All such lumps included in the sample must be thoroughly broken up before quartering.

When the tanks are discharged by a belt-conveyor, or similar mechanical device, samples may be taken, at equal intervals of time, as the material passes a certain point; or an equal portion may be taken once or twice during each revolution of the belt.

Precautions as to Residue Samples. — When residue samples have been taken after cyanide treatment, it is essential to place them in a vessel which will not absorb liquid, and which is not affected by cyanide in any way. An ordinary metal bucket is not suitable for this purpose, but an enameled iron bucket in good condition may generally be used. The samples should be kept covered as much as possible, especially if a moisture determination is to be made.

Mixing, Dividing, and Quartering Samples. — For assay purposes, all samples taken as above described require to be reduced in bulk. The general principles of this process are the same as those underlying the reduction of ore samples, already discussed, the essential point being that the average composition of the sample must not be altered. In all cases a thorough mixing is necessary. The sample is spread out on a smooth surface of suitable material, and sufficiently large to allow of the sample being turned over freely with a shovel. When the material is not very wet, a cement floor answers the purpose. All lumps must be carefully broken up, and in many cases it is very desirable to put the whole sample through a coarse sieve, after which it is reduced by coning and quartering. (See above.) It has already been remarked that any lumps of slimes and such-like

material tend to roll to the bottom of the cone, so that it is as
well not to rely exclusively on this process for mixing the sample.
When the mixing is considered to be sufficient, the conical or
hemispherical heap is divided crosswise into four quarters. Two
opposite quarters are dug away and rejected. The remaining
quarters are mixed together, any lumps further broken up, and
again quartered, this operation being repeated until the required
amount is obtained for the determinations needed. After the
weight of the sample has been sufficiently reduced, the mixing
may conveniently be done by rolling on a sheet of oil-cloth or
rubber, taking care to alter the direction of rolling as often as
possible, so as to avoid the formation of bands of concentrates.
In the case of ordinary sand samples a very satisfactory mixture
may be obtained by rolling, provided the surface be sufficient and
the operation be repeated several times.

Samples containing a moderate amount of moisture and
fairly uniform in composition may be divided by spreading out
in a uniform layer a few inches thick, and selecting with a horn
scoop, spatula, or other instrument, approximately equal portions
at regular intervals all over the layer. It is a good plan to rule
two sets of parallel lines at right angles, so as to divide the layer
into a considerable number of small equal squares, and to take
equal small portions from each square. Before doing this, however, the sample should have been well mixed by coning, rolling,
or otherwise.

Moisture Samples. — It is frequently necessary to make a
determination of the amount of moisture in charge or residue
samples in order to calculate the dry weight of material under
treatment. Thus, the dry contents of a tank are known if we
know the average weight of the truckloads of wet sand required
to fill it, the number of such loads, and the percentage of moisture.
All such "moisture samples" are taken with as little delay as
possible. Probably the most satisfactory method is to spread
out a large quantity of the roughly mixed sample, take out portions at regular intervals in the way described above, and weigh
on a good balance 500 or 1000 grams of the sample. This is
placed in a porcelain or enameled-iron dish and dried at the
same temperature that is used for drying the ordinary assay
samples. When cool, the dried sample is again weighed, the
difference indicating the amount of moisture. In the case of

very wet material (slimes and similar samples) it is best to weigh dish and contents together, and determine the weight of the empty dish after cleaning and drying.

Drying Assay Samples. — After the sample has been reduced by successive quarterings or otherwise to about 5 lbs. (or, say, 2 kg.), it may conveniently be dried by heating in a suitable vessel over a moderately hot fire. In this operation the nature of the material to be treated must be carefully considered. Where pyrites is present, the heat must not be sufficient to cause any roasting. Drying at a temperature of 100° or 110° C. would in general be far too slow for ordinary assay work, though some such standard is usually adopted for samples required for analysis. An effective arrangement for some time used by the writer consisted of a wide chamber about a foot high, with wooden and canvas sides, the whole being heated above and below by means of a coiled steam-pipe, maintaining a temperature of about 90° C. About a dozen ordinary samples, containing, say, 5 per cent. moisture, could be dried if left over night in this apparatus. Samples dried in this and similar ways probably always retain a small percentage of hygroscopic moisture; this, however, does not seriously affect the assay, nor does it affect the accuracy of any calculations of the values per ton of material, provided the moisture sample has been dried at the same temperature.

Drying samples over an open fire in iron pans, as is frequently done, is not to be recommended. Unless constantly watched, there is danger of the sample being overheated; also, in removing the sample from the pan, pieces of metal or rust from the latter are apt to scale off. Samples containing dissolved gold and silver, such as the residues from cyanide-treated tailings, should on no account be dried in unprotected metal vessels. Porcelain or enameled-iron vessels are the most suitable, but when these are used the temperature must be carefully regulated.

Reduction of Dried Sample. — When quite dry, the entire sample must be passed through a sieve of suitable mesh, depending on the nature of the material. For this first sifting a screen of 4 holes to the linear inch may be used, to break up any lumps of slime, etc. The sample is then turned out on an oilcloth or rubber sheet, and thoroughly mixed by rolling. All lumps not passing the sieve are ground up in a mortar or on the buckboard. The softer lumps may be broken by simply rubbing on the sieve with a smooth block of wood or cork. The use of metal weights and similar objects for this purpose is objectionable, as they tend to wear the wires and enlarge the holes of the sieve.

It is essential that the whole sample be passed through, as the portions remaining on the sieve are almost invariably of different average composition from the remainder. After sifting and mixing, the sample is quartered, either by hand or by some mechanical device. Many different types of apparatus have been brought out for this purpose, mostly based on the principle of cutting out a given fraction (say $\frac{1}{4}$ or $\frac{1}{8}$) from a falling stream of sand. The fraction retained as the sample is then passed through a finer sieve, all particles which do not at once pass the sieve being crushed to the requisite degree of fineness. The sample is then again mixed and further divided. This process is repeated, using successively finer sieves, until the assay sample is obtained. This should be crushed at least to 60-mesh; in some cases it is desirable to use a much finer mesh, say 120 or 200. The finer crushing is generally done on a buckboard, as described above.

Size of Grains at Different Stages. — The maximum size of lumps allowable at any stage of the process depends on the weight of the sample at that stage. This subject is discussed above in reference to rock samples, and some examples are given from different writers, but no absolute rule can be given. The point should be determined, in the case of any particular class of material, by making separate assays of each pair of opposite quarters in one or more lots of the material. The following, frequently used by the writer, may be taken as an example in dealing with ordinary sand charges and residues:

Entire sample passed (wet) through $\frac{1}{2}$-inch screen. From this 2000 grams are roughly weighed out after mixing, dried, crushed to $\frac{1}{4}$ in., and halved.

1000 grams crushed to 10-mesh and halved
 500   "   "   30-   "   "   "
 250   "   "   60-   "   "   "
 125   "   "   80-   "   for sample.

The last step is taken only in the case of fairly rich material. The final sample allows of 4 assays of 1 assay ton each; or if the division is stopped at 250 grams of 4 assays of 2 assay tons each.

Weight of the Final Sample. — Sufficient material should be retained to allow of the assay being repeated three or four times if necessary. Generally speaking, the poorer the sample, the larger must be the quantity taken for each assay. The assay charge commonly taken varies from 30 to 60 grams, but in special cases it may be as low as 15 grams or as high as 250 grams; the assay sample may therefore vary from 60 to 1000 grams.

Sampling Slimes. — L. Ehrmann [1] gives the following method for sampling "battery slimes." The slimes are put through a small filter-press after the pulp has been thoroughly mixed; then two, three, or more cakes are made, according to the volume of the original sample, the cakes weighing about $1\frac{1}{2}$ kg. wet or 1 kg. dry. Now from each fresh cake two opposite triangular parts are cut and the portions so selected are made into a pulp with water and again put through the press, until only one cake is obtained. This cake will be an average of the lot, and will weigh, say, 35 assay tons. From this, triangular opposite parts are cut so as to get 16 to 17 assay tons. This may then be dried for the assay sample.

Samples Containing Dissolved Values. — When samples contain soluble gold and silver it is best to make a thorough extraction of the soluble matter, and assay the extract and the washed residue separately, then calculating the results on the weight of the original sample. It has been supposed that losses of precious metal take place in drying finely divided samples containing cyanide solution, but the evidence on this point is not very conclusive. [2]

Estimation of Undissolved Gold in Slime Pulp Undergoing Cyanide Treatment. — It is often necessary to know the exact amount of gold (or silver) remaining undissolved at a given stage of treatment. A sample taken out, and washed with water to remove cyanide and dissolved values, does not yield accurate results, because the action continues in the pulp sample under more favorable conditions than in the bulk of the charge, and the amount of undissolved gold shown is lower than the truth.

H. A. White (Journ. Chem. Met. and Min. Soc. of South Africa, Sept. 1911, p. 89) has shown that the action of cyanide may be instantly arrested, without precipitation of any dissolved gold,

[1] "Proc. Chem., Met. and Min. Soc. of South Africa," II, p. 697.
[2] *Ibid.*, II, 372.

by mixing the sample with a sufficient amount of permanganate solution. Bleaching powder, formaldehyde and alkaline picrates have the same effect, but in the case of the two latter the action is not instantaneous.

The method recommended is as follows: To a bucket half filled with water, enough saturated permanganate solution is added to leave a decided pink color after addition of the slime sample. At a definite time during the treatment, a sample of the pulp to be tested is taken by means of a tube with a flap valve at the bottom, or by other suitable means, and allowed to run into the prepared bucket under the surface of the solution, and with as little exposure to the air as possible. Without delay, the bucket is filled up with water and allowed to settle. The solution is then poured off and discarded. The operation of filling with water, well mixing, and allowing to settle, is repeated four times, in each case with the addition of limewater to hasten settlement. Finally the sample is dried for assay. In performing the assay it is very necessary in some cases to secure a large button of lead. "Washing" with litharge and reducer should not be neglected, and the fusion should be completed at a high temperature.

SECTION II

ASSAY OF A TYPICAL SILICEOUS ORE

(A) Crucible Fire Assay

Object of Assaying. — The process of assaying aims at extracting the whole of the gold and silver in a pure and weighable form from a weighed portion of the ore or other material to be assayed, which portion is assumed to be of the same average composition as the entire mass which it represents.

Quantity of Ore to be Used for Assay. — The amount to be weighed out for the fusion will be determined by the probable richness of the sample.

Where results have to be reported in ounces or pennyweights per ton, it is convenient to use what is known as the "assay ton" system. An assay ton is a weight so chosen that it contains as many milligrams as there are ounces (troy) in a ton.

For the ton of 2000 lb. this weight is $29\frac{1}{6}$ grams ($29166\frac{2}{3}$ mg.)

For the ton of 2240 lb. the assay ton is $32\frac{2}{3}$ grams ($32666\frac{2}{3}$ mg.).

If this weight be taken for an assay, every milligram of metal found will represent 1 oz. (or 20 dwt.) per ton of the material assayed; if 2 assay tons be taken for an assay, every milligram found will represent $\frac{1}{2}$ oz. (or 10 dwt.); or every 0.1 mg. will represent 1 dwt. per ton of the material assayed.

The abbreviations "1 a. t.," "2 a. t.," etc., are commonly used to represent "one assay ton," "two assay ton," etc. When not otherwise stated, the "ton" referred to in this work is the ton of 2000 lbs.

Assays are often made, however, on an even number of grams (say 30, 60, or 100); in this case, if the result is to be reported on the "ounce per ton" or "pennyweight per ton" system, a table for converting the results must be used.

Where the metric system is used throughout, 1 mg. per kilogram represents 1 gram per ton of 1000 kilograms. In this case a convenient quantity for assay is 25 or 50 grams:

1 mg. on 25 grams = 40 grams per ton.
1 mg. on 50 "   = 20 "   per ton.

401

Fluxes. — The "crucible" or "fusion" assay of an ore is carried out in several stages, the first of which consists in forming an alloy of lead with the precious metals; other substances are added at the same time to form fusible compounds with the remaining constituents of the ore, so that every particle of precious metal may readily sink through the mass and alloy with the lead. In order to effectively secure every particle of gold and silver, it is found necessary to use a mixture which will generate metallic lead in a state of very fine division; for this purpose an oxide of lead mixed with some reducing agent is necessary; when this is sufficiently heated, the latter acts upon the oxide, producing minute globules of metallic lead in all parts of the mass; these gradually sink, at the same time combining with and carrying down with them any particles of gold and silver which they encounter.

The remaining constituents of the ore fall into one of two groups, generally described as "acid" or "basic." In order to effectively separate these from the metallic particles, substances must be added to form a fusible compound with each of these classes. Acid substances will require a basic flux, and basic substances an acid flux. We require, therefore:

1. An oxide of lead (litharge, red lead);
2. A reducing agent (charcoal, flour, argol);
3. A basic flux (carbonate of soda, bicarbonate of soda, carbonate of potash);
4. An acid flux (borax, silica).

Other substances, such as niter, fluor-spar, sulphur, metallic iron, etc., are occasionally added for special purposes (see below).

Proportions of Ingredients in Flux. — The nature of the flux necessary and the proportions and total quantity of each ingredient should be ascertained by experiment for each particular class of sample which has to be dealt with. The information given on this subject in different text-books is contradictory and often misleading, but by bearing in mind the object of the different classes of fluxing materials, the proper proportions may generally be discovered after a few trials, and where many samples of similar material have to be assayed, a large stock of flux may be prepared for use.

Standard Flux for Siliceous Ores. —The following may be taken as applicable to a large number of samples of what may fairly be called siliceous or quartzose ores. They were adopted by the writer after an extensive series of experiments on an ore consisting mainly of silica, but containing a fair amount of oxide of iron, and smaller quantities of the silicates of aluminium, magnesium, etc.:

Charge of ore	1 assay ton	2 assay tons
Carbonate of soda	32 grams	57 grams
Fused borax	5 "	8 "
Litharge	57 "	120 "
Charcoal	1.1 "	2 "
Size of crucible required......	G	H
Weight of lead produced	25 grams	40 grams

Quality of Fluxes. — When large quantities of flux are prepared, the ingredients should be carefully mixed, all lumps broken up, and the whole sifted through a coarse screen kept for this special purpose. The flux should be preserved in a suitable box or tin, kept in a dry place, and not exposed to dust from ore samples, etc.

The proportions of borax and soda to be used will vary considerably with the quality of these reagents, which sometimes contain a considerable percentage of moisture and other ingredients. It is not by any means unusual to find commercial "dry" borax with over 30 per cent. of moisture. If used in the flux, about double the quantities given above must be taken, but such material should be avoided as it swells enormously on heating, and causes violent bubbling in the crucible during the early stages of the fusion. There seems to be no object in using bicarbonate of soda in place of the carbonate; the Na_2O being the effective constituent, the use of bicarbonate merely implies loss of fuel and time in expelling the extra quantity of carbonic acid.

Proportions of Litharge and Reducer. — The litharge and charcoal should be so regulated that the button of lead obtained at the end of the fusion is of convenient size for cupellation, and may be reasonably supposed to have taken up practically the whole of the values in the material fused. Some assayers insist on reducing the whole of the lead from the litharge used in the flux; others are equally emphatic as to the necessity of using a large excess of litharge; in the writer's experience it is best to leave a *small* excess of litharge in the slag. The weight of lead button found to be most generally suitable is about 18 to 23 grams for an assay on 1 assay ton of ore, and 25 to 30 grams for 2 assay tons of ore.

In some cases, better results are obtained by increasing both litharge and reducing agent so as to bring down a large button of lead, which is afterwards reduced, as described below, to a convenient size for cupellation.

The special modifications of the process for different classes of ores containing base metals, etc., will be discussed in a later section.

Crucibles. — The crucibles generally used are fire-clay pots of different shapes and sizes. The kinds marked "G" and "H" are those most generally useful for assays on 1 assay ton and 2 assay tons respectively. Previous to use, they should be stored in a warm dry place and carefully cleaned before charging. Covers of the same material are used, which serve to prevent the entrance of coal or other foreign matter into the crucible during the fusion. The same crucible may, with care, be used for five or six successive fusions, but must be carefully examined for cracks before recharging.

Weighing the Assay Charges. — It is advisable to have the portions of flux needed for each assay ready before weighing out the ore charges. The flux need not be weighed with great accuracy; in fact, for all practical purposes the necessary amount may be measured out by means of a small metal cup or other convenient instrument which allows of approximately equal quantities being selected.

The ore is best spread out on a sheet of smooth rubber in front of the pulp balance or ore balance used for weighing. It is assumed that the sample has already been thoroughly mixed, but as an additional precaution the assay charge may be chosen by removing small amounts with a spatula at regular intervals all over the layer of ore. The charge is placed on the left-hand pan of the balance and the assay-ton weight, or other weight required, on the right. The weight for assay may then be adjusted by adding or taking off small portions of ore until it is correct to about 0.1 gram. In an ordinary assay it is a waste of time to adjust more closely than this, as it is obvious that a difference of 0.1 gram on a charge of, say, 30 grams would introduce an error of only $\frac{1}{15}$ dwt. in an assay of a 20-dwt. ore; whereas a balance sensitive to $\frac{1}{100}$ mg. can only detect a difference equivalent to $\frac{1}{100}$ oz., or $\frac{1}{5}$ dwt., in an assay of 20 dwt.

Mixing with Flux. — The weighed portions of an ore are

commonly mixed with the flux on a sheet of glazed paper, using a spatula to obtain a uniform mixture. The charge may then be readily transferred to a clay crucible of suitable size, and the paper swept clean by a small flat brush. Another, perhaps preferable, method is to mix flux and ore in a smooth porcelain mortar, large enough to enable the mixture to be stirred freely. In many assay offices the ore and flux, after mixing, are wrapped in thin tissue paper (Japanese copying-paper answers the purpose), and the little bundle placed in a crucible which has been previously heated. The fusion in this way is probably more rapid than when the charge is added to a cold crucible; but there are objections to the method, as will be shown later.

The Fusion Process. — The assay fusions are generally carried out in small square furnaces (8 in. to 12 in. square, inside measurement), capable of holding from 2 to 6 crucibles at a time, and provided with horizontal or inclined sliding covers. The furnaces and such parts of the flue as are exposed to strong heat are lined with fire-brick. The crucibles should be firmly supported on fire-bricks resting on the bars at the bottom, but arranged in such a way that they may be uniformly heated all round; and the mouths of the crucibles should not be at a higher level than the opening of the flue. When charcoal is used, the crucibles, after putting on the fire-clay lids, may be entirely covered by a layer of fuel. With coke it is hardly necessary to use a support beneath the crucibles. The bars of the furnace must be clear of ash before setting the crucibles in their places.

The operation of fusion usually takes about 40 minutes. For the first 10 minutes or so the temperature should be kept down to avoid loss through the violent expulsion of carbonic acid, water vapor, etc. At this stage, signs of fusion generally begin to appear at the sides, and the lids had better be removed, provided the fuel is burning quietly and there is no danger of lumps of charcoal, etc., being projected into the pots. A semi-fused crust sometimes forms on the surface, which may occasionally be raised nearly to the mouth of the crucible by the action of gases beneath, but if the proper conditions be observed this crust soon sinks down and no projection of the contents from the crucible takes place. As soon as the action begins to moderate, the temperature may be raised (by opening the damper more fully and closing the covers of the furnace), and the heat is continued until all signs of ebullition have ceased. The lids are then replaced for the last five to ten minutes, after which the contents should be ready to pour. A fairly bright red heat is needed for the fusion, but very intense heat is neither necessary nor desirable.

Fusions in the Muffle. — The fusions are sometimes made by placing the crucibles in a large muffle, and with care good results can be obtained; but the method has the disadvantage that the actions going on cannot be easily observed and controlled; also, with poor ores, either a very large muffle must be used to accommodate the large crucibles which are necessary, or a number of assays on each sample must be combined, which is laborious and troublesome.

Pouring. — When ready for pouring, the contents of the crucible should be quite tranquil; the crucibles are then lifted out of the furnace by means of iron tongs, bent at the points so as to grip the edge of the crucible firmly, and the molten contents poured into iron molds, previously cleaned and slightly warmed. The fused mass should pour freely, like oil, and should leave no metal or lumps of any kind in the crucible, which, when cool, should show internally only a thin layer of slag.

Hammering Lead Buttons. — The molds are left about five minutes to cool, then inverted, and the slag removed from the buttons of lead by one or two blows with a hammer. No shots or detached particles of lead should appear in the slag, but in all cases it should be carefully examined, and any detached lead added to the main button. The color and appearance of the slag will depend largely on the nature of the ore; also to some extent on the time and temperature of fusion. When the ore consists chiefly of silica, with small quantites of oxide of iron, as in the case we are considering, the color will be light green; with large amounts of iron it will be dark brown, almost black. When copper is present a characteristic reddish-brown color is produced (due to cuprous oxide). The lead should be easily detached from the slag, and should not leave a scale or film behind on breaking off; this takes place sometimes when the slag is excessively hard and stony, as may be the case if too much borax or acid flux has been used. Often the slag is very brittle and flies to pieces with some violence on merely touching with a hammer. The hammering should be done on a steel block, mounted on a firm wooden support and surrounded on three sides by a frame to retain flying particles. Hammer and anvil faces should be kept carefully cleaned. The detached lead buttons are held in steel forceps and struck on the edges before hammering the upper and lower surfaces, otherwise particles of slag may be hammered into the block of lead, causing spitting and loss in cupellation. By hammering at the edges and brushing the lower surface (where the button was in contact with the slag) with a wire brush, the lead may be effectually cleaned. The block of lead is then beaten into a roughly cubical shape, and placed in a compartment of a cupel tray.

Arrangement of Assays. — In working a batch of assays it is very necessary to arrange them systematically and to adhere to the same order of arrangement at every stage of the process. Each sample is first marked with a number, and the same number entered in the assay book, with sufficient description for identifying the sample with certainty. In placing the crucibles in the fusion furnace, the lead buttons on the cupel trays, or in the muffle for cupellation, etc., these numbers are taken in the order of the words on a page of a book. Thus, 12 assays in 2 fusion furnaces might be arranged:

Furnace No. 1			Furnace No. 2		
1	2	3	7	8	9
4	5	6	10	11	12

When subsequently transferred to the cupel tray they might be placed thus:

1	2	3	4
5	6	7	8
9	10	11	12

If this principle be strictly adhered to, no confusion or ambiguity can occur, even if the crucibles, cupels, etc., be not actually marked.

It is a good plan to arrange the samples so that the lower-grade assays (*e.g.*, cyanide residues) are made first. The same crucibles may then be used for the richer samples without danger of "salting" should any lead remain in the crucible from a previous imperfect fusion, as this could hardly contain a weighable quantity of gold.

Fusion in Hot Crucibles. — The method of adding fresh charges immediately to the hot crucibles, whether these charges are wrapped in paper or not, is not to be recommended, as it results in very violent and rapid action in the early stages, giving rise very probably to loss of material through ejection of particles from the crucible. It is also possible that the lead particles will form and sink to the bottom of the crucible before the siliceous matter is properly fused, and that particles of precious metal will therefore remain in suspension in the slag without ever coming in contact with lead. When the temperature is kept low in the early stages the action is more moderate; the lead formed sinks gradually through the mass and has a better chance of securing all the particles of precious metal.

(B) CUPELLATION AND PARTING

Cupellation. — This process depends upon the fact that when an alloy of gold and silver with lead is heated, with free access of air, in contact with certain porous materials, such as finely powdered bone-ash, the lead oxidizes, this oxide being partly volatilized and partly absorbed by the substance on which it rests; while the gold and silver undergo little or no oxidation, but, after all the lead has disappeared, remain behind in the form of a nearly spherical bead of fine metal.

Muffle-Furnaces. — The operation is generally carried out in a small fire-clay oven, called a "muffle," consisting of an oblong floor surmounted by an arched roof and closed at one end. This is supported within an outer furnace ("muffle-furnace"), so arranged that fuel may be placed below, above, and on both sides of the muffle. The latter is at least partially open in front, and the semicircular end is usually provided with a slit to secure a draft. Slits are also sometimes made in the sides of the arch. Coal, charcoal, gas, or oil are used as fuel.

Cupels. — The "cupels" in general use are small cylinders of porous material with a hollow at one end for the reception of the lead button. Formerly, they were always made of bone-ash, but at the present day various refractory materials, such as magnesia and magnesium borate, have been applied with some success in the manufacture of cupels.

The efficiency of bone-ash cupels depends largely on the amount of water added, the method of molding and pressing, the fineness and purity of the bone-ash, and the time of drying the cupels before use. Some assayers add small quantities of alkaline carbonates and other ingredients, but the general opinion seems to be that with pure bone-ash the best results are obtained by moderately fine crushing, using only distilled water. The amount of water should be only just sufficient to cause the material to cohere, and not sufficient to make it pasty. The cupels are then made by means of a mold consisting of a hollow metal cylinder, which is filled with the moist bone-ash; the plunger is then pushed in and struck once or twice with a mallet. The mold is then inverted and the cupel turned out. Better results can in general be obtained by the use of manufactured cupels, which are much more uniform in quality than can be the case with those made by hand on the spot. Previous to use, the cupels should be dried for a long time in a moderately warm place.

Charging into Muffle. — The cupels are placed, empty, in the muffle, in rows of 3 or 4, as may be convenient, with the hollow uppermost. After about 10 minutes, with a good fire, they will begin to show a red heat, and the lead buttons may then be added by means of suitably shaped cupel tongs, each button being carefully placed in the hollow of its corresponding cupel. It is a good plan to begin by adding the front row (*i.e.*, the row nearest the mouth of the muffle), as these are generally cooler than the rest, and require more time to complete the cupellation, The other rows are then added in succession from front to back. The number of assays to be cupeled simultaneously will depend on the size of the furnace and of the cupels, but it is not a good plan to crowd too many into the muffle at one time; if the furnace be heated sufficiently to cupel the front row in a full muffle, some of the back rows may be much overheated, leading to heavy losses by volatilization.

Size and Form of Cupels. — The size of cupels to be used will depend on the weight of the lead buttons. It is generally assumed that a bone-ash cupel will absorb its own weight of lead. For buttons of 20–25 grams, the size known as No. 7 is convenient; for 15 grams, No. 5 may be used. The depth of the hollow is also of some importance. If too shallow, the molten lead may overflow, although there may be sufficient absorbent material; if too deep, the oxidation proceeds very slowly and the losses of gold and silver by volatilization may be heavy. A layer of bone-ash should be placed on the floor of the muffle, as a precaution against accidents and to keep the litharge from attacking the fire-clay if it should soak through the bottoms of the cupels.

Process of Cupellation. — When the lead buttons are placed on the cupels, in a few moments the lead melts, and the surface shortly afterward brightens and begins to exhibit a rotary movement. This bright appearance continues until the operation is complete, provided a sufficient temperature be maintained. When the operation is finished the movement ceases, but the bead of gold and silver may remain molten for some time. When drawn forward to a cooler part of the muffle, it suddenly solidifies, emitting a bright glow of light; if cooled too rapidly particles of metal may be projected onto the edges of the cupel or entirely lost. This action is generally ascribed to the expulsion of absorbed oxygen as the silver solidifies. Large buttons exhibit a play of iridescent interference colors a few moments before the finish of the operation. At a suitable temperature, a button of lead weighing 25 grams will be cupeled in about half an hour.

Effects of Varying Temperature. — If the temperature of the muffle during cupellation sinks below the melting-point of litharge, a crust forms on the buttons and the movement ceases. This is spoken of as "freezing." When this has taken place it requires a much higher temperature to re-start the cupellation, and there is always a possibility of loss in doing so. The temperature required to start cupellation and also to complete the final stages is considerably higher than that required to maintain the operation during the earlier part, after it has once started. Losses both of gold and silver occur in all cases, by volatilization and by absorption in the cupels, and as these losses are increased by rise of temperature, an accurate assay, especially where the determination of silver is of importance, is always conducted at the lowest temperature which can be used without danger of freezing.

When the cupellation has been carried out at a fairly low temperature, the bead at the finish is surrounded by a ring of feathery yellowish crystals of litharge.

Losses of Gold and Silver in Cupellation. — According to Hillebrand and Allen,[1] "when cupellation takes place at a low temperature, with formation of considerable feather litharge, the loss by volatilization is practically negligible, or at any rate

[1] Bull. No. 253, U. S. G. Survey (1905).

is perhaps compensated by retention of lead; but when the cupellation is made at a higher temperature, the loss is considerable." The following average losses were determined by them in samples of telluride ore, by assaying the slags from the original fusions and the cupels respectively. The results give the amounts of gold and silver together per assay.

Sample	In Slags: Milligram	In Cupels: Milligram
1	.055	.085
2	.08	.10
3	.145	.16
4	.05	.18

This cupellation loss is the amount absorbed, and does not include the loss by volatilization. The latter could only be determined by making a check assay with known quantities of gold and silver under conditions of temperature, etc., as nearly as possible identical with those of the actual assay. Experiments made by Hillebrand and Allen[1] showed that, contrary to the usual opinion, the losses of *gold* in cupellation are not negligible, especially with rich ores. Tests made by cupeling various known weights of gold and silver with 25 grams of lead in each case, to correspond with an ordinary assay button, led to the following conclusions:

(1) That the losses of gold and silver increase progressively from front to back of muffle.

(2) That, for equal amounts of gold, the loss is greater, with 25 grams of lead than with 5 grams.

(3) That the percentage of loss of gold is greater with small than with large amounts of gold, especially in the hotter part of the muffle.

(4) That the loss of gold by absorption is much greater than the loss by volatilization; when no silver is present the loss is chiefly due to absorption.

(5) That the loss of gold is about the same whether silver is present or not.

They insist on the necessity for corrected assays where accurate results are required, and state that the most exact results are obtained when feather litharge is still abundant at the time of brightening.

[1] *Ibid.*

Miller and Fulton [1] state that the absorption of silver by the cupel increases with the size of button, but the amount absorbed per gram of lead button cupeled diminishes regularly as the lead increases. They find also that the loss in diminishing a lead button by scorification is irregular, but usually greater than by cupeling direct. The corrections never account for the whole of the loss of silver.

F. P. Dewey [2] remarks: "While the correction for slag and cupel loss is easily made, and ought always to be made when accurate statistics are kept, there is yet the volatilization loss to correct, and some means of doing this is very desirable. While a check assay answers very well for bullion, it would hardly be possible to construct check material for the varying characters of ores and products ordinarily met with."

W. P. Mason and J. W. Bowman [3] give a table showing a number of results of losses of gold and silver on weighed quantities, when scorified and cupeled in a Battersea F muffle, the conditions under which the assays were made, such as heat of muffle, draft, and manipulations in general, being such as would obtain in careful, practical work. The average of their results show:

TABLE IV. — SCORIFICATION AND CUPELLATION LOSSES IN A
BATTERSEA F MUFFLE

| | PERCENTAGE LOSS | |
	Silver	Gold
In Scorification	0.55	0.574
In Cupellation	1.99	0.296
For entire process................	2.54	0.87

The methods of assaying slags and cupels will be discussed in a later section.

Cleaning and Weighing Fine Metal. — When the cupels are sufficiently cool, the beads of fine metal (gold and silver) are removed by means of a small pair of pliers, pressed tightly, and if large enough brushed with a wire scratch brush, then hammered

[1] "Sch. Mines Quart.," XVII, p. 160 (1896).
[2] "Journ. Am. Chem. Soc.," XVI, p. 505.
[3] "Journ. Am. Chem. Soc.," XVI, p. 313 (Oct. 6, 1893).

on a smooth clean anvil with a clean smooth-faced hammer. The flattened beads may conveniently be transferred to small porcelain crucibles, arranged on a cupel tray or wooden frame, in the same order as the cupels. They are then weighed on the assay balance. For ordinary work it is quite sufficient to weigh the fine metal within .05 or even .1 mg., as the variations in the losses of silver in different parts of the muffle probably exceed this amount.

Inquartation. — In cases where the amount of silver present is less than $2\frac{1}{2}$ times that of the gold, the latter protects the silver to some extent from the action of the nitric acid in the subsequent operation of parting. When this is the case it will be necessary to add a small quantity of silver foil, a milligram or two in excess of the amount theoretically required, wrap the bead and extra silver in a small piece of lead-foil, and cupel in a small cupel (say No. 3 size). A better plan, however, is to make two assays of the material — one without adding silver, and the other with the necessary silver for parting added to the flux before transferring to the crucible. It is advisable to examine the silver-foil used and make sure that it is absolutely free from gold.

Parting. — This operation may be carried out either in porcelain crucibles or in small long-necked glass flasks. The former method is quite satisfactory, and if carefully carried out requires less manipulation and trouble than the method with parting-flasks.

The acid required for parting the beads from ordinary ore assays is nitric acid, 20 per cent. by volume — *i.e.*, a mixture of 20 parts pure acid (say 1.42 sp. gr.) with 80 parts distilled water. From 10 to 15 c.c. of acid are required for each bead. The crucibles are 4 to 4.5 cm. in diameter, and when this quantity of acid is added are filled about two-thirds full. The crucibles may be conveniently arranged on a perforated metal plate supported over a uniform and moderate source of heat; an oil-stove with large wicks can be used, but the heat is not really uniform in all parts. The acid must be heated to boiling, but not allowed to boil too violently, as in some cases the beads may break up and pieces of gold may be projected from the crucibles and lost. The heating must be continued until no signs of red fumes can be observed, and until no further evolution of small bubbles takes place. Care must be taken not to allow the liquid to evaporate to dryness.

With ordinary assay beads a second parting is hardly necessary. With large beads, however, it is advisable to pour off the first (weak) acid and add an equal amount of a stronger acid (50 per cent. by volume). This is heated to boiling and again poured off when the action appears to be finished. The crucible is then filled up with distilled water, which is carefully decanted; if the gold has broken up, as will be the case when a large excess of silver was present, this operation requires close attention, and it is best to pour the water off along a glass rod into a porcelain basin. After washing once or twice in this way, the crucibles are dried slowly to avoid spurting: generally this may be safely done by setting them on the perforated frame in a slightly inclined position and turning the lamp down. They are finally heated until the gold acquires its natural yellow color. When cool it is ready for weighing.

When parting-flasks are used they are generally supported on a specially constructed frame so that they rest in a slightly inclined position; this is heated with the necessary precautions against bumping and spurting. After pouring off the acid the flask is filled to the top with distilled water and inverted into a small crucible of unglazed porous material (annealing-cup). This may easily be done without spilling any liquid by placing the crucible over the mouth of the flask before inverting the latter. The gold then sinks to the bottom of the annealing cup; the inverted flask is gradually raised, allowing the cup to fill with water, the flask is then withdrawn by a quick lateral movement and the surplus water poured off. For small beads, test-tubes may preferably be used instead of parting-flasks. Generally speaking, this method is more troublesome than parting in crucibles. A glazed porcelain crucible may of course be used instead of the porous annealing-cup, and is perhaps preferable, as small particles are liable to chip off the edges of the latter and might occasion errors in weighing broken gold, but greater care is necessary in drying the gold in a glazed crucible.

Purity of Parting Acid. — Care must be taken that the nitric acid used is free from chlorides, free chlorine, hydrocyanic acid, or any impurity which could precipitate silver in acid solution or dissolve gold, either by itself or in conjunction with nitric acid. Silver nitrate added to the diluted parting acid should give no turbidity. If any milkiness be observed, it is best to add a certain quantity of silver nitrate to all the parting acid made up from that lot of strong acid, and to allow the turbidity to settle before using. Nitrous acid has generally been considered to be a solvent for gold;[1] but the researches of Hillebrand and Allen[2] would appear to show that this acid has no such action. Nitrous acid is shown by a reddish brown color and by the dilute acid liberating iodine from potassic iodide (test with starch solution). The same writers also found no evidence of the solution of gold by prolonged boiling, in pure nitric acid. There is, however, some reason for believing that, under conditions not clearly understood, a quite perceptible quantity of gold may be dissolved, so that the assayer will do well to use the purest acid obtainable and avoid prolonging the process of boiling after the action on the silver has entirely ceased.

The distilled water used must of course also be free from any impurities which might precipitate silver, act on the gold, or otherwise interfere with the process.

Adjustment of Assay Balance. — Before weighing a batch of gold assays it is essential to verify the adjustment of the assay balance. The case should be free from dust, inside and out, and the plate on which the balance rests should be quite level, usually shown by two small spirit-levels placed at right angles inside the balance case. If necessary, it must be leveled by raising or lowering the supporting screws. If the balance is much out of adjustment, it is rectified by moving the vane or adjusting screw in the required direction. Small variations in adjustment are best corrected by means of a rider. If the right arm of the balance be used for the weights, as is generally the case, the adjusting rider may conveniently be placed on the left arm, which is kept permanently one- or two-tenths of a mg. lighter than the right arm. Any small variation of adjustment occurring during the course of weighing (due to differences of temperature, etc.) can thus be corrected by moving the left-hand rider, without any disturbance of the balance.

Weighing Parted Gold. — The crucibles or annealing-cups containing the parted gold are arranged in their proper order on a cupel tray, or on any suitable frame which allows them to be moved simultaneously. The gold is then transferred from

[1] G. H. Makins, in " Journ Chem. Soc.," XIII, p. 97 (1861).
[2] Bull. No. 253, U. S. G. Survey (1905).

the crucible to the balance pan by means of a needle or very fine camel's-hair brush; sometimes a slight tap may be needed to move it. The pan should be placed on a sheet of glass or a smooth card, so that any loose particles which might be lost in transfer may be at once seen. A good balance will determine the weight within .01 mg., which is sufficient for most purposes, this amount corresponding to 4.8 grains per ton (20 cents gold value) on an assay of 1 assay ton, or 2.4 grains (10 cents) on an assay of 2 assay tons.

Most balances are so constructed that weights less than 0.05 mg. and weights lying between 0.9 and 1.0 mg. cannot be conveniently weighed with the rider alone. In such cases a counterpoise of known weight (say 0.5 mg.) may be placed in the pan, or the left-hand rider may be moved one or more divisions, so that the zero-point in weighing is no longer at the center of the beam, care being taken to return the rider to its normal position after each such weighing.

After weighing, the gold should be immediately transferred to a small cup or other vessel placed in the balance case, so that the accumulated beads may be recovered, and also to avoid any risk of their finding their way into other samples.

Reporting Results. — A convenient method of recording the assay results is to use the headings here shown:

No. of Sample	Description	Date	WEIGHT FOUND		ASSAY OF SAMPLE	
			Gold and Silver: mg.	Gold: mg.	Gold: Dwt.	Silver: Dwt.

The weights of fine metal in milligrams are noted in the column headed "Gold and Silver," and the weights of gold are subsequently entered in the next column, headed "Gold: mg." The difference of these entries gives the weight of silver in milligrams. If the assay be on 1 assay ton, the weights of gold and silver in milligrams must be each multiplied by twenty to obtain the assay in pennyweights. If the assay be on 2 assay tons, they must be multiplied by ten.

(C) Sources of Error in Assaying Ore Samples

Robert Dures (Journ. Chem. Met. and Min. Soc. of South Africa, June, 1913, p. 608) gives the results of investigations on various points connected with the assay of mine samples. The tests were all made in open pots in a reverberatory furnace, the flame only in contact, and on charges of 2 A.T.

Fineness of Crushing. — Nine assays on portions of a sample crushed to 40 mesh showed an average assay of 40.1 dwt., with an extreme difference of 12.4 dwt. Nine assays on portions of the same sample crushed to 100 mesh showed an average of 37.6 dwt., with an extreme difference of 2.8 dwt. The lower value in the latter case is probably due to the loss of some of the "metallics" in the process of sifting. A possible source of error in fine crushing is the clogging of the meshes of the 100-mesh sieve, with the liability of salting a subsequent sample.

Varying Litharge and Soda in Flux. — Using the same amount of charcoal, the quantity of soda could be varied within wide limits, when the amount of litharge was increased as the soda diminished. In each case 1 gram of charcoal was used, and a cover of 15 grams borax. The fusions occupied 40 minutes. The weight of lead obtained varied widely, but the assay result varied no more than in the case of duplicates run under identical conditions on the same sample. The figures given by Dures are as follows:

No.	Soda Carbonate gm.	Litharge gm.	Weight of Lead Button gm.	Assay dwt. per Ton
1	100	40	34	31
2	95	50	42.8	28.9
3	90	60	50.3	30
4	85	70	54	30
5	80	80	56.2	31.4
6	75	90	57.3	31.3

Varying Charcoal. — Using a fixed charge of soda, litharge and borax, viz.:

Soda................ 90 gm.
Litharge .... 70 "
Borax cover .... 15 "

the charcoal was varied from 0.5 to 2 grams, bringing down lead buttons from 48 to 61.2 grams without affecting the assay result within limits of error in duplicates.

Variation of Duplicates. — Twenty assays made in duplicate on different samples showed an average difference of 4.38 per cent., the differences varying from 0 to 12.5 per cent. The low-grade samples showed the higher proportional differences, this being probably due to the gold being mainly associated with rich pyritic material. The low-grade samples containing only a little of this, it was practically impossible to distribute it so uniformly as to give concordant results, whereas in the high-grade samples it was present in such quantity that it was easy to get a fairly even distribution of this rich material throughout the sample.

Errors in Cupellation. — When the cupellation was finished at too low a temperature, so that the beads had a " frosted " appearance (spongy or pitted, owing to freezing near the finish of the operation), the results were about 9 per cent. too high. On re-cupelling such beads a loss of 9.96 per cent. was observed; deducting the normal cupellation loss of 1.2 per cent., this shows the original weight of the frosted beads to have been 8.75 per cent. too high.

Tests made by cupelling approximately 2.5 mgr. gold with 2 to 7 times its weight of silver showed an average loss in the total weight of fine metal amounting to 3.1 per cent., this loss being practically all in the silver.

SECTION III

SPECIAL METHODS OF ASSAY FOR PARTICULAR ORES AND PRODUCTS

(A) THE SCORIFICATION ASSAY

Variable Nature of Material for Assay. — The ores treated, and the materials obtained as by-products in the treatment of gold and silver ores by the cyanide process, are of such varied nature that no single method of assaying is applicable in all cases. When these ores or materials are of high assay value, or present difficulties in the ordinary method of assaying as described in the previous section, special care and experience are required in dealing with them. In doubtful cases it is always well to try two or more distinct methods on the same sample. The agreement of duplicate assays by the same method cannot be accepted as a proof of correctness, as experience often shows that a different method may yield a totally different result.

Scorification. — The most generally applicable method for many different classes of material is scorification, and this will be first considered. As a rule the method can only be conveniently applied to comparatively rich ores or products, as the amount taken for each assay of the substance is necessarily small, an ordinary charge being 5 to 8 grams. In exceptional cases as much as 15–20 grams may be taken, but this involves the use of large scorifiers. The scorifiers in general use are small fire-clay dishes varying in size from 1¼ in. up to 3 or 4 in. in diameter. The substance to be assayed is mixed with pure lead in the form of small grains, and placed in the scorifiers. The general custom is to mix half the lead with the material to be scorified and add the remainder as a cover, together with a very small amount of borax or other flux. In some cases, litharge may be added with advantage instead of, or as well as, metallic lead.

The following five charges are given as examples for ordinary

419

cases; the quantities given are in grams, unless otherwise designated:

	Charge 1	Charge 2	Charge 3	Charge 4	Charge 5
Weight of material	$\frac{1}{4}$ a. t.	5	5	5	$\frac{1}{4}$ a. t.
Grain lead	50	60	30	—	27
Litharge	—	—	35	70	22
Fused borax	0.2	0.25	0.25	0.25	0.5
Silica	—	—	—	—	0.4
Diam. of scorifier........	$2\frac{1}{2}$ in.	$2\frac{3}{4}$ in.	$2\frac{3}{4}$ in.	3 in	3 in.

The last (No. 5) is suitable for pyritic ore or similar material. The amount of silica may be varied according to the quantity of pyrites present in the ore. In this case the litharge and silica are to be mixed with the ore and the lead and borax added as a cover.

Charges for Material Containing Copper. — The following is suitable for material containing small amounts of copper and low-grade in precious metals:

Charge 6

Material for assay...........	$14\frac{7}{12}$ grams = $\frac{1}{2}$ a.t.
Grain lead	60 "
Fused borax	0.05 "
Scorifier	$3\frac{1}{2}$ in.

For highly cupriferous material:

Charge 7

Substance for assay	7.23	grams = $\frac{1}{4}$ a.t.
Grain lead	75	"
Borax	0.2 to 0.3	"
Scorifier	$3\frac{1}{2}$ in. or 4 in.	

Charges for Rich Silver Ores. — For rich silver-bearing material one of the following may be used:

	Charge 8	Charge 9
Substance for assay.........	$\frac{1}{10}$ a.t.	$\frac{1}{20}$ a.t.
Grain lead	30–70 grams	40–60 grams
Borax	0.3– 3 "	0.5 "
Scorifier	$2\frac{1}{2}$ in. to $2\frac{3}{4}$ in.	$2\frac{1}{2}$ in.

Zinc-Box Precipitate. — The following is given by R. W. Lodge [1] for the assay of zinc-box precipitate:

[1] *Trans.* A. I. M. E., XXXIV, p. 432 (October, 1903).

Charge 10
Precipitate .... 0.05 a.t.
Grain lead .... 65 grams
Borax glass .... 10 "
Scorifier .... 3 in. or 4 in.

35 grams of the lead to be mixed with the material and 30 grams to be used as cover.

Lead Required for Scorifying Different Kinds of Material. — As will be seen by the above figures, the amounts of grain lead required to scorify various kinds of material are very different, according to the nature of the material. The following table gives some of the usually accepted figures on this subject:

TABLE V. — GRAIN LEAD REQUIRED WITH VARIOUS MATERIALS

Nature of Material Taken	To be Added per Gram of Material Assayed	
	Lead: Grams	Borax: Grams
Galena	5– 8	0.1–0.15
Quartzose ores	8–10	nil
Basic ores	8–10	0.25–1
Pyritic ore	10–15	0.1–0.2
Blende..............		
Graphite	12	0.05
Fahl ore..............	12–20	0.1–0.15
Arsenical ore	16	0.1–1
Antimonial ore		
Telluride ore	16–18	0.1 (with litharge cover)

Mode of Procedure. — In an ordinary scorification assay, the materials, mixed as directed, are placed in the hollow of the scorifier, which is then introduced into the muffle at a low red heat. In many cases it is preferable to add the borax little by little, as the action proceeds. The temperature is then raised gradually to bright redness, keeping the muffle door closed, say, for 15 minutes, by which time the mass should be completely fused. The door is then opened sufficiently to give a good current of air, when litharge is rapidly formed, and may be seen forming a ring round a bright central spot or "eye." This spot gradually diminishes until it is completely covered by the layer of molten slag. The temperature is now again raised, by closing the door or increasing the draft. At this stage it is sometimes a good plan to add a small quantity of a reducing agent, say 0.5 to 0.8 part of anthracite powder to every 100 parts of lead originally taken. This is wrapped in tissue paper and placed quickly on top of the slag; it serves to reduce any silicates of silver that may have been formed. When all action appears to have ceased and the contents of the scorifier are quite liquid, the latter is removed from the muffle by means of "scorifier" tongs, consisting of two arms, the upper one of which rests across the top of the scorifier while the lower one is branched in U-shape, so as to hold the bottom of the scorifier firmly. While the scorifier is held in this way its contents are poured into a dry, clean mold. When cool, the slag is detached and the lead button cleaned for cupellation. The process must not be continued too long after the "eye" has closed, i.e., after the lead is completely covered by the layer of slag, as the size of the lead button gradually diminishes, and when too small much of the gold and silver may pass into the slag.

Remelting Slag. — In any case it is advisable, with all samples that are at all rich, to remelt the slag with fresh flux. This may be done by grinding the slag in a mortar and fusing in a clay crucible with 20 grams litharge, 2 grams sodium carbonate, and 1 gram charcoal. The resulting lead button is cupeled and the result obtained added to that from the scorification assay; or the two buttons may be cupeled together.

Theory of Scorification Assay. — The theory of the process is described by J. Daniell [1] as follows:

"The oxidation of lead produces litharge, or litharge is added with the charge. With the silica (of the scorifier or of the material assayed) this forms the extremely fusible silicate Pb_2SiO_4. By sulphides, the oxide is reduced back to metallic lead with formation of sulphur dioxide in presence of a current of air, and a metallic oxide which is dissolved by the excess of litharge. In fact, fused litharge is an extremely powerful solvent, and there are very few substances which are not attacked and held in 'igneous solution' by it. The noble metals are thus concentrated in the unoxidized metallic lead and are subsequently obtained by cupellation."

Inconveniences of Scorifying in Muffle. — With ordinary muffle furnaces there is a tendency, already noted, for the back of the muffle to become much hotter than the front; hence the

[1] "Proc. Chem., Met. and Min. Soc. of South Africa," II, 277.

scorifications at the back are apt to finish first, and where large numbers are assayed simultaneously it may be troublesome to remove these at the proper time without upsetting or interfering with those in front.

Scorification Assay in Fusion Furnace. — A modified form of the process, described by Daniell,[1] will meet this difficulty. The operation is carried out in a "Cornish fire."

"Duplicate trials of about 3 grams of ore were taken, mixed with 25 grams of granulated assay lead and a small quantity of a flux composed of equal weights of carbonate of soda and borax. Another quantity of 25 grams granulated lead was then put in and covered with a further quantity of flux. The crucibles were placed in a full fire at a low heat [small wide-mouthed crucibles were used instead of scorifiers], with damper closed, 4, 6 or more at a time, depending on the size of the furnace. When melted, the bricks were opened, and damper raised, causing a current of air to play upon the crucibles, ensuring rapid oxidation of the lead and a thorough scorification of the ore. . . . The use of the Cornish fire in this way has been thoroughly tested and gives perfectly satisfactory results."

Cautions. — T. Kirke Rose[2] remarks, in reference to scorification: "Effervescence and spurting may occur, especially if the scorifier has not been well dried by warming before it is used."

Daniell[3] says: "There are, of course, a few precautions which have to be taken. For instance, pyritic material, especially arsenical pyrites, has a tendency to decrepitate, throwing out minute particles which burn with a characteristic sparkle and entail loss. This can be obviated by careful attention to the heat at the commencement of the operation. Again, carbonaceous matter in presence of litharge generates gas, which, rising through the melted lead, may occasion loss by projection of particles."

Reduction of Large or Impure Buttons. — It sometimes happens on pouring a scorification assay that the button of lead obtained is too large to be conveniently cupeled, and in other cases it may be hard, brittle, or coppery in appearance. In such cases the slag should be removed and the button replaced in the same scorifier, fresh lead being added in the case of hard or impure buttons. The scorification is continued until the button is reduced to the proper size for cupellation and appears quite clean and malleable. According to Rose[4] (*loc. cit.*) less loss of the precious metals is

[1] *Ibid.*
[2] "Metallurgy of Gold," p. 383.
[3] "Proc. Chem., Met. and Min. Soc. of South Africa," II, 277.
[4] Compare Miller & Fulton, "Sch. Mines Quart.," XVII, p. 160.

incurred by scorifying lead than by cupeling it, and consequently it is better to reduce any lead button weighing more than 20 grams by rescorification before cupellation.

Assay of Graphite Crucibles. — The following method is given by T. L. Carter [1] for assaying graphite crucibles (*i.e.*, old pots that have been used for smelting operations):

" A quantity not exceeding 6–7 grains (say .4 to .5 grams) of the finely powdered material is taken; if poor, several buttons are combined. The bottom of a large scorifier is rubbed with silica, then 40 grains (say 3 grams) of pure litharge is introduced, followed by the plumbago and 4 grains (say .25 gram) of niter. These materials are mixed together, 30 grains (*i.e.*, 2 grams) litharge added,. and finally a covering of borax. The temperature is kept very low for a few minutes, then gradually raised to a white heat till the mass fuses completely. As a check, 3 assays are made by scorification and 3 by pot fusion."

Acid-Treated Residues. — Scorification is also used sometimes for assaying the residues from ores and other products which are largely soluble in acids, as certain cupriferous materials, etc. This will be referred to later.

Limitations of Scorification Assay. — Dr. Loevy [2] observes: " For Rand ores and for all kinds of tailings and slimes, scorification is entirely out of the question; but for a certain class of material, such as graphite crucibles, certain slags, antimonial ores, lead ores, etc., the scorification assay is the only reliable one."

This statement appears a little too sweeping, especially as regards antimonial and lead ores. (See below.)

(B) Crucible Fusions for Various Classes of Ore

Classes of Ore for Assay. — As already remarked, the flux to be used must always be adjusted to the nature of the ore, and no fixed rules can be given. The following examples may, however, be of some use as a guide. . The classes which will be here considered are:

Class   I. Siliceous (Quartzose) Ores.
Class II. Basic (Oxidized) Ores.
Class III. Pyritic Ores.

[1] " Eng. and Min. Journ.," Aug. 5, 1899 (p. 155).
[2] " Proc. Chem., Met. and Min. Soc. of South Africa," **II**, 206.

Class IV. Lead-Zinc Ores.   Class VI. Antimonial Ores.
Class V. Arsenical Ores.   Class VII. Cupriferous Ores.
Class VIII. Telluride Ores.

Methods of Assay. — The methods of assay employed may be grouped under the following heads:

(*a*) Direct fusion with ordinary fluxes (*i.e.*, soda, borax, litharge and charcoal, or substitutes for these).

(*b*) Direct fusion with special oxidizing fluxes (niter, manganese dioxide, etc.).

(*c*) Direct fusion with a desulphurizing agent — generally metallic iron.

(*d*) Fusion after preliminary roast or oxidizing process.

(*e*) Fusion after preliminary wet chemical treatment.

Class I. — Siliceous Ores

Consisting mainly of quartz, with very small quantities of oxide of iron, iron pyrites, calcium and magnesium silicates and carbonates, etc. The quantity taken for assay will vary from $\frac{1}{2}$ assay ton up to 4 assay tons, according to richness, but on a cyanide plant it is seldom desirable to use less than 1 assay ton.

Fluxes for Free-Milling Ores. — We give here a number of fluxes adapted from those recommended by different writers, so as to be applicable to charges of 1 assay ton. The quantities are given in grams:

	Flux No. 1	No. 2	No. 3	No. 4	No. 5	No. 6
Carbonate of soda	30–46	33	40	40	33	32
Borax glass	7–15	10	6	cover	nil	5
Litharge	15–30	40	45	98	49.55	57
Charcoal	1–1.5	as required	—	1	0.45	1.1
Flour	—	—	2.4	—	—	—

Nearly all writers on assaying give bicarbonate of soda in the fluxes recommended. If bicarbonate ($NaHCO_3$) is to. be used, the amount of carbonate is increased in the proportion of 3 to 2; thus 40 grams Na_2CO_3 are equivalent to 60 grams $NaHCO_3$.[1]

Of the above fluxes, No. 1 is based on the recommendations of T. Kirke Rose.[2] The quantities there given are "From 1 to $1\frac{1}{2}$

---

[1] Flux used in assay office of Creston Colorada Co., Sonora, Mexico.

Carbonate of soda	46 grams
Borax	13 "
Litharge	40 "
Flour	1.2 "

[2] " Metallurgy of Gold," 4th edition, p. 470.

assay tons of soda carbonate and from $\frac{1}{4}$ to $\frac{1}{2}$ assay ton of borax to 1 assay ton of ore." And with regard to litharge he remarks: " Litharge or red lead is added in the proportion of one or two parts to two of ore; if too much litharge is used the slags are not clean, as a slag containing lead may mean a loss of silver and gold." [The amount of litharge here recommended appears to be much less than that which the general experience of assayers shows to be desirable. — J. E. C.]

No. 2 is given as applicable to ordinary quartz ores on the Rand, and is recommended by B. W. Begeer.[1]

No. 3 is adapted from the flux given by C. and J. Beringer.[2] Assuming the reducing power of charcoal to be twice that of flour, we might substitute 1.2 grams charcoal for the 2.4 grams flour above.

No. 4 is a flux calculated on the lines of that given by C. H. Fulton[3] for the production of a " sesquisilicate " slag, with a charge of 0.5 assay ton. The amount of $NaHCO_3$ is doubled, and calculated to its equivalent of Na_2CO_3, and the litharge is calculated on the assumption that 76 grams are required for the slag and 22 grams to give a 20-gram lead button.

No. 5 is a flux recommended by G. B. Hogenraad[4] for the ore at Redjang Lebong, Sumatra, which is practically a free-milling ore, but contains small amounts of pyrites, manganese peroxide, copper, and selenium. Borax is omitted, as it was thought to carry part of the silver into the slag.

No. 6 is the flux adopted by the writer for free-milling ores of ordinary types. (See Section II, A.)

The following fluxes are adapted in a similar way for charges of 2 assay tons:

	Flux No. 1	No. 2	No. 3	No. 4	No. 5	No. 6
Carbonate of soda	67	67	78	73	57	83
Fused borax	30	20	12	–	8	24
Litharge	45	80	88	89.7	120	70
Charcoal	–	as required	–	0.3	2	–
Flour	–	–	4.7	–	–	2.2
Argol	2.5	–	–	–	–	–
Glass	–	20	–	–	–	–

The quantities are given in grams.

No. 1 is a flux formerly used in the assay office of the Robinson G. M. Company of Johannesburg; it was found to give lead buttons

[1] " Metallurgy of Gold on the Rand," 1898.

[2] " Text Book of assaying," 9th edition, p. 138.

[3] " A Manual of Fire Assaying," 1907, p. 66.

[4] " Journ. Chem., Met. and Min. Soc. of South Africa," VIII, 73 (1907).

of about 20 grams, with a clear green slag. The borax seems excessive.

No. 2 is given by B. W. Begeer[1] for tailings on the Rand.

No. 3 is adapted from a flux given by C. and J. Beringer,[2] the quantities there given for 50 grams being increased in the proportion 58⅓ : 50. We might substitute 2.35 grams of charcoal for the 4.7 grams of flour above given.

No. 4 is a flux recommended by G. B. Hogenraad for the 2 assay ton charges at Redjang Lebong. It is said to give lead buttons of 30 grams, with a clear green slag. With pure siliceous ores more charcoal would be necessary.

No. 5 adopted by the writer. (See Section II, A.)

No. 6 used in the assay office of Creston Colorada Co., Sonora, Mexico.

The following flux, suitable for cyanide tailings is adapted from one given by C. and J. Beringer,[3] for a charge of 3 assay tons.

Carbonate of soda .... 90 grams
Borax 22   "
Litharge 135   "
Charcoal as required.

The writer has used the following for charges of 4 assay tons, the fusion being made in a Battersea No. 12 fluxing pot.

Carbonate of soda .... 108 grams ⎫
Fused borax.............. 15.7   "   ⎬ Total, 267 grams
Litharge 141   "   ⎪
Charcoal................. 2.3   "   ⎭

Probably the total quantity of flux per charge might be increased with advantage, but in that case a larger crucible would be necessary.

General Remarks

With certain classes of ore the amount of litharge in the flux may be reduced to some extent, if the amount of soda be correspondingly increased, and vice versa. As already pointed out, carbonate of soda is for various reasons preferable to bicarbonate. (Section II, A.) An excessive quantity of borax should be avoided, as this gives rise to a very hard stony slag to which the lead adheres very tenaciously. On hammering such a slag it is liable to fly to pieces with explosive violence, carrying with it portions of the lead. There is also a possibility that some gold

---

[1] " Metallurgy of Gold on the Rand," 1898.
[2] " Text Book of Assaying," 9th edition, p, 138.
[3] *Ibid.*, p. 140.

and silver may be retained chemically in slags high in borax. With little or no borax the slag is usually thick and not very fusible.

Excess of soda gives a very deliquescent slag, which is liable to attack the crucibles.

Excess of charcoal gives a very infusible black slag, and may cause heavy losses in gold and silver.

The use of salt, sometimes recommended as a cover, is chiefly to moderate effervescence during the expulsion of CO_2 in the early stages of the fusion, and to prevent oxidation of lead by exposure to air. It is held by many assayers that the addition of salt causes losses, especially of silver, hence where such a cover is needed it is better to use borax. The writer sometimes uses a cover of soda.

Class II. — Basic Ores.

The general principle in assaying material of this class is to increase the borax and reducing agent, and to somewhat diminish the alkaline flux (soda, litharge). Ferric iron should be converted as much as possible into ferrous. For very basic ores it is advantageous to add silica or other acid flux in addition to borax.

The following fluxes are suggested for assays on 1 assay ton of ore:

	Flux. No. 1	No. 2
Carbonate of soda	18	20
Borax	12	20
Litharge	40	45
Charcoal	2.5	2

No. 1 is adapted from a flux given by E. A. Smith,[1] the litharge and charcoal being unchanged, but the soda and borax reduced proportionally to the amount of ore taken. No. 2 is adapted from a flux given by C. and J. Beringer (loc. cit., p. 139), substituting the equivalent of carbonate for the bicarbonate used by these authors.

For assays on 2 assay tons of ore:

	No. 1	No. 2	No. 3
Carbonate of soda	60	36	12–18
Borax	30	24	36
Litharge	45	40	50
Charcoal	–	2.5	4–6
Sand or glass	–	–	12–24
Argol	3	–	–

[1] "Assaying of Complex Gold Ores," in Trans. I. M. M., IX, p. 320 (1901).

Flux No. 1 was used on the Rand for chlorination tailings and similar material that was mainly siliceous, but contained some ferric oxide. No. 2 is adapted from a flux given by E. A. Smith (*loc. cit.*), the soda and borax being increased proportionally to the amount of ore, leaving the litharge and charcoal unchanged. No. 3 is similarly adapted from a flux given by Smith for very basic ore.

The following are adapted from C. and J. Beringer (*loc. cit.*, p. 139):

	No. 1	No. 2	No. 3
Charge of ore........	2.5 assay tons	50 grams	2 assay tons
Soda (carbonate)	50 grams	20–33 "	24–40 grams
Borax	75 "	30 "	36 "
Litharge	120 "	50 "	50 "
Silica	– "	10–25 "	12–30 "
Charcoal	4 "	– "	– "
Flour	– "	7 "	7 "

No. 1 is for low-grade material such as chlorination tailings; to be assayed in a large crucible (size I). No 2 is for ore consisting chiefly of hematite. It is very similar to the flux given by Smith for very basic ore, except as regards the amount of soda.

Some assayers use metallic iron with this class of ore, but it would seem to be unnecessary if sufficient reducing agent be used. Finely powdered gas-carbon has been recommended as a very efficient reducing agent; it has a higher reducing power, weight for weight, than charcoal, and being less rapidly oxidized acts for a longer time and at a higher temperature during the course of the fusion.

Class III. — Pyritic Ores

Methods of Assaying Pyritic Ores. — The principal methods used in assaying ores containing larger or smaller amounts of iron pyrites are:

(a) Roasting.

(b) Fusion with excess of litharge and little or no reducing agent.

(c) Fusion with metallic iron.

(d) Fusion with an oxidizing agent, generally niter.

These will be considered in the order given.

(a) *Assay by roasting.*— According to the richness of the ore, 1 or 2 assay tons are weighed out and placed in a shallow clay dish

4 to 5 in. in diameter and spread out in a thin layer. Rectangular dishes made to fit inside the muffle are sometimes used. For highly pyritic ores it is better to place a layer of silica (15–20 grams) on the dish before adding the ore. This serves to prevent the roasted material from adhering to the dish and also aids as a flux in the subsequent fusion. The roasting-dish is placed at the mouth of a large muffle, which is kept for some time at a very low heat. To prevent "caking," or the formation of hard lumps during this process, the contents of the dish are stirred occasionally, but not too much, with a thin iron rod, bent and flattened at the end. After some time, if much pyrites be present, a blue flame of burning sulphur will be observed over the dish. The temperature is raised gradually to dull redness and the dish pushed toward the back of the muffle. After all sulphur fumes have ceased and there are no perceptible sparks on stirring, the temperature may be further raised to complete the roast. The material when roasted "dead" should have a uniform appearance, and should emit no smell whatever of sulphur, etc. Arsenical and antimonial ores require special treatment in roasting, which will be described later. When the roasting is complete, the dish and contents are allowed to cool (under a cover, if left for some time), and the roasted ore swept off the dish into a mixture of suitable flux by means of a small flat brush. It is then well mixed with the flux and assayed in the ordinary way.

When roasted without silica, the ordinary pyritic ores, consisting mainly of siliceous materials and containing not more than 25–30 per cent. iron pyrites, may be assayed by one of the first four fluxes given for basic ores; if very pyritic (over 30 per cent. FeS_2) the flux for very basic ores or for hematite should be used for the roasted material, since in roasting the sulphides of iron are converted into oxides. The process takes place in several stages, but the ultimate result may be represented thus:

$$2FeS_2 + 11O = Fe_2O_3 + 4SO_2$$

When the roasting process is properly carried out, the heat kept low enough in the early stages and not hurried, and when the proper flux is used for the fusion of the roasted material, the results are generally at least as good as those obtained by any direct fusion method. However, the time occupied in the roasting, often an hour or more, is a disadvantage; hence the method of fusion with excess of litharge is to be preferred for slightly pyritic material.

Assay of Pyritic Concentrates by Roasting. — The method usually adopted by the writer for the assay of pyritic concentrates on the Rand was as follows: About ½ assay ton (say 15 grams) of siliceous sand, free from gold or silver, was spread out on a 5-in. roasting-dish, and 1 assay ton of the concentrates laid in a thin layer on the sand. This was roasted as above described, and when cool mixed thoroughly with

Carbonate of soda	60 grams
Borax	30 "
Litharge	50 "
Gas carbon	1.3 "

and fused. The carbon was varied, if necessary, so as to give a button of 18 to 25 grams.

(b) *Fusion with Excess of Litharge.* — The principle of this method is that lead oxide and pyrites act on one another when the former is in excess, with formation of oxides of iron and sulphur, metallic lead being liberated. The pyrites thus supplies the place of the whole or part of the reducing agent. In the flux used, rather more than the ordinary amount of litharge is taken; the charcoal is reduced as the amount of pyrites increases, and in some cases may be omitted altogether.

The following fluxes may be used for an ordinary type of slightly pyritic ore:

	No. 1	No. 2
Ore taken	1 assay ton	2 assay tons
Carbonate of soda	45 grams	90 grams
Borax	15 "	15 "
Litharge	50 "	70 "
Charcoal	0.8 to nil "	0.8 to nil "

Mitchell gives the following, which could probably be used for most ores of this class:

Ore	1 assay ton
Carbonate of soda	30 grams
Borax	30 "
Litharge	150 "
Salt cover	30 "

Argol, sufficient to give a button of 13 grams.
[A larger button would probably give better results.
— J. E. C.]

Limitations of the Method. — The method is not convenient for highly pyritic ores, as in such cases a large amount of litharge must be added to ensure having an excess, and consequently very large lead buttons are produced, which must be reduced subsequently by scorification. If the litharge be not in excess, the lead button will be brittle, consisting partly of lead sulphide, and more or less of the values will be lost.

(c) *Fusion with Metallic Iron.* — The principle of this method is to convert the pyrites into ferrous sulphide in accordance with the reaction:

$$FeS_2 + Fe = 2FeS$$

The ferrous sulphide passes into the slag without causing any considerable loss of precious metal, and a malleable lead button is formed. Large nails are sometimes used, but a better plan is to bend a piece of hoop-iron in U-shape; this is pushed into the charge in the crucible; when the fusion appears to be complete and the slag is quite fluid, the strip of iron is removed with the tongs, tapped once or twice to detach any adhering slag or lead, removed from the crucible, and the fusion continued about 5 minutes longer before pouring.

E. A. Smith[1] gives the following flux as suitable with this method of assay (adapted for assay of 2 assay tons):

Ore	2 assay tons
Carbonate of soda	48–60 grams
Borax	12–18 "
Red lead	48–60 "
Charcoal	2 to nil
Hoop-iron (thick)	2 or 3 pieces to be added

Aaron[2] gives the following:

Ore	1 assay ton
Carbonate of soda	90 grams
Borax	15 "
Litharge	30 "
Sulphur	3 "
Flour	3 "
Iron nails	3 to be added (with glass, if a more acid flux is needed, and a cover of salt).

[The object of the sulphur is not particularly obvious. — J. E. C.]

This method requires considerable care and experience for

[1] "Assaying of Complex Gold, Ores," in *Trans.* I. M. M., IX, p. 320.
[2] Rose, "Metallurgy of Gold," 4th edition, p. 469.

satisfactory results. The strips of iron when withdrawn from the crucibles are apt to carry small adhering shots of lead, and should be carefully examined.

(d) *Assay with Niter.* — In this method it is presupposed that the quantity of pyrites in the ore is more than sufficient to bring down the required amount of lead. The surplus lead is oxidized by the addition to the flux of a carefully adjusted quantity of niter. [It is more probable that the niter attacks the iron pyrites rather than the lead, as the slag in this process invariably contains potassium sulphate.] An excess of niter causes heavy losses of precious metals, and sometimes causes corrosion of the crucible itself; hence when this method is tried on an unknown ore, a *trial fusion* is always made to determine the "reducing power" of the ore. In such a case, some such mixture as the following is made:

Ore .... ¼ assay ton (7.292 grams)
Litharge .... 60 grams
Carbonate of soda .... 4 "

Miller, Hall, and Falk[1] give the following:

Ore .... 3 grams
Soda bicarbonate .... 10 "
Litharge .... 50 "
Salt cover

The mixture is fused at a bright red heat for from 10 to 15 minutes, then poured, and the resulting lead button weighed. The weight of this button is a measure of the reducing power of the ore. It is generally assumed that 1 gram of niter will oxidize 4 grams of lead. E. A. Smith (*loc. cit.*, p. 335) says 4 to 5 grams, but the amount varies with the nature of the ore, and may range from 3.2 to 5.3.

Formula for Niter Required. — The following is a general formula which can be applied in all cases for calculating the amount of niter required:

Let a = weight of ore taken for trial fusion.
 " A = weight of ore taken for final assay.
 " w = weight of lead button from trial fusion.
 " W = weight of lead button required.
 " n = number of grams lead oxidized by one gram of niter.
 " x = weight of niter required.

Then

$$x = \frac{Aw - Wa}{na}$$

[1] *Trans.* A. I. M. E., XXXIV, p. 387.

In the special case where 1 gram niter oxidizes 4 grams lead, and ¼ assay ton is taken for trial fusion, 1 assay ton for final fusion, and a button of 20 grams is required, we have

$$x = w - 5$$

Oxidizing Power of Niter on Various Minerals. — Smith (*loc. cit.*, p. 327) gives the following table to determine the oxidizing effect of niter, when the nature and amount of the oxidizable material is known:

TABLE VI. — OXIDIZING EFFECT OF NITER

	Parts Niter Required to 1 Part of "Sulphuret"
Iron pyrites......................	2-2½ parts
Copper pyrite, fahlerz, or zinc-blende	1½-2 "
Antimonite.................	1½ "
Galena...........................	¾ "

Final Fusion. — For the final fusion, as a rule, a flux is used containing much litharge and little soda, together with the indicated amount of niter and a cover of salt to modify the effervescence. The violent action is a frequent cause of loss in assays with niter. The following may be given as examples of fluxes for the final fusion:

	No. 1	No. 2
Ore.................	2 assay tons	½ assay ton
Carbonate of soda	30 grams	15 grams
Borax	12 "	– "
Litharge	—	70 "
Red lead	120 "	– "
Niter as indicated.		
Salt cover.		

No. 1 is given by E. A. Smith (*loc. cit.*). No. 2 is given by Miller, Hall, and Falk (*loc. cit.*). These authors found that the use of borax was objectionable, giving an excessively hard and stony slag, with hard brittle lead buttons. [This would probably apply only to certain classes of ore. — J. E. C.] They summarized their experience of the process as follows:

" It is necessary to determine the oxidizing power of niter with that substance and charge with which it is subsequently used; the niter method, as modified, gives accurate results and is neither as long nor as tedious as the roasting method; it also gives a charge which does not boil over, and yields a lead button agree-ing closely with the calculated weight, and a clean slag."

Another method of applying niter in assaying will be described under "Antimonial Ores" (Class VI).

Class IV. — Lead-Zinc Ores

Ores containing galena and zinc-blende are best assayed by fusion with metallic iron, as already described under "Pyritic Ores" (Class III, c). Where galena predominates, the amount of litharge may be reduced; when the zinc mineral predominates, the amount of soda and borax should be increased. One of the two fluxes given above will probably meet most cases.

Hall and Popper[1] give the following as suitable to certain kinds of zinc ore (sphalerite):

Ore	....	$\frac{1}{3}$ assay ton, say 10 grams		
Carbonate of soda	.....1$\frac{1}{3}$	"	" 40	"
Borax glass	.... $\frac{1}{2}$	"	" 15	"
Litharge	.... $\frac{4}{5}$	"	" 24	"
Argol as required.				

Add iron if more than 15 per cent. pyrites is present. [The amount of soda here seems excessive, and in fact the authors remark that the silica of the crucible appeared to be attacked, and the charge was inclined to boil excessively. — J. E. C.]

Roasting may be employed when the ore is chiefly or exclusively zinc-blende, but is unsatisfactory with any but very small quantities of galena.

Class V. — Arsenical Ores

Mode of Occurrence. — The arsenic in gold ores usually occurs as mispickel (FeAsS), occasionally as arsenide of nickel or as realgar and orpiment (sulphides of arsenic).

Roasting with Charcoal. — The best method is to first roast at a low temperature, eliminating part of the arsenic, then allow to cool somewhat; add 5 to 10 per cent. of powdered charcoal, stirring well, to mix with the charge on the roasting-dish, and continue heating at a somewhat higher temperature until a dead roast is obtained. The roasted material is then fluxed as described under basic ores, or pyritic ores after roasting.

[1] "Sch. Mines Quart.," XXV, p. 355 (1904).

Theory of Method. — The object of adding charcoal is to de-compose the arseniates formed in the first stage of roasting. These compounds, if not destroyed, would carry gold and silver into the slag. It has been stated[1] that the addition of a small quantity of potassium cyanide to the flux facilitates the decomposition of any arseniates remaining in the roasted product. It is not advisable to assay ores of this class by means of niter, as this tends to form arseniates.

Class VI. — Antimonial Ores

Various Methods for Antimonial Ores. — The methods usually adopted for assaying ores of this class may be divided as follows:

(*a*) Fusion with niter.

(*b*) Preliminary oxidation with niter or other oxidizer, followed by fusion with a reducing flux.

(*c*) Preliminary acid treatment, to eliminate as much as possible of the antimony, followed by scorification or fusion of the residue.

(*a*) *Fusion with Niter.* — The details of this method have already been given under "Pyritic Ores" (Class III, *d*). E. A. Smith (*loc. cit.*, p. 334) gives the following flux for an ore containing 75 per cent. of stibnite (here adapted for assay on 2 assay tons):

Ore	2 assay tons
Carbonate of soda	50 grams
Borax	12 "
Red lead	120 "
Niter	36–48 "
Salt cover.	

A large crucible must be used, raising the temperature very gradually at first. The slag is very fluid. "When effervescence has ceased and the contents of the crucible have become thoroughly liquid and tranquil, it is advisable to lift the crucible out of the fire and subject it to a rotary motion, in order to thoroughly mix the liquid contents. If this causes effervescence it indicates that the action of the niter is not complete, and the crucible must be returned to the fire and the fusion continued" (Smith).

(*b*) *Preliminary Oxidation with Niter.* — This method requires some previous knowledge of the nature of the ore. The charge of ore is placed in a crucible with niter equal to twice the weight

[1] E. A. Smith, *Ibid.*, p. 332.

of the pyritic material contained in the ore, together with soda and borax, say for 2 assay tons of ore: soda 50 grams, borax 10 to 15 grams, niter (as required) up to 80 or 90 grams. When the action appears to be complete and the fusion is tranquil, the crucible is withdrawn from the fire and allowed to cool until the slag begins to thicken. The following mixture is then added, and fused for 10 to 15 minutes:

Borax ....10 grams
Red lead ....40 "
Charcoal .... 2 "

Several variations of this method have been used, and a somewhat similar method has been proposed by K. Sander for assaying carbonaceous matter containing lead, in which the preliminary oxidation is made with a mixture of niter and sodium peroxide. (See below.)

(c) *Preliminary Acid Treatment.* — The following combined wet and dry method depends on the fact that sulphide of antimony is soluble in strong hydrochloric acid. The native oxides are, however, only partially soluble. "The weighed quantity of ore (30 to 50 grams) is treated with concentrated hydrochloric acid and heated until decomposition is complete; the solution is diluted with a small quantity of water containing a little tartaric acid to prevent precipitation of antimony as oxychloride. The insoluble residue is allowed to settle, and when the solution is clear as much of the liquid as possible is poured off through a filter, without disturbing the residue. The latter is finally transferred to the filter, allowed to drain, and dried. The filter-paper is burnt, and the ash, with the insoluble residue, is scorified with ten times its weight of grain-lead and a borax cover, or fused with the following flux:" E. A. Smith (*loc. cit.*).

Soda carbonate and borax together............30 grams
Red lead ....30 "
Charcoal ....1.5 "

Class VII. — Cupriferous Ores

Methods of Assaying. — The methods usually applied for this class of ore are:

(a) Preliminary acid treatment to remove the bulk of the copper, followed by fusion or scorification of the residue.

(b) Fusion with a large excess of lead oxide, using charcoal or niter according to the nature of the ore.

(a) *Assay with Preliminary Acid Treatment.* — Dr. J. Percy[1] gives the following method (grains are here expressed in the nearest equivalent in grams):

"Six or seven grams of ore are heated with nitric acid, or a mixture of nitric and sulphuric acids. The liquid is then diluted, a little hydrochloric acid or salt added to precipitate silver, after which the solution is filtered. The filtrate is clear, contains most of the copper, and may be rejected. The residue is dried and fused with

Soda and borax ....20 grams
Red lead 20   "
Charcoal 1   "

for 10 to 15 minutes."

In a modification of this method given by E. A. Smith (*loc. cit.*), 30 to 50 grams of the ore are treated with fuming nitric acid till decomposed, heated to expel nitrous acid, then diluted and mixed with a little HCl or salt to precipitate silver. The ore is then stirred, settled, and decanted through a filter. The dried residue, with the filter-paper, is fluxed with

Soda and borax (together) ....30 grams
Red lead 30   "
Charcoal ....1.5   "

or it may be scorified with ten times its weight of lead.

The writer's experience of this method is that the filtration is very tedious unless a long time be allowed for settlement, as with most ores a considerable portion is crushed fine enough to produce slimes, which soon choke the filter. Moreover, the method for precipitating the silver is certainly faulty, for if insufficient HCl or salt be added, the silver will not be completely precipitated, whereas an excess of either, together with the nitric acid present, would dissolve gold. A better plan would be to filter without addition of chlorides and precipitate or titrate the silver separately in the filtrate; or make an entirely independent assay of it by scorification or some other method.

A similar method, due to C. Whitehead,[2] in which the silver

---

[1] "Metallurgy: Silver and Gold," Part I, p. 115.
[2] See R. W. Lodge in *Trans.* A. I. M. E., XXXIV, p. 432.

in the filtrate is precipitated by potassium bromide, was found by Lodge to give low results.

(b) *Direct Fusion with Excess of Litharge.* — In all cases where cupriferous ores are fused without preliminary acid treatment, a large quantity of lead oxide must be used, and the resulting large lead button reduced by scorification before cupeling. When the buttons are hard and have a coppery appearance, they should be still further scorified, with addition of more lead. According to Percy (*loc. cit.*) one part by weight of copper requires about 16 parts of lead for cupellation, and although a smaller quantity of lead will pass the copper into the cupel, yet in that case some considerable amount of gold and silver is also absorbed by the cupel.

The three following fluxes (adapted from Percy) are given by E. A. Smith (*loc. cit.*) for ores containing various amounts of copper (the weights in grams):

	No. 1	No. 2	No. 3
Ore.....................	10–50	10–50	1 assay ton
Soda Carbonate...........	—	20–30	120
Borax	—	15–20	—
Silica	10–25	—	—
Red lead................	200–300	100	150
Charcoal	1.5–3.5	3.5	—
Niter	—	—	18

No. 1 is for ores containing only moderate amounts of copper; No. 2 for ores after roasting, No. 3 for gray antimonial-copper ore containing say 30 per cent. Cu and 150 oz. Ag per ton.

The quantities of charcoal and niter would of course be varied according to circumstances.

Class VIII. — Telluride Ores.

Points to be Observed in Assaying Telluride Ores. — An excellent article on the assaying of gold telluride ores, by W. F. Hillebrand and E. T. Allen, will be found in Bulletin No. 253 of the United States Geological Survey (1905). The essential points in assaying these ores appear to be: (a) Very fine grinding of the sample, say to 150- or 200-mesh; (b) considerable excess of lead oxide; (c) avoidance of roasting, scorification, or any oxidation process, either for the ore or the lead buttons; hence the button produced should not be larger than can be cupeled direct.

Flux for Cripple Creek Tellurides. — Hillebrand and Allen found the following flux to give very satisfactory results on Cripple Creek telluride ores, using a Battersea "F" crucible for the fusion:

Ore	1 assay ton
Bicarbonate of soda	30 grams
Fused borax	10 "
Litharge	120–180 grams
Salt cover.	

This gave lead buttons of about 25 grams weight, without addition of iron or niter. The losses in the slag were very slight; those by absorption in the cupels were more considerable and were corrected by a subsequent assay of the cupels.

Precautions. — With ores of this class great care is necessary in moderating the heat in the early stages of the fusion, and with an ore not previously tested it would always be as well to remelt the slag with fresh flux. In the case of rich tellurides it is better to reduce the charge of ore rather than increase the litharge, so as to avoid producing an unduly large lead button. The losses in oxidizing telluride ores or lead buttons containing tellurium appear to be due to the volatilization of tellurium trioxide, which carries off gold mechanically. Smith (*loc. cit.*, p. 344) states that in presence of lead oxide tellurium is readily converted into telluric acid, which combines with the lead oxide when the latter is in excess:

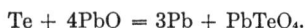

$$Te + 4PbO = 3Pb + PbTeO_4.$$

but if the contrary be the case, the excess of telluric acid is volatilized and telluride of lead is produced:

$$2Te + 3PbO = TePb_3 + TeO_3.$$

Fluxes for Telluride Ores. — Smith gives the following fluxes (adapted):

	No. 1 (for ordinary assay)	No. 2 (for very rich ores)
Ore	2 assay tons	$\frac{1}{3}$ assay ton
Carbonate of soda	60 grams	50 grams
Borax	24 "	— "
Siliceous sand	12–36 "	5–10 "
Red lead	50–200 "	100–150 "
Charcoal	1–2 "	1 "
Salt cover.		

The red lead is varied according to the tellurium present, and charcoal according to the reducing power of the ore. The charcoal may be omitted in some cases.

Combined Wet and Dry Assay. — The following method is given by Hillebrand and Allen (*loc. cit.*). "Fifty-gram portions were mixed with water in large porcelain basins, and strong nitric acid was added by degrees, with constant stirring, till the action had nearly ceased; then bromine water was added to oxidize the gold and sulphur and precipitate the silver, no artificial heat being used." After thorough settling, the residue was filtered and washed repeatedly with water containing Br and HCl, and then washed two or three times with pure water. The combined filtrates were evaporated in the original basins to near dryness, then treated with HCl and covered for a time. When action ceased the covers were removed and evaporation continued, HCl being added two or three times more at intervals. The residues were now digested with dilute HCl and a little bromine water to insure solution of the gold. Considerable calcium sulphate remained undissolved, which was filtered off and found to be practically free from gold. The filtrate was precipitated with ferrous sulphate, allowed to stand for 24 hours, and the precipitated gold collected on a filter, washed, dried, weighed, cupeled, and again weighed (using a check of proof gold for correction). The gold thus thrown down was found to be free from tellurium.

The residue, containing the silver and a small quantity of gold not extracted by the washing, was dried and assayed with the following flux, yielding a lead button of 20 grams:

Carbonate of soda	30 grams
Borax	10 "
Litharge	60 "
Argol	2 "
Salt cover.	

(C) Assay of Certain Refractory Materials

Among the products which the cyanide chemist may occasionally be called upon to assay may be mentioned zinc-gold precipitate (with or without roasting or acid treatment), slags from fusion of zinc precipitate, etc., slags and cupels from the assay of different classes of ore (when exact results are required on rich material), old crucibles, cupel and furnace bottoms, battery screening, scalings from plates, prussian blue, ferric hydrate and other deposits from tanks or precipitation boxes, residues from smelting retorts, sweep from smelting room, old sacking and filter-cloths, and other matters of very varied description. Of course, no general rule can be given for the treatment of such material, but a few remarks may be of use:

Where the material does not admit of fine subdivision no reliable assay can be made, unless a very large sample be taken and the whole of it weighed and subjected to some process for reducing the bulk without loss of values. Thus, samples consisting largely of metal may often be treated by dissolving in some suitable acid, depending on the nature of the material; an assay is then made by scorification or otherwise on the dried residue insoluble in that acid; the residue, after acid treatment, is of course weighed again, unless the whole of it can be dealt with in one operation for extraction of its values. In other cases, where a large part of the sample is inflammable, the whole may be weighed, and then roasted or burnt with suitable precautions and an assay made on the ash.

Samples Containing Metallics. — Material of this class frequently contains metallic particles, such as shots or flakes of bullion, of totally different and usually much higher assay value than the remainder. In all such cases the entire sample should be first weighed. It is then crushed, and as much as possible passed through a sufficiently fine sieve. The portion which passes the sieve, and the metallic and other particles remaining on the sieve, are then weighed and assayed separately.

Formulæ for Assays with Metallics. — The following formula may be used in the ordinary case, where the whole of the metallic portion is treated to recover its entire gold or silver contents, and an assay made in the usual way on the fine portion.

Let M = total weight of metallics in grams.
   F = total weight of fines in grams.
   m = weight in mg. of the metal sought (gold or silver) contained in M.
   a = assay of metallics (in dwt. per ton of 2000 lb.)
   b = assay of fines (in dwt. per ton of 2000 lb.)
   x = assay of entire sample (in dwt. per ton of 2000 lb.)

Then $x = \dfrac{12\,Fb + 7000m}{12\,(M + F)} = \dfrac{Fb + 1750\dfrac{m}{3}}{M + F}$

In the case where an assay is made on a part only of each product,

$$x = \frac{Ma + Fb}{M + F}$$

Some of the substances above mentioned are similar in composition to one or other of the classes of ore previously discussed, and may be assayed by the methods of crucible fusion already given, but scorification is more generally applicable.

Assay of Slags. — Slags (from melting of zinc-gold precipitate, etc.) usually require a considerable amount of litharge and reducer, so as to give a large lead button. Occasionally they are mixed with so much carbonaceous matter that niter must be used instead of reducing agent. Roasting cannot be used for this class of material, as it would fuse and attack the silica of the dish, forming a crust which would not afterwards be separated without loss.

Slags from Assay Fusions frequently have to be remelted with fresh flux to recover the values passing into them in the first fusion. It is usually sufficient to powder the slag coarsely and return it to the original crucible, together with about 50 grams of the ordinary assay flux. This is fused for 10 to 15 minutes, and the small resulting lead button cupeled with the original button, or preferably on a separate cupel, so that the amount recovered from the slag may be noted. The following fluxes may be mentioned, given by different authorities as suitable for remelting the assay slags from various kinds of material:

No. 1.   E. A. Smith (*loc. cit.*, p. 322):

          Soda carbonate .... 10 grams
          Litharge 30   "
          Charcoal ....1.5   "
          Borax added if mixture does not fuse quickly.

No. 2.   Adapted from Percy, *loc. cit.* (fuse for 15 to 20 minutes)

          Soda carbonate .... 2–3.5 grams
          Red lead ....20– 33   "
          Charcoal 1–2   "

No. 3.   For rich silver-lead ore — cerussite (Miller and Fulton[1]). Slag is first remelted with litharge, 1 assay ton, argol, 2 grams, and then the large lead button resulting is scorified. The slag from this scorification is further run down with

[1] " Sch. Mines Quart.," XVII, p. 160.

Carbonate of soda ....1 assay ton
Silica ....½   "   "
Litharge 1   "   "
Argol ....2 grams

No. 4.   For remelting slags from assay of telluride ores (Hille‑
brand and Allen, *loc. cit.*):

Litharge ....1 assay ton
Argol.............................2 grams
Salt cover.

Assay of Bone-ash Cupels.—It is usually only necessary to test
the portion of the cupel which has been stained by absorbed
litharge.   The remainder is scraped off and rejected, and the
portion for assay finely powdered (at least to 80-mesh).   T. Kirke
Rose[1] gives the following as a suitable flux:

Ground cupels ....100 parts
Carbonate of soda 100   "
Borax 50   "
Sand 75   "
Fluor-spar 75   "
Litharge 50   "
Charcoal   4   "

Percy (*loc. cit.*) gives:

Carbonate of soda ....7-17 grams
Borax 7-17   "
Fluor-spar 7-17   "

Charcoal, 4 to 5 parts for every part of litharge in the cupel; if the latter
is insufficient, add red lead, 20 grams, charcoal, 1 gram.
Instead of borax, glass free from lead may be used.

The two following fluxes also are given by (1) Hillebrand and
Allen, and (2) Miller and Fulton:

	No. 1	No. 2
Carbonate of soda	30 grams	30 grams
Borax	45 "	30 "
Litharge	60 "	60 "
Argol	2 "	2 "
Cover of salt.		

For cupels which have been used in the assay of zinc-gold
precipitate the following are given by (1) R. W. Lodge (*Trans.
A. I. M. E.*, XXXIV, p. 432), and (2) W. Magenau ("Min. and
Sci. Press," LXXX, p. 464).   The quantities are given in grams:

[1] "Metallurgy of Gold," 4th edition, p. 481.

	No. 1	No. 2
Cupels	30–40	–
Carbonate of soda	15	40
Borax	10	10
Silica	–	10
Glass	15	–
Litharge	60	20
Argol	2	–
Flour	–	1
Cover of	–	borax

Graphite Crucibles. — A method for assaying graphite crucibles by scorification has already been given in Section III. The following methods of crucible fusion may here be noted.

R. Dures,[1] after sifting out all shots, etc., from the finely-ground material, fluxes the remainder thus:

Finely ground crucible	5 grams
Carbonate of soda	25 "
Borax	15 "
Silica	10 "
Litharge	85 "
Niter	7 "

A. F. Crosse[2] gives a method of assaying this material by preliminary oxidation with manganese peroxide, taking

Powdered crucibles	10 grams
Manganese dioxide	35 "

After heating this mixture to bright redness in an " H " crucible, the temperature is reduced, and the following flux added:

Potassium carbonate	50 grams
Borax	25 "
Salt	25 "
Litharge	50 "
Silica	20 "
Flour	2 "

In some cases the manganese dioxide may be omitted and the material given a preliminary roast with silica, using

Powdered crucibles	1 assay ton
Quartz	2 assay tons

Other assayers, however, obtained very low results by Crosse's method, and Dr. J. Loevy[3] recommends scorifying

[1] " Journ. Chem., Met. and Min. Soc. of South Africa," II, p. 577.
[2] " Proc. Chem., Met. and Min. Soc. of South Africa," II, p. 124.
[3] " Proc. Chem., Met. and Min. Soc. of South Africa," II, p. 205.

Powdered crucibles .... 5 grams
{ Lead ....60 "
{ or Litharge...........................70 "

with a small quantity of borax or glass.

Battery Scrapings (and other coppery material of similar nature). — R. Dures (*loc. cit.*) prefers scorification to nitric acid treatment for this class of material, on the ground that the nitrous acid always present in the commercial acid causes loss of gold. (See Section II, *b*.) Williams and Crosse, however, advocate acid treatment, and the latter adds lead acetate and sulphuric acid to form a precipitate of lead sulphate, which helps to carry down the gold.

Zinc-Box Precipitate. — No. 1. The following scorification charge is given by R. W. Lodge[1]:

Precipitate .... 0.05 assay ton
Grain lead .... 65 grams
Borax glass...................... 10 "

Of the 65 grams of lead, 35 grams are to be mixed with the charge and the remaining 30 grams are to be used as a cover. Use a 3-in. to 4-in. scorifier. The slag from this to be remelted with

Soda carbonate ....15 grams
Glass ....10 "
Litharge ....30 "
Argol .... 2 "

and the cupels assayed as described above.

No. 2. C. H. Fulton and C. H. Crawford[2] give the following:

Precipitate.................0.05 assay ton
Grain lead .... 70 grams
Litharge................... 15 " } added as cover
Borax .... 1 " }

But these writers prefer a "wet and dry" method, as follows:

No. 3. One-tenth assay ton of the precipitate is boiled with a mixture of 20 c.c. conc. H_2SO_4 and 60 c.c. water; then 75 cc. normal NaCl and 20 c.c. lead acetate solution (strength not given) are added. The precipitate is settled for an hour, filtered, washed, and dried, care being taken to wash into the point of the filter. The paper is then burnt at a low heat and the whole residue

[1] *Trans.* A. I. M. E , XXXIV, p. 432 (October, 1903).
[2] "Sch. Mines Quart.," XXII, p. 153 (1901).

scorified with 30 to 40 grams of lead. This gives lower gold results but more silver than the following:

No. 4. W. Magenau's [1] method for crucible fusion:

Zinc precipitate....	$\frac{1}{10}$ assay ton	Litharge..............	70 grams
Dry soda carbonate	5 grams	Flour................	1 "
Borax glass........	2 "	Salt cover.	
Silica	5 "		

Carbonaceous Residues (left in retorts after distillation of Zinc).— K. Sander [2] gives a method for assaying the residue remaining in the muffles after zinc distillation. This residue contains lead and a large amount of carbon, and offers difficulties in the ordinary scorification and crucible assays. He gives a preliminary oxidation as follows: The mixture

Powdered residue......................	20 grams	
Potassium nitrate......................	40	"
Sodium peroxide	10	"

is mixed with an iron spatula and gradually introduced into a red-hot iron crucible. After the violent action has ceased, the following flux is added:

Soda.................	14 parts	Litharge..............	10 parts
Calcined borax........	8 "	Argol	2 "

Pour when tranquil. The results quoted, however, are in general lower than those obtained by scorification.

Assay of Silver Ores from Cobalt. — The following flux is given by J. O. Handy (Intern. Congress of Applied Chem.) for the assay of ordinary Cobalt ores (containing As, Cu, Co, Ni, etc.).

Sodium carbonate	12 gm.	Flour................	1 gm.
Potassium carbonate...	12 "	Ore	0.2 A.T.
Borax.	6 "	Borax cover	8 gm.
Litharge	60 "		

A 30-gram crucible is used; heated moderately for 20 min. in closed muffle, then at bright yellow heat for 30 min.

Cupelled at low temperature. Metallics are scorified with addition of a little borax and ground silica.

High-grade ores are treated in the same way except that more litharge (90 grams) is used, and the metallics are assayed by crucible fusion.

[1] " Min. and Sci. Press," LXXX, p. 464 (1900).
[2] " Zeit. für Angew. Chem.," XV, p. 32 (1902).

SECTION IV

ASSAY OF GOLD AND SILVER BULLION

(A) Fire Assay Method

Sampling: Dip-samples. — The generally accepted opinion is that the most accurate method of sampling bullion is by means of what is known as a "dip-sample." This is taken from the crucible or smelting-furnace while the metal is in a molten condition, immediately before pouring. The bath of metal is first well stirred with an iron bar (rabbling-iron) somewhat flattened at the end, and a small portion of the metal is taken out by means of a plumbago spoon, held by a pair of tongs. The dipper used is often a section cut from an old graphite crucible. When the dip-sample is poured there should be a thin layer of slag over the molten metal to prevent oxidation. The dip-sample may be either poured into a small mold and subsequently flattened and cut into pieces, or it may be granulated by pouring into water.

Drill-Samples. — Samples are also very frequently taken by drilling one or more holes in the bars after casting. These "drill-samples" are usually taken by means of a small vertical machine drill, which is caused to revolve with considerable speed, cutting out fine turnings of the metal. A common plan is to take one drilling near a corner of the bar and the other at the middle of the opposite side, each hole passing half-way through the bar. In bars containing much base metal, there is often a marked difference in the composition of the central and superficial portions, owing to the phenomenon of liquation which occurs during the cooling of the metal.

Cut-Samples. — Samples are also sometimes taken by chipping fragments from the corners or edges of bars, but this procedure can only give correct results when no liquation has occurred. In general it is not to be recommended. The results obtained by Stockhausen[1] showed that drill-samples gave a higher, and cut-

[1] " Proc. Chem., Met. and Min. Soc. of South Africa," II, pp. 46–48.

448

samples from the edges and corners a lower, result than dip-samples from the same bars, and also that the cut-samples exhibit much greater fluctuations than either.

Preparation of Samples. — The samples must be carefully examined before weighing, and any particles of slag or other extraneous matter which may have been accidentally mixed with them are removed. Occasionally it may be necessary to cut them up finer; this may be done by hammering the fragments and cutting with a clean pair of scissors; when the material is brittle, as with some qualities of cyanide bullion, it may be further reduced by grinding in an agate mortar.

Weighing-in. — The sample is next spread out on a clean sheet of paper and the charge weighed. This charge is usually 500 mg., and the bullion taken may be either adjusted to this exact amount, by clipping or filing the fragments, or an approximate amount may be taken and its exact weight determined and recorded. With a good assay balance the correct weight may be determined within 0.01 mg., but this degree of accuracy is commonly not required; for ordinary purposes it is sufficient to weigh to 0.05 mg.

Addition of Lead. — When the portion of bullion for assay has been weighed it is at once transferred by means of a fine brush to a small capsule of lead-foil. The weight of lead to be used depends largely on the nature of the impurities contained in the bullion, as does also the question whether scorification should be employed or not. In an ordinary case about four times the weight of the bullion taken (*i.e.*, about 2 grams of lead-foil) will suffice. With certain kinds of base bullion, particularly such as contain arsenic, antimony, selenium or tellurium, a much larger amount may be necessary, say 15 grams.

Additions of Copper and Silver. — When the bullion contains no copper, it is advisable to add a small amount, say about 20 to 30 mg. in each assay, which prevents "spitting" after the cupellation. When the amount of silver is insufficient for parting the gold (*i.e.*, less than about two and a quarter times the weight of gold present), it is necessary to add a sufficient quantity of silver to the assay to make up this difference. When the bullion is to be assayed for both silver and gold, the most satisfactory method is to make these determinations on separate portions. In many cases, however, sufficiently exact results are obtained by re-cupeling the bead of mixed metal, with the addition of the requisite amount of fine silver and sufficient lead-foil. It is necessary to ascertain that the lead-foil used is free from gold, and its silver contents, if any, must be determined and allowed for.

Checks. — With every batch of bullion assays one or more checks are added. These are made by weighing out exact quantities of pure gold and silver corresponding approximately to the amounts expected to occur in the actual bullion assays. When these are unknown, the approximate composition of the bullion is determined by a preliminary trial. The gold is first adjusted by cutting pieces from a thin sheet of proof gold, to within 0.5 mg. of the theoretical amount; the weight of this quantity is then determined as accurately as possible, say to 0.01 mg. It is then transferred to a small lead capsule (a sheet of lead-foil coiled into a small conical pocket answers the purpose). The amount of silver judged to be present in 500 mg. of the bullion is then weighed out, together with an additional quantity to allow for volatilization and absorption in the muffle; this extra amount depends upon the position the assays are to occupy in cupellation and the temperature at which this operation is to be carried out. The silver is adjusted to within 2 mg. of the amount theoretically required, the exact weight being then determined to within 0.02 mg. It is then added to the capsule containing the gold. It is usual to add also a small quantity of copper, and it is desirable, at least occasionally, to ascertain the effect of adding other ingredients corresponding to the amounts present in the bars to be assayed. In an ordinary case 15 to 20 mg. of copper may be added.

Losses of Silver in Cupellation. — A series of tests were made by the writer on known weights of gold and silver, approximately 400 mg. Ag to 60 mg. Au, with about 15 grams lead-foil in each case, cupeling in an ordinary fire-clay muffle, size " J," with charcoal as fuel, and using 4 cross-rows of 3 cupels each. The extra silver required was shown by these tests to be approximately as follows:

Cupels used Size	Morganite. No. 5 Loss of Silver Mg.	Bone-ash. No. 5 Loss of Silver Mg.
First row (at back)................	6–8	11–14
Second row......................	5–7	9–12
Third row.......................	4–6	7–10
Fourth row (in front).............	3–5	5–8

(See remarks on this subject in Section II, *a*.)

Position of Checks. — The lead packets containing the weighed bullion or check pieces are rolled up and placed in their proper positions in a suitable tray. Generally, two checks with each muffle charge will be sufficient, but where, as in the experiment quoted above, the differences of temperature in different parts of the muffle are considerable, it is as well to use a check with each row, the two remaining places in the row being occupied by duplicates of the same assay sample. In order to equalize the effects of any possible differences of temperature in various parts of the muffle, the position of the checks may be altered with each batch of assays, as shown in the following arrangements, where the numbers 1, 2, etc., represent the bar assays, and the letters *a*, *b*, *c*, *d* the checks:

No. 1			No. 2			No. 3			No. 4		
a	1	1	1	*a*	1	1	1	*a*	1	*a*	1
2	*b*	2	*b*	2	2	2	*b*	2	2	2	*b*
3	3	*c*	3	*c*	3	*c*	3	3	3	*c*	3
4	*d*	4	4	4	*d*	4	*d*	4	*d*	4	4

In an ordinary muffle-furnace it will be found that the temperature is lower and the losses of silver therefore less in the central line of the muffle than at the sides, and the back rows are also considerably hotter than the front.

Cupellation. — The cupels are heated for some time in the muffle before introducing the assay pieces. Claudet[1] recommends baking in the furnace for about an hour, but in most cases a shorter time would seem to be sufficient. The safest plan is to introduce a small piece of pure lead into each, when the cupels appear to be hot enough. When this has melted and begins to cupel freely the assay pieces and checks may be transferred by means of cupel-tongs from the tray to the corresponding cupels in the muffle, each packet being dropped carefully into its proper cupel, where it is at once absorbed in the bath of molten lead. The furnace, where coke or charcoal is used, should be well filled up with fuel, and a moderate red heat maintained throughout the operation. Towards the end the door should be closed as a somewhat higher temperature is required. To secure greater uniformity of temperature, a block of red-hot coke or charcoal

[1] Inst. Min. and Met. Bull No. 27, p. 16 (Dec. 13, 1906).

is sometimes placed at the mouth of the muffle. Another plan is to surround the assay pieces by rows of empty cupels, so that the actual cupellation is carried on only in the central part of the muffle. The time occupied in cupellation will vary from ten minutes to half an hour or more, according to the weight of lead and the temperature at which the operation is conducted. When approaching the finish, a play of iridescent colors is noticed on the surface of the beads; these are interference colors produced by the very thin transparent film of litharge surrounding the beads. The cupels are then withdrawn in their proper order and placed for a few minutes at the mouth of the muffle until the beads have solidified. This is accompanied by a sudden flash of light.

If the cooling has been too sudden, this process may be accompanied by violent projection of particles of silver, by "sprouting" or "vegetating" (i.e., the formation of irregular processes projecting from the surface of the bead). These effects are generally ascribed to the sudden ejection of occluded oxygen absorbed by the molten silver during the cupellation. When the proper conditions are carefully attended to, this will not occur, and the beads when withdrawn from the muffle will show a perfectly smooth, bright surface.

The sizes of cupels from No. 3 to No. 5 are most commonly used for bullion work. Various materials are used, several substitutes for bone-ash having been extensively adopted of late years.

Scorification. — Where the bullion is very base, and especially when it contains ingredients which would be liable to crack or corrode the cupels, a preliminary operation is resorted to. Instead of transferring the lead packets direct to the cupels, they are placed in small previously heated scorifiers (1¼ in. to 1½ in. diameter), with the addition of a very small quantity (say 0.1 gram) of borax. Oxidation of the base metals then takes place, and the molten oxides form fusible silicates with the material of the scorifier itself, producing a slag which floats on the surface of the molten lead. The details of the process have been already described in Section III (a). When the operation is complete, the scorifiers are taken out and their contents poured into small molds previously warmed. The clean lead button is detached when sufficiently cool, and cupeled in the ordinary way. The slag should be uniform; when chiefly consisting of lead silicate, it has a yellowish-brown color, but may be stained dark brown by iron, or green by copper, manganese, etc.

Cleaning and Flattening the Beads. — The beads are detached from the cupels by means of pliers and squeezed with some force to loosen any adhering matter. A convenient method of cleaning the bead is to hold it edgewise with a small pair of steel forceps, striking it once or twice on the edge with a smooth-faced hammer on a smooth, clean anvil. The rough under surface, where the bead was in contact with the cupel, may then be easily cleaned with a wire scratch-brush. The bead is then flattened by one or two blows of the hammer. Small beads containing much silver may quite well be hammered out to about 1 c.m. diameter. The hammer and anvil for this purpose should be kept perfectly clean and polished, and used exclusively for hammering assay beads.

Weighing Fine Metal. — The beads are now weighed, this weight being recorded as "fine metal"; if the operation has been properly conducted, this is the sum of the gold and silver, less whatever losses have taken place by volatilization and absorption in the cupels during cupellation. It is generally sufficient to weigh to 0.02 mg. The variation in the adjustment of the balance may be corrected as described in Section II (*b*), or the method of "substitution" may be used (in which the bead is first placed in the right-hand pan of the balance, and then counterpoised by pieces of metal or rough weights in the left-hand pan). The bullion bead is then removed and accurate weights added on the right until equilibrium is obtained, the final adjustment, of course, being made with the rider.

Annealing and Rolling. — The beads are generally annealed by heating to dull redness on fire-clay tiles, or in a small spirit or gas flame, and rolled into strips about 3 in. in length (fillets) by passing between smooth steel rolls. The most convenient form of rolls have a single adjusting screw, working in such a way that the rolls always remain exactly parallel; when there are two screws, this is difficult to secure. The rolls are first loosened so as to allow the metal to pass through without much force; then they are gradually tightened, so that the strip becomes somewhat longer and thinner each time it is passed through. The edges should be smooth and show no signs of frilling or cracks. After rolling, the fillets are again annealed and loosely coiled, taking care to leave the rougher side, corresponding to the bottom of the bead on the cupel, on the outer side of the roll, so as to be more easily attacked by the acid. Beads containing a large percentage of silver (say 5 to 7 times the weight of the gold) may quite well be parted without rolling out, provided they are well flattened by hammering to at least 1 cm. diameter.

Parting. — This operation is usually carried out in special glass vessels (parting-flasks) having a somewhat lengthened bulb and a long narrow neck. The method of using these is described in Section II (*b*). Equally good results may, with care, be obtained by using porcelain crucibles of somewhat large size (say 4½ cm. diameter by 3 cm. deep). Where large numbers of bullion assays are made at a time, as in mints, etc., it is customary to use a *platinum* parting apparatus, consisting of a tray with a number of perforated recesses ("thimbles"), in which the rolled assay beads ("cornets") are placed. The whole tray is immersed in a platinum vessel containing nitric acid, so that all the cornets receive uniform treatment.

Arthur C. Claudet [1] gives the following description of parting with the platinum apparatus:

"The cornets are placed in a platinum tray, capable of holding sixty cornets, in their respective thimbles. Twenty-five ounces of nitric acid, sp. gr. 1160°, are put into No. 1 platinum boiler, and the tray of cornets immersed in it. The boiling is continued till no more nitrous fumes are observed. The tray is then taken out and drained from the acid liquor, then washed in distilled water and immersed in No. 2 boiler, containing 25 oz. nitric acid, sp. gr. 1250°; boiling is continued for 20 minutes; the tray is then removed from the boiler and the acid liquor drained off; the tray is immersed, without washing, in No. 3 boiler, containing 25 oz. nitric acid, sp. gr. 1250, and boiling continued for 20 minutes. The tray is then removed, drained, washed twice with distilled water, drained again, and dried quietly over a gas-burner covered with wire gauze. The tray is placed in the muffle-furnace on a suitable revolving platform to anneal, and when the cornets are thoroughly red-hot, is withdrawn. The cornets are then of a bright gold color."

In the office of a mine or cyanide plant, where such expensive apparatus is commonly not available, the following method will generally be found to work satisfactorily. The flattened beads or

[1] I. M. M., Bull No. 27, Dec. 13, 1906, p. 17.

cornets are placed in porcelain crucibles of convenient size, supported on a perforated metal plate over a uniform source of heat, such as a large oil-stove, and 15 c.c. of "weak acid" poured over each. The weak acid is prepared by diluting 1 part of pure nitric acid (sp. gr. 1.42) with 7 parts of water (*i.e.*, 12.5 per cent. by volume of concentrated acid). After the cornets have been well boiled, the lamp is turned down, the acid being kept hot, but not actually boiling, for say 15 minutes, after which the liquid is again boiled till quite colorless, and until the bubbles given off no longer show the slightest appearance of brown fumes. If on cooling for a moment a rapid evolution of small bubbles still takes place, the boiling is further continued, but usually about 20 minutes' treatment in the weak acid is sufficient. This acid is then drained off and poured away into the silver residue bottle. The cornets are washed once with distilled water, and 15 c.c. of "strong" acid poured on. The strong acid is prepared by mixing 3 volumes of pure nitric acid (sp. gr. 1.42) with 1 volume of water, and thus contains 75 per cent. by volume of concentrated acid. The liquid is raised cautiously to the boiling-point; the lamp is then turned down for say 25 minutes, thus keeping the crucibles hot but not actually boiling, and adding fresh acid if necessary. At the finish the temperature is again raised to the boiling-point. The time allowed for strong acid treatment is usually about 30 minutes. For good results it is necessary to boil thoroughly in both acids.

Precautions. — Evaporation to dryness must be carefully avoided, especially during the weak-acid treatment, for if the cornets are heated sufficiently at that stage to cause them to acquire a natural gold color, it is afterwards impossible to extract the residual silver by the strong-acid treatment, and the result will be too high.

By the treatment above described, beads containing originally 60 to 80 mg. of gold and 350 to 420 mg. of silver usually remain unbroken. When the proportion of silver is much larger they break to pieces and some care is necessary to avoid loss. Too violent boiling must also be carefully guarded against, as this is liable to cause spurting, especially during the strong acid treatment, and may result in the cornet suddenly breaking in pieces, in which case fragments are likely to be projected out of the crucible.

Solvent Action of Nitric Acid on Gold. — Contradictory opinions are expressed by different writers on this subject. (See Section II, *b*.) Some maintain that neither nitric nor nitrous acids by themselves are capable of dissolving gold. It has been observed, in parting cornets containing a rather high percentage of silver, with nitric acid perfectly free from chlorides, that on diluting the clear liquid containing the dissolved silver, either with cold acid or water, a dark purplish turbidity was formed. On boiling, the liquid became clear again, but a blackish scum was formed which seems to consist, at least partially, of finely divided gold.

Igniting the Cornets. — After pouring off the strong acid, the cornets are washed and the inside surface of the crucible carefully rinsed out with distilled water. This operation is repeated at least twice. The crucibles are then replaced on the perforated frame over the lamp and allowed to dry for a few minutes at a moderate heat. They are then ignited by being placed for a minute or so in a muffle at a dull red heat, or by holding the crucibles in the flame of a blast lamp until they acquire a bright golden color. Care must be taken not to heat so intensely as to fuse the gold.

Weighing Gold Cornets. — When cool, the gold cornets are weighed on the fine assay balance, recording the correct weight to 0.01 mg. A good assay balance will weigh to 0.005 mg., but this degree of accuracy is not usually necessary. The cornets must be a good color; if pale, it is a sign that they contain undissolved silver. The adjustment of the balance must be carefully attended to, and corrected, if necessary, as described under ore assaying. It is advisable to verify the correctness of the adjustment after every two or three weighings by allowing the balance to swing with the pans empty, a special rider on the left arm of the beam being used to correct any inequality in the swings. Errors from this source may also be corrected by the method of substitution above noted.

Accuracy of Assay Weights. — It is necessary to carefully examine the weights used, and to determine their true relative value with reference to the 500 mg. weight used for weighing the bullion-charges, as the sets supplied, even by firms of high standing, are by no means absolutely correct.

Millièmes. — Some assayers use the "millième" system, in which the 500 mg. weight is marked 1000, and its subdivisions numbered in millièmes, or half-milligrams. This has the advantage of giving the "fineness" on a charge of 0.5 gram of bullion directly, without calculation, but is very confusing if the weights are used for other purposes, as the numbers are apt to be mistaken for milligrams.

The fineness of the bullion is generally reported in parts per 1000. As 500 mg. are taken for each assay, it is merely necessary to multiply the result by two to obtain the fineness.

Correction of Results by Checks. — A correction is required to be made for losses of gold and silver in cupellation, and for silver retained in the cornets after parting. In the *gold* assay, the weight actually obtained is the net result of the losses in the first operation and the gains in the second. It is assumed that the fine gold used in the check has undergone the same losses and gains as that from the bar. This is only strictly true when the check is very closely of the same composition as the bullion to be assayed, and has been subjected to precisely similar treatment. If the check, as is usually the case, shows an increase of weight, the amount of this increase is deducted from the weight of gold obtained from the bars adjoining that particular check in the muffle; if the check has diminished, the amount of such decrease is added to the weight of gold from each bar assay.

Amount of Surcharge. — According to A. C. Claudet[1] the amount of surcharge in gold bullion assays may vary from 0.7 to 1.0 per millième (*i.e.*, 0.35 to 0.5 mg. on an assay of 500 mg.) and is always an increase on the amount taken for the check; this amount must therefore be deducted from the weight of the cornet in each bar assay. Professor W. Gowland[2] stated, in the discussion on Claudet's paper, that the surcharge at the Mint of Japan varied during 6 months from 0.5 to 0.7, averaging 0.55. He also remarks that a small surcharge indicates that the cupellation has been carried out at too high a temperature, and that in such a case the assay should be repeated, still more so if there should be no surcharge or a loss on the checks. These figures, however, all probably refer to bullion consisting chiefly of gold. According to T. W. Wood,[3] the surcharge varies according to the relative proportion of gold and silver, and with bullion consisting chiefly of silver the gold checks usually show a loss. The present writer's experience

[1] I. M. M., Bull. No. 27, p. 17.
[2] *Ibid.*, Bull. No. 28, p. 29.
[3] " Proc. Chem., Met. and Min. Soc. of South Africa," II, p. 3.

with bullion averaging about 800 in silver and 60 to 70 in gold was that the surcharge rarely exceeded ± 0.4 per 1000 (*i.e.*, 0.2 mg. on 500 mg.).

Under normal conditions the *silver* always shows a loss, so that the correction indicated by the check has to be added to the silver found. It occasionally happens that the front row in the muffle shows an apparent increase in silver. This is due to retention of lead; in such cases the correction cannot be safely applied and the assays must always be repeated, using a somewhat higher temperature.

Examples of Corrections for Bullion Assays. — (*a*) *Gold Bullion* (*e.g.*, battery gold); 500 mg. taken for assay, in duplicate.

	Gold: mg.	Silver: mg.
Taken for check	284.52	207.85
Found.	284.86	206.42
Correction	—.34	+1.43

Bar assay:	Gold:	
	Found	Corrected
	284.93	284.59
	284.79	284.45

Bar assay:	Silver:	
	Found	Corrected
	206.37	207.80
	206.25	207.68

	Gold	Silver
Mean Fineness.	569.04	415.48

(*b*) *Silver Bullion* (Cyanide). —

	Gold: mg.	Silver: mg.
Taken for check	76.79	397.59
Found.	76.88	391.97
Correction	−.09	+5.62

Bar assay:	Gold:	
	Found	Corrected
	76.70	76.61
	76.68	76.59

Silver:

Found	Corrected
391.80	397.42
392.76	398.38

	Gold	Silver
Mean fineness	153.2	795.8

(B) Volumetric Methods

There are two methods in general use for determining the silver in bullion; several others have been proposed, but these two only need be considered here. They are: (1) Gay-Lussac's method; (2) Volhard's method.

(1) *Gay-Lussac's Method.* — This method depends on the fact that silver is precipitated by solutions of sodium chloride as a white chloride, insoluble in dilute acid solutions; the only other metals precipitated under the same conditions being lead and mercury. Since the chloride coagulates and settles rapidly on agitating the solution, the finishing point of the reaction may be found by noting the point when fresh additions of salt solution fail to produce any further precipitate. It has been found, however, that silver chloride is slightly soluble in the sodium nitrate produced in the reaction:

$$NaCl + AgNO_3 = NaNO_3 + AgCl.$$

Hence the true finishing point occurs slightly before the point at which fresh NaCl ceases to give a precipitate. In fact, at or near the point of exact correspondence the solution will give a precipitate of AgCl if either salt or silver solution be added; the true end-point may be found by determining (1) the point at which NaCl ceases to give a precipitate, (2) the point at which, after a slight excess of NaCl has been added, a corresponding solution of $AgNO_3$ ceases to give a precipitate, and taking the mean of the two results. The amount of AgCl dissolved in this way depends on the temperature and degree of dilution of the solutions.

A simpler method is to determine the finishing point, by means of an accurately prepared salt solution, for different measured quantities of a silver solution of known strength; after a number of such òbservations have been made, a factor can be deduced which will give the proper correction for any reading. The solutions commonly used for assays by this method are:

1. A "standard" salt solution, made by dissolving 5.4202 grams of the purest obtainable dry sodium chloride in distilled water, and making up to a liter; 1 cc. of this solution = .01 gram Ag. The pure salt is best prepared by neutralizing pure sodium carbonate or bicarbonate with hydrochloric acid, evaporating to dryness, and igniting.

2. A salt solution, prepared from the above of exactly one-tenth the strength, known as the "decimal" salt solution; 1 cc. of this solution = .001 gram Ag.

3. A solution of silver nitrate corresponding exactly to the decimal salt solution, that is, containing 1 gram of Ag dissolved in pure nitric acid, heated to expel nitrous fumes and diluted to a liter.

In making a test, the quantity of bullion taken must be such that it contains very slightly more than 1 gram of silver, this being ascertained by a previous approximate assay. This is introduced into a stoppered bottle and dissolved in nitric acid, heating on the water-bath till completely dissolved, after which the stopper is removed and the nitrous fumes dispelled by blowing through a bent tube. Then 100 cc. of the standard salt solution are added by means of an accurately graduated pipette, so that exactly 1 gram of silver is precipitated. The stopper is replaced and the bottle shaken, being kept in the dark as much as possible, until the silver chloride settles clear. The decimal salt solution is then added, a little at a time, as long as a further drop, after agitation and settling, continues to produce a precipitate. A mechanical device is often used for the agitation of the bottles, so arranged that they are kept dark, since silver chloride is rapidly acted upon and decomposed by exposure to light.

With care, this method yields very accurate results, but the necessity of waiting after each addition of the test solution for the settlement of the precipitate renders it very tedious.

(2) *Volhard's Method.* — This depends on the fact that silver in nitric acid solution is precipitated by solutions of a soluble thiocyanate (sulphocyanide), and that if a ferric salt be present the red ferric thiocyanate is decomposed and the color destroyed as long as silver is in excess. The reactions are:

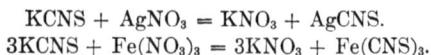

$$KCNS + AgNO_3 = KNO_3 + AgCNS.$$
$$3KCNS + Fe(NO_3)_3 = 3KNO_3 + Fe(CNS)_3.$$

As soon as all the silver is precipitated, the red color remains permanent.

Standard and decimal solutions of potassium or ammonium thiocyanate may be prepared, such that 1 cc. = .01 gram Ag, and 1 cc. = .001 gram Ag, respectively.[1] In this case the solutions cannot conveniently be prepared by direct weighing of the thiocyanate, on account of the deliquescent nature of the salt, but may be adjusted by comparison with a silver solution of exactly known strength.

With this process also there is a difficulty as to the exact finishing point, owing to the fact that the red ferric thiocyanate color disappears very slowly and only after continued agitation when only a small excess of silver is present. Various methods of determining the end-point have been suggested, as follows:

1. A somewhat sharper finish is obtained by adding a slight excess of thiocyanate, shaking thoroughly, then cautiously adding a dilute silver nitrate solution (say 1 cc. = .0005 gram Ag), drop by drop, until the red color just disappears, deducting the equivalent of $AgNO_3$ used from the thiocyanate taken.

2. Precipitating the bulk of the silver as chloride by means of a decimal salt solution (as in Gay-Lussac's method), filtering, and titrating the filtrate only with dilute thiocyanate.

3. Adding a slight excess of thiocyanate, estimating the amount of this excess (or rather its equivalent in terms of silver) by colorimetric comparison with measured quantities of solution of known strength necessary to give a tint of equal intensity, and deducting this from the silver equivalent of the total thiocyanate used. The solution must be filtered before making the comparison. (See E. A. Smith, in I. M. M., Bull. No. 28.)

The method may be used in presence of most of the ordinary metals, such as lead, iron, zinc, copper, etc., which may be present in considerable amount without affecting the result. The presence of copper alters the tint of the solution at the finish, and hence interferes with the colorimetric method of determining the end point.

The writer finds it advisable to avoid filtration. Where the quantity of silver in the solution is considerable (more, say, than 0.1 gram), it is better to add the thiocyanate in slight excess, shake thoroughly, allow to settle for some time, decant the almost clear liquid (tinged red by the ferric indicator and excess of thio-

---

[1] It is more convenient to work with weaker solutions, say 1 cc. = .005 gram and 1 cc. = .0005 gram.

cyanate) into a separate vessel, adding water, agitating, settling and decanting several times, and titrate the whole liquid poured off by means of the dilute standard silver solution until the red color disappears.

PART VIII

ANALYTICAL OPERATIONS

In this part of the present volume, some description will be given of the analytical methods required for the examination of such substances as may occasionally demand analysis in connection with cyanide work. The subject is divided into the following sections:

I. Analysis of ores and similar siliceous material, including slags.

II. Analysis of bullion and other mainly metallic products.

III. Analysis of cyanide solutions after use in ore treatment.

IV. Analysis of commercial cyanide (solid).

V. Analysis of sundry materials (lime, coal, water) used in the process.

ANALYSIS OF ORES AND SIMILAR SILICEOUS MATERIAL

No attempt can be made in a work of this kind to give minute details of analytical methods, but a general outline will be here presented, with references to standard works, where the reader may obtain fuller information. In this section the analysis will be described of such material as might commonly be expected to contain 50 per cent. or more of silica. The constituents whose determination is followed are those which may be of interest and importance in cyanide work, and for convenience of reference they are arranged alphabetically, though of course the actual order of analysis would be determined by the nature of the sample.

PRELIMINARY OPERATIONS

The operations of sampling, crushing, and drying are of at least as much importance in analysis as in assaying; in fact, certain operations which are quite allowable in ordinary assay work would not be admissible in an analysis.

Crushing. — The use of an ordinary iron pestle and mortar might introduce sufficient iron, in the form of metal or oxide, to entirely vitiate the result. The metallic portion may, it is true, be removed with a magnet, but many minerals themselves contain magnetic particles, and the magnet does not remove foreign particles of oxide of iron. The crushing can only be carried out safely by means of a well-polished hard-steel hammer and anvil, provided with a guard to retain flying particles, followed by fine grinding of a well-mixed average portion with an agate pestle and mortar.

Sifting. — Metal sieves are liable to induce contamination, and even sifting through cloth may give rise to organic impurities. In some cases, sifting is entirely unnecessary; in others, judgment must be used as to whether the possible contamination of the sample is sufficient to necessitate a special determination in an unsifted sample of the elements (iron, copper, zinc) which might arise from the wearing of the screen.

Drying. — The presence of moisture in samples for analysis is sometimes overlooked. Many siliceous substances, when finely divided, readily absorb moisture from the air, which adds to their weight without imparting any visible dampness to the sample. This so-called "hygroscopic water," if neglected, will cause unaccountable variations in the determination of the other constituents. On this account it is customary to dry the portion of the sample to be taken for analysis for some hours at a fixed temperature (generally 100° or 110° C.), weighing at intervals in a glass-stoppered bottle, which is of course left open during the drying, until two consecutive weighings agree within sufficiently narrow limits. Another method sometimes employed is to take the crushed mineral in the undried condition, and weigh out, as nearly as possible at the same time, all the portions required for the different determinations, including one for hygroscopic water. The percentages of the various constituents in the dry sample are then obtained by calculation. The drying is carried out in an air-bath or steam oven at constant temperature. The determination of combined water will be described below.

GENERAL OUTLINE OF ANALYSIS

After the preliminary operations above described, the next step, in the treatment of siliceous material, is to "open up" the substance by some method which will render the bases soluble, and the whole of the silica insoluble, in dilute acids. Where conditions allow, this is done by fusion in platinum vessels with alkalis or alkaline carbonates; in other cases the sample is evaporated or boiled with nitric, hydrochloric, or sulphuric acid, according to circumstances. In the filtrate from the insoluble matter, iron and alumina are generally precipitated together by adding excess of ammonia; then, successively, the group containing manganese, zinc, nickel, and cobalt by ammonium sulphide, calcium by ammonium oxalate, and magnesium by an alkaline phosphate. The alkali metals are estimated in a separate portion of the sample. In cases where lead, bismuth, arsenic, antimony, or other heavy metals capable of attacking platinum are present, these must be eliminated before making any fusions in platinum vessels.

ESTIMATION OF VARIOUS INGREDIENTS

ALKALI METALS

1. These are generally estimated by Lawrence Smith's method, as follows: A weighed portion, say 1 gram, of the finely powdered ore is mixed in a mortar with an equal weight of ammonium chloride, then with eight times its weight of pure calcium carbonate, and transferred to a platinum crucible provided with a funnel-shaped prolongation to condense any alkaline chloride which may be volatilized during ignition. The crucible is heated, gently at first, until ammonium salts are decomposed, then kept at redness for, say, three quarters of an hour. When cool the contents are turned into a dish and the crucible washed into the same dish with water (60 to 80 cc.) and boiled. The residue is then filtered, the filtrate mixed with 1.5 gram ammonium carbonate and concentrated somewhat; then more ammonium carbonate and ammonia are added. The liquid is now again filtered into a weighed platinum dish, evaporated to dryness, and ignited cautiously to below redness. When cool the dish is weighed, the increase in weight giving the combined chlorides of the alkali metals. The residue is then dissolved in water, and the chlorine titrated by standard silver nitrate. Deducting the weight of chlorine so found from that of the mixed chlorides, we have the total weight of alkali metals present. (For separation of sodium and potassium see Potassium.)

2. In cases where Lawrence Smith's method is inapplicable, as when the sample contains large quantities of the heavy metals, evaporate to dryness with aqua regia, ignite to render silica insoluble, take up with dilute hydrochloric acid, and precipitate with hydrogen sulphide. Treat the filtrate with ammonia and ammonium sulphide; boil, filter, and wash with dilute ammonium sulphide. Evaporate the filtrate to dryness and ignite to expel all ammonium chloride, etc. Re-dissolve in a little water, add ammonium carbonate and evaporate to dryness on a water-bath. Re-dissolve in warm water and filter into platinum dish. Evaporate on water-bath, ignite gently, and weigh mixed chlorides as above.

ALUMINIUM

This metal is generally precipitated as alumina, together with Fe, Cr, Ti, Zr, and in some cases phosphates and small amounts of SiO_2, in the filtrate from the silica. After determining the total weight of the ignited precipitate, and the separate amounts of the other constituents, the alumina (Al_2O_3) is found by difference. The precipitation is made either by ammonia or by sodium acetate.

1. The filtrate from silica is heated nearly to boiling, and ammonia added until the liquid becomes slightly alkaline. Sufficient ammonium chloride must be present to prevent the precipitation of magnesia, etc.; this is best secured by acidulating the liquid, if not already strongly acid, with additional hydrochloric acid before adding ammonia. After stirring and allowing to settle, wash by decantation with hot water containing a little ammonia; finally transfer to a filter, allow to drain, and wash back the precipitate into the original dish. Re-dissolve in hot dilute HCl, re-precipitate with ammonia from the nearly boiling liquid, again filter, and add the second filtrate to the first. The precipitate is dried carefully in an air-bath, removed from the papers, which should be burnt separately, ignited over an oxidizing blast flame in a platinum crucible, and weighed as alumina, etc.

2. When much manganese is present, it is better to precipitate with sodium acetate, as follows: The acid filtrate from the silica is nearly neutralized with caustic soda, taking care, however, not to produce a permanent precipitate. From 2 to 3 grams of sodium acetate are then added, and the liquid boiled, allowed to settle somewhat, filtered and washed slightly. The precipitate is re-dissolved in nitric acid, transferred to the original vessel, and heated nearly to boiling. It is then precipitated with ammonia, filtered, and washed with a 2 per cent. solution of ammonium nitrate, care being taken to keep the two filtrates separate. By concentrating these filtrates separately, an additional quantity of alumina, etc., is recovered. The first filtrate is then re-filtered, and the second poured on to the same filter, giving a final wash with hot dilute ammonia. The united filtrate thus obtained serves for the estimation of Mn, etc. The additional precipitate recovered is added to the main bulk, dried

in the air-bath, and treated as already described in the first method.

3. The filtrate from the silica is oxidized to bring all iron into the ferric state. It is diluted to 100 cc. and 15 cc. of a saturated solution of microcosmic salt ($Na \cdot NH_4 \cdot HPO_4 \cdot 8H_2O$) is added, followed by ammonia, till a slight permanent precipitate remains. Heat to boiling, acidulate very slightly with HCl, add 25 cc. saturated sodium thiosulphate and 5 cc. glacial acetic acid; boil for 10 minutes. A white precipitate is formed consisting of aluminium phosphate mixed with sulphur. Filter, dry, and ignite with the paper; finally ignite at bright-red heat and weigh as $AlPO_4$.[1]

$$Al_2O_3 \times 0.5303 = Al.$$
$$AlPO_4 \times 0.4185 = Al_2O_3.$$
$$AlPO_4 \times 0.2219 = Al.$$

ANTIMONY

In ores containing antimony, this metal may be brought into solution: (1) By treatment with hydrochloric acid and potassium chlorate. (2) By fusion with an alkali and sulphur. The mass extracted with water contains the antimony as a soluble sulphantimonite. (3) By evaporation with sulphuric and tartaric acids, the latter being added to ensure the conversion of antimony into an antimonious compound; (4) In certain ores, such as antimonite, the metal may be dissolved by simply boiling with hydrochloric acid.

The following precautions are to be noted in manipulating solutions containing antimony: (1) The chloride is very liable to precipitate on dilution, as $SbOCl_3$; this may be prevented by the addition of a little tartaric acid before precipitating the antimony, for instance, with H_2S. (2) Treatment of antimony ores with nitric acid forms antimonic acid, which is soluble with great difficulty. It is therefore preferable to use a mixture of HCl and $KClO_3$. (3) Antimonious chloride, $SbCl_3$, is volatile on evaporation; solutions should therefore be made alkaline before evaporating.

Separation. — 1. Having obtained the metals in HCl solution, the liquid is mixed with excess of caustic soda and a little sulphur; a current of H_2S gas is passed through for some min-

[1] H. T. Walker, Bull. No. 49, I. M. M., Oct. 8, 1908.

utes, and the mixture allowed to stand in a warm place for an hour; it is then filtered and washed, the Sb being in the filtrate as sulphantimonite (together with As and Sn if present). Acidulate filtrate with HCl, which leaves As_2S_3 undissolved on boiling. Filter, dilute, add a little tartaric acid; re-precipitate with H_2S. Filter and wash free from chlorides. Determine Sb as below.

In case tin also is present, transfer the precipitate to a weighed dish and treat cautiously with fuming nitric acid. Evaporate, ignite, and weigh as $xSb_2O_4 + ySnO_2$. The residue in the dish is then transferred to a flask and digested for an hour with concentrated tartaric acid solution, on a water-bath, avoiding undue evaporation. Residue is filtered, washed, ignited, and weighed as $ySnO_2$. The precipitation of tin by H_2S may be prevented by adding a sufficient amount of concentrated oxalic acid solution to the mixed chlorides.

2. The ore is heated in a flask with a mixture of tartaric acid, sulphuric acid, and potassium bisulphate until completely decomposed; all carbon separated is burnt off and the mixture heated till most of the H_2SO_4 is expelled. Allow to cool and dissolve in dilute HCl with addition of tartaric acid. Heat nearly to boiling and pass in H_2S. Filter and wash with H_2S water. Transfer the precipitate to a beaker and warm with potassium sulphide; pass again through same filter, washing with warm dilute K_2S. The Sb (with perhaps As and Sn) is in solution; convert into chlorides by repeating the original treatment with sulphuric acid, etc., finally dissolving in strong HCl. Now pass in H_2S, which, in strongly acid solution, precipitates only the As. Filter, wash with HCl, dilute filtrate with warm water, add oxalic acid if tin is present, and precipitate antimony as Sb_2S_3 by passing H_2S. (For details see A. H. Low, "Technical Methods of Ore Analysis," 3d edition, pp. 32–35.)

Estimation. — 1. The antimony is assumed to have been separated as antimonious chloride, $SbCl_3$, free from other metals; add excess of HCl, boil slightly if necessary to expel SO_2, dilute considerably, and titrate with standard permanganate or bichromate:

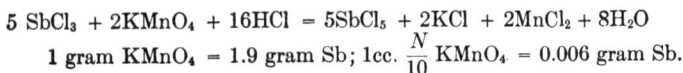

$$5\ SbCl_3 + 2KMnO_4 + 16HCl = 5SbCl_5 + 2KCl + 2MnCl_2 + 8H_2O$$

1 gram $KMnO_4$ = 1.9 gram Sb; 1cc. $\dfrac{N}{10}$ $KMnO_4$ = 0.006 gram Sb.

2. When the antimony has been obtained as antimonic chloride ($SbCl_5$), as in the case where the ore has been treated with HCl and $KClO_3$, the excess of chlorine is boiled off, and the solution cooled, diluted, and mixed with potassium iodide.

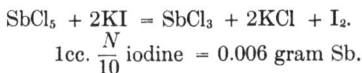

$$SbCl_5 + 2KI = SbCl_3 + 2KCl + I_2.$$
$$1cc. \frac{N}{10} \text{ iodine } = 0.006 \text{ gram Sb.}$$

After standing a few minutes, the liberated iodine is titrated with standard thiosulphate. The thiosulphate must be standardized against a known quantity of antimony, either the pure metal dissolved in HCl + $KClO_3$ or tartar emetic, also heated to convert the antimony into the antimonic condition.

3. Antimonious compounds may also be titrated with a standard solution of iodine in potassium iodide, as described under arsenic.[1]

Detection. — The presence of antimony in an ore may often be detected by heating the extract obtained with hydrochloric acid or aqua regia, in a dish containing a strip of platinum in contact with metallic zinc. A black deposit on the platinum, insoluble in dilute HCl but soluble in HNO_3, indicates antimony.

ARSENIC

Ores containing arsenic, such as mispickel, realgar, orpiment, and some other minerals, are occasionally treated by cyanide; since the presence of arsenic introduces some difficulties in the process, attention may be drawn to the methods of detecting and estimating this element.

Dissolving. — 1. The methods already described for antimony may be applied in general for arsenic.

2. Arsenic may be obtained in the arsenious condition by evaporating the powdered ore with nitric acid, and adding the mixture gradually to a strong solution of alkaline sulphide, warming and filtering, the arsenic being in the filtrate as sulpharsenite.

3. It may also be obtained as sulpharsenite by fusing with sodium carbonate and sulphur.

4. It may be obtained as an alkaline arseniate by digesting the ore with nitric acid, evaporating to dryness, and taking up with an alkali.

[1] Sutton, "Volumetric Analysis," 8th edition, p. 160.

5. As arseniate, it may be obtained by fusing with sodium carbonate and niter, and extracting with water.

6. By suspending the finely powdered ore in a caustic alkali solution and passing in chlorine till saturated.

7. By distilling with zinc and sulphuric acid, which produce hydrogen arsenide; this gas when led through fuming nitric acid forms arsenic acid.

Separation. — 1. Having obtained the arsenic (together with antimony and tin) as a sulpharsenite, the insoluble residue, if any, is collected on a filter and washed with dilute potassium sulphide. The filtrate is acidulated with sulphuric acid and potassium bisulphate, adding tartaric acid if antimony is present. Heat till the free sulphur and most of the acid is expelled, then cool and add strong hydrochloric acid. Warm gently and saturate with H_2S. The arsenic is precipitated as As_2S_3, leaving antimony and tin in solution. Filter and wash with HCl (2: 1).[1]

2. In cases where the arsenic has been converted by preliminary treatment into an arsenious compound, it may be separated from other (non-volatile) ingredients by distillation as follows: The substance is placed in a flask connected with a bulbed U-tube or other suitable condenser, containing a little water. Add to the flask, for each gram of material treated, 30 grams calcium chloride, 15 grams ferric chloride, and 30 cc. hydrochloric acid. Distil for 20 to 30 minutes without evaporating to dryness. The arsenic is present in the distillate as arsenious chloride. In a modification of this method, ferrous sulphate is used instead of ferric chloride, and serves to reduce any arsenic compound that may be present.

Estimation. — 1. Having obtained the arsenic as an arsenious compound by distillation as described above, or by dissolving the sulphide in ammonium sulphide and evaporating with sulphuric acid and potassium bisulphate, the solution is made slightly alkaline with ammonia, then slightly acid with hydrochloric acid. When cool, a sufficient excess of ˙sodium bicarbonate is given, and the mixture titrated with standard iodine and starch indicator. The iodine solution is standardized against known quantities of arsenious acid (As_2O_3) dissolved in hydrochloric acid.[2]

[1] For details see A. H. Low, "Technical Methods of Ore Analysis," pp. 33–41.
[2] C. and J. Beringer, "Text Book of Assaying," pp. 383–384.

$$1 \text{ cc. } \frac{N}{10} \text{ iodine } = 0.00375 \text{ gram As.}$$
$$= 0.00495 \text{ gram As}_2\text{O}_3.$$

2. Gravimetric Method. — When the arsenic has been obtained as an arseniate, the mixture is made alkaline with ammonia, and a mixture of magnesium sulphate and ammonium chloride added (1 part of the former to 2 parts of the latter). The precipitate is allowed to settle, filtered, and washed with a minimum quantity of dilute ammonia, then dried very slowly and carefully to expel ammonium salts, ignited, and weighed as magnesium pyroarseniate, $Mg_2As_2O_7$[1].

$$Mg_2As_2O_7 \times 0.4828 = As.$$

3. Volumetric Method, for Arseniates. — The solution is made slightly alkaline with ammonia, then mixed with sodium acetate and acetic acid till distinctly acid, heated to boiling, and at once titrated with standard uranium acetate, using ferrocyanide as external indicator; the appearance of a brown color in the spots of ferrocyanide marks the end of the titration.[1]

4. After fusing with sodium carbonate and niter (5 parts of each to 1 part of the ore), dissolve the residue in warm water, filter, and wash with cold water. Acidulate slightly with nitric acid, add silver nitrate, then successively ammonia and nitric acid, till the precipitate just dissolves and the liquid is faintly acid. Add sodium acetate in excess; the arsenic is precipitated as silver arsenate, Ag_3AsO_4; heat to boiling, filter, and wash with cold water till free from soluble silver salts, re-dissolve precipitate with dilute nitric acid, and estimate the silver by Volhard's method. (See p. 390.) [Pearce's method.][2]

$$1 \text{ cc. } \frac{N}{10} \text{ thiocyanate } = 0.0025 \text{ gram As.}$$

Detection. — 1. The substance is dissolved in hydrochloric acid, or the arsenic obtained as $AsCl_3$ as described above. A strip of copper is introduced into the liquid, which is then warmed. Small quantities of arsenic give a gray deposit, and larger amounts a black deposit, on the copper.

2. Marsh's Test. — The substance to be examined is introduced into a vessel containing zinc and sulphuric acid, from

---

[1] C. and J. Beringer, *loc cit.*, p. 389.

[2] For details, see A. H. Low, "Technical Methods of Ore Analysis," 3d edition, p. 42.

which a current of hydrogen is being generated. The evolved gases are led through a horizontal tube drawn to a fine point, forming a jet, where the escaping hydrogen may be ignited. By holding a cold porcelain plate in the flame, a black deposit is obtained consisting of reduced metallic arsenic or antimony; the arsenic deposit may be distinguished from that of antimony by its greater solubility in a solution of a hypochlorite. A mirror of arsenic may also be obtained by heating the tube at a certain point; the AsH_3 is decomposed and metallic arsenic deposited in the cooler part of the tube.

3. The substance is mixed with dry charcoal and heated in a bulb tube. A sublimate of arsenic is formed, and when much is present an odor resembling garlic is observed.

BARIUM

When the ore contains the metal as carbonate, it may be extracted by simply boiling with hydrochloric acid. In other cases the ore is treated successively with nitric and sulphuric acids as described under LEAD; barium and strontium remain as sulphates, together with silica, in the portion insoluble in ammonium acetate. They may be separated from silica by treating the residue in a platinum crucible with H_2SO_4 and HF. In the absence of lead, however, it is better to determine barium in a separate portion of the sample; this is treated with hydrofluoric and sulphuric acids, and evaporated repeatedly with sulphuric acid until every trace of fluorine has been expelled. The evaporated residue is digested for some time with 5 per cent. H_2SO_4, boiled if necessary, and filtered. The filtrate contains any other non-volatile metals which may have been present, and may be used for their estimation. The residue, consisting of $BaSO_4$ and $SrSO_4$, with perhaps traces of $CaSO_4$, is ignited with sodium carbonate and extracted with water. The residual carbonates are then dissolved in HCl. If strontium be present, the filtrate is made alkaline with ammonia, a slight excess of acetic acid is added, and potassium chromate, which precipitates the barium as $BaCrO_4$; this is filtered, washed with ammonium acetate, ignited gently and weighed.

$$BaCrO_4 \times 0.542 = Ba.$$

Strontium remains in the filtrate, and may be precipitated by

H_2SO_4. When no strontium is present, the $BaCO_3$ formed by fusion with sodium carbonate is at once dissolved in dilute HCl, diluted, heated to boiling, and precipitated with H_2SO_4. After settling, it is filtered, washed with hot water, ignited, and weighed as $BaSO_4$.

$$BaSO_4 \times 0.5885 = Ba.$$

If any calcium should be present in this precipitate of $BaSO_4$, it may be separated by dissolving the whole in concentrated H_2SO_4, and re-precipitating the $BaSO_4$ by dilution, the $CaSO_4$ remaining dissolved.

BISMUTH

This metal is obtained together with copper in precipitating the acid extract of the ore with H_2S, after removal of lead, etc., with H_2SO_4, as described under copper. In order to separate it, the sulphide precipitate is dissolved in nitric acid and evaporated to dryness, taken up with a little sulphuric acid, warmed and filtered. The filtrate is nearly neutralized with ammonia, ammonium carbonate added in excess, boiled and filtered. The precipitate is then re-dissolved in HNO_3, and re-precipitated from boiling solution by ammonium carbonate, filtered, washed, dried, ignited at low-red heat, and weighed as Bi_2O_3.

$$Bi_2O_3 \times 0.8966 = Bi.$$

CADMIUM

This metal is precipitated by H_2S with copper and bismuth. It remains in solution together with copper when the bismuth is precipitated as carbonate in ammoniacal solution. Cadmium may be detected and estimated by adding cyanide to the blue ammoniacal liquor till it is colorless, then passing in H_2S, which gives a yellow precipitate, leaving the copper in solution. The precipitate is collected on a weighed filter, washed with sulphureted hydrogen water, then extracted with carbon bisulphide to remove free sulphur, dried at 100° C. and weighed as CdS.

$$CdS \times 0.7781 = Cd.$$

CALCIUM

Separation. — 1. Calcium is generally determined in the filtrate from the group containing zinc and manganese; this filtrate is boiled to expel excess of ammonium sulphide, filtered if not quite clear, then mixed with a boiling solution of ammonium oxalate. After settling clear, the precipitate is filtered off and washed with hot water. In cases where much calcium is present the precipitate is re-dissolved in a minimum quantity of HCl and re-precipitated from boiling solution with ammonia and ammonium oxalate. This double precipitation is necessary because calcium oxalate retains both magnesium and sodium salts very tenaciously.

2. A direct separation in the filtrate from silica may be made as follows. The liquid is heated nearly to boiling, excess of ammonia added, then solid oxalic acid till iron is dissolved. Neutralize exactly with ammonia and add oxalic acid until the ferric hydrate just dissolves. On boiling, the calcium is precipitated as oxalate, free from iron. In presence of manganese it may be necessary to add bromine water after the first addition of ammonia; boil, filter, and wash with hot water before proceeding to add oxalic acid.

Estimation. — 1. When calcium has been precipitated as oxalate it may be conveniently and accurately estimated by titration with standard permanganate. The precipitate is washed thoroughly with hot water, dissolved in dilute sulphuric acid, heated to 70° C., and titrated with permanganate till the pink tint remains permanent. The permanganate should be standardized on pure crystallized oxalic acid.

$$1 \text{ part oxalic acid } (C_2H_2O_4 \cdot 2H_2O) = 0.4451 \text{ parts CaO.}$$

Special methods for determining the percentage of caustic lime in commercial samples will be described in a later section.

$$1\text{cc. } \frac{N}{10} \text{ KMnO}_4 = 0.0028 \text{ gram CaO.}$$
$$= 0.0020 \text{ gram Ca.}$$

A convenient standard solution is one containing 1.58 gram $KMnO_4$ per liter, so that

$$1 \text{ cc. } = 0.001 \text{ gram Ca.}$$

The reactions are as follows:

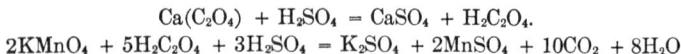

$$Ca(C_2O_4) + H_2SO_4 = CaSO_4 + H_2C_2O_4.$$
$$2KMnO_4 + 5H_2C_2O_4 + 3H_2SO_4 = K_2SO_4 + 2MnSO_4 + 10CO_2 + 8H_2O$$

2. Calcium may also be determined gravimetrically by strong ignition of the precipitate, and weighing as lime, CaO, or by gentle ignition, afterwards moistening the residue with ammo-

nium carbonate, evaporating to dryness, heating just to redness, and weighing as calcium carbonate, $CaCO_3$.

$$CaCO_3 \times 0.4006 = Ca.$$
$$CaCO_3 \times 0.5604 = CaO.$$

CARBONIC ACID

The carbonic acid in ores exists in the form of carbonates, generally insoluble in water but soluble in dilute acids. It may be estimated in one or other of the numerous types of carbonic acid apparatus, of which the general principle is as follows: The apparatus is weighed, first without and afterwards with the portion of ore to be analyzed. The dilute acid (HNO_3 or HCl) which serves to decompose the carbonate is then allowed to come in contact with the ore by turning the tap of the bulb in which it is contained and momentarily opening the stopper. The escaping gas has to pass through a bulb containing sulphuric acid, which retains any moisture. Gentle heat is necessary in some cases for complete expulsion of carbonic acid. When the action is complete, a slow current of air is aspirated through the apparatus, which is then again closed and weighed, the loss of weight giving the CO_2 expelled.

3. For more exact determinations the gas is passed through a series of tubes containing desiccating agents, such as strong sulphuric acid, anhydrous copper sulphate, calcium chloride, etc., and the CO_2 is absorbed by soda-lime contained in a weighed U-tube. The inlet of the flask in which the decomposition of the carbonate takes place is protected by a U-tube of soda lime to absorb any atmospheric CO_2 which might otherwise enter the apparatus, and the exit end of the weighed tube is protected by a U-tube containing a desiccating agent and absorbent of CO_2. Air is then aspirated through the whole apparatus, and the weighed U-tube detached, closed, and weighed. The increase of weight gives the CO_2.

CHLORINE

Chlorine occurs as chlorides in small quantities in certain ores. Silver occurs occasionally as "horn-silver" (AgCl), in which form it is readily extracted by cyanide. When insoluble chlorides are present, the crushed ore is fused with sodium carbonate and chlorine determined in the aqueous extract. Any heavy metals present in the extract are removed by H_2S, the excess of which must be boiled off. The liquid is then made acid with HNO_3 and the chlorine determined.

Estimation. — 1. If present in sufficient quantity, it may be titrated in neutral solution by $AgNO_3$, with potassium chromate (K_2CrO_4) as indicator. The silver solution is added till a permanent reddish tint remains after agitation.[1]

2. In cases where it is inconvenient to neutralize the solution, a measured amount of standard silver nitrate is added, more than sufficient to precipitate the chlorine, and the liquid acidified with HNO_3. The AgCl is then filtered off and the residual $AgNO_3$ determined by titrating with standard potassium or ammonium thiocyanate, using ferric nitrate or sulphate as indicator.[2]

3. Very small quantities are best estimated gravimetrically as AgCl. The liquid, which must be slightly acid with HNO_3, is mixed with a few drops of $AgNO_3$, boiled thoroughly, and allowed to stand for some time. The AgCl should settle in a dense form. It is then washed by decantation, without filtering, transferred to a weighed porcelain crucible, dried and heated to incipient fusion, cooled, and weighed as AgCl.

$$AgCl \times 0.2472 = Cl.$$

CHROMIUM

This metal occurs as chrome iron ore; it is precipitated along with Al and Fe in treating the filtrate from silica with ammonia or with sodium acetate. It is preferable, however, to determine it in a separate portion of the ore. This must be ground extremely fine and fused, either with sodium carbonate and niter, or with sodium peroxide, to convert the chromium into a soluble yellow chromate. The metal is dissolved in water, giving a yellow solution.

Determination. — 1. When only small quantities are present, the chromium may be determined colorimetrically by comparison with a standard solution of K_2CrO_4.

2. If present in larger amount, the solution, after boiling off any excess of sodium peroxide, is mixed with ammonium carbonate to partially neutralize the caustic alkali, filtered, and acidified

---

[1] For details, see Sutton, "Volumetric Analysis," p. 152 (8th edition).

[2] Sutton, *loc. cit.*, pp. 155, 184.

with dilute sulphuric acid. An excess of ferrous ammonium sulphate is then given, by adding a weighed quantity of the salt, and the residual ferrous iron titrated with permanganate.

1 part Fe = 0.3107 parts Cr.

COBALT

The sample is evaporated to dryness with HNO_3 and $KClO_3$. The residue is dissolved in dilute HCl, heated to boiling, and precipitated with H_2S. The filtrate from H_2S precipitate is boiled to expel excess of H_2S, a few drops of HNO_3 added, and the iron precipitated by ammonia. The filtrate is then made strongly acid with acetic acid, adding ammonium acetate and heating to about 70° C. On again passing in H_2S, cobalt, nickel, and zinc are precipitated, leaving manganese in solution. Filter, wash with warm water, and re-dissolve the precipitate in HCl + HNO_3. Add a considerable excess of ammonium chloride, evaporate to dryness, and ignite to volatilize zinc chloride; re-dissolve the residue, which now contains only Co and Ni, in HCl + HNO_3.

Determination of Cobalt. — The extract containing Co and Ni is made alkaline with caustic alkali, acidified with acetic acid, and mixed with a concentrated solution of potassium nitrite. On standing in a warm place (best for 24 hours) the cobalt separates as a yellow precipitate (double nitrite of cobalt and potassium). This is filtered off, washed with potassium acetate solution and finally with alcohol, dried, heated in a porcelain crucible with sulphuric acid, ignited at a very low red heat and weighed as $2 CoSO_4 + 3 K_2SO_4$.

Weight of residue × 0.1416 = Co.

COPPER

The presence of copper in ores to be treated by cyanide introduces such difficulties that the detection and estimation of the metal may become a matter of considerable importance.

Detection. — In most cases the presence of copper is readily detected by boiling a small quantity of the ore in strong nitric acid, diluting and neutralizing the acid with ammonia. On allowing to settle, the presence of copper is shown by a blue color in the supernatant liquid. It may be confirmed by filtering a little of the fluid, rendering just acid with hydrochloric or acetic acid and adding a drop of ferrocyanide solution, which gives a reddish-brown coloration. In order to determine whether the copper in an ore is in a form readily attacked by cyanide, a little of the crushed ore may be shaken with cyanide solution and filtered; if the filtrate be acidulated and a drop of ferrocyanide added, the characteristic reddish-brown color is obtained if any copper has been dissolved.

Separation. — 1. In cases where lead has also to be determined, the copper may be estimated in the filtrate from the lead sulphate precipitate, after the alcohol has been boiled off. It may be precipitated as sulphide, best from a hot solution, by sulphureted hydrogen or by sodium thiosulphate, or it may be precipitated on aluminium as described below, in the form of finely divided metallic copper.

2. A portion of the crushed ore (0.5 gram to 2 or 3 grams, according to richness in copper) is evaporated nearly to dryness with HNO_3. It is then boiled with HCl and finally with H_2SO_4, heating till white fumes are freely evolved. Cool, dilute, and filter. Wash with hot water. The copper in the filtrate is now precipitated as sulphide by passing a current of H_2S gas through the nearly boiling liquid; when saturated, allow to settle for a few minutes, filter while still hot, and wash with H_2S water. In presence of much iron, the sulphide precipitate must be re-dissolved in HNO_3, again boiled with H_2SO_4, diluted and re-precipitated with H_2S. Dissolve the precipitate in concentrated HNO_3, adding a little potassium chlorate if arsenic or antimony is present; boil thoroughly, filter off any undissolved sulphur. If any undissolved precipitate remains on the filter-paper, dry and ignite the latter, dissolve the ash in boiling HNO_3, and add to the bulk of the solution. The copper is in solution as nitrate, with perhaps Bi, Cd, or some other metals. In presence of Bi, add excess of ammonia, boil and filter.

3. Proceed as in method No. 2, until white fumes are given off on heating with H_2SO_4. Then cool, dilute and heat to boiling. Filter off any precipitate of $PbSO_4$, etc. Wash with hot water, dilute to (say) 75 cc., add one or more sheets of aluminium, bent so as not to lie flat in the beaker. Boil about ten minutes, till the copper is precipitated in a spongy form. Decant the liquid through a filter, and wash the copper on to the filter with H_2S water till all iron, etc., is removed. The copper adhering to the

aluminium is then dissolved in hot 50 per cent. nitric acid, which is afterwards poured over the precipitate in the funnel, together with a little bromine or $KClO_3$; collect the filtrate in a flask, wash with hot water, boil thoroughly to expel Br or Cl, and add ammonia.

4. Certain slags contain all or nearly all the silica in a form soluble in HCl, the copper being present as sulphide, insoluble in that acid. In such cases the sample is simply treated with hot dilute HCl and filtered. Dry, ignite, and treat the residue with HNO_3 and a little HCl. Boil off red fumes, dilute, add ammonia, boil, and filter.

Estimation. — 1. If the copper has been obtained in nitric acid solution, neutralize with sodium carbonate, and add ammonia till a clear-blue solution is obtained. Filter if necessary (*e.g.*, if Bi is present). If the copper contents are approximately known, prepare a similar solution containing about the same amounts of copper, nitric acid, and ammonia, using pure copper foil. This solution is used as a standard. Titrate the blue solutions as nearly as possible at the same time with a solution of potassium cyanide 0.5 to 2 per cent. KCN, according to circumstances, running in the solution until a faint violet color of equal intensity remains in each. When approaching the finish, the cyanide solution should be run in a little at a time, in quantities of say 0.5 cc., waiting some time before making a fresh addition, as the reaction takes place slowly.

2. The nitric acid solution of the copper is neutralized with sodium carbonate, and acetic acid added until the precipitate dissolves. (Ammonia may be used instead of sodium carbonate.) Cool, and add about 10 parts of potassium iodide for every part of copper supposed to be present. The following reaction takes place:

$$2Cu(CH_3CO_2)_2 + 4KI = Cu_2I_2 + I_2 + 4 (KCH_3CO_2).$$

The liberated iodine is then titrated at once with standard sodium thiosulphate, adding starch solution towards the finish.[1]

$$1cc. \frac{N}{10} \text{ thiosulphate } = 0.0063 \text{ gram Cu.}$$

In this process care must be taken that no iron or nitrous acid are present, and the liquid must not be too dilute; if these

[1] Sutton, *loc. cit.*, p. 201. Beringer, *loc. cit.*, p. 199.

conditions are neglected, the blue color of the iodine returns on standing for a few moments and no definite end-point can be obtained. Bismuth interferes with the process, as it gives a color with iodine; if present, it may be separated by boiling the solution with carbonate of ammonium and filtering, before adding the acetic acid.

3. When only small quantities of copper are present, they may be determined colorimetrically, by comparing the tint of the ammoniacal solution with that of a liquid in a similar vessel containing amounts of nitric acid and ammonia approximately corresponding to those in the assay, adding standard copper nitrate until the colors are alike. Still smaller quantities may be determined colorimetrically by using a standard ferrocyanide solution.

4. The copper is obtained in slightly acid HCl solution, and is reduced by sodium sulphite and precipitated with thiocyanate. The cuprous thiocyanate is then boiled, filtered, and washed thoroughly with hot water. It is next dissolved off the filter with boiling caustic soda, and the resulting sodium thiocyanate titrated with permanganate after acidulating with H_2SO_4. The permanganate solution should be standardized on pure copper under the conditions of the assay, as cuprous thiocyanate is slightly soluble in the reagents used.

5. The writer has found that small amounts of copper may be readily and accurately estimated by a method nearly identical in principle with that of T. Moore for the estimation of nickel.[1] (See below.) The copper is obtained in a nitrate or sulphate solution free from other metals. Sufficient caustic soda is added to precipitate the whole of the copper, avoiding a large excess of alkali; then standard cyanide solution, in measured amount, until a perfectly colorless liquid is obtained. Potassium iodide is then added and the excess of cyanide titrated with silver nitrate. The presence of ammonia or ammonium salts vitiates the test. A constant excess of cyanide should be used in all cases, as the amount of copper indicated increases slightly with increased excess of cyanide. Satisfactory results were obtained with solutions containing from 1 to 50 mg. Cu. The results were on a par with the iodide method in point of accuracy, and superior to the ordinary cyanide method.

[1] "Chem. News," LXXII, 92.

FLUORINE

Fluorine occurs in ores chiefly as fluor-spar (CaF_2) and as cryolite (Na_3AlF_6). It may be determined [1] by mixing the finely-ground ore with an excess of powdered silica and distilling with concentrated sulphuric acid for two hours at a gentle heat, whereby the fluorine is expelled as gaseous silicon fluoride:

$$2CaF_2 + 2H_2SO_4 + SiO_2 = 2CaSO_4 + 2H_2O + SiF_4.$$

During the distillation a current of air is made to pass slowly through the apparatus, and the silicon fluoride, which should be cooled by leading through a bulb-tube immersed in water, is passed into an alcoholic solution of potassium chloride (30 parts KCl in 100 parts alcohol and 100 water), by which it is decomposed as follows:

$$3SiF_4 + 2H_2O + 4KCl = SiO_2 + 2K_2SiF_6 + 4HCl.$$

The potassium silicofluoride remains undissolved in the alcoholic solution. By titrating the HCl with standard alkali and lacmoid indicator, the amount of fluorine may be determined.

$$1 \text{ equivalent HCl} = 3 \text{ equivalents F.}$$
$$1 \text{cc.} \frac{N}{10} \text{ HCl} = 0.0057 \text{ gram F.}$$

IRON

Separation. — The estimation of iron is generally made in the filtrate after separation of silica. The methods of separation with ammonia and with sodium acetate are described under aluminium. In a few cases the precipitate of alumina, etc., formed under these circumstances, can be ignited and weighed, and afterwards completely dissolved by boiling with concentrated sulphuric acid. In general, however, the ferric oxide, after ignition, resists the action of acids very strongly and can only be got completely into solution by long-continued fusion (2 to 4 hours) with potassium bisulphate. The fusion should be conducted at as low a temperature as possible, and the bisulphate should have been previously fused by itself, to expel water-vapor and excess of acid. The molten mass is allowed to cool, dissolved in water to which a little sulphuric acid is added, boiled and filtered. Any residual silica is here collected and estimated.

[1] Penfield, "Chemical Engineer," III, 65.

Estimation. — 1. The liquid obtained as above is treated with a reducing agent, cooled and titrated with permanganate. The reduction may be carried out in many ways. The writer prefers boiling with clean aluminium turnings until a drop of the liquid no longer reacts with a drop of thiocyanate solution. Other methods of reduction commonly employed are with zinc, with sulphureted hydrogen, and with sodium sulphite.[1] Care must be taken to avoid oxidation of the iron in the solution during the cooling process. When much iron is present, it may be necessary to dilute with boiled, cooled, distilled water before titrating with permanganate. The permanganate solution is standardized on pure iron wire dissolved in pure 20 per cent. sulphuric acid, or on a known weight of pure dry ferrous ammonium sulphate, $FeSO_4(NH_4)_2SO_4 \cdot 6H_2O$.

2. The liquid, reduced in the manner described, or (more rapidly) by addition of a small excess of stannous chloride — which excess is afterwards removed by cautious addition of mercuric chloride — is titrated with a standard solution of potassium bichromate, using ferricyanide of potassium, in spots on a white porcelain plate, as external indicator. For the estimation of iron in slags, 0.5 grams of the sample may be dissolved in 25 cc. boiling water, with addition of 20 cc. HCl. The mixture is stirred vigorously and boiled to expel H_2S. Then add $SnCl_2$ in slight excess, cool, and add 20 cc. of a saturated solution of mercuric chloride. Titrate with bichromate solution, 4.392 grams $K_2Cr_2O_7$ per liter.[2]

3. Very small amounts of iron are best estimated by the colorimetric method. The ferric hydrate or basic acetate precipitate is dissolved in HCl, a sufficient excess of thiocyanate solution added, and the tint compared with that produced under similar conditions by measured quantities of standard ferric chloride solution, using the same amount of thiocyanate in each case. The standard solutions may be conveniently adjusted so that 1 cc. = 0.01 or 0.001 gram Fe, according to the amount to be estimated.

LEAD

Separation. — In some ores, such as pure galena, the lead may be completely extracted by boiling with strong hydrochloric

---

[1] C. and J. Beringer, Text Book of Assaying, 9th edition, p. 235.

[2] H. T. Waller, Bull. I. M. M., No. 49, Oct. 8, 1908.

acid; in other cases nitric acid is used, with addition of a few drops of HCl. After boiling for some time, the flask is allowed to cool, and sufficient sulphuric acid added to precipitate the whole of the lead as PbSO$_4$, together with barium and perhaps some other metals. Evaporate until dense white fumes of sulphur trioxide are given off. Cool, dilute carefully, and boil. Again cool, add alcohol, stir and settle thoroughly (over night if possible). Decant the clear liquid and wash by decantation with dilute sulphuric acid containing about 10 per cent. of alcohol, finally with pure alcohol, leaving the residue as much as possible in the beaker. The filtrate may serve for the estimation of copper, etc. To the residue add a strong solution of ammonium acetate, with a slight excess of acetic acid, to dissolve PbSO$_4$, leaving SiO$_2$, BaSO$_4$, etc., undissolved. Moisten the filter with ammonia and pass the liquid through into a clean flask or beaker. Wash first with dilute ammonium acetate, then with hot water.

Estimation. — 1. Gravimetric Method. Acidulate the filtrate, obtained as above, with sulphuric acid. Allow to settle for some time and filter. The precipitate is separated as much as possible from the paper, ignited carefully, and the ash of the paper, separately ignited and treated successively with HNO$_3$ and H$_2$SO$_4$, is added. Weigh as PbSO$_4$. It is preferable to collect the final precipitate of PbSO$_4$ on a weighed perforated porcelain crucible with asbestos filter. This is washed with absolute alcohol, dried, ignited, and again weighed.

$$PbSO_4 \times 0.6829 = Pb.$$

2. Several volumetric methods are in use,[1] which are applicable where the quantity of lead is considerable. (*a*) The lead sulphate is dissolved in ammonium acetate, and titrated in hot solution with standard ammonium molybdate, using tannin as indicator, which gives a brown tint with excess of molybdate. A correction is required for the amount of molybdate solution necessary to develop this color. (*b*) Chromate of potassium is added in excess; the excess is then determined either colorimetrically or by means of a ferrous iron solution. (*c*) The lead dissolved as acetate is titrated with standard ferrocyanide at about 60° C., using uranium acetate as indicator.

[1] See Beringer, *loc. cit.;* also Irving C. Bull. "Sch. of Mines Qly.," XXIII, No. 4; "Chem. News," LXXXVII, 40.

MAGNESIUM

1. After precipitation of calcium as oxalate, the filtrate is made strongly alkaline with ammonia, and mixed with an excess of sodium or ammonium phosphate, well stirred without touching the sides of the vessel with the stirring-rod, and allowed to settle for a considerable time (2 or 3 hours). The precipitate is then filtered off and washed with 2.5 per cent. ammonia. In exact analysis it should be re-dissolved in HCl and re-precipitated with ammonia and phosphate. In washing, the minimum necessary amount of wash-water should be used, and the first portions of the precipitate are best washed on to the filter, by means of a small quantity of the filtrate. When free from chlorides, the precipitate is dried, ignited, moistened with nitric acid, cautiously evaporated to dryness, then again ignited and weighed as magnesium pyrophosphate ($Mg_2P_2O_7$).

$$Mg_2P_2O_7 \times 0.2188 = Mg.$$

2. Magnesium may also be estimated volumetrically by dissolving the precipitate of magnesium ammonium phosphate in decinormal sulphuric acid and titrating the excess of acid with standard alkali and methyl orange.[1]

$$1cc. \frac{N}{10} H_2SO_4 = 0.001218 \text{ gram Mg.}$$
$$= 0.002018 \text{ gram MgO.}$$

MANGANESE

Separation. — 1. Manganese is contained in the filtrate obtained after precipitating iron, alumina, etc., with ammonia or with sodium acetate. Its presence is generally indicated at an earlier stage by the greenish, sometimes purplish, color of the melt obtained with the ore and sodium carbonate; this color disappears on treatment with hydrochloric acid. Manganese may be precipitated, along with some other metals, by treating the alkaline filtrate from Al, etc., with H_2S or ammonium sulphide solution, and allowing to stand, covered, for a considerable time. A more convenient method, which at once separates the manganese from accompanying metals, is to make the filtrate just acid with sulphuric acid. It is then warmed to about 70° C. and precipitated by a brisk current of H_2S gas. This throws down zinc,

---

[1] Handy, "Journ. Am. Chem. Soc.," XXII, 31.

nickel, cobalt, and also copper and any other of the heavy metals which may be present, together with any traces of platinum which have passed through from previous operations. The precipitate is filtered off and washed with dilute H_2SO_4 containing H_2S. The manganese is contained in the filtrate.

2. The ore is decomposed by treatment with hydrochloric and nitric acids in the usual way. It is then heated to white fumes with sulphuric acid, cooled somewhat, and diluted with water, warmed till iron has dissolved, and then mixed with an emulsion of zinc oxide or pure precipitated zinc hydrate. This precipitates iron, leaving manganese in solution. Filter, add a saturated solution of bromine in water, and 2 to 3 grams of sodium acetate.[1] On boiling, the manganese is precipitated as hydrated manganese dioxide. Boil and wash with hot water.

Estimation. — 1. When manganese has been separated in H_2SO_4 solution, as in the first method, the filtrate is made alkaline with ammonia, which, with the excess of H_2S, causes the manganese to precipitate as sulphide. The flask is stoppered and set aside over night, or preferably for 24 hours. The clear liquid is then decanted through a filter, the precipitate washed with dilute ammonium sulphide, then dissolved in HCl and H_2SO_4, boiled to expel excess of HCl, and the manganese determined by any suitable gravimetric or volumetric method. Perhaps the simplest in ordinary circumstances is to add carbonate of soda in slight excess to the manganese solution, which is previously boiled in a porcelain or platinum vessel. The manganese, precipitated as carbonate, is filtered off, washed with hot water till free from alkali, dried, ignited cautiously in an open crucible, and weighed as Mn_3O_4.

2. Where manganese has been precipitated by bromine as MnO_2, it may be conveniently estimated by adding an excess of oxalic acid, together with some dilute sulphuric acid, heating nearly to boiling, diluting with water, and titrating the warm solution with standard permanganate, to determine the excess of oxalic acid.

126 parts crystallized oxalic acid = 55 parts Mn or 87 parts MnO_2.

Reduction by oxalic acid in presence of dilute sulphuric acid affords an easy means of determining the amount of actual MnO_2

[1] For details see A. H. Low, " Technical Methods of Ore Analysis," 3d edition, p. 146.

in a manganese ore when other oxides are present, as the latter do not oxidize oxalic acid under the conditions of the test.

3. Very small quantities of manganese are best determined by the colorimetric method;[1] the ore is oxidized by boiling with nitric acid and lead peroxide, and the tint of the resulting solution compared with that of a standard permanganate solution containing a known quantity of manganese.

For further details on the estimation of manganese see Beringer, "Text Book of Assaying," pp. 299–306, and Sutton, "Volumetric Analysis, 8th edition, pp. 255–266."

MERCURY

Gold and silver ores occasionally contain cinnabar, and metallic mercury is frequently present in such products as tailings and concentrates, so that its determination may be of importance in connection with the cyanide process. Whenever possible, several pounds of the sample should be carefully panned, in cases where the presence of metallic mercury is suspected. If floured, the metal may generally be obtained in a single globule by the addition of cyanide or ammonium chloride or metallic sodium.

Estimation. — 1. The following rough method frequently suffices. The powdered sample, mixed with soda-lime, is placed in a porcelain crucible covered with a disk of gold or silver shaped like a watch-glass, the hollow side uppermost and filled with water. The mixture is cautiously heated; the cinnabar or other ore is decomposed, and the volatilized mercury condenses on the under side of the disk. When the action is complete, the disk is removed, washed with alcohol, and dried in a desiccator. It is then weighed, the increase of weight giving the amount of mercury deposited.

2. The mineral may also be mixed with lime and heated in a combustion tube closed at one end. The closed end is previously filled with powdered magnesium carbonate. The open end is bent and drawn to a point, dipping under water. The carbonic acid evolved from the magnesium carbonate drives out the last portions of mercury vapor; the sublimed mercury is condensed by the water, collected, dried, and weighed.[2]

[1] Beringer, *loc. cit.*, p. 306.
[2] Beringer, *loc. cit.*, p. 172.

NICKEL

Separation. — Proceed as described under cobalt, until a solution is obtained containing only these metals. In cases where a separation of the two metals is necessary, precipitate the cobalt as described, with potassium nitrite, filter, precipitate the nickel from a boiling solution, by caustic alkali, filter, and wash with hot water.

Estimation.[1] — Dissolve the precipitate of nickel hydroxide in dilute ammonia. Add a measured quantity of a solution of potassium cyanide which has been accurately standardized on a silver nitrate solution of known strength, using potassium iodide indicator. If the nickel precipitate does not dissolve readily in ammonia, add a little ammonium chloride. When sufficient cyanide has been added to give a perfectly clear solution, a moderate excess is added beyond this point, then potassium iodide is added, and the excess of cyanide titrated by means of the standard silver solution. Cobalt and nickel may be estimated together by this method, but the presence of cobalt causes a darkening of the solution. The cyanide solution should also be standardized on a nickel solution of known strength.

OXYGEN

In most cases the amount of oxygen in an ore can only be determined by difference, after all other ingredients have been estimated. Sulphates and carbonates are separately determined and their oxygen contents calculated. The nature and degree of oxidation of the metallic oxides present may sometimes be ascertained by examining uncrushed specimens of the ore. In a few cases oxygen may be directly determined (*e.g.*, in cassiterite) by igniting the ore in a stream of hydrogen and collecting the water formed by means of a suitable absorbent in a weighed tube. The hydrogen must of course be previously dried and freed from any admixed oxygen, and any combined water present in the mineral must be previously expelled, or allowed for in the determination. In cases where oxides of a given metal in various degrees of oxidation are present in an ore, methods may be employed for determining one or other of these separately and deducing the amount of the other from the total quantity

[1] T. Moore, "Chem. News," LXXII, 92.

of the given metal. Thus in an ore containing ferrous and ferric oxides, we may determine the total iron, and the iron present as ferrous oxide, by separate tests. A simple calculation then gives the amount of each oxide present.

PHOSPHORIC ACID

Separation. — Ores frequently contain small quantities of phosphoric acid in the form of metallic phosphates, and calcium phosphate is found pure or mixed with other minerals. Most phosphates are soluble in hydrochloric acid.

1. Boil with hydrochloric acid, filter, ignite the insoluble residue in a platinum crucible and add it to the extract. Evaporate the whole to dryness. Take up the residue with nitric acid, concentrate by boiling, dilute, filter, and wash with water. Precipitate the warm filtrate with a warm clear solution of ammonium nitro-molybdate, which gives a yellow precipitate on standing.

2. In cases where arsenic is present, take up the evaporated residue with HCl instead of HNO_3, precipitate the As with H_2S, filter, boil off excess of H_2S, then add HNO_3 and proceed as above, adding ammonium nitro-molybdate.

3. The substance is treated with HCl as already described, evaporated to dryness, taken up with HCl, diluted and precipitated with H_2S. The precipitate is filtered off and the filtrate boiled to expel excess of H_2S and treated with HNO_3. Ferric chloride is then added, if sufficient iron is not already present, to convert the whole of the phosphoric acid into ferric phosphate; the solution is then just neutralized with ammonia, sodium acetate and acetic acid added, boiled slightly, and filtered. The precipitate, after washing, is transferred to a flask and treated with ammonia and H_2S to remove the iron. It is then filtered, the filtrate containing the phosphoric acid as ammonium phosphate.

Estimation. — 1. When the phosphate has been separated with nitro-molybdate, the precipitate is collected on a weighed filter-paper, and washed, first with dilute nitric acid, then with alcohol, dried at 110° and weighed as ammonium phospho-molybdate $(NH_4)_3 \cdot 12MoO_3 \cdot PO_4$.[1]

Weight of precipitate $\times 0.0165$ = P.

[1] "Journ. Am. Chem. Soc.," XIX, 614.

2. The molybdate precipitate is collected on an unweighed filter, washed with acid ammonium sulphate, then dissolved in ammonia. The solution is reduced with powdered zinc and sulphuric acid, and titrated at 40° C. with permanganate, which oxidizes Mo_2O_3 to $2MoO_3$.

3. The molybdate precipitate is washed with 10 per cent. ammonic nitrate, dissolved in dilute ammonia, and precipitated with "magnesia mixture" ($MgSO_4$, NH_4Cl, and NH_4OH). The precipitate, after standing for some time, is filtered and washed with dilute ammonia and dissolved in HCl, made just alkaline with ammonia, mixed with acetic acid and sodium acetate till distinctly acid, and titrated with standard uranium acetate, with ferrocyanide indicator exactly as in the case of arsenic. (See above.)

4. The phosphate solution, after precipitation as ferric phosphate and removal of iron as described in separation method No. 3, is acidulated with HCl and titrated with uranium acetate as above.

POTASSIUM

The methods of separating the alkali metals from other elements have already been described. (See above.)

Estimation. — 1. In cases where considerable amounts of both sodium and potassium are present, the quantity of potassium may be deduced by calculation, after weighing the mixed chlorides and titrating the chlorine, by means of the formula:

$$K = 2.44\ M - 4.024\ Cl, \text{ when}$$

K = weight of potassium in portion analyzed,
M = weight of mixed chlorides found,
Cl = weight of chlorine found.

The sodium is of course found by difference.

2. In cases where an exact estimation of potassium is necessary, the final residue after ignition is not titrated for chlorine, but dissolved in water and mixed with a solution of platinic chloride (H_2PtCl_6) containing 3 parts of platinum for every part of the mixed sodium and potassium chlorides present.[1] It is then evaporated on a water-bath nearly to dryness, moistened again with the platinum solution, covered with alcohol, and washed by rotating the dish. After settling, the liquid is de-

[1] Tatlock's method. See also A. H. Low, "Technical Methods of Ore Analysis," 3d edition, p. 184.

canted through a filter and the precipitate washed with alcohol until the filtrate is perfectly colorless. The residue in the dish is then washed on to the filter, using as little alcohol as possible, finally washing once or twice on the filter by means of a small wash-bottle containing alcohol and provided with a fine jet. The precipitate is then dried on the paper in the water-oven and transferred to a weighed crucible. It may be weighed directly as potassium platino-chloride (K_2PtCl_6). The paper is ignited and weighed as Pt + 2KCl + ash.

$$K_2PtCl_6 \times 0.30561 = KCl.$$
$$K_2PtCl_6 \times 0.1612 = K.$$
$$(Pt + 2KCl) \times 0.2276 = K.$$

3. In cases where the sodium largely exceeds the potassium, Finkener's method, as modified by Dittmar and MacArthur, is preferable.[1] Having obtained the residue of mixed chlorides as before, add platinic chloride (H_2PtCl_6) containing at least 3.2 parts of platinum for every part of potassium present. Add a little water and heat till the precipitate just dissolves, evaporate on a water-bath to a small bulk and stir to prevent formation of large crystals. Cool, add 10 cc. absolute alcohol, then after some time 5 cc. of ether; stir, allow to stand, covered, for some hours till the precipitate is thoroughly settled. Decant through a filter, and wash with a mixture of 1 vol. ether and 2 vols. alcohol. In some cases it may be necessary to re-dissolve the precipitate in water and re-crystallize by evaporation and addition of alcohol, as before. When washed free from platinum salts (Na_2PtCl_6) transfer the precipitate to a conical flask, add 1 cc. water for every 5 mg. of platinum estimated to be present in the precipitate. Immerse the flask containing the precipitate in a water-bath at 90° C. and pass in a current of purified hydrogen until the solution becomes perfectly colorless and the whole of the platinum is precipitated; thus:

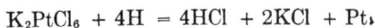

$$K_2PtCl_6 + 4H = 4HCl + 2KCl + Pt.$$

Filter, wash with hot water, ignite and weigh the platinum.

$$Pt \times 0.402 = K.$$

SELENIUM

Selenium and tellurium are not infrequently associated with gold or silver in ores, and are obtained, together with sulphur,

[1] W. Dittmar, "Exercises in Quantitative Analysis," pp. 25–30, 310–313.

by the methods given under that heading, as selenates and tellurates. These compounds are less perfectly precipitated than sulphates by addition of barium chloride.

Detection. — The ore is extracted with nitro-hydrochloric acid, or fused with sodium carbonate and niter and extracted with HCl. The filtrate is evaporated on a water-bath several times with HCl, adding a little NaCl in the first case, until all nitrates are expelled. On adding sodium sulphite to the solution and heating to boiling, selenium gives a red precipitate, becoming black on continued boiling. Tellurium gives at once a black precipitate.

Separation. — Having obtained a hydrochloric acid extract free from nitric acid, pass a current of purified SO_2 gas into the boiling liquid, until the precipitate comes down in a dense form. It is then collected on a weighed filter-paper, washed rapidly with dilute SO_2 water, dried at 100° C., and weighed as xSe + yTe. To separate selenium and tellurium the dry precipitate is mixed with about 10 times its weight of powdered potassium cyanide and fused in an atmosphere of hydrogen. The melt is extracted with water, filtered if necessary, and a current of air passed through the filtrate for some time. Tellurium alone is precipitated, being present as potassium telluride, which gradually decomposes under the influence of air. Selenium remains in solution as KSeCN. After collecting the tellurium on a weighed filter, the liquid is acidulated with HCl and Se precipitated.

Estimation. — Selenium is determined by collecting the precipitate obtained as above on a weighed filter paper, drying at 100° C. and weighing as Se. Small quantities may be estimated with sufficient accuracy by comparing the tint of the emulsion with that obtained by boiling measured quantities of a sodium selenite solution of known strength with HCl and a sulphite.

SILICA

A portion (in ordinary cases 1 gram) of the crushed sample, previously dried at 100° or 110° C., is weighed in a weighing bottle and transferred to a tolerably capacious platinum crucible. This is placed in an inclined position and heated, cautiously at first, by allowing the oxidizing flame of a Bunsen burner or similar appliance to play on the outside of the crucible at the bottom and round the sides. The crucible is moved from time to time to allow the contents to be uniformly heated and to prevent caking. The object of this operation is to remove all oxidizable and volatile material such as sulphur, arsenic, etc., which might injure the crucible in the subsequent fusion.

After the ignition is complete, the crucible is allowed to cool and its contents intimately mixed with 6 or 7 grams of the purest obtainable anhydrous sodium carbonate, finely powdered. The crucible is then gradually raised to a bright-red heat, maintaining an oxidizing atmosphere until the contents are thoroughly fused and there are no signs of effervescence. It is then allowed to cool, cleaned externally if necessary, and immersed in an evaporating dish, preferably of platinum, containing warm water, to which sufficient hydrochloric acid is added to render the liquid slightly acid. The mass generally disintegrates, and can then be detached from the crucible; the latter is carefully washed and removed, and the liquid evaporated, finally, over a water-bath until nearly dry. The mass is then re-dissolved in dilute HCl, warmed and filtered. The filtrate, which may contain about 1 per cent. of silica, is again evaporated on the water-bath, taken up with dilute acid and filtered, best on a separate filter. The precipitates are thoroughly washed with hot water and the additional silica added to the main portion. The united filtrate is reserved for estimation of alumina, iron, etc. It may still contain 0.1 per cent. of silica, which is separated later.

The precipitate is dried with the filter-papers in an air-bath, then transferred to a platinum crucible, ignited strongly, allowed to cool in a desiccator and weighed in the covered crucible, as the finely-divided silica is very hygroscopic. The substance thus obtained nearly always contains some impurity, and in accurate work should be examined by treating the whole, or an aliquot part, with a mixture of sulphuric and hydrofluoric acids, heating the platinum crucible until the whole of the silica is expelled. The crucible and residual contents are then weighed; this weight, deducted from the previous one, gives the amount of silica in the portion of substance examined.

The method here described is inapplicable in presence of lead or other heavy metals liable to attack platinum, or in presence of fluorides. In these cases the interfering substances must be removed by preliminary operations before the estimation of silica is made.

SODIUM

See ALKALI METALS and POTASSIUM.

STRONTIUM

See BARIUM.

SULPHUR

Sulphur exists in ores usually in the form of metallic sulphides, and to a less extent as sulphates. Its estimation, particularly in concentrates and similar products, may sometimes be important, and in some cases useful information may be obtained by determining the mode of combination in which the sulphur occurs.

Separation. — 1. In certain cases sulphur may be brought into solution by evaporating with nitric acid or aqua regia, which convert it into sulphuric acid or a metallic sulphate. In this connection it is well to remember that when iron pyrites is oxidized with nitric acid a portion of the sulphur is necessarily converted into free sulphuric acid:

$$2FeS_2 + 15O + H_2O = Fe_2(SO_4)_3 + H_2SO_4$$

and may be lost if the evaporation is carried too far, unless some base, *e.g.*, Na_2CO_3, be added.

2. A more general method is to fuse the finely powdered mineral in a covered platinum dish or crucible with an excess of niter, to which sodium carbonate is added if the amount of sulphides is large. The fusion is conducted at a low-red heat, and the mass is extracted with water and filtered, most of the other constituents remaining insoluble.

Estimation. — 1. The extract in either case is freed as much as possible from nitric acid by evaporation with HCl, then diluted considerably, heated to boiling, and precipitated with barium chloride. After standing for some time, the clear liquid is poured off through a filter, the residue washed several times by decantation with boiling water containing a few drops of HCl, filtered, washed, dried, ignited at a moderate red heat, and weighed as $BaSO_4$.

$$BaSO_4 \times 0.1373 = S.$$

2. Sulphur may also be estimated volumetrically. For this purpose the HCl extract is concentrated to a small bulk, mixed with sodium acetate and acetic acid, diluted considerably, and boiled. Standard barium chloride solution is then run into the boiling liquid, contained in a large dish, until a portion of the liquid, taken out and filtered, begins to give a slight turbidity with dilute sulphuric acid.

3. In ores containing sulphate of lead, the latter is extracted by treating with hot concentrated ammonium acetate, and the sulphate collected and determined as described under LEAD. Ores containing $CaSO_4$, $BaSO_4$, or $SrSO_4$ are fused with sodium carbonate and the sulphur determined in the aqueous extract. (See BARIUM.)

TELLURIUM

For methods of separation and estimation, see SELENIUM.

Detection. — The presence of small traces of tellurium may be detected by boiling the substance with nitric acid, then evaporating with strong sulphuric acid until white fumes appear: on cooling somewhat, a strip of tin-foil is added to the liquid, when a fine purple color appears if tellurium be present.

Tellurium or selenium may sometimes be present in the precipitate of $BaSO_4$, obtained in the estimation of sulphur as barium tellurate ($BaTeO_4$) or barium selenate ($BaSeO_4$). In this case the precipitate should be fused with sodium carbonate, and the aqueous extract acidified with HCl and treated with SO_2, as described under SELENIUM.

TIN

Ores such as are treated by the cyanide process rarely contain much tin. When present in minute quantities, the metal is detected and estimated with great difficulty. The method of separating tin from arsenic and antimony is given under ANTIMONY.

Estimation. — 1. Where the amount is sufficient, the following method [1] may be used. The ore is crushed coarsely and concentrated by vanning. About 20 grams of the concentrates are then extracted with aqua regia to remove the bulk of the other metals which may be present, SnO_2 being practically insoluble in this reagent. After washing and filtering, the residue on the filter-

[1] C. and J. Beringer, "Text Book of Assaying," 9th edition, p. 278.

paper is placed in a fire-clay crucible, size E, and calcined, then mixed with its own weight of potassium cyanide. An equal weight of cyanide is added as a cover, and the whole is fused. The pot is then removed from the furnace, tapped, and the contents poured into a mold. The slag is dissolved in water, and the button of metallic tin cleaned and weighed.

2. Tin may also be estimated by fusing the oxidized and roasted mineral with soda under a cover of charcoal, in an iron, nickel, or silver crucible. The molten mass is dissolved in strong hydrochloric acid, and the tin, after reducing to stannous chloride, titrated by means of a solution of iodine in potassium iodide. This method may be used in presence of arsenic and antimony, provided the solution is sufficiently acid.

TITANIUM

This metal, which is frequently present in small quantities in siliceous ores, may be conveniently estimated in the solution obtained during the separation of barium, by treating the residue after evaporation with H_2SO_4 and HF with 5 per cent. sulphuric acid. The titanium in this solution is estimated colorimetrically by comparing the tint produced by the addition of about 2 cc. of hydrogen peroxide, free from fluorine, with that produced under similar conditions with measured quantities of normal solution of titanium sulphate. A dark-brown color appears on addition of the hydrogen peroxide.

WATER

The determination of moisture or "hygroscopic water" has already been described, under PRELIMINARY OPERATIONS. In addition to this, many ores contain *combined water*. In most cases it is sufficient to determine the total quantity of water not driven off at 100° or 110° C. The ore is first carefully dried at this temperature until the weight of the sample in consecutive weighings at intervals of (say) half an hour remains constant. The sample is then heated in a hard glass tube connected with weighed tubes containing sulphuric acid or other absorbents, and arranged so that a current of dry air can be aspirated successively through the heated tube and the absorption apparatus. In cases where other volatile substances are given off which would interfere with the determination, the ore may be mixed with a dry alkali, such as calcined magnesia or anhydrous sodium carbonate. In some cases it is necessary to collect the water in a cooled condenser instead of absorbing in the manner described. In a few cases a white heat is necessary for complete expulsion of combined water.

ZINC

Ores of zinc, such as blende and calamine, occur pretty frequently associated with gold- and silver-bearing minerals. Zinc is also an invariable constituent of slags produced in the smelting of zinc-gold precipitate in the cyanide process.

Separation. — Zinc is usually determined in the filtrate after separation of iron and alumina. In cases where manganese also is present, the latter may be removed by boiling with bromine water in presence of an excess of ammonia, or the zinc may be precipitated as sulphide from a warm solution rendered faintly acid with sulphuric or acetic acid, manganese remaining in solution. An excess of ammonium chloride should be present in either case. When the zinc has been obtained as sulphide, free from other metals, filter, wash with boiling water containing a little ammonium sulphide; finally dissolve the precipitate in HCl, boil to expel H_2S, and filter. Where copper and other heavy metals are present, they are first precipitated by H_2S from a strongly acid (HCl) solution, before separating the zinc.

Estimation. — 1. Where the quantity is considerable, make slightly acid with HCl, dilute to about 250 cc., heat to boiling, cool to 70° C., and add a standard solution of potassium ferrocyanide (about 2 per cent. $K_4FeCy_6 \cdot 3H_2O$) until a drop taken out shows a brown coloration with a spot of uranium acetate or nitrate on a porcelain plate. Add about 5 cc. ferrocyanide in excess. Warm gently for ten minutes, and titrate the excess with a standard zinc chloride solution of corresponding strength, until a color is no longer produced.

2. Where the amount of zinc is small, precipitate as sulphide, filter, dry with the filter-paper at the mouth of a muffle, ignite, first at a low temperature, finally at a red heat, and weigh as ZnO.

3. The following method is given by H. T. Waller [1] for deter-

[1] "Bull. I. M. M.," No. 49, Oct. 8, 1908.

mining zinc in slags. The sample for analysis, 0.5 gram, is moistened with water, 5 cc. HCl added, and rubbed till gelatinous silica separates. It is then evaporated to dryness at a moderate heat, finally on a hot plate, to render silica insoluble. Take up residue with 20 cc. of a saturated solution of $KClO_3$ in HNO_3. Evaporate in a covered dish to complete dryness. By this treatment MnO_2 remains insoluble. Add 7 grams solid NH_4Cl, 20 cc. ammonia, and 25 cc. hot water. Stir, boil, and filter. If much iron is present, re-dissolve in $KClO_3$ and HNO_3 and again evaporate to dryness. In presence of copper add granulated or sheet lead and boil. Dilute filtrate to 75 cc., heat to 70° C. and titrate with ferrocyanide as above.

ADDITIONAL NOTES ON ORE ANALYSIS

Estimation of Copper. — P. S. Harrison (Eng. Min. Journ., *95*, 283, Feb. 1, 1913) gives the following method, among others, for determining copper in mattes and ores containing As, Sb and other impurities.

One gram of the material is boiled with 10 cc. conc. HNO_3 and 10 cc. HCl, then 5 cc. H_2SO_4 added and heated to fumes. After cooling, add 30 cc. water and boil for 10 minutes, allow to stand, hot, until all ferric sulphate has dissolved, filter, wash thoroughly with hot water. Add 25 cc. of saturated sodium thiosulphate, boil 20 minutes, filter, wash twice with hot water and once with alcohol, dry and ignite at a low temperature.

From this point either the electrolytic or the iodide method may be used.

For the former, place ignited residue in beaker, add 8 cc. conc. HNO_3, heat with addition of salt to precipitate Ag, filter, wash with hot water, add 2 cc. conc. H_2SO_4 and electrolyze with ND 100 = 0.2 amp., 1.9 to 2.1 volts for 14 hr., wash cathodes twice with warm water and once with alcohol, dry at 100° C. and weigh.

For the iodide method, transfer ignited residue to a flask, add 0.1 gram $KClO_3$ and 10 cc. conc. HNO_3, heat to dryness, cool, add 50 cc. water, then ammonia in excess, boil off excess, add 10 cc. acetic acid, boil to dissolve copper, cool, add 6 grams KI and titrate as usual with thiosulphate and starch indicator.

These two methods gave practically identical results, but the iodide method is simpler and quicker, and requires less expensive apparatus. When arsenic and antimony are present, they are partly precipitated on the cathode with the copper and give too high a result with other forms of the electrolytic method, but these elements appear to be eliminated by either of the methods here given.

E. C. Kendall (Journ. Amer. Chem. Soc., Dec., 1911) gives a modification of the iodide method which avoids the necessity of evaporating to dryness after addition of HNO_3. Sodium hypochlorite is added, which oxidizes nitrous to nitric acid with liberation of chlorine. After standing 2 minutes, add rapidly 10 cc. of a 5 per cent. colorless solution of phenol, to remove free chlorine, blow out the fumes and wash down sides of flask, add NaOH till a white precipitate forms. Dissolve this in a few drops of acetic acid, add 10 cc. of 30 per cent. KI and titrate as usual.

The hypochlorite solution is adjusted to correspond with $\frac{N}{10}$ $Na_2S_2O_3$, and may be titrated with the latter, after addition of water, KI and dilute HCl. In using this method, the solution, before adding hypochlorite, should measure 50 to 60 cc.; acidity should equal 4 or 5 cc. HNO_3 and temperature about 75° F. Sufficient hypochlorite has been added when the color of solution changes from clear blue to green.

Permanganate Method for Copper. — (See text, p. 482, par. 4.) The following additional particulars of this method are given by A. T. French (Bull. 89, I. M. M.). Having obtained the copper as sulphide, free from other metals, it is dissolved in bromine and HNO_3 and evaporated to dryness with addition of .05 gram of iron. The residue is dissolved in 2 to 3 cc. HCl and diluted to 150 to 200 cc. Add 20 cc. 10 per cent. sodium sulphite and 5 cc. 10 per cent. NH_4CNS. Boil till discolored; add more Na_2SO_3, if necessary. Filter and wash 3 times with hot water. Place funnel over flask and wash with hot 8 per cent. NaOH or KOH, also washing the beaker in which the precipitation was made with the same solution. Wash 3 times with hot water, dilute to 300 cc. with cold water, add 300 cc. cold water, acidify with H_2SO_4, adding 10 cc. in excess, and titrate with permanganate (5.7 grams $KMnO_4$ per liter: 1 cc. = .002 gram Cu).

Estimation of Mercury. — The following volumetric method, due to L. L. Krieckhaus, is described by A. H. Low (Technical Methods of Ore Analysis, 5th edition, p. 181).

Two grams of the ore are digested in the cold for 1 hr., with

2 cc. strong HNO_3 and 10 cc. strong HCl. Dilute with 10 to 15 cc. of water, filter, wash with hot water. Add 60 cc. stannous chloride solution gradually, in the cold, and allow to stand in a corked flask, inclined at 45°, for about 2 hr., or until settled clear. The stannous chloride is prepared by mixing 50 grams $SnCl_2$, 50 cc. HCl and 150 cc. of water, and boiling with metallic tin till clear. The reduced mercury is then washed with dilute sulphuric acid, by decantation, and dissolved in 2 to 3 cc. strong HNO_3, then 75 cc. cold water added, and the liquid titrated with thiocyanate and ferric indicator, as in Volhard's method for silver. The reactions are:

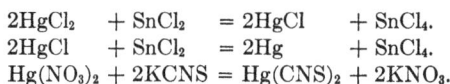

$$2HgCl_2 + SnCl_2 = 2HgCl + SnCl_4.$$
$$2HgCl + SnCl_2 = 2Hg + SnCl_4.$$
$$Hg(NO_3)_2 + 2KCNS = Hg(CNS)_2 + 2KNO_3.$$

With a thiocyanate solution containing 9.722 grams KCNS per liter, 1 cc. = .01 gram Hg, but it is best to standardize on pure silver. The silver value of 1 cc. multiplied by 0.9265 gives the mercury value.

Volumetric Estimation of Mercury. — The writer finds that mercury may be rapidly and accurately estimated as follows. The mercury is brought into the mercuric condition, free from other heavy metals. Free halogens must also be absent. Mercuric oxide is first dissolved in HCl. Make strongly alkaline with caustic alkali, then run in standard KCN (about 1 per cent. strength) until the red precipitate entirely dissolves and a clear colorless solution is obtained. Add, say, 5 cc. of the standard KCN in excess, then KI and titrate excess KCN with standard silver nitrate.

The amount of $AgNO_3$ required in this titration is deducted from the $AgNO_3$-equivalent of the total volume of KCN used. The difference is the equivalent of the mercury present.

1 cc. $AgNO_3$ = .005 gram KCN = .00769 gram Hg.

Estimation of Tin. — Jas. Gray (Journ. Chem. Met. and Min. Soc. of South Africa, March, 1910, pp. 312–315) selects the following as being the most accurate and satisfactory of many methods examined:

"From 0.5 to 2 grams of the finely powdered ore is fused with 4 to 6 times its weight of sodium peroxide in an iron crucible, and, after cooling, the melt is extracted with water. The solution thus obtained is transferred to a conical flask and excess of HCl added. The solution is then heated to 90° C., reduced with iron nails and the reduction continued for 45 to 60 min. after the solution has become colorless. The flask and contents are then rapidly cooled in a stream of running water, a piece of marble added to the solution to maintain a neutral atmosphere, and when cold a little starch paste added and the solution titrated with iodine solution which has previously been standardized against pure tin, until the starch blue appears."

This method was criticized on the ground that an error in the iodine titration is introduced by the use of iron nails, that there is no means of knowing when the reduction to $SnCl_2$ is complete, and that arsenic, tungsten and other elements interfere.

A. H. Low (Technical Methods of Ore Analysis, 5th ed., p. 246) uses a modification of E. V. Pearce's method. After fusing 0.5 gram of the ore in an iron crucible with about 8 grams sodium hydroxide, the melt is poured out into a nickel dish floating in water, and finally dissolved in a casserole with addition of 30 cc. strong HCl, the crucible and cover being washed with hot water and dilute HCl. After adding a further 50 cc. strong HCl and diluting to 200 cc. the solution is reduced by boiling with a coil of metallic nickel, or by the addition of 1 gram powdered antimony, adding marble, or passing a current of CO_2 through the apparatus to avoid oxidation. The reduction is complete in 15 to 20 minutes. Cool flask quickly in running water. Remove any undissolved nickel and rinse the latter with HCl (1:3). Add starch liquor and titrate immediately with point of burette well within the flask and avoiding violent agitation.

Determination of Zinc in Ores. — The following modification of Low's method is described in Journ. Ind. Eng. Chem., 1913, p. 302 (Abs. Eng. Min. Journ., *96*, 836, Nov. 1, 1913) as applicable in presence of impurities of all kinds.

One gram of the ore is dissolved in aqua regia, treated with 1 gram potassium chlorate, evaporated to dryness, the residue treated with 50 cc. of water and 0.5 gram of potassium hydroxide, and then heated nearly to boiling for several minutes with 6 grams of ammonium carbonate, whereby Fe, Al, Mn, Pb and Cd are precipitated. The precipitate is filtered off, washed with a hot 5 per cent. solution of ammonium carbonate, then dissolved in

HCl with a little KNO_3, the solution neutralized with KOH and reprecipitated with ammonium carbonate, the precipitate being washed as before. The filtrates and washings contain copper and zinc. The copper is precipitated as sulphide and the zinc determined by titration with ferrocyanide and uranium indicator.

SECTION II

ANALYSIS OF BULLION AND OTHER MAINLY METAL-LIC PRODUCTS

In this section we shall discuss only the analysis of matter composed entirely or mainly of metallic elements, such as gold and silver bullion, mattes, skimmings, and other by-products. Useful information may sometimes be obtained by an analysis of the zinc-box precipitate, roasted or unroasted, in order to determine the best method of fluxing the same. Occasionally an analysis may be required of such products as the "white precipitate" in the zinc-boxes.

We may classify the ingredients to be determined as follows: (1) Gold and silver. (2) Base metals, more particularly zinc, copper, and lead; in smaller quantities, iron, arsenic, antimony, and occasionally manganese, nickel, and cobalt. (3) Non-metals, chiefly sulphur, also small amounts of carbon, silicon, phosphorus, selenium, and tellurium.

Sampling and Preliminary Operations

Low-grade bullion is best sampled by "dipping" while in the molten condition, just before pouring. (See Part VII.) With high-grade bullion, drill samples are admissible and in some cases preferable. Matte may generally be sampled by crushing and quartering in the manner described for ores. Gold-cyanide precipitates and similar material require extremely careful sampling. Accurate results can only be obtained by passing a large part of the dried precipitate through a fine sieve, and making separate determinations on the portions remaining on and passing the sieve. In some cases the samples contain comparatively large pieces of metal, which may be hammered out or passed between rolls, and finally cut in small pieces with clean scissors. In every case the particles should be reduced to as fine a state of sub-division as possible, and thoroughly mixed before selecting the

504

necessarily small portion taken for the actual analysis. Any particles of slag or other matter obviously not forming a legitimate part of the sample should be carefully picked out. The weight taken for analysis is usually 0.5 gram to 1 gram. Owing to the very varied nature of the materials dealt with it is impossible to give a universally applicable scheme of analysis, and only the more important determinations will be here described. Reference should also be made to Section I (Ore Analysis).

ESTIMATION OF PRECIOUS METALS [1]

GOLD

This is usually made by fire assay (see Part VII), but there are cases where wet methods are desirable, as, for instance, with certain materials heavily charged with copper and other base metals. In any case a wet method of separating gold must be used if other constituents are to be determined. The procedure will depend largely on the amount of gold present.

1. Where the gold constitutes more than (say) 30 per cent. of the material, it may in most cases be dissolved in aqua regia. With alloys containing a certain proportion of silver, however, this is very tedious, as the particles of metal rapidly become coated with a deposit of silver chloride which stops any further action. This may be overcome by alternate treatments with aqua regia and ammonia, but even in this way the process is very slow. A mixture of HCl and $KClO_3$ is more effective than aqua regia, as when sufficiently concentrated it keeps AgCl in solution; the silver may be precipitated as chloride, after all the gold has dissolved, by diluting considerably, heating, and settling.

2. Another method, applicable, however, only in the absence of volatile constituents, is to fuse the bullion in a covered porcelain crucible with five or six times its weight of some very fusible metal, such as cadmium, under a layer of cyanide.[1] The resulting alloy, after cleaning, is then dissolved in dilute and finally in 75 per cent. nitric acid, the insoluble matter being chiefly gold.

3. Where the alloy contains sufficient silver or other metal for parting the gold in nitric acid, it may be dissolved at once in that acid. Two treatments are usually sufficient, using acid of 75 per cent. by volume. The residue is washed several times by

[1] Beringer, "Text Book of Assaying," 9th edition, p. 157.

decantation with hot water, dried, ignited, and weighed. It should then be dissolved in aqua regia, diluted, and allowed to stand and settle clear. The residue is filtered off and its weight determined after careful drying. Any silver present will of course have been converted into chloride, and its amount may be determined by cupeling the dry residue with lead. The proper correction can then easily be calculated, to obtain the true weight of gold in the sample.

4. In cases where the alloy is soluble in aqua regia the gold may be precipitated by oxalic or sulphurous acids. For this purpose the liquid should be concentrated to a small bulk on the water-bath and re-evaporated several times with HCl until practically all HNO_3 is removed. The residue is then diluted, heated to boiling, filtered if necessary, and the precipitant added. A current of purified SO_2 gas is the best for the purpose, as it introduces no substance which can interfere with subsequent operations. A few other elements, such as selenium and tellurium, are, however, precipitated along with the gold by this reagent. After settling, the liquid is decanted and the precipitate washed with hot water, ignited, and weighed.

SILVER

This metal is also usually estimated by fire assay, but it may be determined gravimetrically as chloride, and (perhaps more accurately) by various well-known volumetric methods. (See Part VII.)

1. When the sample has been attacked with aqua regia, the silver, after diluting and settling, remains for the most part as insoluble chloride, possibly mixed with silica, carbon, or other refractory matter. The residue from the aqua regia treatment is washed thoroughly by decantation, dissolved in dilute ammonia, filtered, and the filtrate boiled and re-precipitated with a few drops of HCl, avoiding a large excess. After settling, the AgCl is collected in a weighed porcelain crucible, dried, heated to incipient fusion, and weighed with the crucible.

2. In cases where the sample has been dissolved in nitric acid, the liquid is heated to boiling and the silver precipitated by adding a slight excess of HCl. It is then washed, as much as possible by decantation, with boiling water, and treated as

above, except that in this case it is unnecessary to dissolve in ammonia.

3. Some account of the volumetric methods of Gay-Lussac and Volhard is given in Part VII. In connection with the latter, the writer is accustomed to use the following procedure. In samples consisting chiefly of silver, not more than 0.2 gram need be taken for the estimation. This is dissolved in nitric acid, well boiled, and cooled. The standard thiocyanate is added in slight excess; the excess is then removed by addition of a silver solution equivalent in strength to one-fifth that of the thiocyanate, until the red color of the ferric thiocyanate just disappears.

4. Where silver has been precipitated as chloride, and the quantity is considerable, it may be determined volumetrically by the method of Denigès. The AgCl is dissolved in excess of a standard solution of cyanide, using a measured quantity of the latter. If the last portions of the precipitate dissolve with difficulty, the liquid may be poured off and the residue dissolved in ammonia, the solution so formed being added to the bulk. Potassium iodide is now added, and the residual cyanide titrated with standard $AgNO_3$. The reactions are:

(a) $AgCl + 2KCN = KAg(CN)_2 + KCl$.
(b) $AgNO_3 + KI = AgI + KNO_3$.
(c) $AgI + 2KCN = KAg(CN)_2 + KI$.

ESTIMATION OF BASE METALS

These may generally be determined, after separation of the gold and silver, by the methods described in Section I.

Lead. — In cases where selenium or tellurium is present, the writer has found it desirable to separate these elements as described below, before proceeding to precipitate the lead. After removal of these elements the filtrate, containing the lead as chloride (extracted by boiling water), is evaporated to a small bulk, sulphuric acid is then added, and the liquid boiled till strong white fumes are given off. It is then cooled, diluted, and treated as described in Section I.

Copper is determined in the filtrate from the lead, as already described; the method to be selected will depend on the quantity present.

Iron is generally present in very small amounts (except in the case of zinc-box precipitates, where it may exist as ferrocyanides: see below). It is generally estimated by the colorimetric method. (See Section I.)

Zinc is determined in the filtrate from iron, as described above. (See Section I.)

Calcium. The sample is boiled with HCl or in some cases with $HCl + HNO_3$, adding a considerable excess of HCl. Dilute and make alkaline with ammonia. In presence of manganese, add bromine water. Boil, filter, and wash with hot water. Redissolve residue in HCl and re-precipitate with ammonia, adding the second filtrate to the first. Heat filtrate to boiling, add excess of hot concentrated ammonium oxalate, and proceed as described in Section I.

Manganese, alumina, magnesium, alkali metals, etc. — The methods given in Section I may be applied, having regard to the special circumstances in each case.

Non-Metals and Negative Radicals

Sulphur. — In general, the sample should be first dissolved in nitric acid. Fusion with niter in platinum vessels is usually inadmissible with metallic samples. In cases where HNO_3 cannot be used, dissolve the sample in $HCl + KClO_3$, evaporate and boil with HCl till all free chlorine is expelled, dilute, filter, heat to boiling, and precipitate with $BaCl_2$. Any gold which may possibly remain in solution and be carried down by the $BaSO_4$ precipitate may be determined, after weighing the latter, by scorifying and cupeling, and a correction applied in calculating the sulphur.

Selenium and *tellurium* may occasionally be present in quite considerable amounts in mattes, zinc-gold precipitate, etc., and are tenaciously retained by gold and silver bullion even after careful refining. They may be separated and estimated by the methods given in Section I.

Minute quantities of *carbon, silicon,* and other elements may be found in the residue insoluble in aqua regia, after treatment of the latter with ammonia and ammonium acetate.

Cyanogen. — 1. In most cases, the total cyanogen may be estimated by adding caustic alkali and mercuric oxide, and boiling for some time. The cyanogen compounds are decomposed and yield soluble mercuric cyanide, $HgCy_2$, which unites with the alkali, forming a double cyanide, such as Na_2HgCy_4. Filter and treat while hot with H_2S or Na_2S to remove excess of mercury. The excess of sulphide is removed by agitation with lead carbonate. Filter, add potassium iodide, and titrate the cyanogen with standard silver nitrate.

2. Proceed as above until the double cyanide of mercury is obtained. Then add zinc nitrate dissolved in ammonia; [1] pass in H_2S slowly until a white precipitate (ZnS) begins to appear. Filter, add KI, and titrate with silver nitrate.

3. A weighed portion of the substance is mixed with about four times its weight of granulated zinc, and distilled with dilute sulphuric acid. The mixture of hydrogen and hydrocyanic acid which is given off is led through one or more bulbed U-tubes containing 7 per cent. silver nitrate or (perhaps better) caustic alkali. Boil for a few minutes at the end of the distillation. Sulphuric acid, 75 per cent. by volume, may be used instead of zinc and dilute acid. If the HCN has been collected in silver nitrate, the silver cyanide is filtered off, ignited, and the residual silver weighed. If collected in caustic alkali, the tubes are washed out into a flask, a little KI added, and the cyanogen titrated with silver nitrate.

Ferrocyanides. — 1. In cases where all the iron is present as ferrocyanide, it is sufficient to decompose the sample by any method which will completely destroy or expel the cyanogen and estimate the iron in the residue. This decomposition may generally be carried out by boiling down with nitric acid, adding H_2SO_4 towards the finish, and heating till white fumes are freely given off. In some cases, where the amount of ferrocyanide is small, it is sufficient to boil with HCl and $KClO_3$, without evaporating to a small bulk.

2. E. Donath and B. M. Margoshes [2] give a method depending on the action of brominized caustic soda, which is applicable in cases where part of the iron is present in other forms than ferrocyanide. The substance is first digested with 8 per cent. caustic soda until as much as possible is dissolved, with gentle warming. The whole, or an aliquot part, is then filtered, and the filtrate treated with " brominized caustic soda," prepared by adding 20 cc. of bromine to a little of the 8 per cent. caustic soda. A precipitate of ferric hydrate is thus obtained, repre-

[1] Fresenius, "Quantitative Analysis," Vol. I, p. 376 (7th edition).

[2] "Journ. für prakt. Chem.," LV (1899).

senting only that part of the iron which was originally present as ferrocyanide. This is filtered off, dissolved in HCl, and reprecipitated with ammonia. Its iron contents may then be determined by any convenient method. (See Section I.)

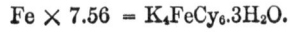

$$Fe \times 7.56 = K_4FeCy_6.3H_2O.$$

SECTION III

ANALYSIS OF CYANIDE SOLUTIONS AFTER USE IN ORE TREATMENT

In general only a few simple chemical tests have to be made on these solutions for the purpose of regulating the daily work of the plant; but there are cases where a complete analysis may be desirable or necessary; we therefore give an outline of the various estimations which might be required. A detailed discussion of this matter will be found in the present writer's "Chemistry of Cyanide Solutions resulting from the Treatment of Ores." [1]

The following list includes all the determinations which would be at all likely to occur in such an investigation, though of course many of the ingredients enumerated would be absent in any particular case.

(a) *Cyanogen Compounds.* — Free cyanide, Total cyanide, Total cyanogen, Hydrocyanic acid, Cyanates, Cyanurates, Haloid cyanides, Ferrocyanides, Ferricyanides, Thiocyanates, Selenocyanates, Tellurocyanates (?).

(b) *Metals.* — The most important in the present connection are: Gold, Silver, Calcium, Copper, Iron, Zinc. Others that may sometimes occur are: Aluminium, Antimony, Arsenic, Barium, Bismuth, Cadmium, Cobalt, Lead, Magnesium, Manganese, Mercury, Nickel, Potassium, Sodium, Strontium, Tin.

(c) *Acid Radicals other than Cyanogen Compounds.* — Bicarbonates, Bromides, Carbonates, Chlorides, Nitrates, Nitrites, Phosphates, Silicates, Sulphates, Sulphides, Tellurides, Thiosulphates.

(d) *Determinations of Alkalinity.* — "Protective" alkali, Total alkali, Alkaline hydrates, Ammonia.

(e) *Organic Matter (excluding cyanogen).* — Among the substances noted by various writers as occurring in cyanide solutions may be mentioned: Azulmates, Formates, Oxalates, Oxamide, Urea.

[1] "Eng. and Min. Journ" (1904).

(*f*) *Special Tests.* — Insoluble suspended matter; total dissolved solids; total reducing power; solvent activity.

Only a few of these determinations can be described here, and where several methods are in use for the same substance, the one given is that which, in the writer's opinion, is of most value to the cyanide worker.

TESTS USED IN DAILY ROUTINE OF CYANIDE PLANT

These are of the simplest description, and generally include only: (*a*) Determinations of free cyanide in the solutions entering and leaving the precipitation boxes, or in the sumps and storage tanks, before and after making up to required strength, with occasional special tests in addition to these. (*b*) Determinations of "protective" alkali (*i.e.*, of the alkalinity of the solution, exclusive of that due to the cyanide itself.) (*c*) Assays of gold (and silver, if necessary) in the solutions entering and leaving the precipitation boxes. These are described below, under METALS.

WORKING TEST FOR FREE CYANIDE

From 10 to 50 cc. of the solution, according to its strength in cyanide, are mixed with a few drops of a strong solution of potassium iodide and titrated with a solution of silver nitrate, adjusted to give the result with little or no calculation. A strength commonly used is 13.04 grams $AgNO_3$ per liter, in which case 1 cc. = 0.01 gram KCN or 0.004 gram CN. The finishing point is usually taken as the first appearance of a permanent *yellowish* turbidity, disregarding a slight white cloudiness which may appear earlier. This probably gives slightly too high a result in presence of K_2ZnCy_4 and its analogues, as some of the cyanogen of these compounds may be also indicated when free alkali is present. This is not, however, of much consequence if the test be made in the same manner for all solutions. In absence of KI the end-point (white turbidity) is reached sooner, but is generally uncertain and indefinite. An intermediate result, and one which, so far as tested, appears to correspond with the actual working strength of the solution, is obtained by neutralizing the protective alkali (as described below), and then adding KI and titrating with $AgNO_3$ in the usual way, taking as the end-point the appearance of a distinct permanent white turbidity.

In making the free cyanide test it is essential to have the solution to be tested perfectly clear. It must be filtered, if necessary, but the addition of lime or other substance for clarifying is generally not admissible. Dilution, addition of alkalis, variation of temperature and other conditions, must also be avoided, as all these affect the reading given by $AgNO_3$, at least in presence of zinc. The tests should be made in perfectly clean flasks and the reaction observed in a good light against a dark background, as the first turbidity, which marks the end-point, is somewhat faint. The result may be calculated in percentage, or in pounds, or kilos per ton, as desired. If 10 cc. of the solution to be examined has been taken, every cc. of $AgNO_3$ (13.04 grams per liter) added will represent:

> 0.1. per cent. KCN or 0.04 per cent CN,
> 2 lb. KCN or 0.8 lb. CN per ton of 2000 lb.,
> 1 kg. KCN or 0.4 kg. CN per metric ton of 1000 kg.

WORKING TEST FOR PROTECTIVE ALKALI

In absence of zinc, add to a measured volume of the solution to be tested sufficient silver nitrate to give a slight permanent turbidity; then without filtering add a few drops of an alcoholic 0.5 per cent. solution of phenol-phthalein, and titrate with standard acid until the pink color just disappears. The amount of standard acid used measures the protective alkali.

In presence of zinc, before titrating with acid, add a sufficient excess (say 10 cc.) of a 5 per cent. solution of potassium ferrocyanide, and proceed exactly as above. The addition of ferrocyanide liberates the alkali, which would otherwise be precipitated as zinc hydrate or carbonate on addition of silver nitrate, the reaction being probably somewhat as follows:

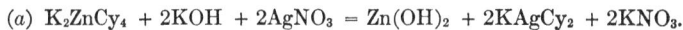

(a) $K_2ZnCy_4 + 2KOH + 2AgNO_3 = Zn(OH)_2 + 2KAgCy_2 + 2KNO_3.$
(b) $2Zn(OH)_2 + K_4FeCy_6 = Zn_2FeCy_6 + 4KOH.$

with analogous reactions in the case of carbonates. In some cases difficulties arise owing to the gradual return of the phenolphthalein color on standing. This may be due to insufficient $AgNO_3$ or ferrocyanide having been added, and may generally be avoided by adding the required amount of $AgNO_3$ to precipitate all the cyanogen as AgCN, adding an excess of ferrocyanide and filtering before titrating the alkali. Any mineral acid (HCl, HNO₃, or H₂SO₄) or oxalic acid (C₂H₂O₄ · 2H₂O) may be used for the standard solution. It is frequently adjusted so as to give the alkalinity in terms of NaOH or CaO without calculation. Thus, a solution of oxalic acid containing 1.575 grams of the crystallized acid per liter may be prepared; in this case 1 cc. of acid = 0.001 gram NaOH.

ESTIMATION OF CYANOGEN COMPOUNDS

(1) *Free cyanide.* (See above.)

(2) *Total cyanide,* defined as "the equivalent in terms of potassium cyanide, of all the cyanogen existing as simple cyanides and easily decomposable double cyanides, such as K₂ZnCy₄" (excluding such bodies as KAuCy₂, HgCy₂, K₄FeCy₆, KCyS, etc.). It is determined by making the solution to be tested strongly alkaline (*e.g.*, with caustic soda), adding KI and titrating the cyanide with AgNO₃ to the appearance of a permanent yellow turbidity. For 50 cc. of the solution to be tested, it is usually sufficient to add 10 cc. of an indicator containing 4 per cent. NaOH and 1 per cent. KI. The following equation may be taken as representing the decomposition of zinc double cyanide in presence of excess of alkali and silver nitrate: (Compare equation *a*, above.)

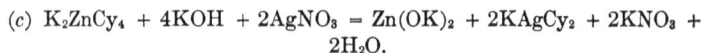

(c) K₂ZnCy₄ + 4KOH + 2AgNO₃ = Zn(OK)₂ + 2KAgCy₂ + 2KNO₃ + 2H₂O.

Total Cyanogen. — The best method of determining the whole of the cyanogen present in a solution appears to be to boil with oxide of mercury in excess, filter, and remove mercury by treatment with an alkaline sulphide. Any excess of sulphide is removed by agitating with lead carbonate, added in small quantities at a time, and filtering. The clear liquid, after addition of KI, is titrated for cyanide in the ordinary way with AgNO₃. By the above treatment practically all cyanogen compounds are converted into HgCy₂, which is subsequently decomposed thus:

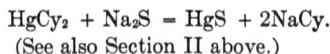

HgCy₂ + Na₂S = HgS + 2NaCy.
(See also Section II above.)

Hydrocyanic Acid. — Add sufficient AgNO₃ to convert free cyanides into double silver salts; in presence of zinc, add also an excess of ferrocyanide. The solution will now appear acid to phenol-phthalein, and the amount of HCy present may be determined by titrating this acidity with standard alkali solution:

$$1cc. \frac{N}{10} \text{ alkali} = 0.0027 \text{ gram HCy.}$$

Cyanates. — These, if present at all, would probably be rapidly decomposed under working conditions. (See Section IV.)

Cyanurates. — See "Notes on Residual Cyanide Solutions," by Chas. J. Ellis.[1]

Haloid Cyanides. — Bromide of cyanogen, as used in the Sulman Teed and Diehl processes (see Part VI), is determined as follows: The solution is acidulated with HCl and KI added. Iodine is liberated as follows:

$$BrCy + HCl + 2KI = HCy + KCl + KBr + I_2.$$

The iodine is estimated in the usual way by titration with standard thiosulphate:

$$1cc. \frac{N}{10} \text{ iodine} = 0.0053 \text{ gram BrCy.}$$

The presence of alkali cyanides does not interfere, as they are converted by the HCl into HCy, which does not affect the reaction.

Ferrocyanides. — 1. In cases where all the iron in the solution exists in the form of ferrocyanide, the latter is best estimated by determining the total iron, after decomposition of the cyanogen compounds. This may be done in some cases by simply adding a powerful oxidizer, such as bromine or HCl + KClO$_3$, and boiling for a few minutes. The iron may then be completely precipitated by adding a slight excess of ammonia to the hot solution, filtering, and determining iron by any suitable method. (See Section I.)

2. Where much ferrocyanide is present it will generally be necessary to evaporate with HNO$_3$ and H$_2$SO$_4$, sometimes more than once, until the liquid, made alkaline and re-acidified with HCl, no longer shows a trace of blue color. The iron is then separated and estimated as above.

3. Where other soluble compounds of iron, such as ferricyanides and nitroprussides, are present, it is perhaps best to determine the total FeCy$_6$, and to estimate the amount of the

[1] "Journ. Soc. Chem. Ind.," February, 1897.

ferricyanides etc., in a separate portion, the ferrocyanide being thus found by difference.

4. Where insoluble iron compounds exist in suspension and cannot readily be removed by filtration, the solution may be agitated with lime and filtered, leaving the soluble cyanogen compounds unaffected. If, however, the suspended matter itself contain insoluble ferrocyanogen compounds, they will be wholly or partially decomposed and a further quantity of ferrocyanogen added to the solution. In many cases the method of Donath and Margoshes may be applied. (See Section II.)

Ferricyanides. — In the absence of other substances capable of liberating iodine from potassium iodide, ferricyanides may be estimated by the method of Lenssen and Mohr.[1] After adding KI and acidulating with HCl, add excess of zinc sulphate, allow to stand some time, and neutralize with $NaHCO_3$. The liberated iodine is then titrated with thiosulphate in the ordinary way.

$$1cc. \frac{N}{10} \text{ thiosulphate } = 0.032941 \text{ gram } K_3FeCy_6.$$

Thiocyanates. — These are most simply estimated by the colorimetric method, merely reversing the procedure for estimating small quantities of iron. (See Section I.) The solution is acidified with HCl and filtered. An excess of ferric chloride or nitrate is then added, and the color compared with that of a similar amount of ferric solution and HCl diluted to a similar volume, to which standard KCyS is added until the tints are alike.

Selenocyanides and *Tellurocyanides.* — Selenium dissolves pretty readily in cyanide, forming compounds analogous in composition to thiocyanates. Tellurium is less easily dissolved, and may possibly be present as telluride, tellurite, or tellurate of an alkali metal. The estimation is made as described below, under *Selenium* and *Tellurium.*

ESTIMATION OF METALS

In most cases these estimations are made, after decomposition of the cyanogen compounds, by the methods already described in Sections I and II. There are, however, certain special methods applicable to cyanide solutions which must be described.

---

[1] Sutton, "Volumetric Analysis," 8th edition, p. 227.

GOLD

1. The standard method is that of evaporation with litharge. A measured quantity of the solution, usually not more than 300 cc., is placed in a porcelain evaporating dish. Litharge (20 to 50 grams) is then sprinkled over the surface of the liquid and the mixture allowed to evaporate at a gentle heat, without boiling. The evaporated residue is then fluxed as an ordinary ore assay.[1]

2 Of the numerous alternative methods that have been proposed, perhaps the best is that suggested originally, about 1896, by S. B. Christy.[2] It consists in acidulating the solution and adding a copper salt together with a reducing agent, such as a soluble sulphite, whereby a precipitate of cuprous cyanide is formed which carries down all but a minute trace of the gold. It was at first considered necessary to heat the solution after adding acid, to expel the bulk of the HCy, but A. Whitby[3] has shown that the reaction is complete in the cold; he adds, first, a sufficient volume of 10 per cent. copper sulphate, then a few cc. conc. HCl. and finally 10 to 20 cc. of 10 per cent. sodium sulphite. After vigorous stirring and settling, the precipitate is filtered off, and the necessary flux added to the filter-paper. The whole is then transferred, without drying, to a small clay crucible and fused like an ordinary assay. Both gold and silver are precipitated together and can be determined in the same assay; the method has the advantage that no heat is required in the preliminary operations. The writer has employed this method, using the following proportions:

Solution to be tested: 10 a. t. = 291⅔cc.
    Copper sulphate (10 per cent. $CuSO_4 \cdot 5H_2O$)......20cc.
    Sodium sulphite (15 per cent. Na_2SO_3) ....20cc.
    Sulphuric acid (10 per cent. H_2SO_4)..............10cc.

These are added in succession, stirring after each addition. A little ferrocyanide added to the solution promotes settlement. After thorough stirring, the mixture is allowed to settle about fifteen minutes, filtered in a large paper, and the filtrate passed back through the funnel once or twice until quite clear. No

[1] For details, see "Chemistry of Cyanide Solutions," p. 114.
[2] "Trans. A. I. M. E." (1896), pp 1–38; "Min. Sci. Press," Dec. 19, 1896.
[3] "Proc. Chem., Met. and Min. Soo f S. A.," III, 15.

water-wash is necessary. When the precipitate is sufficiently drained, 60 grams of the following flux are placed on the same filter: Borax, 30 parts; litharge, 30 parts; charcoal, 1 part.

The paper is then immediately wrapped over the flux and precipitate and the whole placed in a hot E or F clay crucible, fused, and cupeled.

3. H. T. Durant [1] gives a method which, with various modifications, has been adopted by many cyanide workers. The solution is acidulated with H_2SO_4 and boiled. About 5 grams of zinc shavings are then added, in quantities of about 1 gram at a time. More acid is added if required, and the boiling continued till the zinc is apparently dissolved. Remove from the heat and add lead acetate to form lead sulphate, which collects the gold. Filter through a double filter-paper. If the action is slow, owing to the zinc being very pure, add a few drops of copper sulphate at the beginning. The filter, with the precipitate, is placed on a layer of borax in a scorifier, allowed to dry, and char slowly in front of a muffle, then scorified with lead and cupeled in the usual way.

A slight modification of this method is given by N. S. Stines, [2] as follows: 100 cc. of the solution are mixed with 7 cc. of a 10 per cent. lead acetate solution and 1 gram zinc shavings. Heat, without boiling, for say twenty-five minutes. Add 20 cc. HCl and heat till effervescence stops. "The lead is then in such a spongy condition that by the aid of a flattened glass rod it can be pressed into a cake and the clear solution poured off." Transfer the mass to a lead-foil funnel, place in a hot cupel, and complete the assay as usual.

4. Where large numbers of assays have to be made simultaneously, perhaps the simplest and least troublesome method is to evaporate the solution, without boiling, in a dish or other vessel of lead foil; when dry, the whole is scorified and cupeled. For single assays the method is rather slow.

5. A number of colorimetric methods have been suggested for the rapid approximate determination of gold in cyanide solutions, all depending on the oxidation of the cyanide by some powerful reagent, and subsequent addition of stannous chlo-

[1] Private communication to the author.
[2] "Min. Sci. Press," April 28, 1906.

ride to the acidulated solution to obtain the purple of Cassius color.

(a) In Cassel's method,[1] which is the simplest, the preliminary oxidation is obtained with potassium bromate and sulphuric acid, or with $HCl + KClO_3$, $KBr + Na_2O_2$, or $KBr + K_2O_2$. The procedure is as follows: Mix 10 to 50 cc. of the solution to be tested with 0.5 gram $KBrO_3$ and add concentrated H_2SO_4 gradually, with shaking, till the reaction starts. When the action ceases, add drop by drop a saturated solution of $SnCl_2$ till the liquid is just colorless. The purple color now begins to develop and is most intense after about half a minute. The color is compared with that given by a standard gold solution under similar conditions.

(b) In J. Moir's method,[2] 100 cc. of solution are oxidized with 1 to 2 grams Na_2O_2 and boiled. A few drops of lead acetate are added, which produce a brown spot of PbO_2. This re-dissolves immediately if sufficient Na_2O_2 is present. A small quantity of aluminium powder is now added, and a Pb = Al couple forms which throws down the gold. The mixture is filtered and the filtrate rejected. The residue on the filter is dissolved by aqua regia and treated with $SnCl_2$, added drop by drop till the liquid is colorless. Permanent standards for comparison may be made by using a mixture of copper sulphate and cobalt nitrate, adjusting the amounts of each until the tint is the same as that given by a known amount of gold treated as above described. The purple of Cassius color fades on standing, owing to oxidation.

(c) For further modification of this test by Prof. A. Prister (see "Proc. Chem., Met. and Min. Soc. of S. A.," IV, 235, 455, 1904).

SILVER

The methods Nos. 1, 2, and 4 described under gold, serve also for the estimation of silver. In dilute solution, the silver may be separated, free from gold, by precipitating with sodium sulphide after adding a few drops of a solution of some lead compound. The precipitate should be washed with dilute Na_2S, dried, scorified, and cupeled, or it may be converted into bro-

[1] "Eng. and Min. Jour.," Oct. 31, 1903.
[2] "Proc. Chem., Met. and Min. Soc. of S. A.," IV, 298 (September, 1903).

mide by addition of bromine, washed, dried, fused, and weighed as AgBr.

CALCIUM

To 100 cc. of the solution add 10 cc. HCl and boil for ten minutes. Filter, again heat filtrate to boiling, and make slightly alkaline with ammonia. Filter if necessary. To the boiling solution add 10 cc. conc. solution of ammonium oxalate, also boiling, and stir thoroughly. Allow to settle till clear. Filter and wash with hot water till free from soluble oxalates. Wash precipitate back into original flask, add 10 cc. 25 per cent. HCl, heat to boiling, add 50 cc. water and 5 cc. conc. H_2SO_4; heat to 70° C., and titrate with $\frac{N}{10}$ permanganate.

1cc. = 0.001 gram Ca = 0.001 per cent. on 100cc. tested.

The permanganate should be standardized on pure crystallized oxalic acid, under the same conditions as the test.

COPPER

Boil with HCl + $KClO_3$ to decompose cyanogen compounds. In some cases it will be necessary to evaporate with HNO_3 + H_2SO_4 till white fumes of SO_3 are given off. Dilute, add ammonia in excess, boil and filter. Estimate copper by colorimetric method. (See Section I.) When large amounts of copper are present, decompose cyanides as above and precipitate with H_2S in hot dilute acid solution. Determine copper in the precipitate as described in Section I.

IRON

After decomposing cyanogen compounds by boiling with acids and oxidizers, as above, add excess of ammonia and boil. Filter; if the precipitate is considerable, re-dissolve in HCl and re-precipitate with ammonia. Small quantities are determined by the colorimetric test with thiocyanate; larger amounts by reduction and titration with $KMnO_4$ or $K_2Cr_2O_7$. (See Section I.)

ZINC

1. The solution is made strongly alkaline with caustic soda, heated to boiling, and Na_2S added as long as a precipitate forms.

After settling, filter off the precipitate, which may contain Ag, Hg, Pb, and Zn, and wash with hot dilute Na_2S. Dissolve in HCl, boil, dilute, and filter. Titrate filtrate at 70° with standard ferrocyanide and uranium indicator. (See Section I.)

2. Oxidize cyanogen compounds by boiling with HCl + $KClO_3$; precipitate copper, etc., with H_2S, and determine zinc in filtrate by any suitable method.

3. Precipitate zinc as sulphide, as in method No. 1. Wash thoroughly with hot water till free from soluble sulphides; transfer precipitate with filter-paper to a flask. Add $\frac{N}{10}$ iodine in slight excess, together with very dilute HCl. The following reaction occurs:

$$ZnS + 2HCl + I_2 = ZnCl_2 + 2HI + S.$$

Care must be taken to exclude air as much as possible. After shaking, and standing for a few minutes, titrate excess of iodine with $\frac{N}{10}$ thiosulphate.

1cc. $\frac{N}{10}$ iodine or thiosulphate $= 0.00327$ gram **Zn**.

OTHER METALS

The remaining metals, which are commonly of less importance in cyanide solutions, are estimated after complete decomposition of cyanogen compounds by the ordinary methods of analysis. (See Section I.) Where *volatile metals*, such as As and Sb, are to be determined, it is best to destroy the cyanogen by an alkaline oxidizer, such as Na_2O_2 or brominized caustic soda, before acidulating and precipitating with H_2S.

Mercury and *lead* may be precipitated direct from the solution by H_2S. *Lead*, which would in most cases be present as an alkaline plumbate, is shown by a white precipitate on addition of sodium carbonate, which may be filtered off and dissolved in acetic acid. The metal is then detected by the ordinary qualitative tests. It is best estimated by boiling a portion of the original solution with $HNO_3 + H_2SO_4$ till white fumes are freely given off, and proceeding as described in Section I.

Manganese may conveniently be estimated by the colorimetric method. The solution is mixed with concentrated HNO_3 and boiled for some time; then peroxide of lead is added and the boiling continued for a few moments. Dilute to 100 cc. and filter. Reject the first 10 cc. of the filtrate, collect 50 cc. and compare the tint with that of standard permanganate containing nitric acid and diluted to an equal volume. This represents the Mn in half the volume of cyanide solution taken for the test.

Estimation of Acid Radicals (other than Cyanogen Compounds)

Carbonates. — To 100 cc. of the solution add 10 to 15 cc. of a neutral 2 per cent. solution of barium chloride. Agitate and allow to settle for an hour. Filter, passing the filtrate back through the filter-paper till clear. Wash thoroughly with water till washings are free from cyanide and alkali. Transfer paper and precipitate to a flask, add a slight excess of $\frac{N}{10}$ acid (HCl or H_2SO_4), warm to 90° C., and titrate residual acid with $\frac{N}{10}$ alkali and methyl orange.

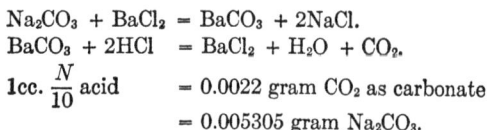

$$Na_2CO_3 + BaCl_2 = BaCO_3 + 2NaCl.$$
$$BaCO_3 + 2HCl = BaCl_2 + H_2O + CO_2.$$
$$1cc. \frac{N}{10} \text{ acid} = 0.0022 \text{ gram } CO_2 \text{ as carbonate}$$
$$= 0.005305 \text{ gram } Na_2CO_3.$$

Bicarbonates. — Add caustic soda free from carbonates to another portion of the solution and apply the test for carbonates as above.[1]

$$NaOH + NaHCO_3 = Na_2CO_3 + H_2O.$$

The increase in the amount of CO_2 or Na_2CO_3 found as compared with the previous test is a measure of the bicarbonate.

Chlorides. — The methods usually given, in which chloride and cyanide are precipitated together by means of silver nitrate, and the cyanide determined separately in another portion of the liquid, are unsatisfactory, because the amount of chloride is usually very small in comparison with the cyanide. The writer has found that the cyanide can be readily decomposed and expelled by boiling the solution with ammonium nitrate or sulphate; the cyanide is volatilized as ammonium cyanide:

$$(NH_4)_2SO_4 + 2NaCy = Na_2SO_4 + 2NH_4Cy.$$

[1] Gerard W. Williams, "Proc. Chem., Met. and Min. Soc. of S. A.," IV, 412.

leaving the chloride unaffected. The solution can then be acidified with HNO_3 and the chloride determined by adding a measured quantity of $AgNO_3$ solution, filtering and determining excess of $AgNO_3$ by titration with thiocyanate.

Nitrates. — Probably the simplest method is to precipitate cyanogen compounds with excess of a soluble silver salt, such as the acetate or sulphate, filter and determine nitrates in the filtrate by distilling with caustic soda and a mixture of granulated zinc and iron filings. The ammonia evolved is collected in a measured volume of standard acid and the residual acid titrated.[1]

$$1cc. \frac{N}{10} \text{ acid consumed} = 0.001401 \text{ gram N} = 0.008506 \text{ gram NaNO}_3.$$

Nitrites. — In absence of ferricyanides and haloid cyanogen compounds, these may be determined by adding an air-free mixture of KI and H_2SO_4 and titrating the liberated iodine with thiosulphate, care being taken to exclude air during the process.

$$H_2SO_4 + KI + KNO_2 = K_2SO_4 + NO + H_2O + I.$$

Phosphates and *silicates* may be estimated by ordinary methods of analysis in the residue after evaporation with acids and ignition to destroy organic matter and render silica insoluble. (See Section I.)

Sulphates are estimated by precipitating with $BaCl_2$, filtering and washing the precipitate till free from cyanides, etc., re-dissolving in hot dilute HCl, filtering and igniting the insoluble residue and weighing as $BaSO_4$.

Sulphides. — 1. These are present usually only in very small quantities and are best estimated colorimetrically by means of a solution of sodium plumbate or alkaline lead tartrate, comparing the tint with that given by a sodium sulphide solution of known strength, to which an equal amount of the lead solution has been added.[2] If the strength of the lead solution is known, that of the sulphide solution may be determined by means of it. A convenient method of standardizing the sodium sulphide solution is to add to a given volume of it an excess of a solution of silver sodium cyanide, filter off the precipitate of Ag_2S, and titrate the liberated cyanide with $AgNO_3$ and KI indicator.

$$1 \text{ gram KCy found} = 0.3 \text{ gram Na}_2\text{S}.$$

[1] Sutton, "Volumetric Analysis," 8th edition, p. 274.
[2] C. J. Ellis, "Journ. Soc. Chem. Ind.," February, 1897.

2. The presence of sulphides in a solution may be detected by adding sodium nitroprusside, which gives a characteristic purple color. A colorimetric method based on this reaction has been devised by Dr. J. Loevy,[1] but the writer finds it less reliable than that depending on the coloration of lead solutions.

Thiosulphates. — The writer finds that thiosulphates may be accurately determined in presence of cyanides and thiocyanates as follows: A drop of methyl orange is added to the solution, and $\frac{N}{10}$ acid run in till the liquid is exactly neutralized. The thiosulphate is then determined by titrating with a standard solution of iodine in potassic iodide, using starch indicator at the finish.

$$2Na_2S_2O_3 + I_2 = 2NaI + Na_2S_4O_6.$$

Total sulphur. — By treating with a powerful oxidizer, such as $HCl + KClO_3$, sulphides, thiosulphates, and thiocyanates are converted into sulphates, and may be determined, along with any sulphates originally present, by boiling off the excess of chlorine, etc., and precipitating the hot solution with barium chloride.

Selenium and *Tellurium.* — On heating a solution containing selenocyanides with excess of HCl, the selenium is thrown down as a scarlet precipitate (mixed with AgCl and certain ferrocyanides in some cases). In absence of cupric ferrocyanide it may be estimated colorimetrically by comparing the tint obtained with that given under similar conditions with a standard solution of sodium selenite or, better, selenocyanide containing a known amount of selenium. The finely divided selenium remains in suspension for a considerable time. In some cases an exact imitation of the tint is obtained by mixing varying amounts of zinc ferrocyanide with the standards.

When both selenium and tellurium are present, the solution is boiled for some time with aqua regia, to form selenates and tellurates, filtered, evaporated with addition of NaCl on a water-bath, and the residue heated with HCl until all HNO_3 has been driven off. The solution is then diluted, heated to boiling, and precipitated with SO_2 gas. The precipitate is collected on a weighed filter, dried at 100° C., and weighed. It may contain gold; for separation see Sections I and II.

---

[1] "Proc. Chem., Met. and Min. Soc. of S. A.," II, 608 (1899); "Chemistry of Cyanide Solutions," p. 92.

ALKALI DETERMINATIONS

Protective alkali. — See above, under "Tests used in daily routine of Cyanide Plant."

Total alkali. — This is defined as the equivalent in terms of NaOH, of all the substances which are alkaline to methyl orange It is generally determined by adding to a measured volume of the solution a known quantity of $\frac{N}{10}$ acid in excess of the required amount, making up to a definite volume and titrating an aliquot part of the solution with $\frac{N}{10}$ alkali, using a few drops of a 0.1 per cent. aqueous solution of methyl orange as indicator.[1]

Alkaline hydrates. — These may be determined by titrating with $\frac{N}{10}$ acid and phenol-phthalein, after precipitating carbonates, as already described, with barium chloride, filtering, and adding silver nitrate to the filtrate until a permanent precipitate is obtained. In presence of zinc it is necessary also to add a ferro-cyanide.[2] The results are not very exact, as the $BaCO_3$ generally carries down some of the hydrate as $Ba(OH)_2$.

Ammonia and *ammonium salts.* — Add excess of $AgNO_3$, *i.e.*, sufficient to precipitate all cyanogen compounds, then a little NaCl to remove excess of $AgNO_3$, filter, wash, evaporate filtrate on a water-bath to a moderate bulk, distil with caustic soda or sodium carbonate, and collect distillate in $\frac{N}{10}$ acid. (For apparatus, see page 442.) Titrate residual acid with $\frac{N}{10}$ alkali and methyl orange. If free ammonia only is to be estimated, use water instead of NaOH or Na_2CO_3 for distilling. In this case also the preliminary evaporation on water-bath must of course be omitted.

ORGANIC MATTER (EXCLUDING CYANOGEN COMPOUNDS)

Many obscure organic compounds may occasionally occur in cyanide solutions, particularly if the material treated has been mixed with decaying animal or vegetable matter. These need

[1] "Chemistry of Cyanide Solutions," p. 62.
[2] L. M. Green, "Trans. I. M. M.," X, 29 (1901).

not be discussed in detail. They may generally be removed, and a clear solution obtained by agitating with lime and filtering; this treatment does not affect ferrocyanides, thiocyanates, and similar bodies. Among the more usual forms of organic matter resulting from the decomposition of cyanide itself under the influence of alkalis and water, may be mentioned oxalates of the alkali metals, and urea.

Oxalates. — C. J. Ellis[1] gives the following method. The solution to be tested is precipitated with calcium chloride in excess. The precipitate, after settling, is filtered off and washed, and dissolved in a small excess of hydrochloric acid. The oxalic acid thus formed is then titrated in warm solution by standard permanganate:

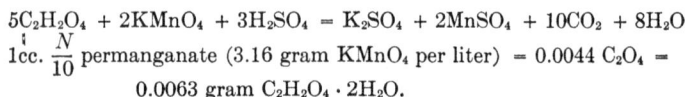

$$5C_2H_2O_4 + 2KMnO_4 + 3H_2SO_4 = K_2SO_4 + 2MnSO_4 + 10CO_2 + 8H_2O$$

$1cc. \frac{N}{10}$ permanganate (3.16 gram $KMnO_4$ per liter) $= 0.0044 \ C_2O_4 =$
0.0063 gram $C_2H_2O_4 \cdot 2H_2O$.

Urea. — If the distillation for ammonia, described above under alkali determinations, be carried out at a moderate temperature without boiling, and without using too large an excess of alkali, the urea will remain practically unchanged in the flask after the operation.[2] It may be determined by decomposing the substance with sodium hypobromite, warming slightly towards the finish.[3] $CO(NH_2)_2 + 3NaBrO = 3NaBr + 2H_2O + CO_2 + N_2$

The nitrogen evolved is collected and measured in a suitable graduated tube, and the necessary corrections for temperature and pressure applied, the CO_2 being absorbed by some alkaline liquid.

SPECIAL TESTS

Solids in Suspension. — The solutions dealt with in cyanide work are very frequently turbid with solid matter in suspension, and it may become a matter of importance to determine the amount and nature of this suspended matter. A measured volume of the liquid is filtered through an ashless paper, previously dried at 100° C. and weighed in a weighing tube. The residue collected on the filter is washed free from soluble matter, dried at

---

[1] "Journ. Soc. Chem. Ind.," XVI, 115 (February, 1897).
[2] C. J. Ellis, *loc. cit.*
[3] Sutton, "Volumetric Analysis," 8th edition, p. 432.

the same temperature as before, and weighed. The residue may then be calcined and examined by the ordinary methods of analysis. Generally, it consists chiefly of silica and compounds of iron, zinc, calcium, etc.

Total Dissolved Solids. — A portion of the liquid is filtered and evaporated to dryness in a weighed dish at a low temperature. In some cases decomposition takes place even at 100° C., and the dry weight of the solids can only be obtained by evaporating in vacuo. When the residue, as is frequently the case, is very deliquescent, it may be transferred when apparently dry from the dish to a weighing-bottle, left for some time at the required temperature, cooled in a desiccator, and weighed in the stoppered bottle.

Reducing Power. — Substances capable of absorbing oxygen are detrimental to the successful working of the process. In the event of bad extractions it is often useful to ascertain whether such substances are present in the solution in large quantity. The test is generally made by acidulating a measured quantity of the solution with H_2SO_4 and running in permanganate of known strength as long as the color disappears.[1]

A better method is to add to the acidulated solution an excess of permanganate, and after standing for some time till the reaction is complete, to determine the excess by adding KI and titrating the liberated iodine with hyposulphite. Dilute standard solutions, $\left(\text{say } \dfrac{N}{100}\right)$ are preferable in most cases. Fixed quantities of acid, solution, and permanganate should be used in all tests. If this determination be made at regular intervals, any variation from normal conditions will be at once detected.

Solvent Activity. — The efficiency of a cyanide solution for dissolving gold obviously depends on a great variety of circumstances besides the strength of the solution in free cyanide, as, for example, on the time of contact, quantity of dissolved oxygen in the solution, temperature, nature of surfaces in contact with liquid, etc.[2] The methods which have been suggested for determining what may be termed the "solvent activity" of a solution generally depend on measuring the quantity of metal dissolved in a given time under given conditions. Weighed pieces of gold

[1] See "Chemistry of Cyanide Solutions," p. 70.
[2] "Chemistry of Cyanide Solutions," p. 54.

or silver are suspended in the liquid, with or without agitation, and weighed after a certain interval of time. This, however, generally gives discordant results, owing to the impossibility of securing uniform conditions of surface, etc.

A more promising method is to prepare equal quantities of precipitated gold in a number of separate and similar vessels by means of gold chloride and sulphurous acid, making faintly alkaline with caustic soda and adding to each a fixed volume of the various solutions to be compared. After agitating a definite length of time, the residual gold is filtered off, dried, cupeled, and weighed. The difference between this weight and that of the gold taken is the measure of the solvent activity of the solution.

ESTIMATION OF SILVER IN CYANIDE SOLUTIONS

The following rapid method is proposed by G. H. Clevenger (Eng. and Min. Journ., *95*, 892, May 3, 1913).

The method depends on precipitation of the silver by zinc-dust, dissolving in nitric acid and titrating with standard thiocyanate.

Five grams of pure zinc-dust are made into an emulsion with distilled water and collected on a Gooch crucible supported over a filtering flask connected with a vacuum pump. As soon as the water is drained off, the solution to be precipitated is run on, preferably from a burette, and allowed to pass through the layer of zinc-dust, about $\frac{1}{8}$ in. thick, which has been formed in the crucible. The surface of the zinc must be kept covered with liquid throughout the operation. Twenty A.T. of solution may be precipitated in this way. Finally wash 3 or 4 times with water, transfer filter-paper and precipitate to a beaker, add 10 to 15 cc. distilled water and add gradually 15 cc. strong nitric acid. Heat till nitrous fumes are driven off, dilute to 100 cc., add ferric alum or ferric nitrate indicator and titrate in the usual manner with standard thiocyanate. In standardizing the latter, add the same quantities of zinc-dust and nitric acid as are used in the test.

SECTION IV

ANALYSIS OF COMMERCIAL CYANIDE

THE extensive and increasing use of the cyanides of the alkali metals in the extraction of gold and silver, and in other industries, renders the question of the accurate analysis of these products one of considerable importance. The mere fact that for several years an impure brand of sodium cyanide, containing 10 per cent. or more of foreign substances, was placed on the market as "98 per cent. potassium cyanide," and was generally regarded as the high-water mark of commercial excellence for this product, ought to demonstrate the necessity for increased attention to this matter.

The following are the chief constituents of the ordinary brands of commercial cyanide.

(1) *Acid Radicals* (generally in combination with the alkali metals): Cyanides, Carbonates, Bicarbonates, Chlorides, Cyanates, Ferrocyanides,[1] Hydroxides, Sulphates,[1] Sulphides,[1] Thiocyanates,[1] Thiosulphates,[1] Silicates.[1]

(2) *Metallic Radicals.* — Sodium, Potassium, Ammonium, Aluminium,[1] Calcium,[1] Iron,[1] Magnesium,[1] Silver,[1] Zinc.[1]

(3) *Moisture and insoluble matter.*

SAMPLING AND PREPARATION

Cyanide is usually consigned in cases with air-tight metallic linings, so that on opening the case it is generally assumed that the contents are in the same condition as on leaving the factory. Probably, however, some decomposition always takes place in transit, as a smell of ammonia is often noticed when the case is opened. Convenient-sized lumps are broken off or picked out from as many different parts as possible, and transferred at once to a jar with a well-fitting stopper. These are crushed to coarse powder in a dry, covered porcelain mortar. Great care must

---

[1] Generally present in traces only.

be taken not to inhale any of the dust produced by the crushing. The powder is at once transferred to a dry clean jar. As the substance is very hygroscopic, and cannot be heated to 100° C. without danger of decomposition, it is best to make all the determinations on the moist sample, and calculate the results to dry weight after estimating moisture. About 100 grams or less will suffice for all the estimations.

Every portion for analysis is weighed out in a small, accurately stoppered bottle of known weight. It is better to take roughly the amount required, transferring with a glass spatula to the weighing bottle. With a little practice it is easy to take out the quantity needed within about 0.1 gram. The exact weight of the portion so taken is then determined on an analytical balance within 0.0001 gram, correcting for weight of bottle.

ESTIMATION OF ACID RADICALS

Cyanogen (as simple cyanides, *e.g.*, KCy, NaCy, CaCy$_2$, NH$_4$Cy). — Weigh about 0.5 gram of the sample and transfer to a 200 cc. conical flask. Dissolve in about 50 cc. of distilled water, rinse out the weighing bottle and add the washings to the same flask. Add 5 cc. of neutral 1 per cent. potassium iodide solution and titrate in the ordinary way with standard silver nitrate to the first appearance of a permanent yellowish turbidity. (See Section III.) The formation of a granular precipitate near the finish generally indicates that the liquid is too concentrated, and in such cases more water may be added. If the sample contains sulphides, there is a darkening of the liquid which interferes with the silver titration. In this case agitate with lead carbonate or preferably a few drops of an alkaline lead solution and filter before titrating with AgNO$_3$.

Carbonates and *Bicarbonates.* —Take 3 to 5 grams of the sample, dissolve in 100 cc. distilled water, and proceed as in Section III.

Hydrates (hydroxides). — It is generally advisable to make a determination of "protective alkali" by adding to a weighed portion of the sample, say 1 gram, dissolved in water, sufficient AgNO$_3$ to give a permanent precipitate, then a few drops of an alcoholic solution of phenol-phthalein, and titrating with standard acid. The result obtained is the equivalent of the hydrate, plus that of half the carbonate, in terms of the standard acid employed. From this the amount of hydrate may be calculated, the percentage of carbonate being already known. Bicarbonates cannot, of course, co-exist with hydrates of the alkali metals, at least in solution.

$$1cc. \frac{N}{10} \text{ acid with phenol-phthalein indicator}$$

= 0.0056 gram KOH = 0.0138 gram K_2CO_3.
= 0.0040 gram NaOH = 0.0106 gram Na_2CO_3.

Cyanates. — 1. The writer has obtained the best results by the method of O. Herting,[1] which is tolerably simple when once the necessary apparatus has been arranged. From 3 to 5 grams of the sample are dissolved in water in an evaporating basin, and sufficient dilute hydrochloric acid (say 30 to 50 cc. of 25 per cent. HCl) are added to completely decompose the cyanide. The mixture is evaporated on a water-bath, care being taken to avoid the access of ammonia vapors. The cyanide is decom-

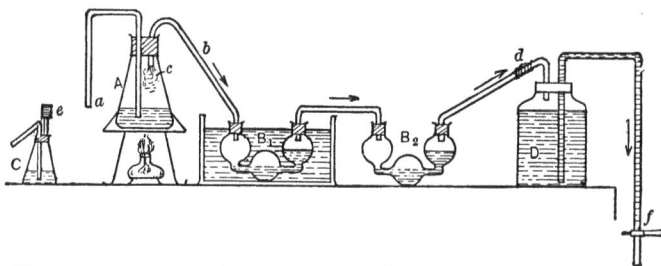

FIG 42. — Apparatus for Estimation of Cyanates and Ammonium.

posed thus:
$$NaCN + HCl = NaCl + HCN$$

while the cyanate gives the reaction
$$NaCNO + 2HCl + H_2O = NaCl + NH_4Cl + CO_2.$$

The evaporated residue is then dissolved in a little water and transferred to a distilling flask, A (see Fig. 42), connected with two bulb U-tubes, B_1 and B_2, or other similar apparatus, each containing, say, 20 cc. of $\frac{N}{10}$ acid and placed in series, the one nearest the distilling flask being kept cool by immersion. The distilling flask, A, is provided with a bent inlet-tube, a, dipping under the surface of the liquid and open at both ends. The mouth of the exit tube, b, is covered loosely with a plug of glass wool, c, which serves to prevent the splashing of any liquid into

[1] "Zeit. für angew. Chem.," XXIV, 585 (1901).

b when the contents of A are boiled. (Instead of this contrivance, the arrangement of two flasks shown in Sutton's "Volumetric Analysis," 8th edition, Fig. 48, p. 274, may be used.) The apparatus is also connected with an aspirator, D, or other means by which a slow current of air may be drawn through uninterruptedly for about thirty minutes. The wash-bottle, C, contains dilute acid and is connected with the tube a by a joint at e at a later stage.

When everything is ready, the current of air is started by opening the clip at f. About 10 cc. of normal NaOH are introduced by holding a vessel containing the alkali under the open end of tube a. The flask C is then connected with a and serves to prevent the introduction of any ammonia vapor from the air of the laboratory. The liquid in A is boiled very gently for fifteen minutes, avoiding violent ebullition or too great concentration. The lamp is then removed and the air current continued for some time longer. The apparatus is then detached at d and the contents of the bulb tubes, B_1 and B_2, together with any liquid condensed in b, are collected and titrated with $\dfrac{N}{10}$ sodium hydrate and methyl orange indicator. The amount of $\dfrac{N}{10}$ acid taken, less that found in the final titration, indicates the amount of ammonia evolved, and hence, by calculation, the amount of cyanate.

When the sample originally also contained ammonium salts, the ammonium must be separately determined (see Section III) and the amount found deducted from the total ammonium given by the test just described. The difference is then calculated to cyanate.

2. An extremely simple method of estimating cyanates has been described by A. C. Cumming and O. Masson,[1] which, if reliable, would seem to be the best method so far announced. A known volume of the solution is first titrated for "total alkali," as described in Section III, using methyl orange or Congo red as indicator. A measured excess of the acid is then added. The mixture is then boiled for a few minutes, when the decomposition given in the previous method takes place, and CO_2 is driven off.

[1] "Proc. Soc. Chem. Ind. of Victoria," July–August, 1903. See also "Chem. News," Jan. 5, 1906.

The solution is cooled and more of the indicator added if necessary. The residual acid is then titrated with standard alkali.

$$1cc. \frac{N}{10} \text{ acid consumed } = 0.003253 \text{ gram NaCNO.}$$

Of course, only that portion of acid is considered which is added after the neutralization of total alkali in the cold. The analyses reported by the authors are satisfactory.

Chlorides. — See Section III.

Ferrocyanides. — Any soluble iron will pretty certainly be present as ferrocyanide. As the amount is commonly very small, a large amount, say 10 grams, of the cyanide is dissolved and filtered. The filtrate is decomposed (under a hood) with HCl, and finally with H_2SO_4, the iron precipitated with ammonia and determined by the colorimetric test. (See Section I.)

Silicates. — Five grams of the sample are evaporated to dryness with excess of HCl, ignited gently to render silica insoluble, taken up with HCl and hot water, boiled and filtered, and washed free from chlorides. The residue is ignited and weighed as SiO_2.

Sulphides. — In the brands of cyanide commonly manufactured at the present day, the amount of soluble sulphide is very small. It may be estimated by the method of C. J. Ellis. (Section III.) When in larger quantities, precipitate a solution of say 10 grams of the sample with alkaline lead tartrate; filter, wash free from cyanide, burn the paper in a weighed crucible, add HNO_3 and H_2SO_4. Evaporate cautiously to dryness, ignite gently, and weigh as $PbSO_4$.

Sulphates and *Thiocyanates.* — See Section III.

Thiosulphates, if present in any considerable quantity, may lead to serious errors in the estimation of cyanide, owing to the solubility of AgCy in alkaline thiosulphates. They may generally be detected by the formation of a white precipitate on addition of a mineral acid; tartaric acid gives no precipitate. For estimation see Section III.[1]

ESTIMATION OF METALLIC RADICALS

Sodium. — This is usually the principal metal in cyanide samples, and is separated together with potassium. A prelim-

---

[1] See also W. Feld, "Journ. für Gasbeleuchtung," XLVI, 561; "Journ. Soc. Chem. Ind.," Sept. 30, 1903.

inary qualitative test is desirable, to determine whether iron, calcium, or other metals are present in appreciable quantities. About 1 gram or more of the sample is dissolved in water in a weighed porcelain (or, better, platinum) dish, excess of HCl added and evaporated to dryness, finishing on a water-bath and finally heating very cautiously, to avoid decrepitation, till the ammonium salts are expelled. Moisten with a few drops of HCl and again evaporate to complete dryness.

1. In the absence of appreciable amounts of metals other than Na and K, or of insoluble matter (SiO_2, etc.), it is sufficient to weigh the dish and contents, after cooling in a desiccator, and to dissolve the residue in water and make up to a definite volume. Take out an aliquot part of this solution, and determine the chlorine by titration with silver nitrate and chromate indicator. Examine another portion of the solution for sulphates; if found, determine by heating to boiling, acidifying slightly with HCl and precipitating with $BaCl_2$. Deduct the equivalent of the total Cl and SO_4 present from the total weight of the ignited residue. The difference gives $xNa + yK$.

Where only chlorides are found, the quantities of sodium and potassium may be calculated from the formula given in Section I (under "Potassium"). This method is, however, not reliable when, as is usually the case in cyanide samples, one of the alkali metals is largely in excess of the other.

2. In cases where silica, iron, etc., are present, the sample is first evaporated with HCl, then re-dissolved in a little hot dilute HCl, precipitated with ammonia, and filtered. If the precipitate is considerable, it is re-dissolved and re-precipitated. Metals other than Fe, Al, Ca, and Mg are rarely present in perceptible quantities, so that it is generally sufficient to employ the ordinary methods of separation with ammonium oxalate and carbonate. (See Section I.) No special precautions need be taken to remove sulphates, as it is more convenient to determine the SO_4 remaining at the end of the operation in the final solution, as already described. Finally evaporate and ignite to expel ammonium salts, again moisten with HCl, dry, ignite, cool in desiccator, and weigh dish and contents.

Potassium. — It will generally be necessary to make a separate determination of this metal whenever a complete analysis is required. When the potassium is largely in excess of the sodium, it is perhaps best determined by Tatlock's method [1] (see Section I), p. 421. In cases where the sodium largely exceeds the potassium, as in samples of the sodium cyanide in ordinary use in cyanide treatment, Finkener's method is preferable (see Section I, method No. 3), p. 422.

Ammonium. — Dissolve 3 to 5 grams of the sample, precipitate completely with $AgNO_3$; *i.e.*, add sufficient to convert all cyanides into AgCy, and thiocyanates, ferrocyanides, and cyanates into silver salts, leaving a small excess of $AgNO_3$ in the solution. Filter, add a few drops of HCl or NaCl to remove excess of silver, concentrate to a moderate bulk, filter into a distilling flask, and distil with NaOH as described under "Cyanates" (see above), collecting the distillate in standard acid. Titrate residual acid with standard alkali.

$$1cc. \frac{N}{10} \text{ acid consumed } = 0.0018042 \text{ gram } NH_4.$$

Other metals are determined by the ordinary methods of analysis after decomposing the cyanide.

MOISTURE AND INSOLUBLE MATTER

Moisture. — About 5 grams of the sample are weighed in a rather large weighing bottle and the open bottle placed in a desiccator over strong sulphuric acid for at least three hours, and weighed after replacing the stopper. The bottle is then left, unstoppered, in the desiccator for another hour and again weighed. The weighings are repeated at intervals of an hour until the loss of weight becomes negligible. It is advisable to have some absorbent of carbonic acid also present in the desiccator, as a slight but measurable decomposition of cyanide may otherwise take place, due to atmospheric carbonic acid. A slight loss of cyanogen also generally occurs on drying at 100° C., even in an atmosphere free from carbonic acid. Ammonium cyanide, if originally present in the sample, is volatilized at a temperature far below 100° C.

Insoluble matter. — If any portion of the sample is insoluble in water, it is advisable to weigh out at least 10 grams, dissolve to a liter, collect on a weighed filter, and wash with warm water till free from cyanide. This residue may consist of silica, ferric

[1] For details see A. H. Low, "Technical Methods of Ore Analysis," p. 184.

oxide, free carbon, carbide of iron, etc. After drying at 100° C., it is weighed with the paper, then ignited and again weighed, deducting ash of paper. The amounts of volatile and non-volatile constituents in the insoluble matter are thus ascertained.

CALCULATION OF RESULTS

This is generally done by an arbitrary rule as follows: The whole of the K and NH_4 are calculated as cyanides. NH_4 may exist in other forms, as NH_4Cl, etc., but these are probably completely converted into NH_4Cy on dissolving in water in presence of a large excess of KCy or NaCy. The remaining Cy is calculated as NaCy. Any Na in excess is distributed among the other acid radicals to form carbonates, cyanates, chlorides, etc., according to the amounts of these found in the analysis. Iron in soluble form is calculated to ferrocyanide.

SECTION V

ANALYSIS OF SUNDRY MATERIALS USED IN CONNECTION WITH THE CYANIDE PROCESS

A. LIME

LIME is universally used as a neutralizing agent, and is frequently added in the battery to counteract the effect of soluble acid salts in the mill or mine water; it is also used as a coagulating agent in bringing about the settlement of slimes. It is added to the cyanide solution as a protection against soluble and insoluble cyanicides in the material treated, and also against atmospheric carbonic acid. For the former purpose carbonate of lime may be used, 100 parts of $CaCO_3$ being equivalent to 56 parts of CaO. For direct use in conjunction with cyanide solution, however, only the soluble CaO or $Ca(OH)_2$ is of any practical value. Under certain circumstances, therefore, separate determinations of the total Ca, and of the equivalent of alkali present as CaO or $Ca(OH)_2$, become necessary.

E. H. Croghan[1] notes the following substances as occurring in "burnt lime" supplied in the Transvaal for use in cyanide plants: Lime (CaO) total, as oxide and hydrate; Magnesia (MgO); Ferric oxide (Fe_2O_3); Alumina (Al_2O_3), probably as aluminates; Manganese dioxide (MnO_2); Carbon dioxide (CO_2), in combination as carbonates; Quartz and insoluble matter; Soluble silica (as compound silicates); sulphuric acid (SO_3), in calcium sulphate; Phosphoric acid (P_2O_5) — traces only; Moisture, water of hydration, in $Ca(OH)_2$, etc; Carbon, from decomposition of organic matter. In the unburnt limestone the iron and manganese are probably present as ferrous and manganous carbonates ($FeCO_3$ and $MnCO_3$).

Sampling and Preparation. — Quicklime exposed to the air

[1] E. H. Croghan. *"Journ. Chem. Met. and Min. Soc. of S. Africa,"* VIII (2) August, 1907, pp. 37–41.

537

gradually absorbs moisture and carbonic acid; in sampling and in preparing the sample for analysis, care must therefore be taken to exclude air and moisture as much as possible. The sample is crushed as quickly as possible till fine enough to pass a 60-mesh sieve, and preserved in stoppered bottles.

Estimation of Constituents other than Caustic Lime. — The various metals and acid radicals contained in samples of ordinary quick- or slaked-lime can all be estimated by the methods already described. (See above.) The portion for analysis, in which the metals are to be determined, is dissolved in hydrochloric acid, evaporated to dryness on the water-bath, taken up with dilute HCl, filtered, and the filtrate again evaporated on the water-bath to render silica insoluble. A separate determination should be made of soluble and insoluble silica, and of carbonic acid. (See Section I.)

Estimation of Caustic Lime. — Caustic lime may be taken to mean the equivalent in terms of CaO of the whole of the calcium existing as calcium oxide (CaO) and as calcium hydroxide $Ca(OH)_2$. There are three methods in general use for this determination.

1. The total calcium is determined, and the amounts present as carbonate, sulphate, aluminate, phosphate, etc., are deducted. The remainder is calculated as CaO.

2. A small weighed portion of the sample is mixed with water, heated to boiling, and titrated with standard acid and phenol-phthalein.[1]

3. The portion for analysis is agitated with a solution of sugar in water, allowed to stand for a considerable time, again agitated and allowed to settle, and an aliquot part of the solution titrated with standard acid and phenol-phthalein.[2]

The first method is somewhat complicated, and liable to serious error, owing to the difficulty of determining exactly in what forms the calcium exists in commercial lime. The second method, according to Croghan (*loc. cit*, p. 39), appears to yield satisfactory results in the case of lime containing little or no magnesia, but gives entirely erroneous results with lime containing 24 per cent. or more of MgO, or with the so-called "blue lime," which also contains considerable amounts of MnO_2. It would not be permissible to reckon the magnesia (MgO) as "caustic lime"

---

[1] Sutton, "Volumetric Analysis," 8th edition, p. 76.

[2] G. W. Williams, "Proc. Chem., Met. and Min. Soc. of S. A.," April, 1905.

for our present purpose, as it is practically insoluble in cold water. The third or "sugar" method seems to be, on the whole, the best so far suggested. It is carried out as follows: 2 grams of the sample are transferred to a liter flask, which is then filled up to the mark with a solution containing 20 grams pure cane sugar per liter. The mixture is thoroughly shaken at intervals for an hour, allowed to stand over night, again shaken for a minute or two, allowed to settle, and an aliquot part, say 250 cc., taken out and titrated with normal HCl, using phenol-phthalein as indicator. When methyl orange is used as indicator, too high a result is obtained, as the lime present as aluminate, aluminosilicate, and perhaps some other forms, is also indicated. These compounds appear to be soluble in the sugar solution, but not so to any large extent in pure water.

Practically identical results were obtained by Croghan in some cases by using ordinary distilled water instead of sugar-water, and titrating as above with normal acid and phenol-phthalein, but as lime is more soluble in the sugar-water, the result is obtained more quickly by that method.

B. Coal

The determinations commonly required for estimating the value of coal are: moisture, volatile matter, fixed carbon, ash, calorific power. Occasionally, sulphur, phosphorus, iron, or other constituents have to be determined.

Preparation. — The sample is crushed and quartered as in the case of ores. A portion of the coarsely crushed sample is reserved for the moisture determination. As the finely crushed coal loses or absorbs moisture and is liable to undergo other changes on exposure, the sample must be kept in a well-stoppered bottle.

Moisture. — One gram or more of the coarsely crushed coal is dried for an hour, at about 105° C., in an open crucible, which is then cooled in a dessiccator, covered, and weighed. Another determination of moisture is made on the finely ground portion, for the purpose of correcting the determinations of volatile matter, carbon, and ash which are made on the fine sample.

Volatile Matter (other than moisture). — One gram of the finely powdered sample is heated in a large, tightly covered platinum crucible over a gas flame for seven minutes, gently at first, finally with the full flame of the burner. When no more flame is seen over the cover, cool and weigh. This is an entirely arbitrary test, as by heating at a higher temperature more volatile matter is expelled, but it serves as a ready means of comparing the quality of different samples of coal.

Fixed Carbon. — This is generally found by difference after determining volatile matter, ash, and moisture. There is an error in this, as in many cases a part of the sulphur is not included either in volatile matter or ash.

Ash. — This may be determined either by strongly igniting, in an open vessel, the coke left after determining volatile matter, or (better) by igniting a fresh quantity of the original finely powdered sample in an open crucible till all combustible matter is burnt off. It is necessary to heat gently at first and gradually raise the temperature.

Calorific Power. — This is a determination of the heat given off in the complete combustion of a given weight of the coal. It is carried out by instruments known as "calorimeters," the general principle being that the combustion takes place inside a closed vessel surrounded by water, the heat of combustion being calculated from the rise of temperature of the water, with corrections, depending on the particular instrument, for losses of heat by radiation and in heating the vessel itself. The powdered coal, mixed with some powerful oxidizer, such as potassium nitrate or chlorate, is placed in a copper cylinder or other suitable receiver, ignited, and quickly placed in the larger vessel filled with water of known mass and temperature. When the action is complete, the water is agitated and the temperature of the water again taken.

The calorific power of the coal is expresed in thermal units, each of which is equal to the quantity of heat required to raise unit mass of water at its maximum density through one degree of temperature. The British unit is the heat required to raise 1 lb. av. of water from 39.1° to 40.1° F.

The Calorie (metric unit) is the heat required to raise 1 kg. from 4° to 5° C.

Sulphur is generally determined by Eschka's method,[1] which has been adopted as a standard. An intimate mixture of the

---

[1] For details see A. H. Low, "Technical Methods of Ore Analysis," 3d edition, p. 272.

finely powdered coal with magnesia and sodium carbonate is gradually heated to dull redness, stirring till carbon is burnt off. The residue is treated with bromine water, filtered, and the filtrate boiled with HCl to expel excess of bromine. It is then precipitated with barium chloride and the sulphur estimated in the usual way as $BaSO_4$.

Phosphorus.[1] — This is determined in the ash from (say) 5 grams of the sample. The ash is digested with HCl, filtered, washed, and the filtrate evaporated to dryness. It is then treated with strong HNO_3, filtered and diluted, then precipitated with ammonic nitromolybdate at 40° to 45° C. After agitating thoroughly and allowing to settle, the precipitate is collected on a weighed filter, washed with 2 per cent. nitric acid, till free from soluble matter, and finally with alcohol. It is then dried for twenty minutes at 110° C. and weighed as ammonium phosphomolybdate. (See Section I.)

Iron. — This may also be determined in the ash from 1 to 5 grams of the finely powdered coal. Evaporate to dryness with HCl, filter, add a little HNO_3, dilute, heat to boiling, and precipitate with ammonia. Wash, re-dissolve in dilute H_2SO_4, reduce to the ferrous state, and titrate with permanganate. (See Section I.)

C. WATER

The quality of the water used in a cyanide plant for various purposes, such as for boilers, for making up solutions, etc., is often a matter of very great importance. Water which is very acid or highly charged with mineral salts should be avoided whenever possible, as also water containing large amounts of dissolved organic matter. On the other hand, the water may contain much common salt without being rendered unfit for making up working solutions. Very salt water is used in some plants in Western Australia. In general there is no necessity for a complete analysis. For example, a rough test will suffice for determining the total amount of organic matter. The following are the more important determinations: Acidity or alkalinity; Total dissolved solids; Solids in suspension; Hardness; Metallic radicals (calcium, magnesium, sodium, and potassium: others are less important); Non-metallic radicals (carbonates, sulphates, chlorides: others are less important).

[1] *Ibid.*, p. 274.

Sampling. — Use a large, clean, stoppered bottle. Rinse well with the water to be sampled; if possible, immerse the bottle. Avoid surface water and deposits at the bottom. If sampling from a pump or tap, allow some water to run to waste before collecting sample, so as to avoid taking water which has been standing in the pipe. Fill the bottle completely and empty again so as to expel any gases which might have been in the bottle; then fill again nearly to the stopper. Keep stoppered when not in use.

Preliminary Observations. — Note color, taste, and smell; particularly observe whether there is much suspended matter. Measure or weigh the whole sample. Allow to settle. If possible, decant as much as can be drawn off clear, measure this and keep separate. Filter the remainder through a weighed filter-paper dried at 100° C. For very accurate work the filter-papers are previously washed with HCl until all iron, etc., is removed, then washed thoroughly with distilled water, and dried at 100° C. before use.

Alkalinity or Acidity. — To 100 cc. of the filtered water add one or two drops of a 0.5 per cent. alcoholic solution of phenolphthalein. If a pink color appears, alkaline hydrates or monocarbonates are probably indicated, though in some cases the reaction might be due to sulphides, cyanides, or other alkaline salts, or to free ammonia. Titrate with very dilute $\left(\text{say } \dfrac{N}{100}\right)$ acid until the color just disappears.

To another 100 cc. of the filtered water add one or two drops of a 0.1 per cent. aqueous solution of methyl orange. A pink color indicates acidity. In this case, titrate with $\dfrac{N}{100}$ alkali till the color just disappears. If no pink color is given by methyl orange, titrate with $\dfrac{N}{100}$ acid until the tint just appears.

It frequently happens that the same water may appear neutral or even acid to phenol-phthalein and alkaline to methyl orange. This is owing to the presence of bicarbonates or of free carbonic acid. Phenol-phthalein is affected by monocarbonates of the alkali metals, but not by bicarbonates, and is also sensitive to carbonic acid, whereas methyl orange is affected by both mono- and bicarbonates, which act as alkalis towards it, while it is not sensitive to carbonic acid.[1] The following possible conditions are to be considered: (a) The water may contain bicarbonates and free carbonic acid. (b) The water may contain carbonates and bicarbonates. (c) The water may contain hydrates and carbonates. (d) Carbonates alone may be present. (e) Bicarbonates alone may be present.

(a) *Bicarbonates and Free Carbonic Acid.* — The water is acid to phenol-phthalein and alkaline to methyl orange.

(i) To one portion add phenol-phthalein and run in $\dfrac{N}{100}$ alkali gradually, without violent agitation, till the pink color just appears. This indicates free CO_2, the reaction being $NaOH + CO_2 = NaHCO_3$, or its equivalent.

$$1cc. \frac{N}{100} \text{ alkali } = 0.00044 \text{ gram } CO_2.$$

(ii) To another portion add methyl orange and titrate with $\dfrac{N}{100}$ acid till a pink tint is just produced.

$$NaHCO_3 + HCl = NaCl + H_2O + CO_2.$$
$$1cc. \frac{N}{100} \text{ acid } = 0.00044 \text{ gram } CO_2 \text{ as bicarbonate.}$$
$$= 0.00084 \text{ gram } NaHCO_3.$$

(iii) To confirm this last result, take another portion of the water, boil for five minutes, cool and add phenol-phthalein. The water will now be alkaline to this indicator; free CO_2 has been expelled and bicarbonates decomposed thus:

$$2NaHCO_3 = Na_2CO_3 + H_2O + CO_2.$$

Titrate with $\dfrac{N}{100}$ acid:

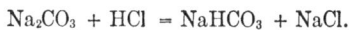

$$Na_2CO_3 + HCl = NaHCO_3 + NaCl.$$
$$1cc. \frac{N}{100} \text{ acid } = 0.00088 \text{ gram } CO_2 \text{ originally as bicarbonate.}$$
$$= 0.00168 \text{ gram } NaHCO_3.$$

(b) *Carbonates and Bicarbonates.* — The water is alkaline to both indicators.

(i) Titrate with $\dfrac{N}{100}$ acid and phenol-phthalein.

[1] See Sutton, "Volumetric Analysis," 8th edition, p. 105.

Reaction:
$$Na_2CO_3 + HCl = NaHCO_3 + NaCl.$$
1cc. $\frac{N}{100}$ acid $= 0.00044$ gram CO_2 as carbonate.
$$= 0.00106 \text{ gram } Na_2CO_3.$$

(ii) Titrate with $\frac{N}{100}$ acid and methyl orange.

Reactions:
$$Na_2CO_3 + 2HCl = 2 NaCl + H_2O + CO_2$$
$$NaHCO_3 + HCl = NaCl + H_2O + CO_2.$$
Let ϕ = no. of cc. required with phenol-phthalein;
μ = no. of cc. required with methyl orange,
using the same volume of water for the test in each case.
Then
$$\phi \times 0.00044 = \text{gram } CO_2 \text{ as carbonate};$$
$$(\mu - 2\phi) \times 0.00044 = \text{gram } CO_2 \text{ as bicarbonate}.$$

(c) *Hydrates and Carbonates.* — The water is alkaline to both indicators.

(i) Titrate with $\frac{N}{100}$ acid and phenol-phthalein.

Reactions:
$$NaOH + HCl = NaCl + H_2O.$$
$$Na_2CO_3 + HCl = NaHCO_3 + NaCl.$$
1cc. $\frac{N}{100}$ acid $= 0.0004$ gram NaOH.
$$= 0.00044 \text{ gram } CO_2 \text{ as carbonate.}$$

(ii) Titrate with $\frac{N}{100}$ acid and methyl orange.

Reaction:
$$NaOH + HCl = NaCl + H_2O.$$
$$Na_2CO_3 + 2HCl = 2NaCl + H_2O + CO_2.$$
$$(2\phi - \mu) \times 0.0004 = \text{grams hydrate, as NaOH.}$$
$$(\mu - \phi) \times 0.00044 = \text{grams } CO_2 \text{ as carbonate.}$$

(d) *Carbonates Alone Present.* — The water is alkaline to both indicators, $\mu = 2\phi$.

$$\phi \times 0.00044 = \text{grams } CO_2 \text{ as carbonate.}$$
$$\mu \times 0.00022 = \text{grams } CO_2 \text{ as carbonate.}$$

(e) *Bicarbonates Alone Present.* — The water is neutral to phenol-phthalein and alkaline to methyl orange.

$$\mu \times 0.00044 = \text{grams } CO_2 \text{ as bicarbonate.}$$

These formulæ are not applicable where other alkalis, such as ammonia, sulphides, or cyanides, are present in the water. The carbonates occurring in ordinary waters are those of the alkali metals (Na and K), and also those of the metals Ca, Mg, Fe, Mn, held in solution by the excess of carbonic acid, and perhaps forming unstable compounds analogous to the alkaline bicarbonates.

Carbonic Acid. — The following method for determining total carbonic acid (free and combined) may be used as a check on the titration given above. A measured volume (100 to 200 cc.) of the water is mixed with a slight excess of ammonia and sufficient calcium chloride solution to precipitate all the CO_2 present as $CaCO_3$. The flask is immersed for two hours in a vessel of boiling water. The precipitate is washed by decantation with hot water containing a little ammonia. The liquid poured off is filtered, and the washing continued till the residue is free from chlorides. Any material collected on the filter is then washed back into the flask, standard acid is added in excess, and the residual acid titrated with $\dfrac{N}{100}$ alkali and methyl orange.

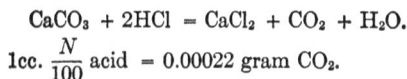

$$CaCO_3 + 2HCl = CaCl_2 + CO_2 + H_2O.$$
$$1 cc. \tfrac{N}{100} \text{ acid } = 0.00022 \text{ gram } CO_2.$$

Solids in Suspension and Solution. — These are determined as described in Section III (for cyanide solutions), except that the solids in solution may generally be evaporated at a higher temperature, first on a water-bath and finally in an air-bath at 150° C. Care must be taken not to heat sufficiently to expel CO_2 from carbonates or to decompose and char any organic matter. After cooling and weighing it is sometimes desirable to determine the weight of residue on ignition at a low temperature. The dish is heated until organic matter is burnt off, without however, decomposing carbonates. The remaining residue may be used for determination of SiO_2, Fe_2O_3, Al_2O_3, CaO, MgO, etc.

Hardness. — The "hardness" of water is estimated by the amount of calcium carbonate contained. One degree of hardness = 1 grain $CaCO_3$ per gallon. It may, therefore, be calculated from the amount of calcium carbonate found in the analysis. For Clark's test with soap solution, see Sutton, " Volumetric Analysis," 8th edition, p. 488. "Permanent hardness" is generally measured by the amounts of calcium and magnesium sulphates present.

Sulphates. — Except in the case of water containing considerable quantities of organic matter, silicates or nitrates, the sulphuric acid may be estimated directly by acidulating a measured volume with HCl, heating to boiling, and precipitating with BaCl₂ in the usual way. Where interfering substances are present, evaporate to dryness with HCl, ignite and re-dissolve residue in dilute HCl before precipitating with BaCl₂.

Chlorides. — If the water is approximately neutral these may be estimated by titration with standard AgNO₃ and neutral chromate indicator. If acid or alkaline, neutralize with standard NaOH or HNO₃ as required before adding AgNO₃. If there be any doubt as to the end-point, pour off half the liquid into another vessel, add another drop of AgNO₃ to one portion, and compare the tints; if any difference is observed the titration may be regarded.as having been finished before adding the extra drop.[1] When the water is strongly alkaline, it is perhaps better to add excess of HNO₃ and AgNO₃, and titrate excess AgNO₃ with thiocyanate and ferric indicator.

Silica is determined by evaporating and igniting, and filtering off the insoluble residue. The latter is then taken up with hot dilute HCl, filtered, and the residue on the filter again strongly ignited and weighed as SiO₂. Silica may exist in suspension as free SiO₂, or in solution as a silicate of an alkali metal.

Calcium and *magnesium* may be determined in the filtrate from the silica by heating to boiling, adding ammonia in slight excess, filtering and proceeding as described in previous sections. (See above.)

Sodium and Potassium. — As a check on the direct estimation of these elements, it is desirable to convert the total evaporated residue into chlorides or sulphates, and weigh as such. Evaporate to dryness and ignite sufficiently to render SiO₂, Fe₂O₃, and Al₂O₃ insoluble. Re-dissolve residue in water, filter, add HCl or H₂SO₄, and• re-evaporate to dryness. Weigh residue. In cases where the water contains no sulphates, hydrochloric acid may advantageously be used. In this case the total metals may be found by simply titrating the chlorine in an aliquot part of

---

[1] See A. H. Low, "Technical Methods of Ore Analysis," p. 71.

the dissolved residue. If the amounts of Ca and Mg are known, these are deducted, together with the total chlorine in the evaporated residue; the balance may generally be taken as xNa + yK.

In cases where sulphates are originally present, it is better to use sulphuric acid. The Ca and Mg are calculated to $CaSO_4$ and $MgSO_4$; the difference gives xNa$_2$SO$_4$ + yK$_2$SO$_4$; from which Na and K can be obtained by calculation. The sulphuric acid is determined in an aliquot part of the dissolved residue, and calculated to SO_4. Deducting this, we obtain the total metals, and deducting Ca and Mg, we obtain the combined weight of Na and K. It is rarely necessary to make separate determinations of Na and K. If required, this can be done as described in Sections I, III, and IV.

Iron and Alumina. — A sufficient quantity of the water is evaporated with HCl, adding a little HNO_3 and igniting residue to destroy organic matter. The residue is then boiled with HCl, filtered, and the filtrate mixed with ammonia in slight excess. After again boiling it is filtered, and the precipitate ignited and weighed as xFe$_2$O$_3$ + yAl$_2$O$_3$. If iron is to be determined separately, instead of igniting the precipitate, dissolve in HCl and estimate colorimetrically with thiocyanate. (See Section I.)

Organic Matter. — Evaporate at 150° C. and weigh residue. Ignite gently to burn off carbonaceous matter without decomposing metallic carbonates, and weigh again. The difference of the two weights gives a rough idea of the amount of organic matter present.

A simple test may also be made as follows:[1] A flask is rinsed out, first with sulphuric acid, then with pure distilled water. A measured volume (100 to 250 cc.) of the water to be tested is run into the flask, which is then stoppered and heated to 80° C. Add 10 cc. sulphuric acid (25 per cent. by vol.) to which enough $KMnO_4$ has previously been added to give a faint permanent pink tint. Add also 10 cc. potassium permanganate (0.395 gram $KMnO_4$ per liter). Allow to stand, stoppered, for four hours, then add a little KI and titrate iodine liberated, representing residual $KMnO_4$, with a corresponding thiosulphate solution, using a little starch indicator at the finish.

1cc. $KMnO_4$ consumed = 0.0001 gram available oxygen.

[1] Sutton, "Volumetric Analysis," 8th edition, p. 518.

A rule [1] based on average analyses of organic carbon in waters gives the following factors:

Organic carbon = O absorbed $\times$ 2.38 (for river water).
$\qquad\qquad$ = O absorbed $\times$ 5.8 (for deep well water).

(D) Standard Method for the Valuation of Caustic Lime

The following method is recommended by the South African Engineering Standards Committee (see abstract of report, Min. Sci. Press, *106*, 652, May 3, 1913).

The sample as delivered for analysis is contained in an air-tight vessel, having been passed through a 30-mesh sieve. It is crushed in a Wedgwood mortar and the whole passed through a 60-mesh sieve, this operation being performed as quickly as possible. It is then placed in a clean, dry, wide-mouthed bottle fitted with a tight-fitting dry glass stopper so as to prevent any access of air. Two grams of the sample are carefully weighed out and agitated with 1 liter of a 2 per cent. cane-sugar solution, or 1 gram with $\frac{1}{2}$ liter of 2 per cent. sugar solution. If a shaking machine be available, 2 hr. continuous agitation should be given; if not, 6 hr. intermittent agitation, every care being taken to prevent coagulation of the lime, and to secure intimate contact of lime and solution. After agitation, the solution is filtered, and aliquot portions titrated with $\frac{N}{10}$ or $\frac{N}{5}$ acid, using rosolic acid as indicator, avoiding delay so as to obviate undue exposure to the atmosphere. Distilled water used in the above determination must be made neutral to rosolic acid, to counteract the effect of dissolved CO_2.

(E) Chemical and Physical Tests to Determine Value of Zinc-dust for Precipitation

The efficiency of zinc-dust depends on its physical condition, particularly the fineness of its particles, as well as upon its composition. W. J. Sharwood (Journ. Chem. Met. and Min. Soc. of South Africa, 12, 332, Feb., 1912) gives a number of measurements of the fineness of particles in various specimens of commercial zinc-dust. An elutriation test was made by agitating a sample of the dust in a cylinder through which a rising stream of

[1] Sutton, " Volumetric Analysis," 8th edition, p. 508.

water was caused to flow with a velocity of about 1 cm. per second. By varying the velocity a number of products of differing fineness are obtained, all passing 200 mesh, and a residue of comparatively coarse particles, some of which remain on 100 and 200 mesh sieves respectively. Good samples of zinc-dust showed not more than 1 per cent. remaining on 100 mesh and not more than 4 per cent. between 100 and 200 mesh. Much of the material was very much finer than 200 mesh, and the minutest particles when examined under the microscope appeared as almost perfect spheres of metallic zinc.

A. M. Merton (Mining World, June 28, 1913, p. 1227) gives the following Chemical and Physical Characteristics of good zinc-dust:

> Specific gravity .... 6.8 to 7.1
> Weight per cubic foot .... 141 lb.

Screen Test:

> 0.5 to 1   per cent. on 100 mesh
> 1.5 to 4.5 per cent. on 200 mesh

Precipitating Efficiency (i.e. lb. silver precipitated per lb. of zinc) 1.15 to 1.65. This represents 35 to 50 per cent. of the theoretically possible precipitation.

Analysis:

> Metallic zinc.................... 75 to 90 per cent.
> Zinc oxide .... 3 to 10 "   "
> Lead........................... 1.5 to 7.5 "   "

Precipitating Efficiency. — Sharwood (*loc. cit.*, p. 334) gives the following test. A solution of potassium-silver cyanide is prepared by dissolving 10 grams of silver cyanide and 5 grams "99 per cent." potassium cyanide in a little water and diluting to a liter, adjusting till the solution indicates .12 per cent. to .15 per cent. free KCN. This is determined by testing with $AgNO_3$, using KI and ammonia as indicator.

The efficiency test is made by placing 0.5 gram zinc-dust in a large beaker, stirring with a little water, and adding 250 cc. of the above silver-potassium cyanide solution. See that all lumps are broken up; stir for 5 minutes, and then occasionally, at least every 10 min., until the end of 2 hr. from the addition of the solution. Filter on an 11 cm. filter, wash thoroughly, scorify with test lead, cupel at a low temperature and weigh the silver.   For this test the present writer uses wide-mouthed bottles closed by a well-fitting rubber stopper, so that all samples simultaneously tested may receive equal and uniform agitation, the bottles being placed on an agitation wheel or similar mechanical contrivance.   It is necessary to leave the bottles open for a few minutes at the start to allow most of the liberated hydrogen to escape.   The theoretical equation is:

$$2KAgCy_2 + Zn = 2Ag + K_2ZnCy_4.$$

but some zinc is always attacked by direct action of the free KCN. From the above equation 1 gram Zn should precipitate 3.303 grams Ag, hence when 0.5 gram zinc-dust is taken, mgr. Ag found $\times$ 0.0606 = precipitating efficiency per cent.

In practice the amount of zinc-dust required is much in excess of that shown by the test.

Analysis. — The effective portion of the zinc-dust is the finely divided metallic zinc, hence it is important to distinguish the zinc present as metal from that present as oxide or carbonate.   Lead is usually present, and is beneficial, up to 5 or 6 per cent., owing to the action of the lead zinc couple.

Analysis of three samples, made at Pachuca, Mexico, by the writer, showed

		A	B	C
Metallic zinc	Zn	84.33	83.29	84.71
Zinc oxide	ZnO	10.70	13.38	14.13
Lead	Pb	0.59	2.44	1.76
Cadmium	Cd	0.52	0.17	1.03
Iron	Fe	0.22	0.05	0.06
Carbon	C	2.04	0.04	0.03
Silica	SiO$_2$	0.33	0.13	0.08
		98.73	99.50	101.80

Metallic zinc is best determined by difference, after determining total zinc, and zinc present as oxide (carbonate, etc.).

Total zinc is determined by the usual ferrocyanide method, but for correct results it is necessary to remove iron and probably lead, a point overlooked both by Sharwood and Merton in the methods described (*loc. cit.*) by them.   The writer's procedure is as follows (for details see Eng. and Min. Journ., April 19, 1913, pp. 793–797):   One gram of the zinc-dust is dissolved in 50 cc. 25 per cent. HCl, and the solution filtered into a 500 cc. flask.   Practically the whole of the Pb, Cd, C and SiO_2 remain in the residue, and only Fe is present with the zinc in the filtrate. Iron is then precipitated by addition of ammonia in slight excess, filtered off, the filtrate made slightly acid with HCl, and an aliquot part titrated in the usual way with ferrocyanide and uranium indicator. A good plan is to heat the solution to be titrated to about 80° C., divide into two approximately equal parts, titrate one part until it shows the uranium reaction, add most of the remainder, and run in the calculated amount of ferrocyanide to complete the reaction. The small reserved portion is then rinsed in with some of the titrated liquid and the process finished drop by drop.

The iodine process, described in the text, may also be used, but gives rather lower results.

Zinc Oxide. — Sharwood's method, slightly modified by the writer, is as follows: A solvent is prepared consisting of water 250 cc., ammonium chloride c.p., 70 grams, ammonia (sp. gr. 0.90) 150 cc.

One gram of the dust is agitated in a stoppered flask with 25 cc. of the above solvent, for 5 min., thrown on to a filter, and washed with hot water till about 100 cc. of filtrate are obtained.

Add litmus paper, neutralize exactly with HCl, then add 5 cc. HCl in excess, heat to 80° C. and titrate zinc with ferrocyanide. .

Zn. found × 1.245 = zinc oxide.

The amount of metallic zinc is now calculated from the following data:

To standardize ferrocyanide: 1 gram pure zinc dissolved in HCl
 and made up to 500 cc.
1 cc. ferrocyanide = .005 gram Zn.
 Zinc as metal = total zinc − zinc as oxide, etc.

Lead. — Sharwood determines this by dissolving 10 grams of the zinc-dust in dilute H_2SO_4 or HCl, treating the final residue with nitric acid, filtering off any carbon, silica, etc. which remain undissolved, evaporating with strong H_2SO_4 to dense fumes. Cool, dilute, filter, wash with 5 per cent. H_2SO_4, redissolve in hot ammonium acetate and titrate at 80° C. with ammonium molybdate, using tannin or ferrocyanide indicator.

The writer obtained satisfactory results from a modification of A. H. Low's method (Technical Methods of Ore Analysis, 5th ed., p. 140): 5 to 10 grams of zinc-dust are dissolved in sufficient

HCl (25 per cent. by vol) to remove the bulk of the zinc. The residue, washed free from chlorides, is dissolved in a small quantity of HNO_3 (50 per cent. by vol.) boiled, filtered and the filtrate exactly neutralized with NaOH, using phenolphthalein indicator. Acetic acid is added till the color disappears and any precipitate dissolves. Potassium chromate is added in slight excess, boiled 5 minutes, filtered and washed with hot water containing a few drops of acetic acid till soluble chromates are removed. The $PbCrO_4$ is then dissolved in HCl (50 per cent. by vol.), potassium iodide added and the liberated iodine titrated with thiosulphate until the blue starch color changes to green.

$$1 \text{ cc. } \frac{N}{10} \text{ iodine or thiosulphate} = 0.0069 \text{ gram Pb.}$$

Cadmium. — A portion of the zinc-dust is dissolved in dilute H_2SO_4 and lead separated as in Sharwood's method. The filtrate containing the cadmium as sulphate must be dilute and not too acid. It is precipitated by H_2S, the sulphide filtered off, redissolved in hot 50 per cent. HCl, evaporated in a crucible with H_2SO_4, ignited and weighed as $CdSO_4$.

When sufficient is present, the sulphide may be dissolved in hot 50 per cent. HCl, diluted, heated to 80° C. and titrated with ferrocyanide in the same way as zinc.

$$1 \text{ cc. standard ferrocyanide} = .0086 \text{ gram Cd.}$$

Iron. — This is precipitated with ammonia as described under total zinc When sufficient is present, it may be determined by dissolving in dilute H_2SO_4, reducing to the ferrous condition, and titrating with $KMnO_4$. Small amounts are best determined by colorimetric method with thiocyanate (see text).

Carbon and Silica. — These are determined in the residue remaining after successive treatment with HCl and HNO_3. This is collected on a weighed filter-paper, dried in a steam oven and weighed. The paper and contents are then ignited and the residue again weighed.

$$\text{The first weight} = C + SiO_2 + \text{filter-paper.}$$
$$\text{The second weight} = SiO_2 + \text{ash of filter-paper.}$$

PART IX

METALLURGICAL TESTS

WE will now describe certain operations which are frequently necessary in controlling the work of a cyanide plant, and in determining the proper conditions of treatment. These consist of various tests and measurements based on well-known mathematical, physical, and chemical principles, but distinct from ordinary analytical processes, and may be roughly described as metallurgical tests. These will be considered under the following heads:

I. Density determinations.
II. Acidity and alkalinity tests.
III. Grading and screening tests.
IV. Concentration and amalgamation tests.
V. Cyanide extraction and consumption tests.
VI. Tables and formulæ.

SECTION I

DENSITY DETERMINATIONS

In cyanide treatment it is very essential to have as exact a knowledge as practicable of the quantities of material treated per day or per charge, in order to regulate operations and to check the actual recovery of values. Differences between "theoretical" and "actual" extraction may often be traced to inaccurate estimates of the quantities treated. In the case of crushed ore, moderately dry tailings and similar material, this may be done by direct weighing, but in many cases, such as slime pulp, it is practically impossible to weigh the charge in bulk, and the weight can only be calculated indirectly after ascertaining the densities of the pulp and of the dry material.

(a) To find weight of a given volume of sand or similar material.

In cases where it is not convenient to weigh the entire charge direct, the wet and dry weight of a charge is often ascertained by allowing a vessel of known capacity, say a large cubical or cylindrical box, to be filled at the same time and in the same manner as the tank or other container in which the material is to be treated. When the charge is completed, the surface is leveled off without pressing down and the box and contents weighed on a platform scale or suitable balance. The entire charge is also leveled and its volume computed from the known dimensions of the containing vessel. In the case of a cylindrical tank,

V (volume) $= \pi\, r^2 h$.

    where r = radius of tank.

        h = average depth charged with material.

        $\pi = 3.14159 \left(= \dfrac{22}{7} \text{ approximately} \right)$.

In the case of a conical vessel,

$$V = \frac{\pi r^2 h}{3}$$

    where r = radius of circle forming upper surface of mass of material filling cone.

    h = depth from surface to apex of cone.

        $\pi = 3.14159 \left(= \dfrac{22}{7} \text{ approximately} \right)$.

After weighing the sample, a sufficient quantity, say 1 or 2 kg. (4 or 5 lb.) is weighed separately, spread out in a thin layer on a metal tray, and dried at a moderate heat.   After drying the tray and contents are re-weighed and the percentage of moisture calculated.

The *wet weight* of the charge is then easily calculated as follows:

Let W = wet weight of charge.
  v = volume of sample.
  w = wet weight of sample.
  h = depth charged in tank (average).
  r = radius of tank.

Then $v: \pi r^2 h: : w: W$.

or $$W = \frac{\pi r^2 h w}{v}$$

The value of h is deduced from the average of a number of measurements taken at different parts of the surface to determine the depth left empty between the top of the tank and the charge.

The *dry weight* of the charge, D, is calculated from p, the percentage of moisture, as follows:

$$D: W: : 100 - p: 100$$
$$\text{or } D = \left(\frac{100 - p}{100}\right) W$$

(*b*) To find the weight of a given volume of wet pulp (*e.g.*, a charge of slime).

In cases where the material is of such a nature that all the interstices between the particles are filled with liquid, as in a charge of slime mixed with water or cyanide solution, the total dry weight may be calculated from the following data:

  V = total volume of charge.
  P = density (specific gravity) of pulp.
  S = density (specific gravity) of the dry material.

The *total volume* (V) is determined as already described, from the known dimensions of the containing vessels.   We may define the *density* or specific gravity of a substance as the ratio between its weight and the weight of an equal volume of water at a temperature of 4° C.   At that temperature the weight of 1 cc. of water is 1 gram; hence the density of any liquid or solid is the weight in grams of a measured quantity, divided by the number of cubic centimeters which it occupies.

The *density of the pulp* is determined by filling a vessel of exactly known volume, for instance a liter flask, with an average sample of the pulp and weighing its contents. The volume of the containing vessel need not be known, if the weight of water filling it to the mark be ascertained. In this case a correction is required for the expansion of water if the determination be made at any other temperature than 4° C., though this is negligible for most practical purposes.

The *density of the dry substance* of which the pulp is formed is generally found by weighing a certain amount of a carefully dried average sample, placing it in a vessel of known capacity and filling up to the mark with distilled water. The weight of the vessel filled with water alone is also determined. Before weighing the vessel containing the material and water, it is necessary in some cases to immerse it for one or two hours in boiling water up to the neck, leaving the stopper out, to allow the expulsion of any air contained in the dry powder. It is then cooled and weighed. The density of the dry material may then be calculated as follows:

Let m = weight of dry substance taken.
a = weight of water filling vessel to mark when no substance is added.
b = weight of substance together with amount of water required to fill vessel to mark after substance has been added.
c = weight of residual water.
d = weight of water displaced by dry material.
s = density of dry material.
Then $b = m + c$
$d = a - c$
$$s = \frac{m}{d} = \frac{m}{a - c} = \frac{m}{m + a - b}$$
Or let W_1 = weight of vessel when filled with water to mark.
W_2 = weight of vessel with dry material, filled up with water to mark.
W_0 = weight of empty vessel
Then $a = W_1 - W_0$
$b = W_2 - W_0$
and
$$s = \frac{m}{m + W_1 - W_2}$$

The *weight of dry material* in the total charge may then be calculated from the known values of P and S.

Let $P =$ density of pulp.
$S =$ density of dry material
$V =$ total volume of pulp
$m =$ total weight of dry material in pulp
$c =$ weight of water in pulp
$w =$ weight of a unit volume of water

Then (1) $P = \dfrac{m + c}{Vw}$ (2) $S = \dfrac{m}{Vw - c}$

From (1) $c = VwP - m$ From (2) $c = \dfrac{Vw\,S - m}{S}$

Hence $m = \dfrac{VwS(P - 1)}{S - 1}$

This formula is subject to the same correction for expansion of water which is involved in calculating the value of P.

(c) Useful data deducible from the above formulæ.[1]

(1) *Percentage of dry material in a charge of pulp.*

Let $p =$ percentage of dry material
Then $p : 100 : : m : m + c$

Hence $c = \dfrac{(100 - p)m}{p}$

From (1) and (2) above: $Vw = \dfrac{m + c}{P} = \dfrac{m + cS}{S}$

or $c = \dfrac{m(S - P)}{S(P - 1)}$; hence $\dfrac{100 - p}{p} = \dfrac{S - P}{S(P - 1)}$

and $p = \dfrac{100\,S(P - 1)}{P(S - 1)}$

(2) *Percentage of solution in a charge of pulp.*

$q =$ percentage of solution

$$q = 100 - p = \dfrac{100(S - P)}{P(S - 1)}$$

(3) *Ratio of solution to dry material* (R).

$$R = \dfrac{q}{p} = \dfrac{S - P}{S(P - 1)}$$

(4) *Volume of unit weight of pulp* (v).

$$v = \dfrac{V}{m + c} = \dfrac{1}{wP}$$

(5) *Weight of unit volume of pulp* (w_1).

$$w^1 = \dfrac{m + c}{V} = wP$$

[1] For a full decussion of this matter see note by W. A. Caldecott, "Proc. Chem., Met. and Min. Soc. of S. A.," II, p. 102, 153, and 837.

(6) *Weight of dry material per unit volume of pulp* (m_1).

$$m^1 = \frac{m}{V} = \frac{wS(P-1)}{S-1}$$

(7) *Volume of pulp containing unit weight of dry material* (v_1).

$$v^1 = \frac{V}{m} = \frac{S-1}{wS(P-1)}$$

(8) *Total weight of solution in charge of pulp* (c).

$$c = \frac{Vw(S-P)}{S-1}$$

(*d*) Application of the above formulæ to particular cases.

(1) *Weights in pounds avoir. Volumes in cubic feet.* — 1 cu. ft. of water = 62.5 lb.

Hence $w = 62.5$

$$m = \frac{62.5 \; VS(P-1)}{S-1}$$

$$v = \frac{1}{62.5 \; P} = \frac{0.016}{P}$$

$w^1 = 62.5 \; P =$ (weight of a cu. ft. of pulp)

$$m^1 = \frac{62.5 \; S(P-1)}{S-1} = \text{(weight of dry material per cu. ft. of pulp)}$$

$$v^1 = \frac{0.016 \; (S-1)}{S(P-1)} = \begin{array}{l} \text{cu. ft. of pulp containing 1 lb. of dry} \\ \text{material} \end{array}$$

$$c = \frac{62.5 \; V(S-P)}{S-1} = \text{lb. of solution in } V \text{ (cu. ft. of pulp)}$$

(2) *Weights in tons of* 2000 *lb. Volumes in cubic feet.* — 1 ton of water = 32 cu. ft.

Hence

$$w = \frac{1}{32}$$

$$v = \frac{32}{P}$$

$$w^1 = \frac{P}{32}$$

$$m^1 = \frac{S(P-1)}{32(S-1)}$$

$$v^1 = \frac{32\;(S-1)}{S(P-1)}$$

$$c = \frac{V(S-P)}{32(S-1)}$$

$$m = \frac{VS(P-1)}{32(S-1)}$$

(3) *Weights in metric tons of* 1000 *kilograms. Volumes in cubic meters.* — 1 ton of water = cu. meter.

$$w = 1$$

$$v = \frac{1}{P}$$

$$w^1 = P$$

$$v^1 = \frac{S-1}{S(P-1)}$$

$$m^1 = \frac{S(P-1)}{S-1}$$

$$c = \frac{V(S-P)}{S-1}$$

$$m = \frac{VS(P-1)}{S-1}$$

SECTION II

ALKALI CONSUMPTION TESTS

IT is often desirable to ascertain by preliminary experiments the quantity of alkali which will be needed to neutralize acidity in any particular ore. In this connection a distinction is sometimes made between "free" and "latent" acidity.

The *free acidity* is that due to substances soluble in water, and is determined by agitating a known weight (say half a pound or 200 grams) in a closed bottle with a sufficient quantity of water (say 4 or 5 parts water to 1 of ore) for half an hour or so. After settling, filter off a sufficient portion of the liquid, take an aliquot part (say $\frac{1}{4}$ or $\frac{1}{10}$ of the whole liquid), and determine the acidity by titrating with standard alkali. In some cases this acidity may be determined by direct titration of the filtrate with standard caustic soda and phenol-phthalein indicator, but the results are usually indefinite owing to the precipitation of metallic salts which obscure the end-point. A better method is to add a decided excess of alkali; filter, and titrate the filtrate or an aliquot part of it with standard acid, using one or two drops of a 0.1 per cent. solution of methyl orange as indicator.

The *latent acidity* may be taken to mean the consumption of alkali due to the combined effect of substances insoluble in water. Thus many ores contain insoluble sulphates of iron, etc., which are decomposed by lime, caustic soda, etc., and thus neutralize alkali.

The *total acidity* (free and latent together) is determined by agitating a weighed quantity of the substance with an excess of standard caustic soda or clear lime water, using a measured quantity of known strength. After agitation for half an hour, or one hour, the liquid is filtered and a measured portion titrated with standard acid, using methyl orange as indicator. A convenient proportion for this test is two parts liquid to one of ore. The standard alkali solution may be so adjusted that the consumption

561

per ton of ore may be obtained with little or no calculation. Thus, if 200 grams ore be used, it is agitated with 400 cc. of 0.1 per cent. NaOH or CaO, containing, therefore, 0.4 gram of the alkali. Say, for example, that 100 cc. of the filtrate are titrated and that V cc. of standard acid are required, equivalent to V cc. of standard alkali; then the entire 400 cc. of liquid would have required 4 V cc., equivalent to 0.004 V gram alkali remaining unconsumed. The alkali consumption is therefore $0.4 - 0.004$ V gram on 200 grams of ore, or 4 $(1 - 0.01\ V)$ lb. per ton of 2000 lb.

The latent acidity may be found, if required, by taking the difference between free and total acidity. It should be remembered that the determination must generally be made on a moist sample of material, as drying vitiates the test by changing the conditions. It must therefore generally be accompanied by a moisture determination, made on a separate portion of the material, as described in Section I, and the results calculated to dry weight. It is also necessary to point out that various soluble salts (ferrous sulphate, magnesium sulphate, alum, etc.) which are not in themselves acid to test paper, may nevertheless constitute the whole or part of the acidity when determined as described.

When an approximate idea of the acidity of an ore has been obtained by tests such as those just described, it is often possible to determine more exactly the quantity which would be needed in practice by agitating or percolating a number of equal portions of the ore with water to which different amounts of lime or caustic soda have been added. By this means the minimum alkali necessary to neutralize acidity may be ascertained, the solution drawn off from each test being titrated with standard acid. When lime is used, sufficient time must be allowed for the reaction, owing to its slight solubility in water.

SCREENING AND HYDRAULIC SEPARATION

Screening Tests. — Much useful information may frequently be obtained by making a "grading test" or "screen analysis" of the ore or product under investigation. If such tests be carried out on a sample of ore or tailings which it is proposed to treat with cyanide, and a similar set with the residue after treatment, it is often apparent that one part of the ore is more refractory than another, and it may be seen whether it would be advantageous to separate the finer and coarser particles for special treatment. Thus, if it be found that the extraction is more effective in the finer portions, it may be inferred that re-grinding of the coarser particles would probably result in improved extraction. The efficiency of stamps, ball-mills, tube-mills, and other crushers may be tested with more or less accuracy by making screening tests on the products entering and leaving the mill. (See Part III.)

These tests are made by passing a weighed quantity of the dried sample successively through a number of sieves, and weighing separately the portions retained on each and the portion which passes through the finest sieve. The material is made to pass through the sieve by gentle rubbing, at the same time shaking or tapping the sieve, taking care to avoid any grinding. The rubbing may generally be done effectively by means of a large cork or flat block of smooth wood. If necessary, the various products are separately assayed, and the percentage of the total values contained in each may be calculated. The coarser products should be re-ground before assaying, so that average samples may be taken. In most cases 500 to 1000 grams of the original material will suffice for a set of grading tests, but where the particles are coarse a larger amount must be taken.

The particulars of a test of this kind are conveniently summarized in some such form as the following, which is given merely as an example, and would, of course, vary according to circumstances.

563

Product	Total Weight of Product: Grams	Per cent. of Total Weight	Assay of Product per ton of 2000 lbs.		Per cent. of Total Values	
			Gold: dwt.	Silver: dwt.	Gold	Silver
(a) Remaining on 20-mesh						
(b) Passing 20-mesh and remaining on 30-mesh						
(c) Passing 30-mesh and remaining on 60-mesh						
(d) Passing 60-mesh and remaining on 120-mesh						
(e) Passing 120-mesh and remaining on 200-mesh						
(f) Passing 200-mesh						

Hydraulic Separation Tests. — In some cases it is essential to make the separation by water, as, for example, when it is desired to ascertain how much slime would be obtained from a given ore after crushing in a given manner. Various forms of apparatus are in use for sizing and screening tests by hydraulic separation, consisting generally of conical vessels of glass or other suitable material, arranged so that a regulated jet of water may be introduced at the apex of the cone. The action is in fact an imitation of that taking place in the spitzlutte described in Part III.

A serviceable practical test is to stir a known weight of material in a bucket with 4 or 5 times its weight of water, allow to settle 30 seconds, 1 minute, 2 minutes, etc., according to the nature of the material, and decant the bulk of the liquid and suspended matter into a larger vessel. This operation is repeated as often as the liquid decanted carries any appreciable amount of matter in suspension. The residue in the bucket is then dried and weighed, and assayed if necessary, and the whole, or a measured average part, of the decanted pulp is also carefully dried, weighed, and assayed. If the density of the pulp and of the dry material contained in it be determined, the weight of the whole may be determined as described in Section I, without the necessity of drying more than a small portion of the decanted pulp. Some ores contain very appreciable amounts of soluble matter, of which due account must be taken in calculating results.

SECTION IV

CONCENTRATION AND AMALGAMATION TESTS

Concentration Tests are made with a prospecting pan, batea, vanning shovel or similiar appliance. Considerable skill and practice is required to obtain successful results or to imitate approximately the effect of large-scale machinery. For hand panning tests, the most convenient method is to arrange a large bowl or bucket of enameled iron at a suitable hight, so that the pan may be dipped into it without discomfort. This vessel is partly filled with clean water and a further supply of water from a tap or another bucket should be close at hand. The weighed sample (1 or 2 lb.) which is to be panned is taken, a handful at a time, and stirred in the pan with a little water, care being taken to make any floating particles sink as much as possible by agitating the surface. The pan is held just above the surface of the water in the large bucket, and the lighter particles are washed off by a combined shaking and revolving motion, dipping the pan occasionally under the water, brushing off the "tail" of light sand and gradually increasing the inclination till only heavy mineral remains. The latter is then washed off into a separate vessel, and a fresh handful of the ore taken for panning. When the whole is finished it is generally necessary to pan the concentrates themselves over again with some care. These are then dried and weighed. As the concentrates frequently contain much pyrites or other easily oxidized ingredients, the drying must generally be done at a low temperature.

The results may be conveniently tabulated in much the same way as those of grading tests. The following shows the manner of arranging a combined concentration and hydraulic separation test:

Product	Total Weight of Product: Grams	Per cent. of Total Weight	Assay of Product per ton of 2000 lbs.		Per cent. of Total Values	
			Gold: dwt.	Silver: dwt.	Gold	Silver
Concentrates..............						
Tailings (sands)...........						
Slimes....................						

Amalgamation Tests are designed to imitate the effect of plate amalgamation after crushing with stamps, or of pan amalgamation in Wheeler, Huntington, or other pans. Many methods have been proposed, but it is doubtful if any of them are reliable under all circumstances. One of the simplest methods of obtaining an approximation to the results of *plate amalgamation* is as follows: One kg. of the dry ore is placed in a large wide-mouthed jar together with 500 cc. of water and (say) 5 grams of caustic soda. After agitating and ascertaining that the charge still shows distinct alkalinity to litmus, 200 grams of mercury are added and the mixture agitated for one or two hours by rolling the closed bottle up and down on a table. The contents are then washed out into an enameled bowl or other suitable vessel and the mercury separated by panning. In cases where the latter has been broken up into small globules it may sometimes be collected by adding a considerable quantity of fresh mercury, or by rubbing with a little sodium amalgam, caustic soda, or ammonium chloride. The floured mercury may also be re-united by addition of cyanide, or of nitric acid, but these are obviously not admissible in a test of this kind.

The extraction indicated by the test is best determined by assays of the ore before and after treatment; results based on the gold found by dissolving and distilling the mercury panned off are usually unreliable, as losses are difficult to avoid and the mercury employed, if not specially purified, is very liable to contain gold originally. Attempts are sometimes made to imitate the conditions of plate amalgamation by washing the ore on an amalgamated copper dish, the surface of which is kept bright by repeated applications of cyanide and sal ammoniac, but in the writer's experience this method cannot be safely relied on. Agitation in a bottle with sheets of amalgamated copper has also been tried, but cannot be recommended. The results by either of these methods are liable to be much lower than those obtained with the same ore on a working scale.

A test imitating *pan amalgamation* may be made by grinding in an iron mortar 300 to 500 grams of ore, 90 to 250 cc. of water, 0.75 to 1.5 grams caustic soda, and 30 to 100 grams of mercury.

The above proportions are usually suitable, but may be varied according to circumstances. The grinding is continued for 1, 2, or 3 hours, according to the nature of the material. The mortar is then emptied into a dish and the mercury panned off as in the preceding test, and the residue dried and assayed. In cases where cyanide tests are to be made on the residue after amalgamation, the latter is simply allowed to drain as much as possible without drying over a fire.

CYANIDE EXTRACTION AND CONSUMPTION TESTS

SMALL scale tests are often of the greatest utility in determining the proper working conditions. There are cases where enormous unnecessary outlay has been incurred, unsuitable plants erected or processes adopted, through the neglect of a few simple and comparatively cheap experiments on the material which it was proposed to treat. Before deciding on any system of cyanide treatment, the following points should be elucidated by preliminary tests: (1) The fineness to which the ore should be crushed. (2) The strength of solution in cyanide and alkali most suitable for the conditions. (3) The time of treatment. (4) The best ratio of solution to ore. (5) The necessity or otherwise of water-washing, aeration, oxidation, roasting, or other preliminary or auxiliary treatment. It is supposed that some idea of the nature of the ore has already been obtained by following out the tests described in previous sections, for acidity, concentration, amalgamation, etc.

STANDARD PRELIMINARY CYANIDE TESTS

In most cases the following test will yield useful information, which may be employed as a basis for further investigations. A carefully selected average sample of the ore is crushed to pass a 40-mesh sieve. The following mixture is then charged into a well-stoppered bottle: Ore, 100 grams; Lime, 1 gram; Solution, 100 cc. containing cyanide equivalent to 0.5 per cent. KCy (0.5 gram KCy or 0.377 gram NaCy). For very poor ores, the weights of ore, lime, and solution should be increased in proportion. The bottle is agitated continuously for 16 hours, at the end of which time it is allowed to stand for an hour and a sufficient quantity of the supernatant liquid decanted off and filtered. The filtrate is tested for cyanide and alkali. The residue in the bottle is then washed out into a large enameled-iron bucket, well stirred with water, and washed at least twice by decantation with a large volume of water so as to remove practically the whole of any soluble matter.   If the nature of the material allows, this washing may conveniently be done on a vacuum filter.   In either case the washed residue is dried and assayed.

For agitation tests of this kind a very convenient apparatus may be constructed by cutting holes of the size and shape required to receive the bottles in a large wooden disk.   These holes should be arranged symmetrically and should be shaped so that the stoppers of the bottles are directed towards the center of the disk. Some arrangement must also be provided for securing the bottles firmly in their places.   The disk is mounted so as to revolve in a vertical plane and driven at a moderate speed. Where such apparatus is not available, the bottles may be securely wrapped in cloth and tied to the spokes of a flywheel or similar revolving mechanism.

The results of this test will give an approximate idea of the extraction likely to be obtainable by ordinary methods of treatment and of the probable consumption of chemicals.   The ore is assayed before and after treatment, the difference of the assays giving the value dissolved per ton.   The difference in the percentage of cyanide in the solution, before and after treatment, multiplied by 20, gives the consumption of cyanide in pounds per ton of 2000 lb., since 100 grams of ore and 100 cc. of solution were taken for the test.   The same difference, multiplied by 10, gives the consumption of cyanide in kilos per metric ton of 1000 kilos.

Tests to Determine Conditions of Treatment

Having obtained a general idea of the behavior of the ore towards cyanides by the preceding standard test, the best conditions may be more precisely determined by arranging series of tests in which one condition is varied at a time.   The following are given merely as examples, and would of course be varied according to circumstances.

(1) Varying fineness of ore:

Weight of ore taken: grams	Weight of solution taken: grams	Lime added: grams	Cyanide strength: KCy per cent.	Time of Agitation: hours	in each case
100	100	1	0.5	16	

Tests made by bottle agitation, as described above, on average portions separately crushed to 30, 40, 60, 100, 150, and 200-mesh.

Let it be supposed that the results show little or no advantage in crushing finer than 60-mesh. A further series is then arranged.

(2) Varying strength of cyanide:

Ore	Mesh	Solution	Lime	Agitation	in each case
100 grams	60	100 grams	1 gram	16 hours	

Cyanide strength in different tests per cent., 0.05, 0.1, 0.2, 0.3, 0.4, 0.5, calculated as KCy.

Supposing the extraction to be sufficiently complete with 0.2 per cent., a further series might be made with 0.1, 0.125, 0.15, 0.175, and 0.2 per cent., the results of which might indicate that there is no advantage in increasing the strength above 0.15 per cent. = 3 lb. KCy per ton of ore treated.

(3) Varying alkali:

Ore	Mesh	Solution	Cyanide	Time	in each case
100 grams	60	100 grams	0.15 per cent.	16 hours	

Lime added: 0.1, 0.2, 0.3, 0.4, 0.5, 0.7, 1, 1.5 grams. Results indicate that 0.2 gram (= 4 lb. per ton of ore) is sufficient.

(4) Varying time of treatment:

Ore	Mesh	Solution	Cyanide	Lime	in each case
100 grams	60	100	0.15 per cent.	0.2 gram	

Time of agitation: 6, 9, 12, 16, 20, 24, 36, 48 hours.

Let it be supposed that the tests show an increased extraction up to 36 hours. A further series may be made, giving 24, 27, 30, 33, 36 hours' agitation. Results show no increase of extraction after 30 hours.

(5) Varying ratio of solution to ore:

Ore   Mesh   Cyanide   Lime   Time of treatment ⎤ in each
100 grams   60   0.15 per cent.   0.2 gram   30 hours   ⎦ case

Weight of solution added: 75, 100, 125, 150, 200, 300 grams. Let it be supposed that no increased extraction is obtained by increasing the amount of solution beyond 125 grams. It is thus shown that the most suitable proportion is 1.25 tons solution per ton of ore.

The indications obtained by these tests may then be confirmed by experiments on a larger scale, in which the ore is treated by some system of percolation or agitation, imitating working conditions as closely as practicable.

Percolation Tests are usually made on a scale of 1 to 25 lb. (500 grams to 10 kg.) For this purpose it is very convenient to use stoneware jars having an outlet at the bottom, a wooden frame covered with cotton cloth, and a glass outlet tube passing through a well-fitting cork or rubber stopper. The filter-frame should be arranged so as not to leave too large a space between it and the bottom of the jar, and the space between the edges of the frame and the sides of the jar should be tightly tamped with twine or other packing. By means of a short length of rubber tube and a screw-clip the outflow of solution can be regulated at will. If preferred, wooden buckets may be used instead of jars, but if allowed to dry when not in use, these may give trouble by leakage. A cover is advisable, to prevent access of dust, etc., and to check undue evaporation.

The treatment is as nearly as possible that which would be given in a leaching tank on a working scale. An average sample is well mixed, with addition of the necessary quantity of lime, as indicated by previous experiments. Solutions are added at suitable intervals, using for each solution from one-tenth to one-fifth of the dry weight of ore treated, giving strong and weak solutions as described in Part IV, and regulating the strength of solution and time of treatment in accordance with the results already obtained in the bottle tests. After sufficient contact with the ore, the solutions are drawn off slowly by slightly unscrewing the clip. The outflowing liquor is received in a jar or bottle and should be measured and tested for cyanide and alkali. If a complete record of these solution tests be kept, a very fair estimate may be made of the probable cyanide consumption in practice.

When the treatment is complete, the whole, or an average sample of the residue is dried at a low temperature and assayed. It is often instructive to compare the result of this assay with that of a similar portion of the residue which is thoroughly washed with water before drying. The washing may generally be done on a vacuum filter.

Aeration Tests. — Where compressed air is available, these may be very effectively carried out by using glass vessels which terminate at the bottom in a sharply pointed cone, at the apex of which the air is introduced, the pressure being regulated by a tap or by a screw-clip on a short length of very stout rubber tubing. The conical bottom avoids the possibility of any material settling and escaping agitation. A fairly satisfactory apparatus may be improvised, where compressed air is not available, by using a large jar filled with water as an aspirator, and connecting this with the agitation apparatus in such a way that a current of air is drawn through the latter. The apparatus is of course not continuous, as the aspirator has to be refilled from time to time.

When a test is to be made the mixture of ore and solution is charged into the vessel with conical bottom and the current of air passed at regulated speed through the pulp as long as may be necessary. Comparative tests under varying conditions may be arranged, on similar lines to the bottle tests above described. A comparison between these aeration tests and the agitation tests made in closed bottles will show whether aeration is advantageous or otherwise.

Other Small-scale Tests. — Many other experiments in connection with cyanide treatment are at times required, to elucidate such points as the following: (1) Use of preliminary water-wash or acid treatment. (2) Use of artificial oxidizing agents, either in preliminary treatment or as adjuncts to cyanide. (3) Use of bromocyanide, ferricyanide, or other auxiliary dissolving agent. (4) Necessity or otherwise of roasting previous to cyanide treatment. (5) Combination of cyanide treatment with amalgamation, concentration, chlorination, or other mechanical process.

TABULATION OF RESULTS

The following are the principal data which should be recorded, though of course the details may vary according to circumstances:

Weight of ore taken for test (wet).

Weight of ore taken for test (dry).

Moisture (per cent.).

Mesh to which ore is crushed.

Time of treatment (hours): (a) agitation:

(b) percolation: } total:

(c) settlement:

Proportion of solution to dry ore.   [average KCy per cent.

Strength of solution used in cyanide {   maximum KCy per cent.

Strength of solution in alkali, calculated as NaOH.

Consumption of cyanide: lb. per ton.

Consumption of lime: lb. per ton.

Assay value: (a) original.   (b) residue.

Extraction: (a) dwts. or grams per ton.   (b) per cent.

SECTION VI

USEFUL FORMULÆ

(1) To find weight of cyanide (or other soluble salt) to be added for making up stock solution to required strength.

When the solutions are contained, as is usually the case, in cylindrical or rectangular tanks, the quantity is calculated by measuring the depth in the tank, having previously determined the weight corresponding to a unit of depth.

Let D = total depth of solution in storage tank when filled to required point.

d = depth in storage tank already filled

a = per cent. of cyanide at present in storage tank

b = per cent. of cyanide in sump or reservoir from which liquid is to be drawn in making up.

p = percentage required in final solution

t = weight of solution corresponding to unit depth in storage tank.

Then $D - d$ = depth of solution to be added to storage tank

x = weight of solid cyanide to be added.

$(D - d)t$ = weight of solution to be added

$\dfrac{(D - d)bt}{100}$ = weight of cyanide in solution to be added

$\dfrac{d\,a\,t}{100}$ = weight of cyanide in solution already present in storage tank

$\dfrac{Dp.t}{100}$ = weight of cyanide in solution as finally prepared

Then $\dfrac{Dp.t}{100} = x + \dfrac{d\,a\,t}{100} + \dfrac{(D - d)\,bt}{100}$

$x = \dfrac{Dt\,(p - b) - dt\,(a - b)}{100}$

Where the weight of solution is given in tons of 2000 lb. and the weight of cyanide in lb. av., the formula becomes

$$x = 20\ \{Dt\,(p - b) - dt\,(a - b)\}$$

Where the weight of solution is given in metric tons of 1000 kg. and the weight of cyanide in kg., the formula becomes

$$x = 10\ \{Dt\,(p - b) - dt\,(a - b)\}$$

Illustrations. — (a) The storage tank is of such a diameter that 1 inch in depth corresponds to 2.5 tons of solution. The total depth is 8 feet (*i.e.*, 96 in.), and it is filled with cyanide solution of 0.19 per cent. strength to a depth of 3 ft. 2 in. (= 38 in.). It is required to fill the storage tank from a sump containing 0.08 per cent. solution. What quantity of solid cyanide must be added to give a solution of 0.20 per cent. KCy?

Here $D = 96$; $d = 38$; $t = 2.5$; $p = 0.2$; $a = 0.19$; and $b = 0.08$
Hence $x = 20 \{240 \times 0.12 - 95 \times 0.11\} = 367$ lbs.

(b) The storage tank is of such a size that 1 cm. in depth corresponds to 0.4 ton. The total depth is 2 m. 50 cm. (= 250 cm.). The solution contained is of 0.22 per cent. strength, and it is required to make it up to 0.25 per cent. from another reservoir containing solution at 0.13 per cent. The present depth in the storage tank is 1 m. 10 cm.

Here $D = 250$; $d = 110$; $a = 0.22$; $b = 0.13$; $p = 0.25$; and $t = 0.4$.
Hence $x = 10 \{100 \times 0.12 - 44 \times 0.09\} =$
$= 80.4$ kg.

(2) To find the strength of resulting solution when two solutions of different strength are mixed.

$A =$ weight of first solution (in tons, lbs. kilos, etc.)
$B =$ weight of second solution (in same units)
$a =$ per cent. of cyanide or other ingredient in first solution
$b =$ per cent. of same in second solution
$p =$ per cent. in mixture
Then $\dfrac{Aa}{100} =$ weight of ingredient in first solution

$\dfrac{Bb}{100} =$ weight of ingredient in second solution

$\dfrac{(A+B)p}{100} =$ weight of ingredient in mixture

$$\frac{Aa}{100} + \frac{Bb}{100} = \frac{(A+B)p}{100} \text{ or } p = \frac{Aa + Bb}{A + B}$$

(3) To find weight of one solution which must be added to another to make it up to required strength.
Using same symbols as above,

$$A = \frac{(p-b)}{(a-p)} B \text{ or } B = \frac{(a-p)}{(p-b)} A$$

(4) Formulæ for converting percentages into results per ton.

P = percentage

$20P$ = lb. av. per ton of 2000 lb.

$22.4\,P$ = lb. av. per ton of 2240 lb.

$10P$ = kg. per ton of 1000 kg. = grams per kg.

$$\frac{7000}{24}\,P = \frac{1750}{6}\,P = \text{oz. Troy per ton of 2000 lb.}$$

$$\frac{980\,P}{3} = \text{oz. Troy per ton of 2240 lb.}$$

$$\frac{70000\,P}{12} = \text{dwt. per ton of 2000 lb.}$$

$$\frac{19600}{3}P = \text{dwt. per ton of 2240 lb.}$$

CONVERSION OF METRIC WEIGHTS AND MEASURES

			APPROXIMATELY
1 meter	=	39.370432 inches	= 39.4
"	=	3.280869 feet	= 3.3
"	=	1.093623 yard	= 1.1
1 inch	=	2.539977 centimeters	= 2.54
1 foot	=	0.304797 meter	= 0.3
1 yard	=	0.914392 meter	= 0.9
1 cubic meter	=	35.3156 cubic feet	= 35.3
1 liter	=	61.0254 cubic inches	= 61
"	=	0.035316 cubic feet	= $\frac{1}{29}$
1 cubic foot	=	0.0283161 cubic meters	= $\frac{1}{35}$
1 gram	=	15.43234874 grains	= 15.4
"	=	0.643015 dwt.	= $\frac{2}{3}$
"	=	0.032151 oz. Troy	= $\frac{1}{31}$
1 kilogram	=	2.20462125 lb. av.	= 2.2
1 grain	=	0.06479895 gram	= 0.065
1 dwt.	=	1.555176 gram	= $1\frac{1}{2}$
1 oz. Troy	= 31.10352 grams		= 31
1 lb. Av.	= 453.592652 grams		= 454
1 ton of 1000 kg.	= 2204.621 lb. Av.		= 2205
1 ton of 2000 lb.	= 907.185 kilograms		= 907
1 ton of 2240 lb.	= 1016.048 kilograms		= 1016

1 gram per ton of 1000 kg. = 14 grains per ton of 2000 lb.

1 " " " " " = $\frac{7}{12}$ dwt. " " " "

1 " " " " " = $\frac{7}{240}$ oz. " " " "

1 dwt. per ton of 2000 lb. = $\frac{12}{7}$ gram per ton of 1000 kg.

1 oz. per ton of 2000 lb. = $\frac{240}{7}$ grams per ton of 1000 kg.

MISCELLANEOUS DATA

1 ton of 2000 lb.	= $29166\frac{2}{3}$ oz. Troy
1 lb. Av.	= 14.58333 oz. Troy
1 " "	= $\frac{175}{12}$ oz. Troy = $\frac{1750}{6}$ dwt. = $291\frac{2}{3}$ dwt.

1 lb. Av.	= 7000 grains
1 oz. Troy	= 480 grains
1 oz. Av.	= 437.5 grains

1 lb. per ton of 2000 lb.	= 0.05 per cent.
" " " " "	= 0.5 kilo per ton of 1000 kg.
1 kg. per ton of 1000 kg.	= 0.1 per cent.
" " " " "	= 2 lb. per ton of 2000 lb.

1 lb. Av. of water	= 27.68122 cubic inches
1 gallon of water	= 10 lb. = 276.8122 cubic inches
1 gallon (U. S) of water	= 8.345 lb. = 231 cubic inches
1 cubic foot of water	= 62.425 lb. = 62½ approx.

INTERNATIONAL ATOMIC WEIGHTS FOR 1914

Element	Symbol	Atomic Weight	Element	Symbol	Atomic weight
Aluminum	Al	27.1	Indium	In	114.8
Antimony	Sb	120.2	Iodine	I	126.92
Argon	A	39.88	Iridium	Ir	193.1
Arsenic	As	74.96	Iron	Fe	55.84
Barium	Ba	137.37	Krypton	Kr	82.9
Bismuth	Bi	208.0	Lanthanum	La	139.0
Boron	B	11.0	Lead	Pb	207.10
Bromine	Br	79.92	Lithium	Li	6.94
Cadmium	Cd	112.40	Lutecium	Lu	174.0
Caesium	Cs	132.81	Magnesium	Mg	24.32
Calcium	Ca	40.07	Manganese	Mn	54.93
Carbon	C	12.00	Mercury	Hg	200.6
Cerium	Ce	140.25	Molybdenum	Mo	96.0
Chlorine	Cl	35.46	Neodymium	Nd	144.3
Chromium	Cr	52.0	Neon	Ne	20.2
Cobalt	Co	58.97	Nickel	Ni	58.68
Columbium	Cb	93.5	Niton	Nt	222.4
Copper	Cu	63.57	Nitrogen	N	14.01
Dysprosium	Dy	162.5	Osmium	Os	190.9
Erbium	Er	167.7	Oxygen	O	16.00
Europium	Eu	152.0	Palladium	Pd	106.7
Fluorine	F	19.0	Phosphorus	P	31.04
Gadolinium	Gd	157.3	Platinum	Pt	195.2
Gallium	Ga	69.9	Potassium	K	39.10
Germanium	Ge	72.5	Praseodymium	Pr	140.6
Glucinum	Gl	9.1	Radium	Ra	226.4
Gold	Au	197.2	Rhodium	Rh	102.9
Helium	He	3.99	Rubidium	Rb	85.45
Hydrogen	H	1.008	Ruthenium	Ru	101.7

INTERNATIONAL ATOMIC WEIGHTS FOR 1914 (*continued*)

Element	Symbol	Atomic weight	Element	Symbol	Atomic weight
Samarium	Sa	150.4	Thorium	Th	232.0
Scandium	Sc	44.1	Thulium	Tm	168.5
Selenium	Se	79.2	Tin	Sn	119.0
Silicon	Si	28.3	Titanium	Ti	48.1
Silver	Ag	107.88	Tungsten	W	184.0
Sodium	Na	23.0	Uranium	U	238.5
Strontium	Sr	87.63	Vanadium.........	V	51.0
Sulphur	S	32.07	Xenon	Xe	130.2
Tantalum	Ta	181.5	Ytterbium	Yb	172.0
Tellurium	Te	127.5	Yttrium	Y	89.0
Terbium	Tb	159.2	Zinc	Zn	65.37
Thallium	Tl	204.0	Zirconium	Zr	90.6

INDEX

www.ingramcontent.com/pod-product-compliance
Lightning Source LLC
Chambersburg PA
CBHW050529190326
41458CB00045B/6765/J